Lecture Notes in Computer Sc

T0230008

Edited by G. Goos, J. Hartmanis and J. van

Springer
Berlin
Heidelberg
New York
Barcelona
Budapest
Hong Kong
London
Milan
Paris
Santa Clara
Singapore
Tokyo

Nachum Dershowitz
Naomi Lindenstrauss (Eds.)

Conditional and Typed Rewriting Systems

4th International Workshop, CTRS-94
Jerusalem, Israel, July 13-15, 1994
Proceedings

Springer

Series Editors

Gerhard Goos, Karlsruhe University, Germany

Juris Hartmanis, Cornell University, NY, USA

Jan van Leeuwen, Utrecht University, The Netherlands

Volume Editors

Nachum Dershowitz
Department of Computer Science, University of Illinois
Urbana, Illinois 61801, USA

Naomi Lindenstrauss
Department of Computer Science, Hebrew University
Givat Ram, Jerusalem 91904, Israel

Cataloging-in-Publication data applied for

Die Deutsche Bibliothek - CIP-Einheitsaufnahme

Conditional and typed rewriting systems : 4th international workshop ; proceedings / CTRS-94, Jerusalem, Israel, July 13 - 15, 1994. Nachum Dershowitz ; Naomi Lindenstrauss (ed.). - Berlin ; Heidelberg ; New York ; Barcelona ; Budapest ; Hong Kong ; London ; Milan ; Paris ; Tokyo : Springer, 1995
(Lecture notes in computer science ; Vol. 968)
ISBN 3-540-60381-6
NE: Dershowitz, Nachum [Hrsg.]; CTRS <1994, Yerûšālayim>; GT

CR Subject Classification (1991): D.3,F.4, I.2.3

ISBN 3-540-60381-6 Springer-Verlag Berlin Heidelberg New York

Typesetting: Camera-ready by author
SPIN 10485430 06/3142 – 5 4 3 2 1 0 Printed on acid-free paper

Preface

This volume contains the papers presented at the **Fourth International Workshop on Conditional (and Typed) Rewriting Systems** (CTRS-94). The meeting was held July 13–15, 1994 on the campus of Hebrew University in Jerusalem, Israel, in conjunction with the Twenty-First International Colloquium on Automata, Languages and Programming (ICALP94).

This series is dedicated to the study of conditional term rewriting and other extensions of term rewriting systems, such as typed systems, higher-order rewriting, as well as to term rewriting in general.

Three previous workshops were held:

- Orsay, France, July 1987
  (*Lecture Notes in Computer Science* **308**, Springer-Verlag);
- Montreal, Canada, June 1990
  (*Lecture Notes in Computer Science* **516**, Springer-Verlag);
- Pont-à-Mousson, France, July 1992
  (*Lecture Notes in Computer Science* **656**, Springer-Verlag).

Each paper was reviewed by one or two referees.

Nachum Dershowitz
Naomi Lindenstrauss
Program Chairs

July 1995

Contents

Associative-Commutative Superposition
Leo Bachmair and Harald Ganzinger 1

A Calculus for Rippling
David A. Basin and Toby Walsh 15

Equation Solving in Geometrical Theories
Philippe Balbiani 31

LSE Narrowing for Decreasing Conditional Term Rewrite Systems
Alexander Bockmayr and Andreas Werner 51

Preserving Confluence for Rewrite Systems with Built-in Operations
Reinhard Bündgen 71

Hierarchical Termination
Nachum Dershowitz 89

Well-foundedness of Term Orderings
M. C. F. Ferreira and H. Zantema 106

A New Characterisation of AC-Termination and Application
Jean-Michel Gélis 124

Relative Normalization in Orthogonal Expression Reduction Systems
John Glauert and Zurab Khasidashvili 144

On Termination and Confluence of Conditional Rewrite Systems
Bernhard Gramlich 166

How to Transform Canonical Decreasing HCTRs into Equivalent Canonical TRSs
Claus Hintermeier 186

Termination for Restricted Derivations and Conditional Rewrite Systems
Charles Hoot 206

Rewriting for Preorder Relations
Paola Inverardi 223

Strong Sequentiality of Left-Linear Overlapping Rewrite Systems
Jean-Pierre Jouannaud and Walid Sadfi 235

A Conflict Between Call-by-Need Computation and Parallelism
Richard Kennaway 247

The Complexity of Testing Ground Reducibility for
Linear Word Rewriting Systems with Variables
Gregory Kucherov and Michaël Rusinowitch 262

Coherence for Cartesian Closed Categories: A Sequential Approach
Akira Mori and Yoshihiro Matsumoto 276

Modular Properties of Constructor-Sharing Conditional Term Rewriting Systems
Enno Ohlebusch 296

Church-Rosser Property and Unique Normal Form
Property of Non-Duplicating Term Rewriting Systems
Yoshihito Toyama and Michio Oyamaguchi 316

The Transformation of Term Rewriting Systems
Based on Well-formedness Preserving Mappings
Jan C. Verheul and Peter G. Kluit 332

Abstract Notions and Inference Systems for Proofs by Mathematical Induction
Claus-Peter Wirth and Klaus Becker 353

Author Index 375

Associative-Commutative Superposition*

Leo Bachmair[1] and Harald Ganzinger[2]

[1] Department of Computer Science, SUNY at Stony Brook,
Stony Brook, NY 11794, U.S.A,
leo@cs.sunysb.edu
[2] Max-Planck-Institut für Informatik,
Im Stadtwald, D-66123 Saarbrücken, Germany,
hg@mpi-sb.mpg.de

Abstract. We present an associative-commutative paramodulation calculus that generalizes the associative-commutative completion procedure to first-order clauses. The calculus is parametrized by a selection function (on negative literals) and a well-founded ordering on terms. It is compatible with an abstract notion of redundancy that covers such simplification techniques as tautology deletion, subsumption, and simplification by (associative-commutative) rewriting. The proof of refutational completeness of the calculus is comparatively simple, and the techniques employed may be of independent interest.

1 Introduction

Rewrite techniques are one of the more successful approaches to equational reasoning. In theorem proving these techniques usually appear in the form of completion-like procedures, such as ordered completion (Bachmair, Dershowitz and Plaisted 1989, Hsiang and Rusinowitch 1987), associative-commutative completion (Peterson and Stickel 1981), or basic completion (Bachmair, Ganzinger, Lynch, et al. 1992, Nieuwenhuis and Rubio 1992). Traditionally completion procedures were formulated for sets of equations (unit clauses), but ordered completion has been generalized to arbitrary non-unit clauses, resulting in several variants of paramodulation called superposition (Rusinowitch 1991, Zhang 1988, Bachmair and Ganzinger 1990, Pais and Peterson 1991), and the basic strategy has actually first been developed for first-order clauses. Associativity and commutativity have been built into ordered paramodulation (Paul 1992, Rusinowitch and Vigneron 1991), a calculus (Hsiang and Rusinowitch 1991) that does not generalize completion, but includes similar rewrite techniques; and Wertz (1992) designed an associative-commutative superposition calculus. Unfortunately, the completeness proofs proposed for these calculi are technically involved and quite complicated.

The calculus described in this paper is obtained by applying the technique of extended rules (Peterson and Stickel 1981) to a superposition calculus of Bachmair and

* The research described in this paper was supported in part by the NSF under research grant INT-9314412, by the German Science Foundation (Deutsche Forschungsgemeinschaft) under grant Ga 261/4-1, by the German Ministry for Research and Technology (Bundesministerium für Forschung und Technologie) under grant ITS 9102/ITS 9103 and by the ESPRIT Basic Research Working Group 6028 (Construction of Computational Logics). The first author was also supported by the Alexander von Humboldt Foundation.

Ganzinger (1994). A similar calculus has been discussed by Wertz (1992), and our completeness proof, like Wertz's proof, is based on the model construction techniques we originally proposed in (Bachmair and Ganzinger 1990). The main difference with our current approach is that we use *non-equality* partial models to construct an equality model. These modifications were motivated by our recent work on rewrite techniques for transitive relations in general (Bachmair and Ganzinger 1993), and allow us to more naturally deal with associative-commutative rewriting.

Completion procedures are based on commutation properties (often called "critical pair lemmas") of the underlying rewrite relation. We believe that our approach may be of independent interest in that it provides a general method for extending such procedures from unit clauses to Horn clauses to full clauses. Clauses are interpreted as conditional rewrite rules (with positive and negative conditions); new clauses need to be inferred with suitably designed inference rules, so that this clausal rewriting relation is well-defined and satisfies the required commutation properties. Associative-commutative superposition represents a non-trivial application of this general methodology.

The next section contains basic notions and terminology of theorem proving and rewriting. The associative-commutative superposition calculus is described in Section 3, and proved complete in Section 4.

2 Preliminaries

2.1 Clauses

We consider first-order languages with equality. A *term* is an expression $f(t_1, \ldots, t_n)$ or x, where f is a function symbol of arity n, x is a variable, and $t_1, \ldots, t_n$ are terms. For simplicity, we assume that equality is the only predicate in our theory. By an *atomic formula* (or *atom*) we mean a multiset $\{s, t\}$, called an *equality* and usually written $s \approx t$, where s and t are terms.[3] A *literal* is either a multiset (of multisets) $\{\{s\}, \{t\}\}$, called a *positive* literal and also written $s \approx t$, or a multiset $\{\{s, t\}\}$, called a *negative* literal and written $s \not\approx t$ or $\neg(s \approx t)$.[4] A *clause* is a finite multiset of literals. We write a clause by listing its literals $\neg A_1, \ldots, \neg A_m, B_1, \ldots, B_n$, or as a disjunction $\neg A_1 \vee \cdots \vee \neg A_m \vee B_1 \cdots \vee B_n$. An expression is said to be *ground* if it contains no variables.

A (*Herbrand*) *interpretation* is a set I of ground atomic formulas. We say that an atom A is *true* (and $\neg A$, *false*) *in* I if $A \in I$; and that A is *false* (and $\neg A$, *true*) in I if $A \notin I$. A ground clause is *true* in an interpretation I if at least one of its literals is true in I; and is *false* otherwise. A (non-ground) clause is said to be true in I if all its ground instances are true. The *empty clause* is false in every interpretation. We say that I is a *model* of a set of clauses N (or that N is *satisfied* by I) if all elements of N are true in I. A set N is *satisfiable* if it has a model, and *unsatisfiable* otherwise. For instance, any set containing the empty clause is unsatisfiable.

[3] The symmetry of equality is thus built into the notation.

[4] This somewhat unusual definition of literals as multisets of multisets of terms is motivated by our subsequent definition of an ordering on clauses in Section 2.2.

An interpretation is called an *equality interpretation* if it satisfies the *reflexivity* axiom

$$x \approx x$$

the *transitivity* axiom

$$x \not\approx y \lor y \not\approx z \lor x \approx z$$

and all *congruence* axioms

$$x \not\approx y \lor f(\ldots, x, \ldots) \approx f(\ldots, y, \ldots)$$

where f ranges over all function symbols. (Symmetry is already built into the notation.) We say that I is an *equality model* of N if it is an equality interpretation satisfying N. A set N is *equality satisfiable* if it has an equality model, and *equality unsatisfiable* otherwise.

We will consider the problem of checking whether a given set of clauses has an equality model, and for that purpose have to reason about Herbrand interpretations. Two concepts are useful in this context: rewrite systems and reduction orderings.

2.2 Reduction Orderings

An *ordering* is a transitive and irreflexive binary relation. A *rewrite relation* is a binary relation $\succ$ on terms such that $s \succ t$ implies $u[s\sigma] \succ u[t\sigma]$, for all terms s, t, $u[s\sigma]$, and $u[t\sigma]$ in the given domain.[5] By a *rewrite ordering* we mean a transitive and irreflexive rewrite relation; by a *reduction ordering*, a transitive and well-founded rewrite relation. (A binary relation $\succ$ is *well-founded* if there is no infinite sequence $t_1 \succ t_2 \succ \ldots$ of elements.)

Any ordering $\succ$ on a set S can be extended to an ordering on finite multisets over S (which for simplicity we also denote by $\succ$) as follows: $M \succ N$ if (i) $M \neq N$ and (ii) whenever $N(x) > M(x)$ then $M(y) > N(y)$, for some y such that $y \succ x$. If $\succ$ is a total (resp. well-founded) ordering, so is its multiset extension. If $\succ$ is an ordering on terms, then its multiset extension is an ordering on equations; its twofold multiset extension, an ordering on literals; and its threefold multiset extension, an ordering on clauses.

2.3 Associative-Commutative Rewriting

An *equivalence* (*relation*) is a reflexive, transitive, and symmetric binary relation. A *congruence* (*relation*) is an equivalence such that $s \sim t$ implies $u[s] \sim u[t]$, for all terms s, t, $u[s]$, and $u[t]$ in the given domain. Note that if I is an equality interpretation, then the set of all pairs (s, t) and (t, s), for which $s \approx t$ is true in I, is a congruence relation on ground terms. Conversely, if $\sim$ is a congruence relation, then the set of all ground equations $s \approx t$, for which $s \sim t$, is an equality interpretation. In short, equality interpretations can be described by congruence relations. Rewrite systems can be used to reason about congruence relations.

[5] In our notation for terms we follow Dershowitz and Jouannaud (1990).

By a *rewrite system* we mean a binary relation on terms. Elements of a rewrite system are called (*rewrite*) *rules* and written $s \rightarrow t$. If R is a rewrite system, we denote by $\rightarrow_R$ the smallest rewrite relation containing R. The transitive-reflexive closure of $\rightarrow_R$ is denoted by $\rightarrow_R^*$; while $\leftrightarrow_R^*$ denotes the smallest congruence relation containing R. A rewrite system R is said to be *terminating* if the rewrite relation $\rightarrow_R$ is well-founded.

We assume that some function symbols f are *associative* and *commutative*, i.e., satisfy the axioms

$$f(x, f(y, z)) \approx f(f(x, y), z)$$
$$f(x, y) \approx f(y, x)$$

For the rest of the paper, let AC be a set of such axioms. We write $f \in AC$, if AC contains the associativity and commutativity axioms for f, and also use AC to denote the binary relation containing the pairs $(f(x, f(y, z)), f(f(x, y), z))$ and $(f(x, y), f(y, x))$. We say that two terms u and v are *AC-equivalent* if $u \leftrightarrow_{AC}^* v$.

A reduction ordering $\succ$ is *AC-compatible* if $u' \leftrightarrow_{AC}^* u \succ v \leftrightarrow_{AC}^* v'$ implies $u' \succ v'$, for all terms u, u', v, and v'. If $\succ$ is AC-compatible, we write $u \succeq v$ to indicate that $u \succ v$ or $u \leftrightarrow_{AC}^* v$. We say that a rewrite system R is *AC-terminating* if there exists an AC-compatible reduction ordering $\succ$ such that $s \succ t$, for all rules $s \rightarrow t$ in R.

Associativity and commutativity are built into rewriting via matching. If R is a rewrite system, we denote by $AC\backslash R$ the set of all rules $u' \rightarrow v\sigma$, such that $u' \leftrightarrow_{AC}^* u\sigma$ for some rule $u \rightarrow v$ in R and some substitution σ. The rewrite relation $\rightarrow_{AC\backslash R}$ corresponds to rewriting by R via AC-matching. We say that a term t can be *rewritten* (*modulo AC*), or is *AC-reducible*, if $t \rightarrow_{AC\backslash R} t'$. Terms that cannot be rewritten are said to be in *AC-normal form* or *AC-irreducible* (with respect to R). We write $u \downarrow_{AC\backslash R} v$ if there exist terms u' and v' such that $u \rightarrow_{AC\backslash R}^* u' \leftrightarrow_{AC}^* v' \leftarrow_{AC\backslash R}^* v$. Given an AC-compatible reduction ordering, we also write $u \leftrightarrow_{AC\cup R}^{\preceq t} v$ (resp. $u \leftrightarrow_{AC\cup R}^{\prec t} v$) to indicate that there exists a sequence

$$t_0 \leftrightarrow_{AC\cup R} \cdots \leftrightarrow_{AC\cup R} t_n$$

such that $t \succeq t_i$ (resp. $t \succ t_i$), for all i with $0 \leq i \leq n$.

If R is AC-terminating and we have $u \downarrow_{AC\backslash R} v$, for all terms u and v with $u \leftrightarrow_{AC\cup R}^* v$, then R is called *AC-convergent*. We shall use AC-convergent rewrite systems to describe equality models of AC.

It is a standard result from the theory of term rewriting (see Dershowitz and Jouannaud 1990 for details and further references) that an AC-terminating rewrite system R is AC-convergent if (i) $u \downarrow_{AC\backslash R} v$ whenever $u \leftarrow_{AC\backslash R} t \leftrightarrow_{AC} v$ (a property called *local AC-coherence* of R) and (ii) $u \downarrow_{AC\backslash R} v$ whenever $u \leftarrow_{AC\backslash R} t \rightarrow_R v$ (a property called *local AC-confluence* of R).

Any rewrite system R can easily be extended to a locally AC-coherent system by a technique proposed by Peterson and Stickel (1981). An *extended rule* is a rewrite rule of the form $f(x, u) \rightarrow f(x, v)$, where $f \in AC$, u is a term $f(s, t)$, and x is a variable not occurring in u or v. We also say that $f(x, u) \rightarrow f(x, v)$ is an *extension* of $u \to v$. If R is a rewrite system, we denote by R^e the set R plus all extensions of rules in R. Any rewrite system R^e is locally AC-coherent: if $u \leftarrow_{AC\backslash R^e} t \leftrightarrow_{AC} v$, then $u \leftrightarrow^*_{AC} u' \leftarrow_{AC\backslash R^e} v$, for some term u'.

The local AC-confluence property is satisfied whenever it can be shown to hold for certain "minimal" rewrite sequences $u \leftarrow_{AC\backslash R} t \to_R v$ that are also called "critical overlaps." We only have to consider certain ground rewrite systems R where critical overlaps involve extended rules in R^e. This is summarized in the following lemma.

Lemma 1. *Let R be a ground rewrite system contained in an AC-compatible reduction ordering $\succ$ and suppose that no left-hand side of a rule in R can be rewritten modulo AC by any other rule in R or any extended rule in R^e. Furthermore, let t be a ground term, such that for all ground instances $f(u,u') \to f(u,u'')$ and $f(v,v') \to f(v,v'')$ of extended rules in R^e, where $f(u,u') \leftrightarrow^*_{AC} f(v,v')$ and $t \succeq f(u,u')$, we have $f(u,u'') \downarrow_{AC\backslash R^e} f(v,v'')$. Then $w \downarrow_{AC\backslash R^e} w'$ for all ground terms w and w' such that $w \leftrightarrow^{\preceq t}_{AC \cup AC\backslash R} w'$.*

If the same conditions are only satisfied for ground instances of extended rules with $t \succ f(u,u')$, then $w \downarrow_{AC\backslash R^e} w'$ for all ground terms w and w' such that $w \leftrightarrow^{\prec t}_{AC \cup AC\backslash R} w'$.

The lemma can be proved by standard techniques from term rewriting.

3 Associative-Commutative Superposition

We formulate our inference rules in terms of a reduction ordering. Furthermore, negative literals in a clause may be marked, in which case they are said to be *selected.* We assume that at least one literal is selected in each clause in which the maximal literal is negative. (If the maximal literal is positive, there may be no selected literals.)

Let $\succ$ be a well-founded AC-compatible reduction ordering, such that $s \leftrightarrow^*_{AC} t$ or $s \succ t$ or $t \succ s$, for all ground terms s and t. (Such orderings have been described by Narendran and Rusinowitch (1991) and Rubio and Nieuwenhuis (1993).) We say that a clause $C \vee s \approx t$ is *reductive* for $s \approx t$ (with respect to $\succ$) if there exists a ground instance $C\sigma \vee s\sigma \approx t\sigma$ such that $s\sigma \succ t\sigma$ and $s\sigma \approx t\sigma \succ C\sigma$. By an *extended clause* we mean a reductive clause $C \vee f(x,s) \approx f(x,t)$, where $f \in AC$, s is a term $f(u,v)$, and x is a variable not occurring in C, s, or t. We also say that $C \vee f(x,s) \approx f(x,t)$ is an *extension* of $C \vee s \approx t$. We will see that only extensions of certain reductive clauses without selected literals are needed. (Such extensions are themselves reductive.)

Associativity and commutativity are built into the inference rules via AC-unification. Two terms u and v are *AC-unifiable* if $u\sigma \leftrightarrow^*_{AC} v\sigma$, for some substitution σ. If two terms u and v are AC-unifiable, then there exists a *complete set of AC-unifiers* $CSU_{AC}(u,v)$, such that for any substitution σ with $u\sigma \leftrightarrow^*_{AC} v\sigma$ there exist substitutions $\tau \in CSU_{AC}(u,v)$ and ρ, such that $x\sigma \leftrightarrow^*_{AC} (x\tau)\rho$, for all variables x in u and v. We assume that a function CSU_{AC} is given and call a substitution in $CSU_{AC}(u,v)$ a *most general AC-unifier* of u and v.

3.1 Inference Rules

The calculus $S^{\succ}_{AC}$ consists of the following inference rules. (We assume that the premises of an inference share no common variables. If necessary the variables in one premise need to be renamed.)

AC-Superposition:
$$\frac{C,\, s \approx t \quad D,\, u[s'] \approx v}{C\sigma,\, D\sigma,\, u[t]\sigma \approx v\sigma}$$

where (i) σ is a most general AC-unifier of s and s', (ii) the clause $C\sigma \vee s\sigma \approx t\sigma$ is reductive for $s\sigma \approx t\sigma$ and $C \vee s \approx t$ either contains no selected literals or is an extended clause,[6] (iii) the clause $D\sigma \vee u\sigma \approx v\sigma$ is reductive for $u\sigma \approx v\sigma$ and either contains no selected literals or is an extended clause, and (iv) s' is not a variable.

Negative AC-Superposition:
$$\frac{C,\, s \approx t \quad D,\, u[s'] \not\approx v}{C\sigma,\, D\sigma,\, u[t]\sigma \not\approx v\sigma}$$

where (i) σ is a most general AC-unifier of s and s', (ii) the clause $C\sigma \vee s\sigma \approx t\sigma$ is reductive for $s\sigma \approx t\sigma$ and $C \vee s \approx t$ either contains no selected literals or is an extended clause, (iii) the literal $u \not\approx v$ is selected in $D \vee u \not\approx v$ and $v\sigma \not\succeq u\sigma$, and (iv) s' is not a variable.

Reflective AC-Resolution:
$$\frac{C,\, u \not\approx v}{C\sigma}$$

where σ is a most general AC-unifier of u and v and $u \not\approx v$ is a selected literal in the premise.

AC-Factoring:
$$\frac{C,\, s \approx t,\, s' \approx t'}{C\sigma,\, t\sigma \not\approx t'\sigma,\, s'\sigma \approx t'\sigma}$$

where σ is a most general AC-unifier of s and s', $t'\sigma \not\succ t\sigma$, the literal $s\sigma \approx t\sigma$ is maximal in $C\sigma$, and the premise contains no selected literals.

3.2 Lifting Properties

In proving the refutational completeness of the superposition calculus $S^{\succ}_{AC}$, we shall have to argue about ground instances of clauses and inferences. The connection between the general level and the ground level is usually stated in the form of so-called "lifting" properties.

Let $C_1 \ldots C_n$ be clauses and let

$$\frac{C_1\sigma \ldots C_n\sigma}{D}$$

be a ground inference, i.e., all clauses $C_i\sigma$ and D are ground. (We assume that each clause $C_i\sigma$ is marked in the same way as C_i; that is, $L\sigma$ is selected in $C\sigma$ if and only

[6] In other words, selection is ignored for extended clauses.

if L is selected in C.) We say that this ground inference can be *AC-lifted* if there is an inference

$$\frac{C_1 \ldots C_n}{C}$$

such that $C\sigma \leftrightarrow^*_{AC} D$. In that case, we also say that the ground inference is an *AC-instance* of the general inference.

Lifting is no problem for resolution and factoring inferences, but is more difficult for superposition inferences. For example, if N contains two clauses $f(x, x) \approx x$ and $a \approx b$, where $a \succ b$, then there is a superposition inference from ground inferences

$$\frac{a \approx b \quad f(a,a) \approx a}{f(a,b) \approx a}$$

but no superposition inference from the clauses themselves. To prove completeness, one has to show that all necessary inferences can be AC-lifted. Let us illustrate the conditions under which AC-superposition inferences can be lifted. (The case of negative AC-superposition is similar.)

Let $C \vee s \approx t$ and $D \vee u[s'] \approx v$ be two clauses, each either containing no selected literals or being an extended clause. An AC-superposition inference

$$\frac{C\tau,\ s\tau \approx t\tau \quad D\tau,\ u[s']\tau \approx v\tau}{C\tau,\ D\tau,\ u[t]\tau \approx v\tau}$$

from ground instances of these clauses, where $s'\tau \leftrightarrow^*_{AC} s\tau$, can be AC-lifted if s' is a non-variable subterm of u. (*Proof.* The ordering restrictions require that $s\tau \approx t\tau$ is strictly maximal in $C\tau$ and $s\tau \succ t\tau$, and $u\tau \approx v\tau$ is strictly maximal in $D\tau$ and $u\tau \succ v\tau$. Consider the inference

$$\frac{C,\ s \approx t \quad D,\ u[s'] \approx v}{C\sigma,\ D\sigma,\ u[t']\sigma \approx v\sigma}$$

where σ is a most general AC-unifier of s and s', such that for some substitution ρ we have $x\tau \leftrightarrow^*_{AC} (x\sigma)\rho$, for all variables x occurring in s' or s, and $y\tau = (y\sigma)\rho$, for all other variables y. It can easily be shown that $C\sigma \vee s\sigma \approx t\sigma$ is reductive for $s\sigma \approx t\sigma$ and $D\sigma \vee u\sigma \approx v\sigma$ is reductive for $u\sigma \approx v\sigma$. Also, s' is not a variable, so that the inference is an AC-superposition inference. Moreover, the conclusion of the ground inference is AC-equivalent to an instance of the conclusion of the general inference.)

3.3 Redundancy

Simplification techniques, such as tautology deletion, subsumption, demodulation, contextual rewriting, etc., represent an essential component of automated theorem provers. These techniques are based on a concept of redundancy (Bachmair and Ganzinger 1994), which we shall now adapt to the AC-case.

Let RA denote the set consisting of the reflexivity axiom, F the set of all congruence axioms, and T the set consisting of the transitivity axiom, cf. Section 2. Furthermore, for any ground term s, let $T_{\preceq s}$ be the set of all ground instances

$u \not\approx v \vee v \not\approx w \vee u \approx w$ of the transitivity axiom, for which $s \succeq u$ and $s \succeq v$ and $s \succeq w$; and $T_{\prec s}$ be the set of all ground instances for which $s \succ u$ and $s \succ v$ and $s \succ w$;

Let N be a set of clauses. A ground clause C (which need not be an instance of N) is said to be *AC-redundant* with respect to N (and ordering $\succ$) if there exist ground instances $C_1, \ldots, C_k$ of N, such that $C \succ C_j$, for all j with $1 \leq j \leq k$, and C is true in every model of $AC \cup RA \cup FU T_{\preceq s} \cup \{C_1, \ldots, C_k\}$, where s is the maximal term in C. It can easily be seen that the clauses $C_1, \ldots, C_k$ can be assumed to be non-redundant. A non-ground clause is called AC-redundant if all its ground instances are.

Remark. We emphasize that AC-redundancy is defined with respect to arbitrary interpretations, not just equality interpretations. In particular, there are restrictions on the use of the transitivity axiom. The restrictions are of more theoretical than practical significance, as all the usual simplification techniques can be formalized using AC-redundancy. Wertz (1992) uses a different notion of redundancy, where restrictions are imposed, not on transitivity, but instead on the use of the associativity and commutativity axioms.

For example, consider the (unit) clauses $s[u'] \approx t$ and $u \approx v$, where $u' \leftrightarrow^*_{AC} u\sigma$, for some substitution σ. If $u\sigma \succ v\sigma$ and either $s \succ u\sigma$ or else $t \succ v\sigma$, then $s[u'] \approx t$ is AC-redundant with respect to any set N containing $u \approx v$ and $s[v\sigma] \approx t$. In practice, this allows one to replace $s[u'] \approx t$ by $s[v\sigma] \approx t$ in the presence of the equation $u \approx v$. In other words, the subterm u' can be rewritten modulo AC to a smaller term $v\sigma$.

A ground inference with conclusion B is called *AC-redundant* with respect to N if there exist ground instances $C_1, \ldots, C_k$ of N, such that B is true in every model of $AC \cup RA \cup FU T_{\prec s} \cup \{C_1, \ldots, C_k\}$, where s is the maximal term in C, and such that the C_j are "sufficiently small". In the case of an inference in which both premises are ground instances of extended clauses, sufficiently small means that $\{\{s\}\} \succ C_j$.[7] For any other ground inference it means that $C \succ C_j$, where C is the maximal premise of the inference. A non-ground inference is called AC-redundant if all its ground instances are AC-redundant. One way to render an inference in $S^{\succ}_{AC}$ redundant is to add its conclusion to the set N.

We say that a set of clauses N is *saturated up to AC-redundancy* if all inferences, the premises of which are non-redundant clauses in N or extensions of non-redundant clauses in N, are AC-redundant.

Let us conclude this section by pointing out that extended clauses are redundant, but inferences with them are not. Therefore it is possible to dispense with extended rules altogether, and instead encode inferences with them directly in an "extended" AC-superposition calculus, cf., Rusinowitch and Vigneron (1991).

4 Refutational Completeness

In this section we will show how to define an equality model for saturated clause sets that do not contain the empty clause. The definition of a suitable model uses

[7] $\{\{s\}\} \succ C_j$ is true if and only if all terms in C_j are smaller than s.

induction on $\succ$ and is adapted from Bachmair and Ganzinger (1994). The presence of extended rules causes technical complications, though, and requires certain modifications in the model construction process.

Given a set N of ground clauses, we define a corresponding Herbrand interpretation I using induction on $\succ$. For every clause C in N we define R_C to be the set $\bigcup_{C \succ D} E_D$; and denote by I_C the set $\{u \approx v : u \text{ and } v \text{ ground}, u \downarrow_{AC \backslash R_C^e} v\}$, which we also call the *AC-rewrite closure* of R^e. Furthermore, if C is a clause $C' \vee s \approx t$, where $(s \approx t) \succ C'$ and $s \succ t$, then $E_C = \{s \approx t\}$ if (i) C is false in I_C, (ii) C' is false in the AC-rewrite closure of $(R_C \cup \{s \approx t\})^e$, and (iii) the term s is AC-irreducible by R_C^e. In that case, we also say that C is *productive* and that it *produces* $s \approx t$. In all other cases, $E_C = \emptyset$. Finally, let R be $\bigcup_C E_C$ and let I be the AC-rewrite closure of R^e.

The Herbrand interpretation I is intended to be an equality model of $N \cup AC$, provided N is saturated and does not contain the empty clause. The following lemmas state the essential properties of the interpretations I and I_C.

Lemma 2. *Let C be a ground clause (which need not be in N) and D be a clause in N with $D \succeq C$. If $\neg A$ is a negative literal in C with $A \notin I_D$, then $A \notin I$. As a consequence, if C is true in I_D, then it is also true in I and in any interpretation $I_{D'}$ with $D' \in N$ and $D' \succ D$.*

Proof. Let $\neg A$ be a negative literal in C such that $A \notin I_D$. If $B \in I \setminus I_D$, then $B \succ \neg A$, and therefore $A \notin I$. Also, if A is a positive atom in C and $A \in I_D$, then A is contained in any set $I_{D'}$ with $D' \succ D$ and in I. From this we may conclude that if C is true in I_D, then it is also true in I and in any interpretation $I_{D'}$ with $D' \in N$ and $D' \succ D$.

The next lemma follows immediately from the definition of the interpretations I_C and I.

Lemma 3. *The interpretation I and all interpretations I_C are models of the associativity, commutativity, reflexivity, symmetry, and all congruence axioms.*

We emphasize that transitivity need not be satisfied by all interpretations I_C or I, and consequently these interpretations need not be equality interpretations. However, if the clause set N is saturated, then "sufficiently many" instances of the transitivity axiom are true in each interpretation I_C, so that the final interpretation I does indeed satisfy transitivity, and hence is an equality interpretation. The following lemma is essential in this regard.

Lemma 4. *The interpretation I_C is a model of $T_{\preceq t}$ (resp. $T_{\prec t}$) if and only if for all ground terms u and v with $u \leftrightarrow^{\preceq t}_{AC \cup R_C^e} v$ (resp. $u \leftrightarrow^{\prec t}_{AC \cup R_C^e} v$) we have $u \downarrow_{AC \backslash R_C^e} v$.*

Proof. First let us assume that we have $u \downarrow_{AC \backslash R_C^e} v$ for all ground terms u and v with $u \leftrightarrow^{\preceq t}_{AC \cup R_C^e} v$. We show that I_C is a model of $T_{\preceq t}$. Let u, v, and w be ground terms with $t \succeq u$, $t \succeq v$, and $t \succeq w$. If $u \approx v$ and $v \approx w$ are true in I_C, then $u \downarrow_{AC \backslash R_C^e} v$ and $v \downarrow_{AC \backslash R_C^e} w$. In other words, we have $u \leftrightarrow^{\preceq t}_{AC \cup R_C^e} w$; hence $u \downarrow_{AC \backslash R_C^e} w$ and $u \approx w$ is true in I_C.

On the other hand, if I_C is a model of $T_{\preceq t}$, then we have $u \downarrow_{AC\backslash R^e_C} w$ whenever $u \leftarrow_{AC\backslash R^e_C} v \rightarrow_{AC\backslash R^e_C} w$, for some term v with $t \succeq v$. Using Lemma 1, we may conclude that $u' \downarrow_{AC\backslash R^e_C} v'$ for all ground terms u' and v' with $u' \leftrightarrow^{\preceq t}_{AC\cup R^e_C} v'$.

We now come to the main properties of the model construction.

Lemma 5. *Let N be a set of clauses that is saturated up to AC-redundancy, does not contain the empty clause, but contains an extension of every non-redundant reductive clause in N. Let I be the interpretation constructed from the set of all ground instances of $N \cup RA$. Then for every ground instance C of a clause in $N \cup RA$ we have:*

(1) If s is the maximal term in C, then I_C is a model of $T_{\prec s}$.

(2) If C is an instance $\hat{C}\sigma$ of a clause in N, such that $x\sigma$ is AC-reducible by R^e_C, for some variable x in $\hat{C}$, then C is true in I_C.

(3) If C is an instance of a clause with a selected literal, then it is true in I_C.

(4) If C is non-productive, then it is true in I_C.

(5) If $C = C' \vee A$ produces A, then C is a non-redundant instance of a clause in N with no selected literals and C' is false in I.

Proof. We use induction on $\succ$. Let C be a ground instance of $N \cup RA$ with maximal term s and assume (1)-(7) are already satisfied for all ground instances C' of N with $C \succ C'$.

(1) Since for every ground term t there is at least one clause with maximal term t (namely the ground instance $t \approx t$ of the reflexivity axiom), we may use the induction hypothesis to infer that I_C is a model of $T_{\prec s}$ By Lemmas 1 and 4, it is therefore sufficient to prove that for any two ground instances $f(u, u') \rightarrow f(u, u'')$ and $f(v, v') \rightarrow f(v, v'')$ of extended rules in R^e_C, with $s \leftrightarrow^*_{AC} f(u, u') \leftrightarrow^*_{AC} f(v, v')$, we have $f(u, u'') \downarrow_{AC\backslash R^e_C} f(v, v'')$.

Suppose $u' \approx u''$ is produced by $C' \vee u' \approx u''$ and $v' \approx v''$ is produced by $D' \vee v' \approx v''$. Both clauses are strictly smaller than C, so that by the induction hypothesis, they are true in I_C, whereas C' and D' are false in I. By AC-superposition we obtain the clause $C'' = C' \vee D' \vee f(u, u'') \approx f(v, v'')$. Since this inference can be AC-lifted, we may use saturation up to AC-redundancy, to infer that there exist ground instances $C_1, \ldots, C_k$ of N, such that $\{\{s\}\} \succ C_j$, for all j with $1 \leq j \leq k$, and C'' is true in every model of $AC \cup RA \cup F \cup T_{\prec s} \cup \{C_1, \ldots, C_k\}$. By induction hypothesis, I_C is a model of $AC \cup RA \cup F \cup T_{\prec s} \cup \{C_1, \ldots, C_k\}$, hence C'' is true in I_C. This implies that $f(u, u'') \approx f(v, v'')$ is true in I_C, and thus $f(u, u'') \downarrow_{AC\backslash R^e_C} f(v, v'')$.

(2) Suppose $C = \hat{C}\sigma$ is a ground instance of N, such that $x\sigma$ is AC-reducible by R^e_C, say $x\sigma \rightarrow_{AC\backslash R^e_C} u$. Let σ' be the same substitution as σ, except that $x\sigma' = u$, and let C' be the clause $\hat{C}\sigma'$. Since $C \succ C'$, we may use the induction hypothesis and Lemma 2 to infer that C' is true in I_C. Furthermore, since $x\sigma \approx u$ is true in I_C, a literal $L\sigma$ in C is true in I_C if and only if the corresponding literal $L\sigma'$ in C' is true in I_C. This implies that C is true in I_C.

If C is an instance $\hat{C}\sigma$, where $x\sigma$ is irreducible for all variables x in $\hat{C}$, then any inference with maximal premise C can be AC-lifted. We call C an *AC-reduced ground instance* of N in that case.

(3) Let $C = C' \vee s \not\approx t$ be an AC-reduced ground instance of a clause in which the literal corresponding to $s \not\approx t$ is selected. The assertion is obviously true if $s \approx t$ is false in I_C. So let us assume $s \downarrow_{AC\backslash R_C^e} t$.

(3.1) If $s \leftrightarrow^*_{AC} t$, then C' is a reflective AC-resolvent of C. Resolution inferences can be lifted, so that we may use saturation up to AC-redundancy, to infer that C' is true in I_C.

(3.2) If s and t are not AC-equivalent, then we have either $s \succ t$ or $t \succ s$. We discuss only one case, as the other is similar. Suppose $s \succ t$, in which case s is AC-reducible by R_C^e and can be written as $s[u']$, where $u' \leftrightarrow^*_{AC} u$, for some clause $D = D' \vee u \approx v$ with maximal literal $u \approx v$, and $s[v] \downarrow_{AC\backslash R_C^e} t$. The clause D is either productive or else is a ground instance of an extension of a productive clause. Since $C \succ D$, we use the induction hypothesis to infer that D' is false in I. Also, if D is productive, it is non-redundant and hence must be a ground instance of a clause with no selected literals. The clause $C'' = C' \vee D' \vee s[v] \not\approx t$ can be obtained from C and D by negative AC-superposition. By saturation up to redundancy, C'' must be true in I_C. Since D' and $s[v] \approx t$ are false in I_C, we may infer that C', and thus C, is true in I_C.

(4) Suppose C is false in I_C but non-productive. That is, C violates one of the other two conditions imposed on productive clauses.

(4.1) If condition (ii) is violated, then C is of the form $C' \vee s \approx t \vee s' \approx t'$, where $s \approx t$ is a maximal literal, $s \succ t \succ t'$, $s \leftrightarrow^*_{AC} s'$, and $t \downarrow_{AC\backslash R_C^e} t'$. We obtain the clause $C'' = C' \vee t \not\approx t' \vee s' \approx t'$ from C by AC-factoring. Using saturation up to redundancy, we may infer that C'' is true in I_C. Since $t \approx t'$ is true in I_C, this implies that $C' \vee s' \approx t'$, and therefore C, is true in I_C, which is a contradiction.

(4.2) Otherwise, C must be a clause $C' \vee s \approx t$ with maximal literal $s \approx t$, where $s \succ t$ and s is AC-reducible by R_C^e. Then there exists a clause $D = D' \vee u \approx v$ with maximal literal $u \approx v$, such that s can be written as $s[u']$ and $u' \leftrightarrow^*_{AC} u$. Moreover, D is either productive or is a ground instance of an extension of a productive clause. By the induction hypothesis, the clause D' is false in I and if D is productive it is a non-redundant instance of a clause with no selected literals. Thus, we may obtain from C and D by AC-superposition the clause $C'' = C' \vee D' \vee s[v] \approx t$, which by saturation has to be true in I_C. Since D' is false in I_C, either C' or $s[v] \approx t$ must be true in I_C. If C' is true in I_C, so is C. If $s[v] \downarrow_{AC\backslash R_C^e} t$, then we have $s = s[u'] \rightarrow_{AC\backslash R_C^e} s[v] \downarrow_{AC\backslash R_C^e} t$, which indicates that $s \approx t$, and hence C, is true in I_C—a contradiction.

(5) A productive clause $C = C' \vee A$ is false in I_C, hence is non-redundant and, by part (3), cannot be an instance of a clause with selected literals. The definition of productive clauses also ensures that C' is false in the AC-rewrite closure of $(R_C \cup E_C)^e$, and hence false in I.

The lemma indicates that under the given assumptions I is an equality model of $AC \cup N$. (Any non-productive ground instance C of N is true in I_C, while non-productive ground instances C are true in the AC-rewrite closure of $(R_C \cup E_C)^e$. Furthermore, I is a model of $T_{\preceq t}$, for all ground terms t; and hence is a model of T.)

The proof of the lemma also shows that certain inferences with extended clauses are unnecessary. For instance, AC-superpositions on a proper subterm of an extended clause, i.e., inferences of the form

$$\frac{C,\ s\approx t \quad D,\ f(x,u[s'])\approx f(x,v)}{C\sigma,\ D\sigma,\ f(x,u[t])\sigma\approx f(x,v)\sigma}$$

where σ is a most general AC-unifier of s and s', and the second premise extends the clause $D \vee u[s'] \approx v$, are not needed.

As an immediate corollary of the above lemmas we obtain:

Theorem 6. *Let N be a set of clauses that is saturated up to AC-redundancy with respect to the associative-commutative superposition calculus $\mathcal{S}^{\succ}_{AC}$. Furthermore, suppose N contains an extension of every non-redundant reductive clause in N. Then $N \cup AC$ is equality unsatisfiable if and only if it contains the empty clause.*

Proof. If $N \cup AC$ contains the empty clause it is unsatisfiable. If $N \cup AC$ does not contain the empty clause, let I be the interpretation constructed from the set of all ground instances of $N \cup RA$. By Lemmas 3 and 5, I is an equality model of $N \cup AC$.

This theorem shows that associative-commutative superposition calculi provide a basis for refutationally complete theorem provers. Such a theorem prover has to saturate a given set of input clauses up to AC-redundancy. The empty clause will be generated if the input set, plus AC, is equality unsatisfiable. Saturation of clause sets can be achieved by *fair* application of inference rules. These aspects of the saturation process have been discussed elsewhere, e.g., Bachmair and Ganzinger (1994) or Bachmair, Ganzinger and Waldmann (1994), and apply directly to the present case.

5 Summary

We have presented an associative-commutative superposition calculus and proved its refutational completeness. We have tried to keep the exposition clear and simple, and therefore have not discussed various possible improvements to the calculus, most of which can be derived from redundancy. Some are relatively minor, such as the redundancy of certain inferences involving an extended clause, while others are more important, e.g., the use of redex orderings to achieve a similar effect as critical pair criteria. The main difference of our calculus with other associative-commutative paramodulation calculi is that the associativity and commutativity axioms, but not all instances of the transitivity axiom, are built into our notion of redundancy; whereas other researchers have opted for transitivity and compromised on associativity and commutativity (Wertz 1992, Nieuwenhuis and Rubio 1994). From a more practical perspective our approach is preferrable, as enough instances of transitivity are provided to cover full associative-commutative rewriting, while otherwise associative-commutative rewriting actually needs to be restricted. In addition, our completeness proof is simpler than previous proofs.

We believe that the approach we have outlined for associativity and commutativity can be applied to other equational theories, such associativity and commutativity with identity (Baird, Peterson and Wilkerson 1989, Jouannaud and

Marché 1992); for some work in this direction see Wertz (1992). Recently, associative-commutative calculi have been combined with the basic strategy, or more generally, with constraint-based deduction, and completeness results have been proved by Nieuwenhuis and Rubio (1994) and by Vigneron (1994).

References

L. BACHMAIR, N. DERSHOWITZ AND D. PLAISTED, 1989. Completion without failure. In H. Ait-Kaci, M. Nivat, editors, *Resolution of Equations in Algebraic Structures, vol. 2*, pp. 1–30. Academic Press.

L. BACHMAIR AND H. GANZINGER, 1990. On Restrictions of Ordered Paramodulation with Simplification. In M. Stickel, editor, *Proc. 10th Int. Conf. on Automated Deduction, Kaiserslautern*, Lecture Notes in Computer Science, vol. 449, pp. 427–441, Berlin, Springer-Verlag.

L. BACHMAIR AND H. GANZINGER, 1993. Rewrite Techniques for Transitive Relations. Technical Report MPI-I-93-249, Max-Planck-Institut für Informatik, Saarbrücken. Short version in Proc. LICS'94, pp. 384–393, 1994.

L. BACHMAIR AND H. GANZINGER, 1994. Rewrite-based equational theorem proving with selection and simplification. *Journal of Logic and Computation*, Vol. 4, No. 3, pp. 217–247. Revised version of Technical Report MPI-I-91-208, 1991.

L. BACHMAIR, H. GANZINGER, CHR. LYNCH AND W. SNYDER, 1992. Basic Paramodulation and Superposition. In D. Kapur, editor, *Automated Deduction — CADE'11*, Lecture Notes in Computer Science, vol. 607, pp. 462–476, Berlin, Springer-Verlag.

LEO BACHMAIR, HARALD GANZINGER AND UWE WALDMANN, 1994. Refutational Theorem Proving for Hierarchic First-Order Theories. *Applicable Algebra in Engineering, Communication and Computing*, Vol. 5, No. 3/4, pp. 193–212. Earlier version: Theorem Proving for Hierarchic First-Order Theories, in Giorgio Levi and Hélène Kirchner, editors, *Algebraic and Logic Programming, Third International Conference*, LNCS 632, pages 420–434, Volterra, Italy, September 2–4, 1992, Springer-Verlag.

TIMOTHY BAIRD, GERALD PETERSON AND RALPH WILKERSON, 1989. Complete sets of reductions modulo associativity, commutativity and identity. In *Proc. 3rd Int. Conf. on Rewriting Techniques and Applications*, Lecture Notes in Computer Science, vol. 355, pp. 29–44, Berlin, Springer-Verlag.

N. DERSHOWITZ AND J.-P. JOUANNAUD, 1990. Rewrite Systems. In J. van Leeuwen, editor, *Handbook of Theoretical Computer Science B: Formal Methods and Semantics*, chapter 6, pp. 243–309. North-Holland, Amsterdam.

J. HSIANG AND M. RUSINOWITCH, 1991. Proving refutational completeness of theorem proving strategies: The transfinite semantic Tree method. *Journal of the ACM*, Vol. 38, No. 3, pp. 559–587.

JIEH HSIANG AND MICHAEL RUSINOWITCH, 1987. On Word Problems in Equational Theories. In *Proc. 14th ICALP*, Lecture Notes in Computer Science, vol. 267, pp. 54–71, Berlin, Springer-Verlag.

JEAN-PIERRE JOUANNAUD AND CLAUDE MARCHÉ, 1992. Termination and completion modulo associativity, commutativity and identity. *Theoretical Computer Science*, Vol. 104, pp. 29–51.

PALIATH NARENDRAN AND MICHAËL RUSINOWITCH, 1991. Any Ground Associative-Commutative Theory has a Finite Canonical System. In Ronald V. Book, editor, *Proc. 4th Rewriting Techniques and Applications 91*, Como, Italy, Springer-Verlag.

R. NIEUWENHUIS AND A. RUBIO, 1992. Basic superposition is complete. In *ESOP'92*, Lecture Notes in Computer Science, vol. 582, pp. 371–389, Berlin, Springer-Verlag.

R. NIEUWENHUIS AND A. RUBIO, 1994. AC-superposition with constraints: No AC-unifiers needed. In *Proc. 12th International Conference on Automated Deduction*, Lecture Notes in Computer Science, vol. 814, pp. 545–559, Berlin, Springer-Verlag.

JOHN PAIS AND G.E. PETERSON, 1991. Using Forcing to Prove Completeness of Resolution and Paramodulation. *Journal of Symbolic Computation*, Vol. 11, pp. 3–19.

E. PAUL, 1992. A general refutational completeness result for an inference procedure based on associative-commutative unification. *Journal of Symbolic Computation*, Vol. 14, pp. 577–618.

G. PETERSON AND M. STICKEL, 1981. Complete sets of reductions for some equational theories. *Journal of the ACM*, Vol. 28, pp. 233–264.

A. RUBIO AND R. NIEUWENHUIS, 1993. A precedence-based total AC-compatible ordering. In *Proc. 5th Int. Conf. on Rewriting Techniques and Applications*, Lecture Notes in Computer Science, vol. 690, pp. 374–388, Berlin, Springer-Verlag.

M. RUSINOWITCH, 1991. Theorem proving with resolution and superposition: An extension of the Knuth and Bendix completion procedure as a complete set of inference rules. *J. Symbolic Computation*, Vol. 11, pp. 21–49.

M. RUSINOWITCH AND L. VIGNERON, 1991. Automated deduction with associative-commutative operators. In *Proc. Int. Workshop on Fundamentals of Artificial Intelligence Research*, Lecture Notes in Artificial Intelligence, vol. 535, pp. 185–199, Berlin, Springer-Verlag.

L. VIGNERON, 1994. Associative-commutative dedution with constraints. In *Proc. 12th International Conference on Automated Deduction*, Lecture Notes in Computer Science, vol. 814, pp. 530–544, Berlin, Springer-Verlag.

U. WERTZ, 1992. First-Order Theorem Proving Modulo Equations. Technical Report MPI-I-92-216, Max-Planck-Institut für Informatik, Saarbrücken.

H. ZHANG, 1988. *Reduction, superposition and induction: Automated reasoning in an equational logic.* PhD thesis, Rensselaer Polytechnic Institute, Schenectady, New York.

A Calculus for Rippling

David A. Basin[1] and Toby Walsh[2]
[1] Max-Planck-Institut für Informatik
Saarbrücken, Germany
Email: basin@mpi-sb.mpg.de
[2] IRST, Trento &
DIST, University of Genova, Italy.
Email: toby@irst.it

Abstract. We present a calculus for rippling, a special type of rewriting using annotations. These annotations guide the derivation towards a particular goal. Although it has been suggested that rippling can be implemented directly via first-order term rewriting, we demonstrate that this is not possible. We show how a simple change to subterm replacement and matching gives a calculus for implementing rippling. This calculus also allows us to combine rippling with conventional term rewriting. Such a combination offers the flexibility and uniformity of conventional rewriting with the highly goal-directed nature of rippling. The calculus we present here is implemented and has been integrated into the Edinburgh CLAM proof-planning system.

1 Introduction

Rippling is a special type of rewriting originally developed for inductive theorem proving [Bundy *et al.*, 1993, Bundy *et al.*, 1990b] but since applied successfully to other domains [Basin and Walsh, 1993, Walsh *et al.*, 1992]. Rippling is rewriting tightly restricted by special kinds of annotation. It has many desirable properties. For example, it terminates yet allows rules like associativity to be used in both directions; it is also highly goal-directed applying only those rewrite rules needed to remove the differences between the goal and some hypothesis.

To illustrate the essential ideas behind rippling, consider a proof of the associativity of multiplication using structural induction. In the step-case, the induction hypothesis is,

$$(x \times y) \times z = x \times (y \times z)$$

and the induction conclusion is,

$$(\boxed{s(\underline{x})} \times y) \times z = \boxed{s(\underline{x})} \times (y \times z).$$

The annotations in the induction conclusion mark the differences with the induction hypothesis. Deleting everything in the box that is not underlined gives the *skeleton*; this is identical to the induction hypothesis and is preserved during rewriting. By comparison, simply removing annotations from terms gives the *erasure*; this is the unannotated induction conclusion. The boxed but not underlined terms are *wave-fronts*; these are moved through the term by rippling.

The underlined parts are *wave-holes*; they represent terms in the wave-front we would like to leave unchanged. To remove the wave-fronts, rippling uses annotated rewrite rules called *wave-rules*; these preserve the skeleton but move wave-fronts in some well founded manner (usually towards the top of the term where cancellation can occur). In this example, we need the wave-rules,

$$\boxed{s(\underline{U})} \times V \Rightarrow \boxed{(\underline{U \times V}) + V} \tag{1}$$

$$\boxed{\underline{U} + V} \times W \Rightarrow \boxed{\underline{U \times W} + V \times W} \tag{2}$$

Wave-rules are derived from recursive definitions and lemmas. Here, we use the recursive definition of multiplication and a lemma for distributing multiplication over addition. Rippling on both sides using (1) gives

$$(\boxed{\underline{x \times y} + y}) \times z = \boxed{(\underline{x \times (y \times z)}) + y \times z}$$

and then rippling with (2) gives

$$\boxed{(\underline{(x \times y) \times z}) + y \times z} = \boxed{(\underline{x \times (y \times z)}) + y \times z}.$$

As the wave-fronts are now at the top of each term, we have successfully rippled-out both sides of the equality. We can complete the proof by simplifying with the induction hypothesis.

While the ideas behind rippling are simple and intuitive, implementation is more complex than one might expect. In this paper we show where some of the difficulties lie and why first-order rewriting alone cannot directly implement rippling. We propose a new calculus that redefines the normal notions of substitution and matching and prove that it has the desired properties. The calculus we present has been implemented and integrated into the Edinburgh CLAM proof planning system [Bundy *et al.*, 1990a]. What we describe here is a slightly simplified version of the implemented calculus. To simplify the presentation we restrict ourselves in a minor way by considering only so called "outward oriented" wave-rules (see [Bundy *et al.*, 1993]); this restriction is easily lifted.

2 Properties Required

Rippling may be viewed as first-order rewriting with the addition of annotation to restrict applicability of rewrite rules. In the CLAM system [Bundy *et al.*, 1990a], rippling was originally implemented as first-order rewriting in a meta-level theory in which the signature of the original theory is extended with two new functions that represent the boxes and holes of annotation. Proofs constructed in this annotated meta-theory are translated into proofs in the original theory by erasing annotations. Unfortunately this simple idea of rippling as ordinary first-order rewriting is not adequate. To show why, we define the properties desired of rippling. These in turn require some preliminary definitions: well annotated terms, and the skeleton and the erasure of well annotated terms.

Well annotated terms, or *wats* are terms formed by enriching a term signature with two new unary function symbols *wf* and *wh* (representing wave-fronts and wave-holes). In a well annotated term, every wave-front has at least one proper subterm that is a wave-hole. Terms in wave-holes may be further annotated.

Definition 1 (wats) *Well-annotated terms, (or wats) are the smallest set such that,*

1. *t is a wat for all unannotated terms;*
2. *$wf(f(t_1,\ldots,t_n))$ is a wat iff for at least some i, $t_i = wh(s_i)$ and for each i where $t_i = wh(s_i)$, s_i is a wat and for each i where $t_i \neq wh(s_i)$, t_i is an unannotated term;*
3. *$f(t_1,\ldots,t_n)$ is a wat where $f \neq wf$ and $f \neq wh$ iff each t_i is a wat.*

For example, the annotated term,

$$wf(s(wh(x))) \times wf(s(wh(y)))$$

is a *wat* representing the term,

$$\boxed{s(\underline{x})} \times \boxed{s(\underline{y})}.$$

To aid the reader, we will continue to write annotated terms using boxes and holes, though this is *just* meant to be syntactic sugar for *wf* and *wh*. Note that in our formalization of annotation, wave-holes occur as immediate subterms of the function symbol in the wave-front. As additional syntactic sugar, we will sometimes merge adjacent wave-fronts and wave-holes when displaying annotated terms. For example, we will display the term

$$\boxed{s(\underline{\boxed{s(\underline{x})}})}$$

as the annotated term

$$\boxed{s(s(\underline{x}))}.$$

Insisting that wave-fronts are "maximally split", simplifies the presentation of rippling that follows. It also leads to a simpler implementation of rippling since wave-fronts need not be dynamically split or merged during rippling, as in [Bundy *et al.*, 1993].

The skeleton of an annotated term is the set of unannotated terms formed by deleting function symbols and variables within wave-fronts that are not within wave-holes. The skeleton is a set since wave-fronts can have multiple wave-holes. These arise, for example, when there are multiple induction hypotheses.

Definition 2 (Skeleton) *The function $skel{:}\,wats \rightarrow \mathcal{P}(unats)$ is defined by,*

1. *$skel(x) = \{x\}$ for all variables x;*
2. *$skel(\boxed{f(t_1,\ldots,t_n)}) = \{s \mid \exists i.t_i = \underline{t_i'} \wedge s \in skel(t_i')\}$;*
3. *$skel(f(t_1,\ldots,t_n)) = \{f(s_1,\ldots,s_n) \mid \forall i.s_i \in skel(t_i)\}$.*

For example, the skeleton of the annotated term

$$s(\boxed{\boxed{s(\underline{a})} + (\underline{b \times c})}) \tag{3}$$

is the set $\{s(a), s(b \times c)\}$.

Erasing annotation gives the corresponding unannotated term.

Definition 3 (Erasure) *the function erase: wats → unats is defined by,*

1. $erase(x) = x$ *for all variables* x;
2. $erase(\boxed{f(t_1, \ldots, t_n)}) = f(s_1, \ldots, s_n)$ *where if* $\underline{t_i'}$ *then* $s_i = erase(t_i')$ *else* $s_i = erase(t_i)$;
3. $erase(f(t_1, \ldots, t_n)) = f(s_1, \ldots, s_n)$ *where* $s_i = erase(t_i)$.

For example, the erasure of (3) is $s(s(a) + (b \times c))$.

We now formalize the properties desired of rippling that motivate the definition of our rewrite calculus.

Well-formedness: if s is a *wat*, and s ripples to t, then t is also a *wat*.
Skeleton preservation: if s ripples to t then $skel(t) \subseteq skel(s)$;
Correctness: if s ripples to t then $erase(s)$ rewrites to $erase(t)$ in the original (unannotated) theory;
Termination: rippling terminates.

The first property means that terms manipulated by rippling always stay well-formed. Hence we can always compute their skeleton and erasure. Skeleton preservation ensures that rippling directs the derivation towards (at least) one of the induction hypotheses. Correctness guarantees that we could have performed the corresponding derivation in the underlying object-level theory; annotation merely guides search. Finally, termination is important for practical considerations; it means we can backtrack and try other possibilities (e.g., other inductions) when the derivation fails.

3 Some Problems

Unfortunately, these four aims cannot be achieved by first-order rewriting alone. Consider a wave-rule formed from the recursive definition of multiplication,

$$\boxed{s(\underline{x})} \times y \Rightarrow \boxed{y + \underline{x \times y}}.$$

If we implement rippling directly via first-order rewriting this rule generates terms which are not well annotated. For example, consider

$$\boxed{s(\underline{a})} \times \boxed{s(\underline{b})}.$$

This is a *wat* but rewrites to

$$\boxed{\boxed{s(\underline{b})} + \underline{a \times \boxed{s(\underline{b})}}}\,.$$

This is not a *wat* since the first argument of plus contains a wave-front (a box) directly inside another without an intermediate wave-hole.

Termination also fails using first-order rewriting. Consider, for example, the equation, $h(f(x, s(y))) = s(h(f(s(x), y)))$. This is a wave-rule in the forward direction (see [Basin and Walsh, 1994]) annotated as

$$h(\boxed{f(x, s(\underline{y}))}) \Rightarrow \boxed{s(\underline{h(\boxed{f(s(x), \underline{y})})})}\,.$$

This rule should not lead to non-termination; the intuition is that the skeleton of both terms is $h(y)$ and the wave-rule transforms a wave-front on the left-hand side that is two nested function symbols deep to one on the right-hand side at the same position that is only one function symbol deep and another wave-front at a higher position in the skeleton. This equation also constitutes a wave-rule in the reverse direction with different annotation.

$$\boxed{s(\underline{h(\boxed{f(s(\underline{x}), y)})})} \Rightarrow h(\boxed{f(\underline{x}, s(y))})$$

Again this rule should not lead to non-termination; the intuition here is that the outermost wave-front is rippled "off the top" and disappears. However, these wave-rules together lead to cycling, as the following derivation illustrates:

$$h(\boxed{f(\underline{a}, s(\underline{a}))}) \;\mapsto\; \boxed{s(\underline{h(\boxed{f(s(\underline{a}), \underline{a})})})} \;\mapsto\; h(\boxed{f(\underline{a}, s(\underline{a}))}) \;\mapsto\; \dots .$$

Note that unlike the multiplication example, all the terms involved in this derivation are well annotated and share the same skeleton. With two equations, we can also construct looping derivations.

The problems of improperly annotated terms and non-termination arise when an annotated term t replaces an unannotated term in a wave-front. We therefore define a calculus for rippling based on a new notion of term replacement for annotated terms that erases annotation where appropriate; in this case, it erases the annotations on t. This new notion of term replacement naturally gives a new notion of substitution, and thus matching. By means of these simple modifications, we get a calculus for rewriting annotated terms which is guaranteed to preserve well-formedness of annotation, skeletons, and correctness wrt the underlying theory. Moreover, for unannotated terms and rewrite rules, this calculus performs conventional rewriting.

4 A Calculus for Rippling

Notational Convention: To simplify notation in proofs, in this and following sections we assume that the arguments of an annotated term like $\boxed{f(t_1, \ldots, t_n)}$ *may be partitioned so that the first j arguments are headed by wave-holes, i.e., of the form* $t_i = \underline{t_i'}$ *for* $i \in \{1..j\}$, *and the last* $n - j$ *are unannotated. Hence the term may be written as* $\boxed{f(\underline{t_1}, \ldots, \underline{t_j}, t_{j+1}, \ldots, t_n)}$, *or when there is no risk of confusion* $\boxed{f(\underline{t_1}, \ldots, t_n)}$. *This is without loss of generality as the proofs below do not depend on the order of wave-holes.*

We begin with ground rewriting and extend to first-order rewriting in a straightforward way. We distinguish between two kinds of variables: those in rewrite rules and those in terms. The later kind, "term variables" will be treated as constants here (see, for example, [Dershowitz, 1987] Section II). Hence we shall always consider the rewritten term as ground and by non-ground rewriting we mean equations between terms which may contain (non-term) variables.

We first redefine subterm replacement. Let $s[l]$ represent a *wat* with a distinguished subterm l that is also a *wat*. Let $rep(s[l], r)$ denote subterm replacement of r for the distinguished occurrence of l in the term s; rep is defined identically to the usual subterm replacement of l by r in s except that if l occurs within a wave-front any annotations on r are erased before replacement. For example, replacing a in $\boxed{\underline{b} + h(a)}$ by $\boxed{s(\underline{a})}$ gives $\boxed{\underline{b} + h(s(a))}$, but replacing b by $\boxed{s(\underline{b})}$ gives $\boxed{\underline{\boxed{s(\underline{b})}} + h(a)}$. From now on, we will perform all term replacement (including that occuring during substitution) using this function.

In rewriting terms, we will use just *proper* rewrite rules.

Definition 4 (Properness) *$l \rightarrow r$ is a* proper rewrite rule *iff l and r are wats, $erase(l) \rightarrow erase(r)$ is a rewrite rule, and $skel(r) \subseteq skel(l)$.*

The last two requirements are needed for well-formedness and skeleton preservation. Note also that the requirement that $erase(l) \rightarrow erase(r)$ is a rewrite rule means that $Vars(r) \subseteq Vars(l)$.

Ground rewriting consists of rewriting using proper rewrite rules that contain no (non-term) variables. Let R be a set of rewrite rules that are (for now) ground. If $s[l]$ is a *wat* then ground rewriting with a rule $l \rightarrow r$ in R yields $rep(s[l], r)$; we use $s[l] \mapsto_R s[r]$ to denote such *ground annotated rewriting.* In what follows we will assume a particular rewrite rule set R, and drop subscripted references, e.g., writing simply $s[l] \mapsto s[r]$.

Ground annotated rewriting preserves the well-formedness of annotated terms, and preserves skeletons. In addition, the corresponding (unannotated) rewriting can be performed in the underlying (unannotated) theory.

Theorem 1 *if s is a wat, $l \rightarrow r$ a proper rewrite rule between ground wats l and r and $s[l] \mapsto s[r]$, then*

1. $s[r]$ is a wat,
2. $skel(s[r]) \subseteq skel(s[l])$,
3. $erase(s[l]) \rightarrow erase(s[r])$.

Proof sketch: By structural induction on the *wat* s.

The only non-trivial case is when s is headed by a wave-front, i.e.,

$$s = \boxed{f(\underline{s_1}, \ldots, \underline{s_j}, s_{j+1}, \ldots, s_n)},$$

and l is *strict* subterm of one of the s_i. There are two cases depending on if $i \leq j$. In the first case, s_i is a *wat*; by the induction hypothesis, $s_i[r]$ is a *wat*. Thus, $s[r]$ is a *wat*. Also, by the induction hypothesis $skel(s_i[r]) \subseteq skel(s_i[l])$. As no other subterm is changed, the union of their skeletons is unchanged. Hence $skel(s[r]) \subseteq skel(s[l])$. Finally, by the induction hypothesis, $erase(s_i[l]) \rightarrow erase(s_i[r])$. Again, as all the other subterms are unchanged, their erasures stay the same. Thus, $erase(s[l]) \rightarrow erase(s[r])$. In the second case, s_i is an unannotated term that is part of the wave-front. Thus, when we substitute r for l we will erase annotations on r. Hence, $s_i[r]$ is unannotated, and $s[r]$ is a *wat*. From the definition of the skeleton it follows that term replacement in wave-fronts has no effect on the skeleton, so $skel(s[l]) = skel(s) = skel(s[r])$. Finally s_i is unannotated and $l \rightarrow r$ is a proper equation, $erase(s_i[l]) \rightarrow erase(s_i[r])$. Hence we have $erase(s[l]) \rightarrow erase(s[r])$. □

By Theorem 1 it follows by induction on the number of rewrite steps that the reflexive transitive closure of $\mapsto$ on ground *wats* also preserves well-formedness, skeletons and correctness with respect to the theory.

As a simple example, let $\boxed{h(\underline{a})} \rightarrow a$ be a proper rewrite rule. We can apply this rule only to the first subterm of the *wat*

$$\boxed{f(\underline{\boxed{h(\underline{a})}}, h(a))} \tag{4}$$

and this results in $\boxed{f(\underline{a}, h(a))}$. Alternatively, we could apply the proper rewrite rule $a \rightarrow \boxed{h(\underline{a})}$ to both occurrences of a in (4) resulting in

$$\boxed{f(\underline{\boxed{h(\underline{\boxed{h(\underline{a})}})}}, h(h(a)))}. \tag{5}$$

Note that annotation was erased when substituting $\boxed{h(\underline{a})}$ for the second occurrence of a. We can apply this rewrite rule again to both occurrences of a in (5). We can see that annotated rewriting with proper rewrite rules may be structure preserving (the skeleton of the rewritten term is always the same as the skeleton of the original term), but it is not necessarily terminating.

5 Annotated Matching

We now consider rewriting with variables. Since substitution depends on subterm replacement, our new definition of subterm replacement gives rise to a new kind of substitution; in particular, during substitution terms replacing variables in wave-fronts are erased of annotation. For example, applying $\sigma = \{x/\boxed{s(\underline{a})}\}$ to $t = \boxed{f(x, \underline{x})}$ gives $\boxed{f(s(a), \underline{\boxed{s(\underline{a})}})}$. Unlike regular substitution, our substitution function preserves well-formedness, skeletons and erasure wrt the theory provided the terms being substituted are themselves properly annotated.

Definition 5 (was) $\sigma = \{x_1/t_1, ..., x_n/t_n\}$ *is a* well-annotated substitution *(or* was*) iff σ is a substitution and every t_i is a wat.*

For σ a *was*, we define the *erase*(σ), and *skel*(σ) as the result of applying these functions to each element in the range of the substitution, i.e.,

$$erase(\sigma) = \{X/t' \mid X/t \in \sigma \wedge t' = erase(t)\}.$$

Theorem 2 *If t is a wat and σ a was then*

1. *$\sigma(t)$ is a wat,*
2. *$skel(\sigma(t)) = (skel(\sigma))(skel(t))$,*
3. *$erase(\sigma(t)) = (erase(\sigma))(erase(t))$.*

Proof sketch: By structural induction on the *wat* t.

As before, the only interesting case is when $t = \boxed{f(\underline{t_1}, \ldots, \underline{t_j}, t_{j+1}, \ldots, t_n)}$

(1) follows as

$$\sigma(t) = \boxed{f(\underline{\sigma(t_1)}, \ldots, \underline{\sigma(t_j)}, erase(\sigma(t_{j+1})), \ldots, erase(\sigma(t_n)))}$$

and subterms in wave-fronts, i.e., $\sigma(t_i)$ for $i \in \{1..j\}$ are *wats* by the induction hypothesis and the remaining subterms are unannotated and therefore also *wats*.

(2) follows as

$$\begin{aligned} skel(\sigma(t)) &= skel(\boxed{f(\underline{\sigma(t_1)}, \ldots, \underline{\sigma(t_j)}, erase(\sigma(t_{j+1})), \ldots, erase(\sigma(t_n)))}) \\ &= \cup_{i=1}^{j}\{s \mid s \in skel(\sigma(t_i))\} \\ &= \cup_{i=1}^{j}\{s \mid s \in skel(\sigma)(skel(t_1))\} \\ &= (skel(\sigma))skel(\boxed{f(\underline{t_1}, \ldots, \underline{t_j}, t_{j+1}, \ldots, t_n)}) \end{aligned}$$

Finally (3) follows as

$$\begin{aligned} erase(\sigma(t)) &= erase(\boxed{f(\underline{\sigma(t_1)}, \ldots, erase(\sigma(t_n)))}) \\ &= f(erase(\sigma(t_1)), \ldots, erase(erase(\sigma(t_n)))) \end{aligned}$$

$$= f(erase(\sigma(t_1)), \ldots, erase(\sigma(t_n)))$$
$$= f((erase(\sigma))(erase(t_1)), \ldots, (erase(\sigma))(erase(t_n)))$$
$$= (erase(\sigma))(f(erase(t_1), \ldots, erase(t_n)))$$
$$= (erase(\sigma))(erase(t))$$

□

To perform rewriting we need a notion of annotated matching corresponding to this new notion of annotated substitution.

Definition 6 (Annotated match) *if s and t are wats, then σ is an* annotated match *of s with t iff $Dom(\sigma) = Vars(s)$, σ is a was and $\sigma(s) = t$.*

Observe that even if we restrict σ to variables in s annotated matching is not unique. For example, $\{x/s(0)\}$ and $\{x/\boxed{s(\underline{0})}\}$ are both annotated matches of $\boxed{f(x,\underline{0})}$ with $\boxed{f(s(0),\underline{0})}$. Matches differ only in the amount of annotation which appear on substitutions for variables that occur in wave-fronts but not in skeletons. We can, however, define a notion of minimality of annotation so that annotated matching is unique. If σ and τ are well annotated substitutions for s and t then we write $\sigma \prec \tau$ iff there exists $X/t_1 \in \sigma$ and $X/t_2 \in \tau$ with $t_1 = erase(t_2)$ and all other pairs in σ and τ are identical. We write $\prec^+$ for the transitive (but not reflexive) closure of $\prec$.

Definition 7 (Minimality) *if s and t are wats, then σ is a* minimal match *of s with t iff σ is an annotated match of s with t and there does not exist any annotated match τ with $\tau \prec^+ \sigma$.*

It follows from this definition that if we have a minimal match, we cannot remove any annotation and have the result remain a match. Indeed, minimal matches are unique up to reordering.

Theorem 3 (uniqueness) *if σ is a minimal match of s with t then σ is unique.*

Proof. By contradiction. Let σ_1 and σ_2 be different minimal matches of s with t. Since $Dom(\sigma_1) = Vars(s) = Dom(\sigma_2)$ and $\sigma_1 \neq \sigma_2$ there must be at least one variable x in $Vars(s)$ on which σ_1 and σ_2 disagree. Let $x/s_1 \in \sigma_1$ and $x/s_2 \in \sigma_2$ where $s_1 \neq s_2$. By the well formedness of annotated substitution, $erase(s_1) = erase(s_2)$. Assume x occurs in the skeleton of s. By well formedness again, $skel(s_1) = skel(s_2)$. But this means that $s_1 = s_2$. Hence x can only occur in the wavefront but not in the skeleton. As σ_1 is minimally annotated, s_1 must therefore be unannotated. Similarly, s_2 must be also be unannotated. Thus $s_1 = s_2$. □

We now give an algorithm for computing minimal matches, $amatch(s,t)$ by means of a set of transformation rules. Because term replacement, and hence substitution is dependent on context (*i.e.* whether or not the term to be replaced is in a wave-front), our rules manipulate equations labeled with context information (wf for "in the wave-front" and sk for "in the skeleton"). As notational shorthand, Pos is a meta-variable that matches either wf or sk.

DELETE :
$S \cup \{t = t : Pos\} \;\Rightarrow\; S$

DECOMPOSE :
$S \cup \{f(s_1, \ldots, s_n) = f(t_1, \ldots, t_n) : Pos\} \;\Rightarrow\; S \cup \{s_i = t_i : Pos \mid 1 \leq i \leq n\}$
$S \cup \{\boxed{f(\underline{s_1}, \ldots, \underline{s_j}, s_{j+1}, \ldots, s_n)} = \boxed{f(\underline{t_1}, \ldots, \underline{t_j}, t_{j+1}, \ldots, t_n)} : sk\} \;\Rightarrow$
$\quad S \cup \{s_1 = t_1 : sk, \ldots, s_j = t_j : sk, s_{j+1} = t_{j+1} : wf, \ldots, s_n = t_n : wf\}$

Fig. 1. Transformation rules for $amatch(s, t)$.

Starting with the set containing the match problem $\{s = t : sk\}$, we apply these transformation rules exhaustively. An equation set is *normalized* if no more transformation rules can be applied. A normalized equation set is *compatible* iff every equation is a variable assignment $X = s : Pos$ where s is a *wat*, for each variable X there exists only one equation of the form $X = s : sk$, and if $X = s : sk \in S$ and $X = t : wf \in S$ then $erase(s) = t$. If a normalized equation set is not compatible, matching has failed; otherwise matching has succeeded and we return the answer substitution

$$\{X/s \mid X = s : sk \in S \text{ or } (X = s : wf \in S \text{ and } X = t : sk \notin S)\}.$$

As an example, if we match $\boxed{f(\underline{X}, X)}$ with $\boxed{f(\underline{\underline{\boxed{s(\underline{a})}}}, s(a))}$ then our initial matching problem is

$$\{\boxed{f(\underline{X}, X)} = \boxed{f(\underline{\underline{\boxed{s(\underline{a})}}}, s(a))} : sk\}$$

and after applying DECOMPOSE, we have

$$\{X = \boxed{s(\underline{a})} : sk, X = s(a) : wf\}.$$

This normalized equation set is compatible and yields the answer substitution $\sigma = \{X/\boxed{s(\underline{a})}\}$.

It is easy to see that execution of these rules in any order terminates in time linear in the size of the smaller of s and t. Note that the DELETE and the first DECOMPOSE rule implement regular matching. For unannotated terms, the compatibility check reduces to the requirement that the normalized equation set only contains variable assignments, and each variable has a single substitution. Annotated matching therefore subsumes regular matching. In addition, the extension of annotated matching to richer domains (for example, second order matching, or equational theories) should be relatively straightforward. As with regular matching, we can also add two failure rules for greater efficiency: CONFLICT which causes annotated matching to fail when the outermost function symbols disagree; and INCOMPATIBLE which causes annotated matching to fail immediately if the set of equations of the form $X = s : Pos$ is not compatible. These additional failure rules are not, however, needed for the completeness or correctness of annotated matching.

Theorem 4 (Correctness and completeness) *if s and t are wats then,*

1. *$amatch(s,t) = \sigma$ iff σ is the minimal match of s and t;*
2. *$amatch(s,t)$ fails iff s and t do not have an annotated match.*

Proof sketch: We prove a stronger result: a set of labeled equations S can be transformed to a compatible set of equations from which we extract σ iff for all $s = t : sk \in S$, $\sigma(s) = t$, and for all $s = t : wf \in S$, $(erase(\sigma))(s) = t$. The minimality of the answer substitution extracted holds by construction.

($\Rightarrow$) We use induction on the length of the transformation. The base case is when S is already a normalized compatible set of equations; the result follows directly from the way the answer substitution is computed. Alternatively, in the step case we must apply a transformation rule. The interesting case is when the DECOMPOSE rule is applied to $\boxed{f(\underline{s_1}, \ldots, s_n)} = \boxed{f(\underline{t_1}, \ldots, t_n)} : sk$ giving the set of equations S'. By the induction hypothesis, for all $s = t : sk \in S'$, $\sigma(s) = \sigma(t)$, and for all $s = t : wf \in S'$, $(erase(\sigma))(s) = (erase(\sigma))(t)$. Thus, by the definition of annotated substitution, $\sigma(\boxed{f(\underline{s_1}, \ldots, s_n)}) = \boxed{f(\underline{t_1}, \ldots, t_n)}$ and $\sigma(\boxed{f(\underline{s_1}, \ldots, s_n)}) = \boxed{f(\underline{t_1} \ldots, t_n)}$.

($\Leftarrow$) We use induction on the weight of the largest LHS of an equation. If this is atomic then, possibly after applications of DELETE, the equation set will be compatible. If it is not atomic, then the equation with the largest LHS will be of the form $f(s_1, \ldots, s_n) = t : Pos$, $\boxed{f(\underline{s_1}, \ldots, s_n)} = t : sk$, or $\boxed{f(\underline{s_1}, \ldots, s_n)} = t : sk$. In all three cases, we can apply DECOMPOSE and then appeal to the induction hypothesis.

Note that the normalized equation set is not compatible iff either two occurrences of a variable in the skeleton need a different substitution, or a variable in the wave-front needs a substitution which is not the erasure of the substitution needed by an occurrence in the skeleton, or there is a conflict in function symbols or annotation preventing application of DECOMPOSE. But this occurs iff s and t do have an annotated match. □

6 Non-ground Rewriting

We can now show that rippling with proper equations containing variables is correct and specify sufficient conditions for termination. We do this by lifting the results of ground annotated rewriting to the first-order case. First, we define non-ground rewriting. Let $s[t]$ be a *wat* with a distinguished subterm t and $l \rightarrow r$ be a proper rewrite rule. Further, let $\sigma = amatch(l, t)$. Then $s[t]$ rewrites to $s[\sigma(r)]$, which we write as $s[\sigma(l)] \mapsto s[\sigma(r)]$. Correctness parallels the ground case. The proof relies on the fact that σ is a *was* and we can thus reduce the problem to the correctness of ground-rewriting.

Theorem 5 *If s is a wat, $l = r$ a proper equation, and $s[\sigma(l)] \mapsto s[\sigma(r)]$, then*

1. *$s[\sigma(r)]$ is a wat,*
2. *$skel(s[\sigma(r)]) \subseteq skel(s[\sigma(l)])$,*
3. *$erase(s[\sigma(l)]) \rightarrow erase(s[\sigma(r)])$.*

Proof. Annotated matching ensures that σ is a *was*. By Theorem 2, $\sigma(l)$ and $\sigma(r)$ are *wats*, $skel(\sigma(r)) \subseteq skel(\sigma(l))$ and $erase(\sigma(l)) \rightarrow erase(\sigma(r))$. Now, since $\sigma(l)$ is syntactically identical to a subterm of s, it is ground (no non-term variables). Furthermore $\sigma(r)$ is also ground because since $l \rightarrow r$ is a rewrite rule we have $Vars(r) \subseteq Vars(l)$. Hence the rewriting of $s[\sigma(l)] \mapsto s[\sigma(r)]$ is equivalent to rewriting $s[\sigma(l)]$ with the ground proper rewrite rule $\sigma(l) \rightarrow \sigma(r)$. Thus, by Theorem 1, the three properties hold. □

Let $\overset{*}{\mapsto}$ be the reflexive transitive closure of $\mapsto$. By induction on number of rewrite steps, it follows from Theorem 5 that annotated rewriting is correct. That is, if we erase all annotations, we can perform the same (object-level) rewriting. Annotations merely guide rewriting in a skeleton preserving way.

Theorem 6 (Correctness) *If s is a wat and $s \overset{*}{\mapsto} t$ then*

1. *t is a wat,*
2. *$skel(t) \subseteq skel(s)$,*
3. *$erase(s) \rightarrow^* erase(t)$.*

7 Termination

We can now turn towards questions of termination. A simple way to ensure termination of ground annotated rewriting is to rewrite only with wave rules which are measure decreasing under some well founded order which is monotonic with respect to wats. We define a weak version of monotonicity that ensures monotonicity with respect to replacement of *wats* by *wats*. This is adequate to show termination when our rewrite rules are between *wats*.

Definition 8 (Monotonicity) *An order $>$ is* monotonic with respect to *wats iff for all wats s and for all ground wats l and r, if $l > r$, then $s[l] > s[r]$.*

The orders in [Basin and Walsh, 1994, Bundy *et al.*, 1993], for example, are monotonic wrt wats.

Theorem 7 (Ground termination) *Let R be a set of (not necessarily proper) rewrite rules between ground wats. If $>$ is well-founded and monotonic with respect to wats, and for all $l \rightarrow r \in R$ we have $l > r$ then rewriting with R is terminating.*

Proof. By monotonicity, an infinite rewrite sequence $t_1 \mapsto t_2 \mapsto \ldots$ gives an infinite sequence of *wats* $t_1 > t_2 > \ldots$. But this contradicts the well-foundedness of $>$. □

To lift termination to non-ground rewriting we use a restricted form of stability.

Definition 9 (Stability) *An order $>$ is* stable wrt wats *iff for all wats s, t and any was σ, $s > t$ implies $\sigma(s) > \sigma(t)$.*

The orders in [Basin and Walsh, 1994, Bundy *et al.*, 1993] are stable wrt wats.

Definition 10 (Annotation Reduction order) *An order on annotated terms that is well-founded, monotonic, and stable with respect to wats is an* annotation reduction order.

Note that annotation reduction orders are strictly weaker than normal reduction orders which are monotonic and stable over all terms in the signature, as opposed to just well-annotated ones.

Based on these definitions, we now formally define a *wave-rule.*

Definition 11 (Wave-rule) *Let $>$ be an annotation reduction order. Then a proper rewrite rule $l \rightarrow r$ is a* wave-rule with respect to $>$ *iff $l > r$.*

Theorem 8 (Termination) *for an annotation reduction order $>$ and R a set of wave-rules with respect to $>$, rippling using wave-rules in R is terminating.*

Proof. We again reduce the problem to the ground case. If $s \mapsto t$ using $l \rightarrow r$ then this is equivalent to rewriting with a rewrite rule $\sigma(l) \rightarrow \sigma(r)$ between ground *wats*. Since $>$ is stable with respect to *wats*, $\sigma(l) > \sigma(r)$. By the termination of ground annotated rewriting, we have termination in the general case. □

Note that our proof is similar to the one given in [Dershowitz, 1987] for regular unannotated rewriting (Corollary to Theorem 5) since we need not show that the ordering used is a reduction ordering, but rather only monotonic and stable with respect to possible instances of the rewrite rules.

In [Basin and Walsh, 1994] we define a number of orders useful for applying rippling within inductive theorem proving and other domains as well. These orders are based on measuring the weight (i.e., number of nested function symbols in the wave-front) and depth (relative to the skeleton) of wave-fronts. It is not difficult to establish formally that these orders are annotation reduction orders as defined above.

8 Related Work

Rippling has been proposed not as a rewrite calculus but as first-order rewriting with the strong precondition that "...each wave-front in the expression [being rewritten] is matched with a wave-front of the same kind in the rule" (Definition 5, page 222, [Bundy *et al.*, 1993]). Variables in wave-rules cannot therefore be instantiated with annotated terms. This is sufficient for rippling to preserve well annotated terms, to preserve skeletons, and (as formalized there) to terminate. However, this precondition is an unacceptably large restriction. Indeed, under this restriction, not all of the examples in [Bundy *et al.*, 1993] are correct. For example (see page 222) we cannot rewrite the immediate subterm of

$$even(\boxed{s(\boxed{s(\underline{x})})} + y)$$

with the wave-rule derived from the recursive definition of plus since the left-hand side of this wave-rule is $\boxed{s(\underline{U})} + V$ and there is an extra wave-front in the subterm being rewritten.

Rippling was originally implemented in the CLAM system without the above restriction. Unfortunately, it suffers from the problems described in §3 that arise when first-order rewriting is used to implement rippling directly. In particular, ill-formed terms can appear during rewriting and an auxiliary routine must occasionally "clean-up" annotations (e.g., consider the multiplication example given in §3). Moreover, if one admits as wave-rules rewrite rules under the annotation reduction orders we give in [Basin and Walsh, 1994], it is possible for rippling in CLAM to loop (e.g., §3).

Hutter, in [Hutter, 1990, Hutter, 1991], describes a calculus for rippling implemented in the INKA system [Biundo *et al.*, 1986]. Hutter rigorously develops an algebra of annotated terms, called C-terms. These are terms in an extended signature where functions and variables each carry a "colour" which represents annotation. Hutter's motivations and developments are similar to ours: he defines congruence relations corresponding to equality of terms after erasure, equivalence of skeletons, and develops algorithms to unify and rewrite C-terms that respect these congruences.

The calculus he develops is more general than ours. However, it is more complex, both conceptually, and in implementation. Wave-fronts can be thought of as contexts. In our calculus we augment the signature only as is required to specify these contexts: i.e., we introduce new function symbols so that we may mark the beginning of a context with a wave-front, and the end of the context with wave-holes. In Hutter's calculus, annotation is the primary concept and matching and rewriting of such terms can be understood independently of contexts.

Hutter's idea is based on marking *every* function symbol and variable with annotation. Annotations are colours and any combination of markings is allowed, not just those which generate an analog of our *wats*. Matching and rewriting over ground C-terms is fairly simple: C-terms match only if their colours match and this matching is used to drive rewriting in a straightforward way. On the other hand, if rewrite equations contain variables, Hutter's formalism becomes significantly more complex. The problem is determining the relationship between the colour of a variable and the colour of the term that instantiates it. If we demand that all the function symbols in the instantiating term have the same colour, then rewriting is too restrictive (this would be analogous to the restriction in [Bundy *et al.*, 1993] that variables are instantiated only with unannotated terms). Hutter solves this by introducing "colour variables" which indicate unspecified annotation and these must be properly manipulated and instantiated during unification and rewriting. As an example, the recursive definition for multiplication (1) could be represented as follows in Hutter's calculus.

$$s_g(U_\beta) \times_\alpha V_\gamma = (U_\beta \times_\alpha V_\gamma) +_g V_g$$

Here we are using the colour given by the subscript g to represent the colour of function symbols in wave-fronts. We now might match the right-hand side of this against a term like the following where w is the colour for the skeleton

$$s_g(a_w) \times_w s_g(b_w)),$$

which represents, in our notation, the term

$$\boxed{s(\underline{a})} \times \boxed{s(\underline{b})}.$$

Applying the rewrite rule via Hutter's matching produces instantiations for the set of variables on the left-hand side of the C-equation, i.e.,

$$\{\alpha, \beta, \gamma, U_\beta, V_\gamma\},$$

in particular U_β / a_w, $V_\gamma / s_g(b_w)$, and α / w. In applying the substitution, the domain must be extended to include $V_g / s_g(b_g)$. The result is the term

$$(a_w \times_w s_w(b_w)) +_g s_g(a_g).$$

The result would be identical to that returned by our rewrite calculus (although not [Bundy *et al.*, 1993]).

Hutter's calculus and ours implement similar kinds of rewriting. Essentially, his unification algorithm enforces constraints similar to those in our matching algorithm: annotation must match up and variables can be instantiated with terms that differ only in their annotation. The extension of his domain when applying a substitution plays an analogous role to our redefinition of subterm replacement. Overall, Hutter's calculus is more complex than ours, but also more general. He can, for example, match (or unify) terms which would not be well-annotated in our setting.

Hutter has not addressed termination of rippling in his work. However, with minor restrictions on his calculus, all of our results should carry over, although we have not formally checked this.

9 Conclusions

Our results show that conventional term rewriting systems can be used to perform rippling via a simple modification to the routines for subterm replacement and matching. Moreover, the same routines can be used for annotated and conventional unannotated term rewriting. The calculus has been implemented and is part of the CLAM proof planning system. Many possible directions can now be explored: for example, rippling in an implicit induction setting, Knuth-Bendix completion of annotated terms, simplification of skeletons and wave-fronts using term rewriting. We believe that such combinations will offer the best of both worlds, the flexibility and uniformity of conventional rewriting procedures combined with the highly goal-directed nature of rippling.

Acknowledgments

We thank Leo Bachmair, Alan Bundy, and Michael Rusinowitch for comments on this draft. Andrew Ireland and Ian Green were responsible for integrating our implementation of the calculus presented here into the CLAM system. The first author is funded by the German Ministry for Research and Technology (BMFT) under grant ITS 9102. The second author is supported by a Human Capital and Mobility Postdoctoral Fellowship from the European Union.

References

[Basin and Walsh, 1993] D. Basin and T. Walsh. Difference unification. In *Proceedings of the 13th IJCAI*, pages 116–122. 1993

[Basin and Walsh, 1994] D. Basin and T. Walsh. Termination orderings for rippling. In *Proc. of CADE-12*, Nancy, France, June 1994.

[Biundo *et al.*, 1986] S. Biundo, B. Hummel, D. Hutter, and C. Walther. The karlsruhe induction theorem proving system. In *Proc. of CADE-8*, Oxford, UK, 1986.

[Bundy *et al.*, 1990a] A. Bundy, F. van Harmelen, C. Horn, and A. Smaill. The Oyster-Clam system. In *Proc. of CADE-10*, Kaiserslautern, Germany, 1990.

[Bundy *et al.*, 1990b] A. Bundy, F. van Harmelen, A. Smaill, and A. Ireland. Extensions to the rippling-out tactic for guiding inductive proofs. In *Proc. of CADE-10*, Kaiserslautern, Germany, 1990.

[Bundy *et al.*, 1993] A. Bundy, A. Stevens, F. van Harmelen, A. Ireland, and A. Smaill. Rippling: A heuristic for guiding inductive proofs. *Artificial Intelligence*, 62:185–253, 1993.

[Dershowitz, 1987] N. Dershowitz. Termination of rewriting. In J.-P. Jouannaud, editor, *Rewriting Techniques and Applications*, pages 69–116. Academic Press, 1987.

[Hutter, 1991] D. Hutter. Colouring terms to control equational reasoning. An Expanded Version of PhD Thesis: Mustergesteuerte Strategien für Beweisen von Gleichheiten (Universität Karlsruhe, 1991), in preparation.

[Hutter, 1990] D. Hutter. Guiding inductive proofs. In *Proc. of CADE-10*, Kaiserslautern, Germany, 1990.

[Walsh *et al.*, 1992] T. Walsh, A. Nunes, and A. Bundy. The use of proof plans to sum series. In *Proc. of CADE-11*, Albany, NY, 1992.

Equation solving in geometrical theories

Philippe Balbiani

Institut de recherche en informatique de Toulouse
118 route de Narbonne
F-31062 Toulouse Cedex

Abstract. Incidence geometries of dimension two and three, miquelian geometry and projective geometry are defined through conditional equational axioms. The mechanization of these geometries is done using their associated positive/negative conditional term rewriting systems. To any figure and to any property of the figure are associated two terms t_1 and t_2 such that the figure possesses the property if and only if t_1 and t_2 have a same normal form for the conditional term rewriting system corresponding to the considered geometry.

1 Introduction

A geometry is defined by a set of beings (points, lines, planes, circles, etc) and a set of relations between these beings (collinearity and order between points, congruence between segments, parallelism and orthogonality between lines, etc). The work of the mathematician - more precisely, the logician - is to find postulates sufficient for the formalization of geometries (affine [4], projective [4] [11], euclidian [12] [18], hyperbolic [21], etc). The work of the computer scientist is to define a language for geometrical concepts as well as automated deduction algorithms for mechanical theorem proving in geometries [6] [7] [19] [20] [21]. The algebraic approach to geometrical reasoning consists of the translation of every geometrical sentence (incidence of a point to a line or to a plane, orthogonality of two lines, congruence of two segments, etc) into a polynomial equation and of the reduction of the deduction of geometrical theorems to simple and purely mechanical transformation rules on polynomials. Based on the theory of characteristic sets [20] [21] or on the theory of Gröbner bases [5], these methods are able to prove in a few seconds the most difficult theorems of synthetic geometry (Desargues, Pascal, Simson, etc) [7] [19]. Their main weakness is to destroy from the beginning the geometrical flavor of the figures which properties are considered. The manipulations on polynomials made by Wu method [20] [21] or Buchberger algorithm [5] are hardly interpretable in geometrical terms, and the obtained proofs cannot be easily translated in synthetic terms into a pure geometrical language.

The equational approach we present here uses a first order language of terms for the representation of geometrical beings and for the representation of their relations. In incidence geometry of dimension two, if, for example, the capital letters X, Y and Z denote geometrical beings of type point then the term $I(X,Y,Z)$ will denote a projection of Z onto the line XY. The equation $I(X,Y,Z) = Z$ will be interpreted by the collinearity of the three points X, Y and Z. The function symbol I is of arity three and type point × point × point → point. In miquelian geometry, a function symbol c of arity three and of type point × point × point → circle will be used for the definition of the term c(X,Y,Z) denoting the circle going through the three points X, Y and Z. In projective geometry, another function symbol i of type line × line → point will be used for the definition of terms like i(x,y) which will denote the intersection point of the two lines x and y. In our equational approach to geometrical reasoning, the automated deduction of the properties of the figure we consider will be realized with term rewriting. The problems of figure analysis - which consist, given a figure, to study its properties - will be solved using term rewriting techniques : to any figure and to any property of the figure are associated two terms t_1 and t_2 such that the figure possesses the property if and only if t_1 and t_2 have a same normal form for the rewriting system corresponding to the considered geometry.

We will study, one after the other, incidence geometries of dimension two and three, miquelian geometry and projective geometry. We will associate to these geometries first order languages of terms as well as equational conditional axioms precising under which conditions two terms are equal. These axioms immediatly define conditional term rewriting systems. All the rewriting systems we have considered are terminating and confluent. We show how they can be used for the resolution of analysis problems in their corresponding geometrical theories. The basic strength of our approach is that the geometrical interpretation of the first order terms we manipulate is always possible : at any time, the obtained deduction can be immediately translated into a pure synthetic geometrical language.

2 Two-dimensional incidence geometry

A first order theory with equality is associated to the two-dimensional incidence geometry. The corresponding positive/negative conditional term rewriting system defines a decidable, terminating and confluent relation of reduction. This relation is used for the mechanization of reasoning in incidence geometry. See [1] [2] [9] [10] [13] [16] [17] for an introduction to unification theory and term rewriting systems.

2.1 A geometrical language

The two-dimensional incidence geometry E_L is made of one set of geometrical beings of type points, together with a relation L of arity three : point × point × point. Points will be denoted by capital letters X, Y, Z, etc. The relation of collinearity between points is defined by the following postulates :

E_{L0} $\quad L(X,Y,Z) \Rightarrow L(Y,X,Z)$

E_{L1} $\quad L(X,Y,X)$

E_{L2} $\quad L(X,Y,Z) \wedge L(X,Y,T) \Rightarrow X = Y \vee L(Y,Z,T)$

These axioms are equivalent to the planar incidence axioms of Hilbert [12]. Models of postulates E_{L0} through E_{L2} are incidence planes. That $L(X,Y,Z) \Rightarrow L(X,Z,Y)$ is a theorem of E_L can be proved in the following way :

(1) $L(Y,X,Z) \wedge L(Y,X,Y) \Rightarrow Y = X \vee L(X,Z,Y)$ axiom E_{L2}

(2) $Y = X \Rightarrow L(X,Z,Y)$ axiom E_{L1}

(3) $L(Y,X,Z) \Rightarrow L(X,Z,Y)$ (1) and (2)

(4) $L(X,Y,Z) \Rightarrow L(X,Z,Y)$ (3) and axiom E_{L0}

2.2 A first order theory with equality

A first order theory E_I with equality is associated to E_L in the following way. Its alphabet contains a function symbol I of arity three : point × point × point → point, defining a set of terms denoted by T. The term $I(X,Y,Z)$ denotes a projection of Z onto the line passing through X and Y. The equation between $I(X,Y,Z)$ and Z will denote the collinearity of X, Y and Z. The first order theory E_I is defined by the following postulates :

E_{I0} $\quad I(X,Y,Z) = I(Y,X,Z)$

E_{I1} $\quad X \neq Y \Rightarrow I(X,Y,X) = X$

E_{I2} $\quad X \neq Y \wedge Z \neq T \wedge I(X,Y,Z) = Z \wedge I(X,Y,T) = T \Rightarrow I(X,Y,I(Z,T,U)) = I(Z,T,U)$

E_{I3} $\quad X \neq I(Y,Z,T) \wedge Y \neq Z \wedge I(Y,Z,X) = X \Rightarrow I(X,I(Y,Z,T),U) = I(Y,Z,U)$

The geometrical interpretation of these axioms is not difficult. Axiom E_{I2}, for example, means that if, on one hand, X, Y and Z are collinear and, on the other hand, X, Y and T are collinear then any projection $I(Z,T,U)$ onto the line passing through Z and T is collinear with X and Y. Let * be the function of translation between the languages of E_L and E_I defined by :

$$L(X,Y,Z)^* = (X = Y \vee I(X,Y,Z) = Z).$$

proposition 2.2.1 If F is a statement in the language of E_L and F^* is its translation in the language of E_I then $\vdash_{EL} F$ if and only if $\vdash_{EI} F^*$.

Therefore, the mechanization of reasoning in E_L can be done using equational reasoning techniques in E_I.

2.3 A conditional term rewriting system

Let R_I be the positive/negative conditional term rewriting system (P/N CTRS) made of the following rules :

R_{I1} $\quad X \neq Y \Rightarrow I(X,Y,X) \to X$

R_{I2} $\quad X \neq Y \wedge Z \neq T \wedge I(X,Y,Z) = Z \wedge I(X,Y,T) = T \Rightarrow$

$$I(X,Y,I(Z,T,U)) \to I(Z,T,U)$$

R_{I3} $\quad X \neq I(Y,Z,T) \wedge Y \neq Z \wedge I(Y,Z,X) = X \Rightarrow$

$$I(X,I(Y,Z,T),U) \to I(Y,Z,U)$$

Let C_I be the first order equational theory generated by E_{I0}. Unification and matching under C_I are finitary and constitute decidable classes of problems for which type conformal algorithms exist [17]. Let $(>_n)_{n\geq 0}$ be the sequence of binary relations on T defined in the following way :

$t >_0 t'$ if and only if either t' is a proper subterm of t
or $t = I(X,Y,I(Z,T,U))$ and $t' = I(X,Y,Z)$
or $t = I(X,I(Y,Z,T),U)$ and $t' = I(Y,Z,X)$
or $t = I(X,I(Y,Z,T),U)$ and $t' = I(Y,Z,U)$

$t >_{n+1} t'$ if and only if either $t_0 = t >_n t_1 \ldots >_n t_k = t'$
or $t = u[s]_p$, $t' = u[s']_p$ and $s >_n s'$
or $t' =_{C_I} t''$ and $t >_n t''$
or $t =_{C_I} t''$ and $t'' >_n t'$

Let $> = \cup_{n\geq 0} >_n$. It is straightforward to prove that $>$ is a reduction ordering on T. Moreover, R_I is a reducing P/N CTRS with respect to $>$ (in the sense of Kaplan [14]). The axiom E_{I0} postulates the commutativity of I with respect to its first and second arguments. Since equality classes modulo C_I are finite, then rewriting modulo the commutativity of I is a straightforward generalization of standard rewriting and the following proposition can be considered as a direct consequence of Kaplan's work on positive/negative conditional rewriting [14] [15].

proposition 2.3.1 The rules of R_I define a decidable and terminating relation of reduction $\to$ on T such that $t \to t'$ if and only if there are a position p in t, a rule $Q \Rightarrow l \to r$ in R_I and a substitution σ such that $t/p =_{C_I} l\sigma$, $t' = t[r\sigma]_p$ and $Q\sigma\downarrow$, that is to say : for every equation $u = v$ in Q, $u\sigma\downarrow v\sigma$ and, for every disequation $u \neq v$ in Q, $u\sigma \not\downarrow v\sigma$, the relation $\downarrow$ being defined by $s\downarrow t$ if and only if $s \to^* s' =_{C_I} t' {}^*\!\leftarrow t$, the relation $\to^*$ being the transitive and reflexive closure of $\to$.

A normal form is a term that cannot be reduced. Since $\to$ is terminating, then every term possesses at least one normal form. As an example, a normal form of $I(I(X,Y,Z),I(X,Y,T),U)$ is $I(X,Y,U)$. Confluence is an essential property for the proof that every term possesses exactly one normal form. The proof of next theorem has been detailed by Balbiani et al [3]. It can be performed through a careful examination of the contextual critical pairs between rules of R_I.

theorem 2.3.2 The relation of reduction $\to$ is confluent on T that is to say : for every term s, if $s \to^* t$ and $s \to^* u$ then $t{\downarrow}u$.

As a corollary of theorem 2.3.2, the normal forms of a term are equal modulo C_I. For every term t, $NF_I(t)$ denotes the class modulo C_I of its normal forms. Let M_I be the interpretation of E_I defined by $\to$, that is to say : the domain of M_I is $\{NF_I(t) : t \in T\}$ and the interpretation of the function symbol I is defined by :

$$I^{M_I}(NF_I(X), NF_I(Y), NF_I(Z)) = NF_I(I(X,Y,Z))$$

The confluence of $\to$ implies the correctness of this definition.

proposition 2.3.3 The interpretation M_I is a model of E_I.

This model will be particularly useful for the mechanization of two-dimensional geometry.

2.4 Mechanization of two-dimensional incidence geometry

To mechanize a geometry is to define a decision method for a large subset of formulas of its language. A geometrical fact of the two-dimensional incidence geometry E_L is one of the following two sentences :

$$X \neq X_1 \wedge \ldots \wedge X \neq X_n$$

$$L(X,Y,Z) \wedge Z \neq X_1 \wedge \ldots \wedge Z \neq X_n$$

The fact $X \neq X_1 \wedge \ldots \wedge X \neq X_n$ defines the point X while the fact $L(X,Y,Z) \wedge Z \neq X_1 \wedge \ldots \wedge Z \neq X_n$ defines the point Z. A finite sequence $[F_1,\ldots,F_k]$ of geometrical facts is coherent whenever, for every $i=1,\ldots,k$, if F_i contains the atom $L(X,Y,Z)$ then, for some $j=1,\ldots,i-1$, F_j contains the disequation $X \neq Y$. In the geometrical fact $X \neq X_1 \wedge \ldots \wedge X \neq X_n$, the point X depends on $X_1, \ldots, X_n$. In the geometrical fact $L(X,Y,Z) \wedge Z \neq X_1 \wedge \ldots \wedge Z \neq X_n$, the point Z depends on X, Y, $X_1, \ldots, X_n$. A constructive geometrical proposition is a coherent finite sequence of geometrical facts in which no point depends directly or indirectly on itself.

example The following sequence of facts :

$$X \neq Y$$

$$L(X,Y,Z) \wedge Z \neq Y$$

$$L(Y,Z,T)$$

is a constructive geometrical proposition.

Given a constructive proposition $G = [F_1,...,F_k]$ with points X, Y and Z, we will associate to a geometrical formula like $F_1 \wedge ... \wedge F_k \Rightarrow L(X,Y,Z)$ two terms t_1 and t_2 such that $\vdash_{EI} F_1 \wedge ... \wedge F_k \Rightarrow L(X,Y,Z)$ if and only if $t_1 \downarrow t_2$. Let $G = [F_1,...,F_k]$ be a constructive proposition and E be a quantifier-free equational formula in the language of E_I. Then, G^*E is the equational formula in the language of E_I defined by induction on the length of G in the following way :

(i) $[]^*E$ is E,

(ii) $[F_1,...,F_{k-1},(X \neq X_1 \wedge ... \wedge X \neq X_n)]^*E$ is
$$[F_1,...,F_{k-1}]^*(X \neq X_1 \wedge ... \wedge X \neq X_n \Rightarrow E)$$ and

(iii) $[F_1,...,F_{k-1},(L(X,Y,Z) \wedge Z \neq X_1 \wedge ... \wedge Z \neq X_n)]^*E$ is
$$[F_1,...,F_{k-1}]^*(Z \neq X_1 \wedge ... \wedge Z \neq X_n \Rightarrow E)\sigma,$$

σ being the substitution $\{I(X,Y,Z)/Z\}$.

example Let F_1 be $X \neq Y$, F_2 be $L(X,Y,Z) \wedge Z \neq Y$, F_3 be $L(Y,Z,T)$ and E be $I(X,Y,T) = T$. Then $[F_1,F_2,F_3]^*E$ is the equational formula in the language of E_I obtained in the following way :

$$[F_1,F_2,F_3]^*(I(X,Y,T) = T)$$
$$[F_1,F_2]^*(I(X,Y,I(Y,Z,T)) = I(Y,Z,T))$$
$$[F_1]^*(I(X,Y,Z) \neq Y \Rightarrow I(X,Y,I(Y,I(X,Y,Z),T)) = I(Y,I(X,Y,Z),T))$$
$$(X \neq Y \Rightarrow (I(X,Y,Z) \neq Y \Rightarrow I(X,Y,I(Y,I(X,Y,Z),T)) = I(Y,I(X,Y,Z),T)))$$

proposition 2.4.1 Let $G = [F_1,...,F_k]$ be a constructive proposition and E be a quantifier-free equational formula in the language of E_I. Then, G^*E is logically equivalent to $F_1^* \wedge ... \wedge F_k^* \Rightarrow E$.

Given a constructive proposition $G = [F_1,...,F_k]$ with points X, Y and Z, if E is the equation $I(X,Y,Z) = Z$ then G^*E is equivalent to a formula in the language of E_I of the form $Q \Rightarrow t_1 = t_2$ such that Q is a finite set of disequations. It can be proved that Q is consistent in E_I and that $t_1 \downarrow t_2$ is a necessary and sufficient condition for $F_1 \wedge ... \wedge F_k \Rightarrow L(X,Y,Z)$ to be a theorem of E_L.

proposition 2.4.2 Let $G = [F_1,...,F_k]$ be a constructive proposition and X and Y be two variables such that, for some $i=1,...,k$, F_i contains the disequation $X \neq Y$. Then, for every variable Z, $G^*(I(X,Y,Z) = Z)$ is equivalent in E_I to an equational formula of the form $Q \Rightarrow t_1 = t_2$ such that $Q\downarrow$. Moreover, $t_1 \downarrow t_2$ if and only if $\vdash_{EI} Q \Rightarrow t_1 = t_2$.

proof The disequations in Q are either of the form $W \neq W'$, $I(U,V,W) \neq W'$ or $I(U,V,W) \neq I(U',V',W')$ where W and W' are distinct variables. Consequently, $Q\downarrow$ and Q is true in M_I. We now prove that $t_1 \downarrow t_2$ if and only if $\vdash_{EI} Q \Rightarrow t_1 = t_2$.

$\Rightarrow$. If $\vdash_{EL} F_1 \wedge ... \wedge F_k \Rightarrow L(X,Y,Z)$ then $\vdash_{EI} F_1^* \wedge ... \wedge F_k^* \Rightarrow X = Y \vee I(X,Y,Z) = Z$. Since, for some $i=1,...,k$, F_i contains the disequation $X \neq Y$, then $\vdash_{EI} F_1^* \wedge ... \wedge F_k^* \Rightarrow I(X,Y,Z) = Z$. Therefore, according to proposition 2.4.1, $\vdash_{EI} Q \Rightarrow t_1 = t_2$. Then $Q \Rightarrow t_1 = t_2$ is true in M_I. Since $Q\downarrow$, then $t_1 \downarrow t_2$.
$\Leftarrow$. If $t_1 \downarrow t_2$ then it can be proved by induction on the length of the reduction of t_1 and t_2 into their normal forms that $\vdash_{EI} Q \Rightarrow t_1 = t_2$. Therefore, $\vdash_{EI} F_1^* \wedge ... \wedge F_k^* \Rightarrow X = Y \vee I(X,Y,Z) = Z$ and $\vdash_{EL} F_1 \wedge ... \wedge F_k \Rightarrow L(X,Y,Z)$. **QED**

example Let F_1 be $X \neq Y$, F_2 be $L(X,Y,Z) \wedge Z \neq Y$ and F_3 is $L(Y,Z,T)$. Then, $[F_1,F_2,F_3]^*(I(X,Y,T) = T)$ is equivalent to :

$$X \neq Y \wedge I(X,Y,Z) \neq Y \Rightarrow I(X,Y,I(Y,I(X,Y,Z),T)) = I(Y,I(X,Y,Z),T).$$

Since $I(X,Y,I(Y,I(X,Y,Z),T))\downarrow I(Y,I(X,Y,Z),T)$, then :

$$\vdash_{EL} X \neq Y \wedge L(X,Y,Z) \wedge Z \neq Y \wedge L(Y,Z,T) \Rightarrow L(X,Y,T).$$

example Let F_1 be $X \neq Y$, F_2 be $L(X,Y,Z)$, F_3 be $L(X,Y,T) \wedge T \neq Z$ and F_4 be $L(X,Y,U)$. Then, $[F_1,F_2,F_3,F_4]^*(I(Z,T,U) = U)$ is equivalent to :

$$X \neq Y \wedge I(X,Y,T) \neq I(X,Y,Z) \Rightarrow$$
$$I(I(X,Y,Z),I(X,Y,T),I(X,Y,U)) = I(X,Y,U).$$

Since $I(I(X,Y,Z),I(X,Y,T),I(X,Y,U))\downarrow I(X,Y,U)$, then :

$$\vdash_{EL} X \neq Y \wedge L(X,Y,Z) \wedge L(X,Y,T) \wedge T \neq Z \wedge L(X,Y,U) \Rightarrow L(Z,T,U).$$

3 Three-dimensional incidence geometry

The three-dimensional incidence geometry E_P is an extension of E_L. Its language contains the ternary relation L and a relation P of arity four : point × point × point × point. The relation of coplanarity between points is defined by the following postulates :

E_{P0} $\quad P(X,Y,Z,T) \Rightarrow P(X,Z,Y,T)$

E_{P1} $\quad L(X,Y,T) \Rightarrow P(X,Y,Z,T)$

E_{P2} $\quad P(X,Y,Z,T) \wedge P(X,Y,Z,U) \wedge L(T,U,V) \Rightarrow T = U \vee P(X,Y,Z,V)$

E_{P3} $\quad P(X,Y,Z,T) \wedge P(X,Y,Z,U) \wedge P(X,Y,Z,V) \Rightarrow L(X,Y,Z) \vee P(Z,T,U,V)$

These axioms, together with postulates E_{L0} through E_{L2} are equivalent to the incidence axioms of Hilbert [12]. Models of postulates E_{L0} through E_{L2} and E_{P0} through E_{P3} are incidence spaces. That the formula :

$$P(X,Y,Z,T) \Rightarrow P(Z,T,X,Y)$$

is a theorem of E_P can be proved in the following way.

(1)	$P(X,Y,Z,T) \wedge P(X,Y,Z,X) \wedge P(X,Y,Z,Y) \Rightarrow L(X,Y,Z) \vee P(Z,T,X,Y)$	axiom E_{P3}
(2)	$P(X,Y,Z,X)$	axioms E_{L1}, E_{P1}
(3)	$P(X,Y,Z,Y)$	axioms E_{L0}, E_{L1}, E_{P1}
(4)	$P(X,Y,Z,T) \Rightarrow L(X,Y,Z) \vee P(Z,T,X,Y)$	(1), (2), (3)
(5)	$L(X,Y,Z) \Rightarrow L(Z,X,Y)$	theorem of E_L
(6)	$L(Z,X,Y) \Rightarrow P(Z,T,X,Y)$	axioms E_{P1}, E_{P0}
(6)	$P(X,Y,Z,T) \Rightarrow P(Z,T,X,Y)$	(4), (5), (6)

The alphabet of the first order theory E_p with equality associated to E_P contains the function symbol l and a function symbol p of arity four : point × point × point × point → point, defining a set of terms denoted by T. The term $p(X,Y,Z,T)$ denotes a projection of T onto the plane passing through X, Y and Z. The first order theory E_p is defined by the postulates of E_l and the following postulates :

E_{p0} $\quad p(X,Y,Z,T) = p(Y,X,Z,T)$

E_{p1} $\quad p(X,Y,Z,T) = p(X,Z,Y,T)$

E_{p2} $\quad X \neq Y \wedge l(X,Y,Z) \neq Z \Rightarrow p(X,Y,Z,X) = X$

E_{p3} $\quad X \neq Y \wedge l(X,Y,Z) \neq Z \wedge T \neq U \wedge l(T,U,V) \neq V \wedge p(X,Y,Z,T) = T \wedge p(X,Y,Z,U) = U \wedge p(X,Y,Z,V) = V \Rightarrow p(X,Y,Z,p(T,U,V,W)) = p(T,U,V,W)$

E_{p4} $\quad X \neq Y \wedge l(X,Y,Z) \neq Z \wedge T \neq U \wedge p(X,Y,Z,T) = T \wedge p(X,Y,Z,U) = U \Rightarrow p(X,Y,Z,l(T,U,V)) = l(T,U,V)$

E_{p5} $\quad X \neq Y \wedge l(X,Y,p(Z,T,U,V)) \neq p(Z,T,U,V) \wedge Z \neq T \wedge l(Z,T,U) \neq U \wedge p(Z,T,U,X) = X \wedge p(Z,T,U,Y) = Y \Rightarrow p(X,Y,p(Z,T,U,V),W) = p(Z,T,U,W)$

E_{p6} $\quad X \neq Y \wedge l(X,Y,l(Z,T,U)) \neq l(Z,T,U) \wedge Z \neq T \wedge p(Z,T,X,Y) = Y \Rightarrow p(X,Y,l(Z,T,U),V) = p(Z,T,X,V)$

Let * be the function of translation between the languages of E_P and E_p defined by :

$$P(X,Y,Z,T)^* = (X = Y \vee l(X,Y,Z) = Z \vee p(X,Y,Z,T) = T).$$

proposition 3.1 If F is a statement in the language of E_P and F^* is its translation in the language of E_p then $\vdash_{E_P} F$ if and only if $\vdash_{E_p} F^*$.

Let R_p be the P/N CTRS made of the rules of R_l and the following rules :

R_{p2} $X \neq Y \wedge I(X,Y,Z) \neq Z \Rightarrow p(X,Y,Z,X) \rightarrow X$

R_{p3} $X \neq Y \wedge I(X,Y,Z) \neq Z \wedge T \neq U \wedge I(T,U,V) \neq V \wedge$
$p(X,Y,Z,T) = T \wedge p(X,Y,Z,U) = U \wedge p(X,Y,Z,V) = V \Rightarrow$
$p(X,Y,Z,p(T,U,V,W)) \rightarrow p(T,U,V,W)$

R_{p4} $X \neq Y \wedge I(X,Y,Z) \neq Z \wedge T \neq U \wedge p(X,Y,Z,T) = T \wedge$
$p(X,Y,Z,U) = U \Rightarrow p(X,Y,Z,I(T,U,V)) \rightarrow I(T,U,V)$

R_{p5} $X \neq Y \wedge I(X,Y,p(Z,T,U,V)) \neq p(Z,T,U,V) \wedge Z \neq T \wedge$
$I(Z,T,U) \neq U \wedge p(Z,T,U,X) = X \wedge p(Z,T,U,Y) = Y \Rightarrow$
$p(X,Y,p(Z,T,U,V),W) \rightarrow p(Z,T,U,W)$

R_{p6} $X \neq Y \wedge I(X,Y,I(Z,T,U)) \neq I(Z,T,U) \wedge Z \neq T \wedge$
$p(Z,T,X,Y) = Y \Rightarrow p(X,Y,I(Z,T,U),V) \rightarrow p(Z,T,X,V)$

Let C_p be the first order equational theory generated by E_{I0}, E_{p0} and E_{p1}. Unification and matching under C_p are finitary and constitute decidable classes of problems for which type conformal algorithms exist. Consequently, the rules of R_p define a decidable and terminating relation of reduction $\rightarrow$ on T such that $t \rightarrow t'$ if and only if there are a position p in t, a rule $Q \Rightarrow l \rightarrow r$ in R_p and a substitution σ such that $t/p =_{C_p} l\sigma$, $t' = t[r\sigma]_p$ and $Q\sigma\downarrow$. Moreover, it can be proved that the relation of reduction $\rightarrow$ is confluent on T. See Balbiani et al [3] for details.

4 Miquelian geometry

Miquelian geometry is made of two sets of geometrical beings of types points and circles, together with two incidence relations of arity two : point × circle and circle × point. Circles will be denoted by lower case letters x, y, z, etc. The relations of incidence between points and circles are defined by the following postulates :

E_{M0} X is incident with x if and only if x is incident with X

E_{M1} Three pairwise distinct points are together incident with exactly one circle

E_{M2} If a point X is incident with a circle x then, for every point Y not incident with x, there is exactly one circle incident with Y whose only common point with x is X

E_{M3} If each cyclically adjacent pair of four circles have a pair of common points, forming altogether eight distinct points, and if four of these points, one from each pair, are incident with a circle, then the remaining four points are incident with a circle too

Models of postulates E_{M0} through E_{M3} are miquelian planes [8]. The alphabet of the first order theory E_m with equality associated to the theory E_M defined by E_{M0} and E_{M1} contains a function symbol c of arity three : point × point × point → circle and a function symbol p of arity two : point × circle → point, defining a set of terms denoted by T. The term c(X,Y,Z) denotes the circle passing through X, Y and

Z. The term $p(X,x)$ denotes a projection of the point X onto the circle x. The first order theory E_m is defined by the following postulates :

E_m0 $c(X,Y,Z) = c(Y,X,Z)$

E_m1 $c(X,Y,Z) = c(X,Z,Y)$

E_m2 $X \neq Y \wedge X \neq Z \wedge Y \neq Z \Rightarrow p(X,c(X,Y,Z)) = X$

E_m3 $p(p(X,x),x) = p(X,x)$

E_m4 $X \neq Y \wedge X \neq p(Z,x) \wedge Y \neq p(Z,x) \wedge p(X,x) = x \wedge$
$p(Y,x) = x \Rightarrow c(X,Y,p(Z,x)) = x$

Let * be the function of translation between the languages of E_M and E_m defined by : $(X \text{ is incident with } x)^* = (x \text{ is incident with } X)^* = (p(X,x) = X)$.

proposition 4.1 If F is a statement in the language of E_M and F^* is its translation in the language of E_m then $\vdash_{E_M} F$ if and only if $\vdash_{E_m} F^*$.

Let R_m be the P/N CTRS made of the following rules :

R_m2 $X \neq Y \wedge X \neq Z \wedge Y \neq Z \Rightarrow p(X,c(X,Y,Z)) \rightarrow X$

R_m3 $p(p(X,x),x) \rightarrow p(X,x)$

R_m4 $X \neq Y \wedge X \neq p(Z,x) \wedge Y \neq p(Z,x) \wedge p(X,x) = x \wedge$
$p(Y,x) = x \Rightarrow c(X,Y,p(Z,x)) \rightarrow x$

Let C_m be the first order equational theory generated by E_m0 and E_m1. Unification and matching under C_m are finitary and constitute decidable classes of problems for which type conformal algorithms exist. Consequently, the rules of R_m define a decidable and terminating relation of reduction $\rightarrow$ on T such that $t \rightarrow t'$ if and only if there are a position p in t, a rule $Q \Rightarrow l \rightarrow r$ in R_m and a substitution σ such that $t/p =_{C_m} l\sigma$, $t' = t[r\sigma]_p$ and $Q\sigma\downarrow$. Moreover, it can be proved that the relation of reduction $\rightarrow$ is confluent on T. See Balbiani et al [3] for details.

5 Projective geometry

Projective geometry E_l is made of two sets of geometrical beings of types points and lines, together with two incidence relations of arity two : point × line and line × point. Lines will be denoted by lower case letters x, y, z, etc. The relations of incidence between points and lines are defined by the following postulates :

E_l0 X is incident with x if and only if x is incident with X

E_l1 Two distinct points are together incident with exactly one line

E_l2 Two distinct lines are together incident with exactly one point

Models of postulates E_{l0} through E_{l2} are projective planes. The alphabet of the first order theory E_i with equality associated to the theory E_l contains a function symbol l of arity two : point × point → line and a function symbol i of arity two : line × line → point, defining a set of terms denoted by T. The term $l(X,Y)$ denotes the line passing through the points X and Y. The term $i(x,y)$ denotes the intersection point of the lines x and y. It also contains a function symbol p of arity two : point × line → point and a function symbol q of arity two : line × point → line. The term $p(X,x)$ denotes a projection of the point X onto the line x such that if X is incident with x then $p(X,x)$ is equal to X. The term $q(x,X)$ denotes a projection of the line x onto the point X such that if x is incident with X then $q(x,X)$ is equal to x. The first order theory E_i is defined by the following postulates :

E_{i0} $\quad l(X,Y) = l(Y,X)$

E_{i1} $\quad i(x,y) = i(y,x)$

E_{i2} $\quad X \neq i(x,y) \wedge x \neq y \wedge p(X,x) = X \Rightarrow l(X,i(x,y)) = x$

E_{i3} $\quad x \neq l(X,Y) \wedge X \neq Y \wedge q(x,X) = x \Rightarrow i(x,l(X,Y)) = X$

E_{i4} $\quad X \neq p(Y,x) \wedge p(X,x) = X \Rightarrow l(X,p(Y,x)) = x$

E_{i5} $\quad x \neq q(y,X) \wedge q(x,X) = x \Rightarrow i(x,q(y,X)) = X$

E_{i6} $\quad X \neq Y \Rightarrow p(X,l(X,Y)) = X$

E_{i7} $\quad x \neq y \Rightarrow q(x,i(x,y)) = x$

E_{i8} $\quad x \neq y \Rightarrow p(i(x,y),x) = i(x,y)$

E_{i9} $\quad X \neq Y \Rightarrow q(l(X,Y),X) = l(X,Y)$

E_{i10} $\quad p(p(X,x),x) = p(X,x)$

E_{i11} $\quad q(q(x,X),X) = q(x,X)$

E_{i12} $\quad p(X,q(x,X)) = X$

E_{i13} $\quad q(x,p(X,x)) = x$

Let * be the function of translation between the languages of E_l and E_i defined by : $(X \text{ is incident with } x)^* = (p(X,x) = X)$ and $(x \text{ is incident with } X)^* = (q(x,X) = x)$.

proposition 5.1 If F is a statement in the language of E_l and F^* is its translation in the language of E_i then $\vdash_{E_l} F$ if and only if $\vdash_{E_i} F^*$.

Let R_i be the P/N CTRS made of the following rules :

R_{i2} $\quad X \neq i(x,y) \wedge x \neq y \wedge p(X,x) = X \Rightarrow l(X,i(x,y)) \to x$

R_{i3} $\quad x \neq l(X,Y) \wedge X \neq Y \wedge q(x,X) = x \Rightarrow i(x,l(X,Y)) \to X$

R_{i4} $\quad X \neq p(Y,x) \wedge p(X,x) = X \Rightarrow l(X,p(Y,x)) \to x$

R_{i5} $\quad x \neq q(y,X) \wedge q(x,X) = x \Rightarrow i(x,q(y,X)) \to X$

R_{i6} $\quad X \neq Y \Rightarrow p(X,l(X,Y)) \to X$

R_{i7}	$x \neq y \Rightarrow q(x,i(x,y)) \rightarrow x$
R_{i8}	$x \neq y \Rightarrow p(i(x,y),x) \rightarrow i(x,y)$
R_{i9}	$X \neq Y \Rightarrow q(I(X,Y),X) \rightarrow I(X,Y)$
R_{i10}	$p(p(X,x),x) \rightarrow p(X,x)$
R_{i11}	$q(q(x,X),X) \rightarrow q(x,X)$
R_{i12}	$p(X,q(x,X)) \rightarrow X$
R_{i13}	$q(x,p(X,x)) \rightarrow x$

Let C_i be the first order equational theory generated by E_{i0} and E_{i1}. Unification and matching under C_i are finitary and constitute decidable classes of problems for which type conformal algorithms exist. Consequently, the rules of R_i define a decidable and terminating relation of reduction $\rightarrow$ on T such that $t \rightarrow t'$ if and only if there are a position p in t, a rule $Q \Rightarrow l \rightarrow r$ in R_i and a substitution σ such that $t/p =_{C_i} l\sigma$, $t' = t[r\sigma]_p$ and $Q\sigma\downarrow$. Moreover, it can be proved that the relation of reduction $\rightarrow$ is confluent on T. See Balbiani et al [3] for details. We present now the mechanization of projective geometry. A geometrical fact is one of the following four sentences :

(X is incident with x)
(X is the only point incident with x and y)
(x is incident with X)
(x is the only line incident with X and Y)

In the geometrical fact (X is incident with x) which defines X, the point X depends on the line x. In the geometrical fact (X is the only point incident with x and y) which defines X, X depends on x and y. In the geometrical fact (x is incident with X) which defines x, x depends on X. In the geometrical fact (x is the only line incident with X and Y) which defines x, x depends on X and Y. A constructive geometrical proposition is a finite sequence of geometrical facts in which no point depends directly or indirectly on itself. Let $G = [F_1,\ldots,F_k]$ be a constructive proposition and E be a quantifier-free formula in the language of E_i. Then, G^*F is the equational formula in the language of E_i defined by induction on the length of G in the following way :

(i) $[F_1,\ldots,F_{k-1},$(X is incident with x)$]^*E$ is
$[F_1,\ldots,F_{k-1}]^*E\sigma$ where σ is the substitution $\{p(X,x)/X\}$

(ii) $[F_1,\ldots,F_{k-1},$(X is the only point incident with x and y)$]^*E$ is
$[F_1,\ldots,F_{k-1}]^*(x \neq y \Rightarrow E\sigma)$ where σ is the substitution $\{i(x,y)/X\}$

(iii) $[F_1,\ldots,F_{k-1},$(x is incident with X)$]^*E$ is
$[F_1,\ldots,F_{k-1}]^*E\sigma$ where σ is the substitution $\{q(x,X)/x\}$

(iv) $[F_1,\ldots,F_{k-1},$(x is the only line incident with X and Y)$]^*E$ is
$[F_1,\ldots,F_{k-1}]^*(X \neq Y \Rightarrow E\sigma)$ where σ is the substitution $\{I(X,Y)/x\}$

Next proposition associates to every constructive geometrical proposition $G = [F_1,\dots,F_k]$ and to every variable X, x, an equational formula $Q \Rightarrow t_1 = t_2$ logically equivalent to $F_1^* \wedge \dots \wedge F_k^* \Rightarrow$ (X is incident with x)*.

> **proposition 5.2** Let $G = [F_1,\dots,F_k]$ be a constructive proposition. Then, $G^*(p(X,x) = X)$ is equivalent to an equational formula of the form $Q \Rightarrow t_1 = t_2$ logically equivalent to $F_1^* \wedge \dots \wedge F_k^* \Rightarrow$ (X is incident with x)*.

The proof that $t_1 \downarrow t_2$ is a necessary and sufficient condition for $F_1 \wedge \dots \wedge F_k \Rightarrow$ (X is incident with x) to be a theorem of E_I needs the following notion of reduction with respect to a set of disequations. For every set S of disequations, there is a decidable and terminating relation of reduction $\rightarrow_S$ such that, for every term t and t', $t \rightarrow_S t'$ iff there are :

- **(i)** a position p of t,
- **(ii)** a substitution σ and
- **(iii)** a variant $Q \Rightarrow l \rightarrow r$ of a rule such that
- **(iv)** $t|_p =_{CI} l\sigma$,
- **(v)** $t' = t[r\sigma]_p$ and
- **(vi)** $Q\sigma\downarrow_S$,

that is to say : for every u=v in Q, $u\sigma\downarrow_S v\sigma$ and, for every $u \neq v$ in Q, there is $u' \neq v'$ in S such that $u\sigma\downarrow_S u'$ and $v\sigma\downarrow_S v'$, the relation $\downarrow_S$ being defined by $u\downarrow_S v$ if there are two terms w and w' such that $w =_{CI} w'$, $u \rightarrow_{S^*} w$ and $v \rightarrow_{S^*} w'$, the relation $\rightarrow_{S^*}$ being the reflexive and transitive closure of $\rightarrow_S$. This reduction relation can be defined inductively. It can be proved that if S is consistent then $\rightarrow_S$ is confluent. This is a consequence of the fact that for every term w, if $w \rightarrow w'$ then $\vdash_{Ei} S \Rightarrow w = w'$. See Balbiani et al [3] for details. The function denoted by the symbols I and i are only partial, that is to say : they are not properly defined when their two arguments are equal. A set S of equations and disequations is saturated if, for every of its subterm of the form I(u,v) or i(u,v) it contains a disequation $u' \neq v'$ such that $u\downarrow_S u'$ and $v\downarrow_S v'$.

> **proposition 5.3** Let $G = [F_1,\dots,F_k]$ be a constructive proposition. Then, $G^*(p(X,x) = X)$ is equivalent to an equational formula of the form $Q \Rightarrow t_1 = t_2$ such that the set $Q \cup \{t_1 = t_2\}$ is saturated.

Let S be any saturated set of disequations. Then, for every subterm w of S, if $w \rightarrow w'$ then $w \rightarrow_S w'$ (this can be proved by induction on the complexity of w). Now suppose that S is also consistent. Let $u \neq v$ be in S. If $u\downarrow_S v$ then $\vdash_{EI} S \Rightarrow u = v$, a contradiction with the consistency of S. As a consequence, $u \not\downarrow v$. Now, we can prove the following proposition which relates the consistency of $F_1 \wedge \ldots \wedge F_k$ in E_l with $Q\downarrow$.

proposition 5.4 Let $G = [F_1,\ldots,F_k]$ be a constructive proposition. Then, $G^*(p(X,x) = X)$ is equivalent to an equational formula in the language of E_l of the form $Q \Rightarrow t_1 = t_2$ such that $F_1 \wedge \ldots \wedge F_k$ is consistent in E_l if and only if $Q\downarrow$.

As a consequence, there is a necessary and sufficient condition for $F_1 \wedge \ldots \wedge F_k \Rightarrow$ (X is incident with x) to be a theorem of E_l and this is : either $Q\not\downarrow$ or $t_1 \downarrow t_2$.

proposition 5.5 Let $G = [F_1,\ldots,F_k]$ be a constructive proposition. Then, $G^*(p(X,x) = X)$ is equivalent to an equational formula in the language of E_l of the form $Q \Rightarrow t_1 = t_2$ such that $\vdash_{E_l} F_1 \wedge \ldots \wedge F_k \Rightarrow$ (X is incident with x) if and only if either $Q\not\downarrow$ or $t_1 \downarrow t_2$.

example Let :

F_1 be (X is incident with a),
F_2 be (Y is incident with a),
F_3 be (z is the only line incident with X and Y) and
F_4 be (Z is incident with z).

Then $[F_1,F_2,F_3,F_4]^*(p(Z,a) = Z)$ is equivalent to $p(X,a) \neq p(Y,a) \Rightarrow p(p(Z,l(p(X,a),p(Y,a))),a)=p(Z,l(p(X,a),p(Y,a)))$. Since $p(X,a)$ and $p(Y,a)$ do not have a same normal form, then $F_1 \wedge F_2 \wedge F_3 \wedge F_4$ is consistent. Since $p(p(Z,l(p(X,a),p(Y,a))),a)\downarrow p(Z,l(p(X,a),p(Y,a)))$, then the formula

$$F_1 \wedge F_2 \wedge F_3 \wedge F_4 \Rightarrow (Z \text{ is incident with } a)$$

is a theorem of E_l.

example Let :

F_1 be (a is incident with X),
F_2 be (b is incident with Y),
F_3 be (Z is the only point incident with a and b) and
F_4 be (z is the only line incident with X and Y).

Then $[F_1,F_2,F_3,F_4]^*(p(Z,z) = Z)$ is equivalent to $q(a,X) \neq q(b,Y) \wedge X \neq Y \Rightarrow p(i(q(a,X),q(b,Y)),l(X,Y)) = i(q(a,X), q(b,Y))$. Since $q(a,X)$ and $q(b,Y)$ do not have a same normal form, then $F_1 \wedge F_2 \wedge F_3 \wedge F_4$ is consistent. Since $p(i(q(a,X),q(b,Y)),l(X,Y)) \not\downarrow i(q(a,X), q(b,Y))$, then the formula

$$F_1 \wedge F_2 \wedge F_3 \wedge F_4 \Rightarrow (Z \text{ is incident with } z)$$

is not a theorem of E_I.

6 Conclusion

We have presented a family of incidence geometries together with their first order axiomatizations with equality. The conditional equational axioms of these theories defined P/N CTRS. The termination and the confluence of these systems are an essential element of their use for the mechanization of geometrical reasoning. In our work, geometry is not defined with relations (incidence, order, congruence, etc) as it is the case with most of the geometers but with functions. A figure is then represented by a first order term and a figure property is represented by an equation between two terms.

This presentation of geometry is justified by the constructive aspect of theorem proving in synthetic geometries. To show that a figure possesses such and such property is to prove that two terms are equal in some first order theory with equality. Problems of figure analysis becomes rewriting problems : reduce two terms in their normal forms and prove that this normal forms are equal. During the computation of these normal forms, some negative conditions are evaluated. They correspond, for example, to the unicity of the line passing through two points (the two points should be distinct) or to the unicity of the plane passing through three points (the three points should not be collinear). These negative conditions can be directly translated in geometrical terms. Every rewriting step on the first order terms of our geometries can also be interpreted immediatly in a purely geometrical deduction step.

The extension of our method to more conceptual geometries (geometries with order or congruence, euclidean or hyperbolic geometries, etc) has been partially realized and presented by Balbiani et al [3]. Nevertheless, we are far away from a complete mechanization of geometry and our work should be pursued. A next step could be the extension of our conditional equational axiomatizations to spatial projective geometry. The mechanization of this geometry being still an open problem for mathematicians [21].

Acknowledgement

Special acknowledgement is heartly granted to Luis Fariñas del Cerro and Harald Ganziger for helpful comments on a preliminary version of this paper.

References

[1] F. Baader and J. Siekmann. Unification theory. In D. Gabbay, C. Hogger and J. Robinson (Eds) : Handbook of Logic in Artificial Intelligence and Logic Programming. Oxford University Press, Oxford, England, to appear.

[2] L. Bachmair. Canonical Equational Proofs. Birkhaüser, Boston, Massachusetts, 1991.

[3] P. Balbiani, V. Dugat, L. Fariñas del Cerro and A. Lopez. Eléments de géométrie mécanique. Hermès, Paris, France, 1994.

[4] L. M. Blumenthal. A Modern View of Geometry. Freeman, San Francisco, California, 1961.

[5] B. Buchberger. Gröbner bases : an algorithmic method in polynomial ideal theory. In : N Bose (Ed.), Recent Trends in Multidimensional Systems Theory. Reidel, Dordrecht, Netherlands, 1985.

[6] B. Buchberger, G. Collins and B. Kutzler. Algebraic methods for geometric reasoning. In : Ann. Rev. Comput. Sci., volume 3, pages 85-119, 1988.

[7] S.-C. Chou. Mechanical Geometry Theorem Proving. Reidel, Dordrecht, Netherlands, 1988.

[8] H. Coxeter. The finite inversive plane with four points on each circle. In : L. Mirsky (Ed.), Papers in Combinatorial Theory, Analysis, Geometry, Algebra and the Theory of Numbers. Academic Press, London, Great Britain, 1971, pp. 39-51.

[9] N. Dershowitz and J.-P. Jouannaud. Rewrite systems. In J. van Leeuwen (Ed) : Handbook of Theoretical Computer Science, volume B, Formal Models and Semantics, pages 243-320. Elsevier, Amsterdam, Netherlands, 1990.

[10] H. Ganzinger. A completion procedure for conditional equations. In S. Kaplan and J.-P. Jouannaud (Eds) : Conditional Term Rewriting Systems, 1st International Workshop, Orsay, France, July 1987, Proceedings, pages 62-83. Lecture Notes in Computer Science 308, Springer-Verlag, Berlin, Germany, 1988.

[11] A. Heyting. Axiomatic Projective Geometry. North-Holland, Amsterdam, Netherlands, 1963.

[12] D. Hilbert. Foundations of Geometry. Open Court, La Salle, Illinois, 1971.

[13] J.-P. Jouannaud and C. Kirchner. Solving equations in abstract algebras: a rule based survey of unification. In J.-L. Lassez and G. Plotkin (Eds) : Computational Logic, Essays in Honor of Alan Robinson. MIT Press, Cambridge, Massachusetts, pages 257-321.

[14] S. Kaplan. Positive/negative conditional rewriting. In : S. Kaplan, J.-P. Jouannaud (Eds), Conditional Term Rewriting Systems, 1st International Workshop, Orsay, France, July, 1987, Proceedings. Lecture Notes in Computer Science 308, Springer-Verlag, Berlin, Germany, 1988, pp. 129-143.

[15] S. Kaplan and J.-L. Rémy. Completion algorithms for conditional rewriting systems. In H. Aït-Kaci and M. Nivat (Eds) : Resolution of Equations in Algebraic Structures, volume 2, Rewriting Techniques, pages 141-170. Academic Press, San Diego, California, 1989.

[16] M. Rusinowitch. Démonstration automatique : techniques de réécriture. InterEditions, Paris, France, 1989.

[17] J. Siekmann. Unification theory. In : C. Kirchner (Ed.), Unification. Academic Press, London, Great Britain, 1990, pp. 1-68.

[18] A. Tarski. What is elementary geometry ? In L. Henkin, P. Suppes and A. Tarski (Eds) : The Axiomatic Method, with Special Reference to Geometry and Physics, pages 16-29. North-Holland, Amsterdam, Netherlands, 1959.

[19] D. Wang. A new theorem discovered by computer prover. In : Journal of Geometry, volume 36, pages 173-182, 1989.

[20] W.-T. Wu. Basic principles of mechanical theorem proving in elementary geometries. In : Journal of Automated Reasoning, volume 2, number 3, pages 221-252, 1986.

[21] W.-T. Wu. Mechanical Theorem Proving in Geometries. Springer-Verlag, Wien, Austria, 1994.

Annex

proposition 2.2.1 If F is a statement in the language of E_L and F^* is its translation in the language of E_I then $\vdash_{E_L} F$ if and only if $\vdash_{E_I} F^*$.

proof $\Rightarrow$. The proof is done by induction on the proof of F in E_L. We consider the case where F is the axiom E_{L2}. In this case F^* is the equational formula $(X = Y \vee I(X,Y,Z) = Z) \wedge (X = Y \vee I(X,Y,T) = T) \Rightarrow X = Y \vee (Y = Z \vee I(Y,Z,T) = T)$ which is a theorem of E_I since if $X \neq Y$, $I(X,Y,Z) = Z$, $I(X,Y,T) = T$ and $Y \neq Z$ then $I(X,Y,Y) = I(Y,X,Y) = Y$ and $I(Y,Z,T) = I(Y,I(X,Y,Z),T) = I(X,Y,T) = T$.

$\Leftarrow$. Let M be a model of E_L and D be its domain. Let M^* be the interpretation of E_I which has the same domain as M and which is defined by : for every element X, Y and Z in D, if X and Y are distinct and if $L(X,Y,Z)$ is true in M then the value of $I(X,Y,Z)$ is Z otherwise this value is any point collinear with X and Y. It is straightforward to see that M^* is a model of E_I. For any variable U and for any element A in D, $M[U=A]$ denotes the model assigning the value A to U. The same definition applies for $M^*[U=A]$ too. It should be remarked that $(M[U=A])^* = M^*[U=A]$. Let it be proved by induction on the complexity of F that, for every model M of E_L, if M is a model of F then M^* is a model of F^*.

first case : F is $L(X,Y,Z)$. If M is a model of F then either the interpretations of X and Y in M^* are equal or the interpretations of $I(X,Y,Z)$ and Z in M^* are equal. Therefore, M^* is a model of F^*.
second case : F is $\neg L(X,Y,Z)$. If M is a model of F then the interpretations of X and Y in M^* are distinct and the interpretations of $I(X,Y,Z)$ and Z in M^* are distinct. Therefore, M^* is a model of F^*.
third case : F is the formula $F_1 \wedge F_2$. If M is a model of F then M is a model of F_1 and M is a model of F_2. Thus, by induction hypothesis, M^* is a model of F_1^*, M^* is a model of F_2^* and M^* is a model of F^*. The same argument applies if F is the formula $F_1 \vee F_2$.
fourth case : F is the formula $\forall U\ F_1$. If M is a model of F then, for every element A in D, $M[U=A]$ is a model of F_1. Thus, by induction hypothesis, for every element A in D, $M^*[U=A]$ is a model of F_1^*. Therefore, M^* is a model of F^*. The same argument applies if F is the formula $\exists U\ F_1$. **QED**

proposition 2.4.1 Let $G = [F_1,\dots,F_k]$ be a constructive proposition and E be a quantifier-free equational formula in the language of E_I. Then, G^*E is logically equivalent to $F_1^* \wedge \dots \wedge F_k^* \Rightarrow E$.

proof The proof is done by induction on the length of the constructive proposition. Let $G = [F_1,\dots,F_k]$ be a constructive proposition and E be a quantifier-free equational formula in the language of E_I. We consider the case where F_k is $L(X,Y,Z) \wedge Z \neq X_1 \wedge \dots \wedge Z \neq X_n$. In this case G^*E is $[F_1,\dots,F_{k-1}]^*(Z \neq X_1 \wedge \dots \wedge Z \neq X_n \Rightarrow E)\sigma$ where σ is the substitution $\{I(X,Y,Z)/Z\}$ and, by induction hypothesis, G^*E is logically equivalent to $F_1^* \wedge \dots \wedge F_{k-1}^* \Rightarrow (Z \neq X_1 \wedge \dots \wedge Z \neq X_n \Rightarrow E)\sigma$. Thus, G^*E is logically equivalent to $F_1^* \wedge \dots \wedge F_{k-1}^* \wedge I(X,Y,Z) \neq X_1 \wedge \dots \wedge I(X,Y,Z) \neq X_n \Rightarrow E\sigma$. Since, for some $j=1,\dots,k-1$, F_j contains the disequation $X \neq Y$, then G^*E is logically equivalent to $F_1^* \wedge \dots \wedge F_{k-1}^* \wedge (X = Y \vee I(X,Y,Z) = Z) \wedge Z \neq X_1 \wedge \dots \wedge Z \neq X_n \Rightarrow E$. **QED**

proposition 5.2 Let $G = [F_1,\dots,F_k]$ be a constructive proposition. Then, $G^*(p(X,x) = X)$ is equivalent to an equational formula of the form $Q \Rightarrow t_1 = t_2$ logically equivalent to $F_1^* \wedge \dots \wedge F_k^* \Rightarrow (X \text{ is incident with } x)^*$.

proof The proof is done by induction on the length of the constructive sequence. If $k=0$ then Q is true and $(X \text{ is incident with } x)^*$ is the equational formula $(p(X,x) = X)$. Now, suppose that $F_1^* \wedge ... \wedge F_k^* \Rightarrow (X \text{ is incident with } x)^*$ and $Q \Rightarrow t_1=t_2$ are logically equivalent in E_i and consider F_0 such that $[F_0,F_1,...,F_k]$ is a constructive sequence. We consider the case when F_0 is (U is the only point incident with u and v). In this case F_0^* is equivalent to the equational formula $u \neq v \wedge U = i(u,v)$. Let σ be the sustitution $\{i(u,v)/U\}$. Consequently, $u \neq v \wedge U = i(u,v) \wedge F_1^* \wedge ... \wedge F_k^* \Rightarrow (X \text{ is incident with } x)^*$ and $u \neq v \wedge Q\sigma \Rightarrow t_1\sigma=t_2\sigma$ are logically equivalent. **QED**

proposition 5.3 Let $G = [F_1,...,F_k]$ be a constructive proposition. Then, $G^*(p(X,x) = X)$ is equivalent to an equational formula of the form $Q \Rightarrow t_1 = t_2$ such that the set $Q \cup \{t_1=t_2\}$ is saturated.

proof The proof is done by induction on the length of the constructive sequence $[F_1,...,F_k]$. If $k=0$ then Q is empty. Now, suppose that the set $Q \cup \{t_1=t_2\}$ associated to $[F_1, ..., F_k]$ is saturated and consider a fact F_0 such that $F_0, F_1, F_2, ..., F_k$ is a constructive sequence of geometrical facts. We consider the case where F_0 is (U is the only point incident with u and v). Let σ be the substitution $\{i(u,v)/U\}$. Since $Q \cup \{t_1=t_2\}$ is saturated then $Q\sigma \cup \{u \neq v\} \cup \{t_1\sigma=t_2\sigma\}$ is saturated too. **QED**

proposition 5.4 Let $G = [F_1,...,F_k]$ be a constructive proposition. Then, $G^*(p(X,x) = X)$ is equivalent to an equational formula in the language of E_i of the form $Q \Rightarrow t_1 = t_2$ such that $F_1 \wedge ... \wedge F_k$ is consistent in E_I if and only if $Q\downarrow$.

proof $\Rightarrow$. Let M be a model of projective geometry in which $F_1 \wedge ... \wedge F_k$ is true and D be its domain. Let M^* be an interpretation of E_i which has the same domain as M and which is defined by, for every element X and Y in D, the value of $I(X,Y)$ is the line of M on X and Y and, for every element x and y in D, the value of $i(x,y)$ is the point of M on x and y. It is straightforward to prove that M^* is a model of E_i. Let j be in $\{0,...,k\}$ and Q_j be the set of disequations associated to $[F_{j+1},...,F_k]^*(p(X,x) = X)$. The proof is done by induction on j that Q_j is true in M^*. If $j=k$ then Q_j is empty. Let j be in $\{1,...,k\}$ such that Q_j is true in M^*. We consider the case where F_j is (U is the only point incident with u and v). In this case Q_{j-1} is $u \neq v \wedge Q_j\sigma$ where σ is the substitution $\{i(u,v)/U\}$. Since the interpretations of u and v in M^* are distinct and the interpretations of U and

$i(u,v)$ in M^* are equal, then Q_{j-1} is true in M^*. Consequently, Q is consistent and, since Q is saturated, then $Q\downarrow$.
$\Leftarrow$. If $Q\not\downarrow$ then Q is true in M_i, the quasi-initial model of E_i. Let M^* be an interpretation of projective geometry which domain is the set of congruence classes modulo C_i of the normal forms generated by the reduction relation $\rightarrow$ and which relations of incidence are defined by NF(X) is incident with NF(x) if and only if $p(X,x)\downarrow X$ and NF(x) is incident with NF(X) if and only if $q(x,X)\downarrow x$. It is straightforward to prove that M^* is a model of projective geometry. Let the interpretation in M^* of a geometrical being appearing in F_1, ..., F_k be the congruence class modulo C_i of the normal form of the proper subterm of t_1 it corresponds to. Let j be in {1,...,k}. We consider the case where F_i is (U is the only point incident with u and v). In this case there is a substitution σ such that U, u and v respectively correspond to proper subterms $i(w\sigma,w'\sigma)$, $w\sigma$ and $w'\sigma$ of t_1. Since $p(i(w\sigma,w'\sigma),w\sigma)\downarrow i(w\sigma,w'\sigma)$ then F_j is true in M^*. **QED**

proposition 5.5 Let $G = [F_1,...,F_k]$ be a constructive proposition. Then, $G^*(p(X,x) = X)$ is equivalent to an equational formula in the language of E_i of the form $Q \Rightarrow t_1 = t_2$ such that $\vdash_{EI} F_1 \wedge ... \wedge F_k \Rightarrow$ (X is incident with x) if and only if either $Q\not\downarrow$ or $t_1\downarrow t_2$.

proof $\Rightarrow$. If $Q\downarrow$ and $t_1 \not\downarrow t_2$ then $Q \wedge t_1 \neq t_2$ is true in M_i, the quasi-initial model of E_i, and $Q \Rightarrow t_1 = t_2$ is not a theorem of E_i, neither is $F_1^* \wedge ... \wedge F_k^*$ $\Rightarrow$ (X is incident with x)* and, consequently, $F_1 \wedge ... \wedge F_k \Rightarrow$ (X is incident with x) is not a theorem of E_I.
$\Leftarrow$. If $Q\not\downarrow$ or $t_1\downarrow t_2$ then $\vdash_{Ei} Q \Rightarrow t_1=t_2$, $\vdash_{Ei} F_1^* \wedge ... \wedge F_k^* \Rightarrow$ (X is incident with x)* and, consequently, $\vdash_{EI} F_1 \wedge ... \wedge F_k \Rightarrow$ (X is incident with x). **QED**

LSE Narrowing for Decreasing Conditional Term Rewrite Systems*

Alexander Bockmayr[1], Andreas Werner[2]

[1] MPI Informatik, Im Stadtwald, D-66123 Saarbrücken, bockmayr@mpi-sb.mpg.de
[2] SFB 314, Univ. Karlsruhe, D-76128 Karlsruhe, werner@ira.uka.de

Abstract. LSE narrowing is known as an optimal narrowing strategy for arbitrary unconditional canonical term rewrite systems without additional properties such as orthogonality or constructor discipline. In this paper, we extend LSE narrowing to confluent and decreasing conditional term rewrite systems.

1 Introduction

Narrowing is a universal unification procedure for equational theories defined by canonical term rewrite systems. It is also the operational semantics of various logic and functional programming languages (Hanus 1994). Narrowing was introduced in (Fay 1979; Hullot 1980) for the unconditional case, and generalized in (Kaplan 1984b; Dershowitz and Plaisted 1985; Hußmann 1985) to the conditional case. In its original form, it is extremely inefficient. Therefore, many optimizations have been proposed during the last years. In (Bockmayr, Krischer, and Werner 1992, 1993) we introduced a new narrowing strategy for unconditional term rewrite systems, LSE narrowing, which is complete for arbitrary canonical systems and improves all other strategies with this property. LSE narrowing is optimal in the sense that two different LSE narrowing derivations cannot generate the same narrowing substitution. Moreover, all narrowing substitutions computed by LSE narrowing are normalized. Empirical results demonstrating the benefits of this strategy have been given in (Bockmayr, Krischer, and Werner 1993), an efficient implementation has been described in (Werner, Bockmayr, and Krischer 1993, 1994).

LSE narrowing works for arbitrary canonical term rewrite systems and does not require additional properties such as orthogonality or constructor discipline. For restricted classes of term rewrite systems, more efficient strategies may be possible. For example, in (Antoy, Echahed, and Hanus 1994) a narrowing strategy for inductively sequential term rewrite systems is developed, which is optimal in the sense that only needed narrowing steps are performed, the narrowing derivations have a minimal length, and the solutions found in different derivations are independent. However, these properties no longer hold if more general systems, for example with overlapping left-hand sides, are considered.

* The first author's work was supported by the German Ministry for Research and Technology (BMFT) under grant ITS 9103 and the ESPRIT Working Group CCL (contract EP 6028). The second author's work was supported by the Deutsche Forschungsgemeinschaft as part of the SFB 314 (project S2).

The aim of this paper is to extend LSE narrowing to the case of confluent and decreasing *conditional* term rewrite systems. The main tool is the calculus of conditional rewriting without evaluation of the premise (Bockmayr 1990). It is used to associate with each conditional rewriting derivation a conditional narrowing derivation. Conditional LSE narrowing derivations then correspond to a special form of leftmost-innermost conditional rewriting derivations. We show that most of our previous results can be generalized to the conditional case. However, some of them hold only for successful derivations and also minor additional hypotheses may be necessary. The organization of the paper is as follows. After some preliminaries in Sect. 2, we present in Sect. 3 conditional rewriting without evaluation of the premise and relate it to ordinary conditional rewriting and to conditional narrowing. Conditional LSE narrowing is introduced in Sect. 4 and its main properties are proven. In Sect. 5, we extend the results to the case of normalizing conditional LSE narrowing. Finally, in Sect. 6, we relate normalizing conditional LSE narrowing to normalizing conditional left-to-right basic narrowing, which is the basis for an efficient implementation of our strategy.

2 Basic Definitions

Throughout this paper, we use the standard terminology of term rewriting (Huet and Oppen 1980; Dershowitz and Jouannaud 1990; Klop 1992). A *conditional term rewrite system* R is a set of rules π of the form $P \Rightarrow l \to r$, where the premise P is a possibly empty *system of equations* $t_1 \doteq u_1 \wedge \ldots \wedge t_n \doteq u_n, n \geq 0$. Variables in P or r that do not occur in l are called *extravariables*. For two terms s, t the *conditional rewrite relation* $\to_R$ is defined by $s \to_R t$ if there is a rule $t_1 \doteq u_1 \wedge \ldots \wedge t_n \doteq u_n \Rightarrow l \to r$ in R, a substitution σ, and an occurrence $\omega \in Occ(s)$, such that $s/\omega = \sigma(l), t = s[\omega \leftarrow \sigma(r)]$ and $\sigma(t_1) \downarrow_R \sigma(u_1), \ldots, \sigma(t_n) \downarrow_R \sigma(u_n)$. Here $t \downarrow_R u$ iff there exists v such that $t \xrightarrow{*}_R v$ and $u \xrightarrow{*}_R v$. A *conditional rewriting derivation* $t_0 \to_R^* t_n$ is a sequence of rewriting steps $t_0 \to_R t_1 \to_R \ldots \to_R t_n, n \geq 0$. A term t is $\to$*-irreducible* if there exists no term t' with $t \to_R t'$. Otherwise t is $\to$*-reducible*. A substitution σ is *normalized* iff $\sigma(x)$ is $\to_R$-irreducible for all variables x. A conditional term rewrite system R without extravariables is *decreasing* iff there exists a well-founded ordering $>$ such that (a) $t > u$, if u is a proper subterm of t, (b) $s > t$, if $s \to_R t$, and (c) $\sigma(l) > \sigma(t_1), \sigma(u_1), \ldots, \sigma(t_n), \sigma(u_n)$ for any rule $t_1 \doteq u_1 \wedge \ldots \wedge t_n \doteq u_n \Rightarrow l \to r$ in R and any substitution σ.

Narrowing allows us to find complete sets of R-unifiers for conditional equational theories $\equiv_R$ that can be defined by a confluent and decreasing conditional term rewrite system R. Let R be a conditional term rewrite system. A system of equations G is *narrowable* to a system of equations G' with *narrowing substitution* δ, $G \rightsquigarrow_{[v,\pi,\delta]} G'$, iff there exist a non-variable occurrence $v \in Occ(G)$ and a rule $\pi : P \Rightarrow l \to r$ in R such that G/v and l are syntactically unifiable with most general unifier δ and $G' = \delta(P) \wedge \delta(G)[v \leftarrow \delta(r)]$. We always assume that G and $P \Rightarrow l \to r$ have no variables in common. A *narrowing derivation* $G_0 \rightsquigarrow_\sigma^* G_n$ with *narrowing substitution* σ is a sequence of narrowing steps $G_0 \rightsquigarrow_{\delta_1} G_1 \rightsquigarrow_{\delta_2} \ldots \rightsquigarrow_{\delta_n} G_n$, where $\sigma \stackrel{\text{def}}{=} (\delta_n \circ \ldots \circ \delta_1)\mid_{Var(G_0)}$ and

$n \geq 0$. The narrowing substitution leading from G_i to G_j, for $0 \leq i \leq j \leq n$, will be denoted by $\lambda_{i,j} \stackrel{\text{def}}{=} \delta_j \circ \ldots \circ \delta_{i+1}$. In particular, $\lambda_{i,i} = id$, for $i = 0, \ldots, n$.

In order to treat syntactical unification as a narrowing step, we add to the rewrite system R the rule $x \doteq x \rightarrow true$, where x denotes a variable. Then $t \doteq t' \rightsquigarrow_\delta true$ holds iff t and t' are syntactically unifiable with most general unifier δ. A system of equations is called *trivial* if it is of the form $true \wedge \ldots \wedge true$. Given an arbitrary system of equations $H : s_1 \doteq t_1 \wedge \ldots \wedge s_n \doteq t_n, n \geq 1$, there exists a trivial system $Triv$ with $H \stackrel{*}{\rightarrow}_R Triv$ iff $s_1 \downarrow_R t_1, \ldots, s_n \downarrow_R t_n$.

3 Conditional Rewriting Without Evaluation of the Premise

If the rewrite system R is confluent, then a substitution μ is an R-unifier of a system of equations G iff there is a rewriting derivation $\mu(G) \stackrel{*}{\rightarrow}_R true \wedge \ldots \wedge true$. The idea of narrowing is to lift this derivation to a corresponding narrowing derivation $G \rightsquigarrow^* true \wedge \ldots \wedge true$. While in the unconditional case there is a direct correspondence between rewriting and narrowing steps, this is no longer true in the presence of conditions. A conditional rewrite step may be performed only if its premise can be evaluated to true. In a conditional narrowing step, however, the premise is not evaluated but simply added to the next goal. In order to get the correspondence between rewriting and narrowing steps needed to establish the correctness and completeness of the narrowing procedure, we use the calculus of *conditional rewriting without evaluation of the premise* (Bockmayr 1990). In a first step, we associate with each conditional rewriting derivation a conditional rewriting derivation without evaluation of the premise, in a second step this derivation is lifted to the narrowing level.

Definition 1. Let R be a conditional term rewrite system. For systems of equations T, T' we define the *conditional term rewriting relation without evaluation of the premise* $\multimap_R$ by $T \multimap_R T'$ if there exists an occurrence ω in T, a rule $P \Rightarrow l \rightarrow r$ in R, and a substitution τ such that $\tau(l) = T/\omega$, and $T' = \tau(P) \wedge T[\omega \leftarrow \tau(r)]$. A term t is *$\multimap$-reducible* if there exists an occurrence ω in t, a rule $P \Rightarrow l \rightarrow r$ in R, and a substitution τ such that $\tau(l) = t/\omega$. Otherwise t is *$\multimap$-irreducible.*

There is a close relation between ordinary conditional rewriting and conditional rewriting without evaluation of the premise. Concerning reducibility to trivial systems $true \wedge \ldots \wedge true$, both rewriting relations have the same power. The next results summarize the basic properties of the relation $\multimap$ (Bockmayr (1993) gives proofs in the more general case of conditional rewriting modulo a set of equations).

Theorem 2. *Let R be a conditional term rewrite system and T be a system of equations. Then there exists a trivial system $Triv$ with $T \stackrel{*}{\rightarrow}_R Triv$ if and only if there exists a trivial system $Triv'$ with $T \stackrel{*}{\multimap}_R Triv'$.*

Corollary 3. *Consider a conditional term rewrite system R and two systems of equations T, T'. If $T \overset{*}{\rightarrow}_R T'$ then $T \mathrel{-\!\!\circ\!\!\!\rightarrow}^{*}_R Triv \wedge T'$, for some empty or trivial system $Triv$.*

Proposition 4. *For a decreasing conditional term rewrite system R the relation $\mathrel{-\!\!\circ\!\!\!\rightarrow}_R$ is noetherian.*

The confluence of $\rightarrow$ does not carry over to $\mathrel{-\!\!\circ\!\!\!\rightarrow}$.

Example 5. The conditional rewrite relation $\rightarrow$ defined by the rules

$$x \geq 0 \doteq \mathtt{true} \Rightarrow |x| \rightarrow x, \quad x \geq 0 \doteq \mathtt{false} \Rightarrow |x| \rightarrow -x,$$

together with some appropriate rules for $\geq$, is confluent. But, $\mathrel{-\!\!\circ\!\!\!\rightarrow}$ is not confluent because

$$\begin{array}{ccccc} & \nearrow^{\circ} & s(0) \geq 0 \doteq \mathtt{true} \;\wedge\; s(0) \doteq s(0) & \not\searrow^{*}_{\circ} & \\ |s(0)| \doteq s(0) & & & & \bullet \\ & \searrow_{\circ} & s(0) \geq 0 \doteq \mathtt{false} \wedge -s(0) \doteq s(0) & \not\nearrow^{*}_{\circ} & \end{array}$$

However, Theorem 2 implies that if R is confluent, then a substitution μ is an R-unifier of G iff $\mu(G) \mathrel{-\!\!\circ\!\!\!\rightarrow}^{*}_R Triv$, for some trivial system $Triv$. Our aim is to lift this $\mathrel{-\!\!\circ\!\!\!\rightarrow}$-derivation to the narrowing level. The next theorem generalizes the classical lifting lemma of Hullot (1980) to the case of conditional term rewrite systems. In contrast to the unconditional case, it is not possible to get for each $\mathrel{-\!\!\circ\!\!\!\rightarrow}$-reduction $\mu(G) \mathrel{-\!\!\circ\!\!\!\rightarrow}^{*} H$, where μ is normalized, a corresponding narrowing derivation. The reason is that $\mu(x)$ may be $\mathrel{-\!\!\circ\!\!\!\rightarrow}$-reducible for some variable x, although $\mu(x)$ is $\rightarrow$-irreducible for all variables x. On the other hand, requiring that $\mu(x)$ is $\mathrel{-\!\!\circ\!\!\!\rightarrow}$-irreducible, for all variables x, is a restriction which would be too strong.

Example 6. Consider the rule $\pi : x \doteq p(0) \Rightarrow s(x) \rightarrow 0$, the substitution $\mu = [y \mapsto s(0)]$, and the systems $G_0 : y \doteq p(0)$ and $H_0 = \mu(G_0) : s(0) \doteq p(0)$. We have $H_0 : s(0) \doteq p(0) \mathrel{-\!\!\circ\!\!\!\rightarrow}_{[1,\pi]} 0 \doteq p(0) \wedge 0 \doteq p(0)$, but it is not possible to perform a narrowing step at occurrence 1 in G_0, since $G_0/1 = y$ is a variable. Note that μ is normalized, but $\mu(y)$ is $\mathrel{-\!\!\circ\!\!\!\rightarrow}$-reducible.

However, it is still possible to obtain for each successful $\mathrel{-\!\!\circ\!\!\!\rightarrow}$-reduction a corresponding narrowing derivation. A $\mathrel{-\!\!\circ\!\!\!\rightarrow}$-derivation $T \mathrel{-\!\!\circ\!\!\!\rightarrow}^{*} T'$ or a narrowing derivation $G \rightsquigarrow^{*} G'$ are called *successful* iff they lead to a trivial system $T' = Triv$ or $G' = Triv$ respectively.

Theorem 7. *Let R be a conditional term rewrite system without extravariables and let G be a system of equations. If μ is a normalized substitution and V a set of variables such that $Var(G) \cup Dom(\mu) \subseteq V$, then for every successful $\mathrel{-\!\!\circ\!\!\!\rightarrow}$-derivation*

$$H_0 \overset{\text{def}}{=} \mu(G) \mathrel{-\!\!\circ\!\!\!\rightarrow}_{[v_1, P_1 \Rightarrow l_1 \rightarrow r_1, \pi_1]} H_1 \ldots \mathrel{-\!\!\circ\!\!\!\rightarrow}_{[v_n, P_n \Rightarrow l_n \rightarrow r_n, \pi_n]} H_n, \qquad (1)$$

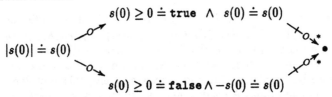

to a trivial system $H_n = Triv$ there exist a normalized substitution λ and a conditional narrowing derivation

$$G_0 \stackrel{\text{def}}{=} G \rightsquigarrow_{[v_1, P_1 \Rightarrow l_1 \to r_1, \delta_1]} G_1 \quad \ldots \quad \rightsquigarrow_{[v_n, P_n \Rightarrow l_n \to r_n, \delta_n]} G_n \quad (2)$$

using the same conditional rewrite rules at the same occurrences such that $\mu = \lambda \circ \delta_n \circ \ldots \circ \delta_1$ $[V]$ and

$$H_i = (\lambda \circ \delta_n \circ \ldots \circ \delta_{i+1})\,(G_i), \text{ for all } i = 0, \ldots, n. \quad (3)$$

Conversely, if $\mu \stackrel{\text{def}}{=} \lambda \circ \delta_n \circ \ldots \circ \delta_1$ then there exists for any narrowing derivation (2) and any substitution λ a $\multimap$-derivation (1) such that (3) holds.

Proof. The proof is by induction on n. First assume that we are given the $\multimap$-derivation (1) of length n. If $n = 0$, the theorem holds with $\lambda = \mu$. So suppose $n > 0$ and $\mu(G) \multimap_{[v_1, P_1 \Rightarrow l_1 \to r_1, \tau_1]} H_1 \multimap \ldots \multimap H_n$.

Assume v_1 is not a non-variable occurrence in G. Then $\mu(G)/v_1 = \mu(x)/v$ for some variable $x \in Var(G)$ and some occurrence $v \in Occ(\mu(x))$. From $H_1 = \tau_1(P_1) \wedge \mu(G)[v_1 \leftarrow \tau_1(r_1)] \stackrel{*}{\multimap} Triv$ we get by Theorem 2 that $\tau_1(P_1) \stackrel{*}{\to} Triv'$, for some trivial system $Triv'$. Together with $\mu(x)/v = \mu(G)/v_1 = \tau_1(l_1)$ this implies $\mu(x) \to_{[v, P_1 \Rightarrow l_1 \to r_1, \tau_1]} \mu(x)[v \leftarrow \tau_1(r_1)]$ in contradiction to the assumption that μ is normalized. Hence, v_1 is a non-variable occurrence in G and $\mu(G)/v_1 = \mu(G/v_1)$.

Since we may assume that V and l_1 have no variables in common and that $Dom(\tau_1) \subseteq Var(l_1)$, the substitution $\phi \stackrel{\text{def}}{=} \tau_1 \uplus \mu \stackrel{\text{def}}{=} \begin{cases} \mu(x), & \text{if } x \in Dom(\mu) \\ \tau_1(x), & \text{if } x \in Dom(\tau_1) \end{cases}$ is well-defined and $\phi(G/v_1) = \phi(l_1)$. This means that ϕ is a syntactic unifier of G/v_1 and l_1. Let δ_1 be a most general syntactic unifier of G/v_1 and l_1 with $Dom(\delta_1) \subseteq Var(G/v_1) \cup Var(l_1)$. Then there exists a substitution ρ with $Dom(\rho) \subseteq (((Dom(\tau_1) \cup Dom(\mu)) \setminus Dom(\delta_1)) \cup Im(\delta_1)$ such that $\phi = \rho \circ \delta_1$ $[V \cup Var(l_1)]$. It follows $G \rightsquigarrow_{[v_1, P_1 \Rightarrow l_1 \to r_1, \delta_1]} G_1$ with $G_1 \stackrel{\text{def}}{=} \delta_1(P_1) \wedge \delta_1(G[v_1 \leftarrow r_1])$.

Next we show that the substitution ρ is normalized. Suppose there exists a variable $x \in Dom(\rho)$ such that $\rho(x)$ is $\to$-reducible. Since μ is normalized we get $x \in (Dom(\tau_1) \setminus Dom(\delta_1)) \cup Im(\delta_1)$.

- If $x \in Dom(\tau_1) \setminus Dom(\delta_1)$ then $x \in Var(l_1)$ and since $\delta_1(x) = x$, we get $x \in Var(\delta_1(l_1))$.
- If $x \in Im(\delta_1)$, then it follows from $Dom(\delta_1) \subseteq Var(G/v_1) \cup Var(l_1)$ that x occurs in $\delta_1(l_1)$ or $\delta_1(G/v_1)$.

But since $\delta_1(l_1) = \delta_1(G/v_1)$, in both cases x must occur in $\delta_1(G/v_1)$. So there exists a variable $y \in G$ such that x occurs in $\delta_1(y)$. Then $\rho(x)$ is a subterm of $(\rho \circ \delta_1)(y)$. This implies that $(\rho \circ \delta_1)(y) = \mu(y)$ is $\to$-reducible in contradiction to the fact that μ is normalized.

Since $Var(P_1) \cup Var(r_1) \subseteq Var(l_1)$ we have $\rho(G_1) = \rho(\delta_1(P_1) \wedge \delta_1(G[v_1 \leftarrow r_1])) = (\rho \circ \delta_1)(P_1 \wedge G[v_1 \leftarrow r_1]) = \phi(P_1 \wedge G[v_1 \leftarrow r_1]) = \phi(P_1) \wedge \phi(G)[v_1 \leftarrow \phi(r_1)] = \tau_1(P_1) \wedge \mu(G)[v_1 \leftarrow \tau_1(r_1)] = H_1 \multimap_{[v_2, P_2 \Rightarrow l_2 \to r_2, \tau_2]}$

$H_2 \mathrel{-\!\!\circ\!\!\rightarrow^*} H_n$, with ρ normalized. Let $V' \stackrel{\text{def}}{=} V \cup Im(\delta_1)$. Then by the induction hypothesis there exists a substitution λ and a narrowing derivation $G_1 \rightsquigarrow_{[v_2, P_2 \Rightarrow l_2 \to r_2, \delta_2]} \cdots \rightsquigarrow_{[v_n, P_n \Rightarrow l_n \to r_n, \delta_n]} G_n$ such that $\rho = \lambda \circ \delta_n \circ \ldots \circ \delta_2\ [V']$ and $H_i = (\lambda \circ \delta_n \circ \ldots \circ \delta_{i+1})(G_i)$, for $i = 1, \ldots, n$. By the disjointness of V and $Var(l_1)$ we get $\mu = \rho \circ \delta_1\ [V]$. From $V' = V \cup Im(\delta_1)$ and $\rho = \lambda \circ \delta_n \circ \ldots \circ \delta_2\ [V']$ we conclude $\rho \circ \delta_1 = \lambda \circ \delta_n \circ \ldots \circ \delta_1\ [V]$. Together this implies $\mu = \lambda \circ \delta_n \circ \ldots \circ \delta_1\ [V]$ and in particular $H_0 = \mu(G_0) = (\lambda \circ \delta_n \circ \ldots \circ \delta_1)(G_0)$.

The reverse direction is again proved by induction. Note that we will not use the assumption that the rules do not contain extravariables. Let λ be a substitution and let $G \rightsquigarrow_{[v_1, P_1 \Rightarrow l_1 \to r_1, \delta_1]} G_1 \rightsquigarrow_{\delta_2} \cdots \rightsquigarrow_{\delta_n} G_n$ be a narrowing derivation. The case $n = 0$ is trivial. So assume $n > 0$. Define the substitution $\nu \stackrel{\text{def}}{=} \lambda \circ \delta_n \circ \ldots \circ \delta_2$. Then it follows from the induction hypothesis that $\nu(G_1) = H_1 \mathrel{-\!\!\circ\!\!\rightarrow} \ldots \mathrel{-\!\!\circ\!\!\rightarrow} H_n$ with $H_i = \lambda \circ \delta_n \circ \ldots \circ \delta_{i+1}(G_i)$, for $i = 1, \ldots, n$. From $G \rightsquigarrow_{[v_1, P_1 \Rightarrow l_1 \to r_1, \delta_1]} G_1$ we get $\delta_1(G/v_1) = \delta_1(l_1)$ and $G_1 = \delta_1(P_1) \wedge \delta_1(G[v_1 \leftarrow r_1]) = \delta_1(P_1) \wedge \delta_1(G)[v_1 \leftarrow \delta_1(r_1)]$. This means $\delta_1(G) \mathrel{-\!\!\circ\!\!\rightarrow}_{[v_1, P_1 \Rightarrow l_1 \to r_1, \delta_1|_{Var(l_1)}]} G_1$. Since $\mathrel{-\!\!\circ\!\!\rightarrow}$ is stable under substitutions we obtain $\mu(G) = (\nu \circ \delta_1)(G) \mathrel{-\!\!\circ\!\!\rightarrow} \nu(G_1) = H_1 \mathrel{-\!\!\circ\!\!\rightarrow^*} H_n$. This proves the theorem. □

Lemma 8. *Let R be a conditional term rewrite system without extravariables. Let G_0 and G'_0 be two systems of equations which are identical up to variable renaming, that is there exist substitutions τ_0, τ'_0 such that $\tau_0(G_0) = G'_0$ and $\tau'_0(G'_0) = G_0$. If $G_0 \rightsquigarrow_{[v, P \Rightarrow l \to r, \delta]} G_1$ and $G'_0 \rightsquigarrow_{[v, P \Rightarrow l \to r, \delta']} G'_1$, then G_1 and G'_1 are also identical up to variable renaming.*

Proof. Without loss of generality, we can assume $Dom(\tau_0) \subseteq Var(G_0)$ and $Dom(\tau'_0) \subseteq Var(G'_0)$. By definition, $G_1 = \delta(P) \wedge \delta(G_0[v \leftarrow r])$ and $G'_1 = \delta'(P) \wedge \delta'(G'_0[v \leftarrow r])$ with a most general unifier δ of G_0/v and l and a most general unifier δ' of G'_0/v and l. Since $G_0/v = \tau'_0(G'_0)/v$ and $G'_0/v = \tau_0(G_0)/v$ we can conclude from $Dom(\tau_0) \cap Var(l) = Dom(\tau'_0) \cap Var(l) = \emptyset$ that $\delta(\tau'_0(l)) = \delta(l) = \delta(G_0/v) = \delta(\tau'_0(G'_0)/v) = \delta(\tau'_0(G'_0/v))$ and similarly $\delta'(\tau_0(l)) = \delta'(l) = \delta'(G'_0/v) = \delta'(\tau_0(G_0)/v) = \delta'(\tau_0(G_0/v))$. Hence $\delta \circ \tau'_0$ unifies G'_0/v and l and $\delta' \circ \tau_0$ unifies G_0/v and l. Since δ, δ' are most general unifiers, there exist substitutions τ_1, τ'_1 with $\tau_1 \circ \delta = \delta' \circ \tau_0$ and $\tau'_1 \circ \delta' = \delta \circ \tau'_0$. It follows $\tau_1(G_1) = \tau_1(\delta(P \wedge G_0[v \leftarrow r])) = \delta'(\tau_0(P \wedge G_0[v \leftarrow r])) = \delta'(\tau_0(P) \wedge \tau_0(G_0)[v \leftarrow \tau_0(r)]) = \delta'(P \wedge G'_0[v \leftarrow r]) = G'_1$ and similarly $\tau'_1(G'_1) = G_1$. This shows that G_1 and G'_1 are identical up to variable renaming. □

4 Conditional LSE Narrowing

By Theorem 7, *any* rewriting derivation $\mu(G) \mathrel{-\!\!\circ\!\!\rightarrow^*} Triv$, with a normalized substitution μ, can be lifted to a corresponding narrowing derivation $G \rightsquigarrow^*_\sigma Triv$ such that σ is more general than μ. A naive narrowing procedure would enumerate all these narrowing derivations, which is extremely inefficient. To ensure completeness it is enough to generate for each normalized substitution μ the narrowing derivation corresponding to one special rewriting derivation. In LSE

narrowing, we select a unique $\dashrightarrow$-derivation $\mu(G) \multimap\!\!\rightarrow^* Triv$, called *left reduction*, and then use reducibility tests such that only narrowing derivations corresponding to left reductions are generated.
In the sequel, we assume that the rules in R are ordered by a total well-founded ordering $<$.

Definition 9. A rewriting step $G \rightarrow_{[v,\pi,\tau]} G'$ or $G \multimap\!\!\rightarrow_{[v,\pi,\tau]} G'$ is called a *left reduction step* iff

- all subterms G/ω with ω strictly left of v are $\rightarrow$-irreducible ("leftmost")
- all proper subterms of G/v are $\rightarrow$-irreducible ("innermost")
- G cannot be $\rightarrow$-reduced at occurrence v by a rule π' with $\pi' < \pi$ ("minimal rule").

A conditional rewriting derivation is called a *left reduction* iff all steps are left reduction steps.

Note that even for a $\multimap\!\!\rightarrow$-reduction step we require $\rightarrow$-irreducibility. The following example illustrates the difference.

Example 10. Consider the rules $\pi_1 : x \doteq p(0) \Rightarrow s(x) \rightarrow 0$ and $\pi_2 : x \doteq s(0) \Rightarrow x + y \rightarrow s(y)$. The step $s(0) + 0 \doteq s(0) \multimap\!\!\rightarrow_{\pi_2} s(0) \doteq s(0) \wedge s(0) \doteq s(0)$ is a left $\multimap\!\!\rightarrow$-reduction because it is $\rightarrow$-innermost, although it is not $\multimap\!\!\rightarrow$-innermost. Note that the $\multimap\!\!\rightarrow$-innermost step $s(0) + 0 \doteq s(0) \multimap\!\!\rightarrow_{\pi_1} 0 \doteq p(0) \wedge 0 \doteq s(0)$ does not lead to a trivial system.

Proposition 11. *Let R be a confluent and terminating conditional term rewrite system without extravariables. Then for all terms t there exists a unique left $\rightarrow$-reduction to the normal form $t\!\downarrow$ of t.*

Proof. We prove the theorem by noetherian induction on $\rightarrow$. If t is $\rightarrow$-irreducible, then the theorem holds trivially. If t is $\rightarrow$-reducible, then there exists a unique first left reduction step $t \rightarrow t'$, since the ordering $\sqsupset$ on $Occ(t)$ defined by $v \sqsupset v'$ iff v is strictly left of or strictly below v' and the ordering $>$ on rules are total and well-founded. By induction hypothesis, there is a unique left derivation $t' \xrightarrow{*} t'\!\downarrow$. If we join the two derivations together, we get the unique left reduction $t \rightarrow t' \xrightarrow{*} t'\!\downarrow = t\!\downarrow$. □

For left $\rightarrow$-reductions that lead to a trivial system we now construct a unique corresponding left $\multimap\!\!\rightarrow$-reduction.

Theorem 12. *Let R be a confluent and decreasing conditional term rewrite system and let T be a system of equations. If $T \xrightarrow{*}_R Triv$ for some trivial system $Triv$ then there is a unique left $\multimap\!\!\rightarrow$-reduction $T \multimap\!\!\xrightarrow{*}_R Triv'$ to some trivial system $Triv'$.*

Proof. We prove the theorem by noetherian induction on $\multimap\!\!\rightarrow_R$, which by Proposition 4 is noetherian. Assume $T \xrightarrow{*}_R Triv$ to some trivial system $Triv$. If T is trivial then it has the form $true \wedge \ldots \wedge true$ and the theorem trivially holds. So suppose that T is not trivial.

First we construct a left $-\!\!\circ\!\!\to$-reduction to a trivial system. Since $T \xrightarrow{*}_R \mathit{Triv}$, by Proposition 11 there exists a unique left $\to$-reduction $T \to_{[v,\pi,\tau]} T' \xrightarrow{*} \mathit{Triv}_2$ to some trivial system Triv_2. From $T \to T'$ we get $T -\!\!\circ\!\!\to_{[v,\pi,\tau]} \tau(P) \wedge T'$ where $\tau(P) = \emptyset$ or $\tau(P) \xrightarrow{*}_R \mathit{Triv}_3$ to some trivial system Triv_3. Hence, $\tau(P) \wedge T' \xrightarrow{*}_R \mathit{Triv}_4$ to some trivial system Triv_4. By induction hypothesis there is a unique left $-\!\!\circ\!\!\to$-reduction $\tau(P) \wedge T' -\!\!\circ\!\!\xrightarrow{*}_R \mathit{Triv}_5$ to some trivial system Triv_5. Since $T \to T'$ is a left $\to$-reduction step, $T -\!\!\circ\!\!\to \tau(P) \wedge T'$ is a left $-\!\!\circ\!\!\to$-reduction step. If we join the two derivations together, we get a left $-\!\!\circ\!\!\to$-reduction $T -\!\!\circ\!\!\to \tau(P) \wedge T' -\!\!\circ\!\!\xrightarrow{*}_R \mathit{Triv}_5$.

Now we show that there is no other left $-\!\!\circ\!\!\to$-reduction to a trivial system. Assume, $T -\!\!\circ\!\!\to_{[v',\pi',\tau']} \tau'(P') \wedge T'' -\!\!\circ\!\!\xrightarrow{*}_R \mathit{Triv}_6$ is an arbitrary left $-\!\!\circ\!\!\to$-reduction to some trivial system Triv_6. Then $T \to_{[v',\pi',\tau']} T''$. Since $T -\!\!\circ\!\!\to_{[v',\pi',\tau']} \tau'(P') \wedge T''$ is a left $-\!\!\circ\!\!\to$-reduction step, $T \to_{[v',\pi',\tau']} T''$ is a left $\to$-reduction step. By Proposition 11, $v' = v$ and $\pi' = \pi$ and since the rules do not contain extravariables $\tau'(P') \wedge T'' = \tau(P) \wedge T'$. By induction hypothesis, $\tau(P) \wedge T' -\!\!\circ\!\!\xrightarrow{*}_R \mathit{Triv}_5$ is the only left $-\!\!\circ\!\!\to$-reduction to a trivial system. So we we conclude that the arbitrary left $-\!\!\circ\!\!\to$-reduction coincides with the left $-\!\!\circ\!\!\to$-reduction constructed before. □

Now we lift left $-\!\!\circ\!\!\to$-reductions to the narrowing level.

Definition 13. In a conditional narrowing derivation

$$G_0 \rightsquigarrow_{[v_1,\pi_1,\delta_1]} G_1 \rightsquigarrow_{[v_2,\pi_2,\delta_2]} \cdots G_{n-1} \rightsquigarrow_{[v_n,\pi_n,\delta_n]} G_n$$

the step $G_{n-1} \rightsquigarrow_{[v_n,\pi_n,\delta_n]} G_n$ is called *LSE* iff for all $i \in \{0,\ldots,n-1\}$:

(Left-Test) The subterms of $\lambda_{i,n}(G_i)$ which lie strictly left of v_{i+1} are $\to$-irreducible.

(Sub-Test) The proper subterms of $\lambda_{i,n}(G_i/v_{i+1})$ are $\to$-irreducible.

(Epsilon-Test) The term $\lambda_{i,n}(G_i/v_{i+1})$ is not $\to$-reducible at occurrence ϵ by a rule π with $\pi < \pi_{i+1}$.

A conditional narrowing derivation is *LSE* iff any single narrowing step is LSE.

There is a 1-1 correspondence between left $-\!\!\circ\!\!\to$-reductions and conditional LSE narrowing derivations.

Proposition 14. *Let R be a conditional term rewrite system without extravariables. Consider a system of equations G and a normalized substitution μ. If*

$$H_0 \stackrel{\text{def}}{=} \mu(G_0) -\!\!\circ\!\!\to_{[v_1,\pi_1]} H_1 -\!\!\circ\!\!\to_{[v_2,\pi_2]} \cdots H_{n-1} -\!\!\circ\!\!\to_{[v_n,\pi_n]} H_n, \qquad (4)$$

with a trivial system H_n, is a left $-\!\!\circ\!\!\to$-reduction, then the corresponding narrowing derivation

$$G_0 \rightsquigarrow_{[v_1,\pi_1,\delta_1]} G_1 \rightsquigarrow_{[v_2,\pi_2,\delta_2]} \cdots G_{n-1} \rightsquigarrow_{[v_n,\pi_n,\delta_n]} G_n \qquad (5)$$

is a conditional LSE narrowing derivation. Conversely, given a conditional LSE narrowing derivation (5), *the corresponding rewriting derivation* (4), *with $H_i \stackrel{\text{def}}{=} \lambda_{i,n}(G_i)$, for $i = 0,\ldots,n$, is a left $-\!\!\circ\!\!\to$-reduction.*

Proof. By Theorem 7 there exists a substitution λ such that $H_i = (\lambda \circ \lambda_{i,n})(G_i)$, for $i = 0, \ldots, n$. We have to show that none of the reducibility tests detects a redundancy. Suppose that the step $G_{m-1} \rightsquigarrow_{[v_m, \pi_m, \delta_m]} G_m$ is not LSE, for some $m \in \{1, \ldots, n\}$. Then there exists $i \in \{0, \ldots, m-1\}$ such that either

1. $\lambda_{i,m}(G_i)$ is $\rightarrow$-reducible at an occurrence v strictly left of v_{i+1} or
2. $\lambda_{i,m}(G_i)$ is $\rightarrow$-reducible at an occurrence v strictly below v_{i+1} or
3. $\lambda_{i,m}(G_i)$ is $\rightarrow$-reducible at occurrence v_{i+1} with a rule smaller than π_{i+1}.

Since $H_i = (\lambda \circ \lambda_{m,n} \circ \lambda_{i,m})(G_i)$ and $\rightarrow$ is stable under substitutions this implies that one of the properties (1) to (3) must hold with H_i in place of $\lambda_{i,m}(G_i)$. But this means that $H_i \multimap_{[v_{i+1}, \pi_{i+1}]} H_{i+1}$ is not a left $\multimap$-reduction step in contradiction to our assumption.

Conversely, suppose that the derivation (4) is not a left $\multimap$-reduction. Then there exists $i \in \{0, \ldots, n-1\}$ such that $H_i = \lambda_{i,n}(G_i)$ is $\rightarrow$-reducible either

1. at an occurrence v strictly left of v_{i+1} or
2. at an occurrence v lies strictly below v_{i+1} or
3. at occurrence v_{i+1} with a rule $\pi < \pi_{i+1}$.

But this implies that the narrowing step $G_{n-1} \rightsquigarrow_{[v_n, \pi_n, \delta_n]} G_n$ is not LSE in contradiction to our assumption. Note that for this part of the proof we do not need the assumption that there are no extravariables. □

Using this result, we can prove our main theorems on the completeness and optimality of conditional LSE narrowing.

Theorem 15. *Conditional LSE narrowing is complete for confluent and decreasing conditional term rewrite systems.*

Proof. Let the substitution μ be an R-unifier of a system of equations G. We have to show that there exists a conditional LSE narrowing derivation $G = G_0 \rightsquigarrow^*_\sigma G_n, n \geq 0$, to a trivial system $G_n = Triv$ such that the narrowing substitution σ is more general than μ. More precisely, this means that there exists a substitution λ with $\lambda(\sigma(x)) \equiv_R \mu(x)$, for all $x \in Var(G)$.

Let $\mu^\downarrow$ be the normal form of μ, that is $\mu^\downarrow(x) \stackrel{\text{def}}{=} \mu(x)\downarrow$, for all variables x. Then $\mu^\downarrow$ is also an R-unifier of G. Therefore there exists a $\rightarrow$-derivation $\mu^\downarrow(G) \xrightarrow{*} Triv'$ to some trivial system $Triv'$. By Theorem 12, there is a left $\multimap$-reduction $\mu^\downarrow(G) \overset{*}{\multimap} Triv$ to some trivial system $Triv$. By Proposition 14 and Theorem 7 there exists a corresponding conditional LSE narrowing derivation $G \rightsquigarrow^*_\sigma G_n$ together with a substitution λ such that $\mu^\downarrow = \lambda \circ \sigma\ [Var(G)]$ and $\lambda(G_n) = Triv$. Since $\lambda(G_n)$ can be trivial only if G_n is trivial and since $\mu^\downarrow \equiv_R \mu$, this implies the theorem. □

Theorem 16. *Let R be a confluent and decreasing conditional term rewrite system. Consider two conditional LSE narrowing derivations*

$$G = G_0 \rightsquigarrow_{[v_1, \pi_1, \delta_1]} G_1 \rightsquigarrow_{[v_2, \pi_2, \delta_2]} G_2 \ldots \rightsquigarrow_{[v_n, \pi_n, \delta_n]} G_n$$
$$G = G'_0 \rightsquigarrow_{[v'_1, \pi'_1, \delta'_1]} G'_1 \rightsquigarrow_{[v'_2, \pi'_2, \delta'_2]} G'_2 \ldots \rightsquigarrow_{[v'_m, \pi'_m, \delta'_m]} G'_m$$

with $n \leq m, \sigma \stackrel{\text{def}}{=} \lambda_{0,n} = \delta_n \circ \ldots \circ \delta_1$ *and* $\sigma' \stackrel{\text{def}}{=} \lambda'_{0,m} = \delta'_m \circ \ldots \circ \delta'_1$. *If there are conditional LSE narrowing derivations* $G_n \rightsquigarrow^*_{\tau} \mathit{Triv}$ *and* $G'_m \rightsquigarrow^*_{\tau'} \mathit{Triv}'$, *for trivial systems Triv and Triv', such that* $\tau \circ \sigma$ *and* $\tau' \circ \sigma'$ *are identical up to variable renaming on* $\mathit{Var}(G)$, *that is there exist substitutions* ρ, ρ' *such that* $\rho \circ \tau \circ \sigma = \tau' \circ \sigma'$ *and* $\rho' \circ \tau' \circ \sigma' = \tau \circ \sigma$ *on* $\mathit{Var}(G)$, *then* $\pi_i = \pi'_i$ *and* $v_i = v'_i$ *for* $i = 1, \ldots, n$. *If additionally* σ *and* σ' *are identical up to variable renaming, then the conditional narrowing derivation*

$$G'_n \rightsquigarrow_{[v'_{n+1}, \pi'_{n+1}, \delta'_{n+1}]} G'_{n+1} \cdots \rightsquigarrow_{[v'_m, \pi'_m, \delta'_m]} G'_m$$

is a left $\multimap$*-reduction up to variable renaming.*

Proof. By Proposition 14 the $\multimap$-derivations

$$\begin{array}{l} (\tau \circ \sigma)(G) \multimap_{[v_1, \pi_1]} \cdots \multimap_{[v_n, \pi_n]} (\tau \circ \lambda_{n,n})(G_n) \multimap^* \mathit{Triv} \\ (\tau' \circ \sigma')(G) \multimap_{[v'_1, \pi'_1]} \cdots \multimap_{[v'_m, \pi'_m]} (\tau' \circ \lambda'_{m,m})(G'_m) \multimap^* \mathit{Triv}' \end{array}$$

are both left $\multimap$-reductions. Since $\tau \circ \sigma$ and $\tau' \circ \sigma'$ coincide on $\mathit{Var}(G)$ up to variable renaming, the systems $(\tau \circ \sigma)(G)$ and $(\tau' \circ \sigma')(G)$ are identical up to variable renaming. By the uniqueness of left $\multimap$-reductions, which follows from Theorem 12, this implies $\pi_i = \pi'_i$ and $v_i = v'_i$ for $i = 1, \ldots, n$.

If σ and σ' are identical up to variable renaming we can conclude by Lemma 8 and by induction that G_i and G'_i resp. $\lambda_{i,n}(G_i)$ and $\lambda'_{i,m}(G'_i)$ are identical up to variable renaming, for $i = 1, \ldots, n$. Since $\lambda_{n,n}(G_n) = G_n$, this implies that G'_n and $\lambda'_{n,m}(G'_n)$ are identical up to variable renaming. The narrowing derivation starting from G'_n uses the same rules at the same occurrences as the $\multimap$-derivation starting from $\lambda'_{n,m}(G'_n)$. Therefore the narrowing substitutions δ'_i restricted to the variables in G'_{i-1}/v'_i have to be renaming substitutions, for $i = n+1, \ldots, m$. This shows that up to variable renaming the narrowing derivation $G'_n \rightsquigarrow^* G'_m$ is a left $\multimap$-reduction. □

If we assume that narrowing derivations starting from the same goal and using the same rules at the same occurrences produce the same narrowing substitution (in any practical implementation, this will be the case), we get the following corollary.

Corollary 17. *If conditional LSE narrowing enumerates two solutions* σ *and* σ' *which coincide up to variable renaming, then* $\sigma = \sigma'$ *holds and the two derivations coincide.*

Proof. We use the same notation as in Theorem 16. Then $G_n = G'_m = \mathit{true} \wedge \ldots \wedge \mathit{true}$ implies $n = m$. □

In contrast to the unconditional case, Theorem 16 holds only for successful narrowing derivations.

Example 18. Consider the two rules $\pi_1 : x \doteq 0 \Rightarrow x + y \rightarrow y$ and $\pi_2 : x \doteq s(0) \Rightarrow x + y \rightarrow s(y)$. Starting with the equation $G : x + x \doteq s(s(0))$ there are two conditional LSE narrowing steps with the empty narrowing substitution. But the narrowing step using rule π_1 cannot be extended to a successful derivation.

However, like in the unconditional case, conditional LSE narrowing generates only normalized narrowing substitutions.

Proposition 19. *Let R be a conditional term rewrite system such that variables are not $\to$-reducible. Then for any conditional LSE narrowing derivation*

$$G_0 \rightsquigarrow_{[v_1,\pi_1,\delta_1]} G_1 \rightsquigarrow_{[v_2,\pi_2,\delta_2]} \cdots G_{n-1} \rightsquigarrow_{[v_n,\pi_n,\delta_n]} G_n$$

the narrowing substitution $(\delta_n \circ \ldots \circ \delta_1)|_{Var(G_0)}$ is normalized.

Proof. Let x be a variable of G_0 such that $\lambda_{0,n}(x)$ is $\to$-reducible. Suppose x is instantiated for the first time in the i-th narrowing step, $i \in \{1,\ldots,n\}$. Then there must be an occurrence of the variable x in G_{i-1} which lies strictly below the non-variable narrowing occurrence v_i. More formally, there exists an occurrence $v \neq \epsilon$ such that $G_{i-1}/v_i.v = x$. Then $\lambda_{i-1,n}(G_{i-1}/v_i.v) = \lambda_{i-1,n}(x) = \lambda_{0,n}(x)$ is $\to$-reducible in contradiction to the Sub-Test. □

5 Normalizing Conditional LSE Narrowing

One of the most important optimizations of naive narrowing is *normalizing narrowing* where after each narrowing step the goal is normalized with respect to the given canonical rewrite system.

Definition 20. Let R be a confluent and terminating conditional term rewrite system and let G be an $\to$-irreducible system of equations. A *normalizing conditional narrowing step*

$$G \rightsquigarrow^{\downarrow}_{[v,l\to r,\delta]} G'\!\downarrow$$

is given by a narrowing step $G \rightsquigarrow_{[v,l\to r,\delta]} G'$ followed by a normalization $G' \xrightarrow{*}_R G'\!\downarrow$ with $G'\!\downarrow$ $\to$-irreducible. In a normalizing conditional narrowing derivation

$$G_0\!\downarrow \rightsquigarrow_{[v_1,\pi_1,\delta_1]} G_1 \xrightarrow{*} G_1\!\downarrow \rightsquigarrow \cdots \rightsquigarrow_{[v_n,\pi_n,\delta_n]} G_n \xrightarrow{*} G_n\!\downarrow$$

the step $G_{n-1}\!\downarrow \rightsquigarrow G_n \xrightarrow{*} G_n\!\downarrow$ is called a *LSE step* iff the following three conditions are satisfied for all $i \in \{0,\ldots,n-1\}$:

(Left-Test) The subterms of $\lambda_{i,n}(G_i\!\downarrow)$ which lie strictly left of v_{i+1} are $\to$-irreducible.
(Sub-Test) The proper subterms of $\lambda_{i,n}(G_i\!\downarrow/v_{i+1})$ are $\to$-irreducible.
(Epsilon-Test) The term $\lambda_{i,n}(G_i\!\downarrow/v_{i+1})$ is not $\to$-reducible at occurrence ϵ by a rule π with $\pi < \pi_{i+1}$.

A normalizing conditional narrowing derivation is called a *normalizing conditional LSE narrowing derivation* iff all steps are LSE steps.

It is not possible to associate with each $-\!\!\circ\!\!\!\rightarrow$-derivation a corresponding normalizing conditional narrowing derivation where the same rules are applied at the same occurrences. However, for any $-\!\!\circ\!\!\!\rightarrow$-derivation $\mu(G) \stackrel{*}{-\!\!\circ\!\!\!\rightarrow} Triv$, where μ is normalized and $Triv$ is trivial, we can construct a $(-\!\!\circ\!\!\!\rightarrow \circ \stackrel{*}{\rightarrow})$-derivation $\mu(G)\,(-\!\!\circ\!\!\!\rightarrow \circ \stackrel{*}{\rightarrow})^*\, Triv'$ for which a corresponding normalizing conditional narrowing derivation exists. Moreover, we can assume that the rewriting steps on $\mu(G)$ corresponding to narrowing steps on G are left $-\!\!\circ\!\!\!\rightarrow$-reduction steps. This implies that the narrowing derivation is even a LSE derivation.

Theorem 21. *Let R be a confluent and decreasing conditional term rewrite system. Consider an $\rightarrow$-irreducible system of equations G, a normalized substitution μ and a set of variables V such that $Var(G) \cup Dom(\mu) \subseteq V$. If there exists a $\rightarrow$-derivation $H \stackrel{\text{def}}{=} \mu(G) \stackrel{*}{\rightarrow} Triv^*$ to some trivial system $Triv^*$, then there exist a trivial system $Triv$, and a $(-\!\!\circ\!\!\!\rightarrow \circ \stackrel{*}{\rightarrow})$-derivation*

$$H = H_0' -\!\!\circ\!\!\!\rightarrow_{[v_1,\pi_1]} H_1 \stackrel{*}{\rightarrow} H_1' -\!\!\circ\!\!\!\rightarrow \ldots -\!\!\circ\!\!\!\rightarrow_{[v_n,\pi_n]} H_n \stackrel{*}{\rightarrow} H_n' = Triv,$$

with left $-\!\!\circ\!\!\!\rightarrow$-reduction steps $H_i' -\!\!\circ\!\!\!\rightarrow_{[v_{i+1},\pi_{i+1}]} H_{i+1}$, $i = 0,\ldots,n-1$, such that there exists a normalizing conditional LSE narrowing derivation

$$G = G_0\!\downarrow \rightsquigarrow_{[v_1,\pi_1,\delta_1]} G_1 \stackrel{*}{\rightarrow} G_1\!\downarrow \rightsquigarrow \ldots \rightsquigarrow_{[v_n,\pi_n,\delta_n]} G_n \stackrel{*}{\rightarrow} G_n\!\downarrow = Triv$$

which uses the same rules at the same occurrences. Moreover, there exists a normalized substitution λ such that

- $\lambda \circ \delta_n \circ \ldots \circ \delta_1 = \mu\ [V]$
- $H_i = (\lambda \circ \delta_n \circ \ldots \circ \delta_{i+1})\,(G_i), i = 1,\ldots,n$
- $H_i' = (\lambda \circ \delta_n \circ \ldots \circ \delta_{i+1})\,(G_i\!\downarrow), i = 0,\ldots,n.$

Proof. By noetherian induction on the relation $-\!\!\circ\!\!\!\rightarrow \circ \stackrel{*}{\rightarrow}$. To see that this relation is noetherian note that $-\!\!\circ\!\!\!\rightarrow$ is noetherian by Proposition 4 and $T \rightarrow_{[v,P\Rightarrow l\rightarrow r,\tau]} T'$ implies $T -\!\!\circ\!\!\!\rightarrow_{[v,P\Rightarrow l\rightarrow r,\tau]} \tau(P) \wedge T'$. Therefore an infinite $(-\!\!\circ\!\!\!\rightarrow \circ \stackrel{*}{\rightarrow})$-derivation would give us an infinite $-\!\!\circ\!\!\!\rightarrow$-derivation, which is not possible.

If $H = \mu(G)$ is trivial, then G is trivial and the theorem holds with $\lambda = \mu$ and $Triv = H$. If H is not trivial, there exists by Theorem 12 a left $-\!\!\circ\!\!\!\rightarrow$-reduction $H\ -\!\!\circ\!\!\!\rightarrow_{[v_1,\pi_1]}\ H_1 \stackrel{*}{-\!\!\circ\!\!\!\rightarrow} Triv'$ to some trivial system $Triv'$. By Theorem 7, the first step in this reduction can be lifted to a conditional narrowing step $G \rightsquigarrow_{[v_1,\pi_1,\delta_1]} G_1$. Moreover, there exists a normalized substitution ψ with $\mu = \psi \circ \delta_1\ [V]$ and $H_1 = \psi(G_1)$. Let $G_1 \stackrel{*}{\rightarrow} G_1\!\downarrow$. By the stability of $\rightarrow$ under substitutions we get a corresponding $\rightarrow$-derivation $H_1 = \psi(G_1) \stackrel{*}{\rightarrow} \psi(G_1\!\downarrow) = H_1'$ and $H\,(-\!\!\circ\!\!\!\rightarrow \circ \stackrel{*}{\rightarrow})\,H_1'$. By Theorem 2, there is a $\rightarrow$-derivation from H_1 to some trivial system. Since $\rightarrow$ is confluent and noetherian, this holds also for H_1'. Let $V' \stackrel{\text{def}}{=} V \cup Im(\delta_1)$. By applying the induction hypothesis to H_1', we obtain a $(-\!\!\circ\!\!\!\rightarrow \circ \stackrel{*}{\rightarrow})$-derivation

$$H_1' -\!\!\circ\!\!\!\rightarrow_{[v_2,\pi_2]} H_2 \stackrel{*}{\rightarrow} H_2' -\!\!\circ\!\!\!\rightarrow \ldots -\!\!\circ\!\!\!\rightarrow_{[v_n,\pi_n]} H_n \stackrel{*}{\rightarrow} H_n' = Triv,$$

with left reduction steps $H'_i \multimap\!\!\to_{[v_{i+1},\pi_{i+1}]} H_{i+1}$, $i = 1,\ldots,n-1$, and a corresponding normalizing conditional narrowing derivation

$$G_1{\downarrow} \rightsquigarrow_{[v_2,\pi_2,\delta_2]} G_2 \overset{*}{\to} G_2{\downarrow} \rightsquigarrow \ldots \rightsquigarrow_{[v_n,\pi_n,\delta_n]} G_n \overset{*}{\to} G_n{\downarrow} = Triv.$$

Moreover, there is a normalized substitution λ such that

- $\lambda \circ \delta_n \circ \ldots \circ \delta_2 = \psi\ [V']$
- $H_i = (\lambda \circ \delta_n \circ \ldots \circ \delta_{i+1})\,(G_i), i = 2,\ldots,n$
- $H'_i = (\lambda \circ \delta_n \circ \ldots \circ \delta_{i+1})\,(G_i{\downarrow}), i = 1,\ldots,n.$

From $\mu = \psi \circ \delta_1\ [V]$ and $\psi = \lambda \circ \delta_n \circ \ldots \circ \delta_2\ [V']$, we get $\lambda \circ \delta_n \circ \ldots \circ \delta_1 = \mu\ [V]$.

Finally, we have to show that none of the LSE reducibility tests detects a redundancy. Suppose that the step $G_{m-1}{\downarrow} \overset{\downarrow}{\rightsquigarrow}_{[v_m,\pi_m,\delta_m]} G_m{\downarrow}$ is not LSE, for some $m \in \{1,\ldots,n\}$. Then there exists $i \in \{0,\ldots,m-1\}$ such that either

1. $\lambda_{i,m}(G_i{\downarrow})$ is $\to$-reducible at an occurrence v strictly left of v_{i+1} or
2. $\lambda_{i,m}(G_i{\downarrow})$ is $\to$-reducible at an occurrence v strictly below v_{i+1} or
3. $\lambda_{i,m}(G_i{\downarrow})$ is $\to$-reducible at occurrence v_{i+1} with a rule smaller than π_{i+1}.

Since $H'_i = (\lambda \circ \lambda_{m,n} \circ \lambda_{i,m})(G_i{\downarrow})$ and $\to$ is stable under substitutions this implies that one of the properties (1) to (3) must hold with H'_i in place of $\lambda_{i,m}(G_i{\downarrow})$. But this means that $H'_i \multimap\!\!\to_{[v_{i+1},\pi_{i+1}]} H_{i+1}$ is not a left reduction step, which is a contradiction. □

The next proposition gives a converse of Theorem 21.

Proposition 22. *Let R be a confluent and terminating conditional term rewrite system. Let*

$$G = G_0{\downarrow} \rightsquigarrow_{[v_1,\pi_1,\delta_1]} G_1 \overset{*}{\to} G_1{\downarrow} \rightsquigarrow \ldots \rightsquigarrow_{[v_n,\pi_n,\delta_n]} G_n \overset{*}{\to} G_n{\downarrow}$$

be a normalizing conditional LSE narrowing derivation. Then in the corresponding $(\multimap\!\!\to \circ \overset{*}{\to})$-derivation

$$H = H'_0 \multimap\!\!\to_{[v_1,\pi_1]} H_1 \overset{*}{\to} H'_1 \multimap\!\!\to \ldots \multimap\!\!\to_{[v_n,\pi_n]} H_n \overset{*}{\to} H'_n,$$

where $H_i \overset{\text{def}}{=} \lambda_{i,n}(G_i)$, for $i = 1,\ldots,n$ and $H'_i \overset{\text{def}}{=} \lambda_{i,n}(G_i{\downarrow})$, for $i = 0,\ldots,n$, the steps $H'_i \multimap\!\!\to_{[\pi_{i+1},v_{i+1}]} H_{i+1}$ are left $\multimap\!\!\to$-reduction steps, for all $i = 0,\ldots,n-1$.

Proof. Similar to the last part of the proof of Proposition 14 with H'_i instead of H_i and $G_i{\downarrow}$ instead of G_i. □

We now extend the completeness and optimality results for conditional LSE narrowing to normalizing conditional LSE narrowing.

Theorem 23. *Normalizing conditional LSE narrowing is complete for confluent and decreasing conditional term rewrite systems.*

Proof. Similar to the proof of Theorem 15. Since G and $G{\downarrow}$ have the same set of R-unifiers we can assume that G is in normal form. If μ is a normalized R-unifier of G, then there exists a $\rightarrow$-derivation $\mu(G) \xrightarrow{*} Triv^*$. By Theorem 21, this implies the existence of a $(\multimap \circ \xrightarrow{*})$-derivation $\mu(G)\,(\multimap \circ \xrightarrow{*})^*\, Triv$ which can be lifted to a normalizing conditional LSE narrowing derivation $G \rightsquigarrow_{\sigma}^{*\downarrow} G_n{\downarrow} = Triv$ such that the narrowing substitution σ is more general than μ. □

Theorem 24. *Let R be a confluent and decreasing conditional term rewrite system. Consider two normalizing conditional LSE narrowing derivations*

$$G_0 \rightsquigarrow^{\downarrow}_{[v_1,\pi_1,\delta_1]} G_1{\downarrow} \rightsquigarrow^{\downarrow}_{[v_2,\pi_2,\delta_2]} \cdots \rightsquigarrow^{\downarrow}_{[v_n,\pi_n,\delta_n]} G_n{\downarrow}$$

$$G'_0 \rightsquigarrow^{\downarrow}_{[v'_1,\pi'_1,\delta'_1]} G'_1{\downarrow} \rightsquigarrow^{\downarrow}_{[v'_2,\pi'_2,\delta'_2]} \cdots \rightsquigarrow^{\downarrow}_{[v'_m,\pi'_m,\delta'_m]} G'_m{\downarrow}$$

leading to trivial systems $G_n{\downarrow} = Triv$ and $G'_m{\downarrow} = Triv'$.

Suppose that G_0 and G'_0 respectively $\lambda_{0,n}$ and $\lambda'_{0,m}$ are identical up to variable renaming, that is there exist substitutions $\tau_0, \tau'_0, \rho, \rho'$ such that

- $\tau_0(G_0) = G'_0$, $\tau'_0(G'_0) = G_0$ *and*
- $\lambda_{0,n} = \rho' \circ \lambda'_{0,m} \circ \tau_0\ [Var(G_0)]$, $\lambda'_{0,m} = \rho \circ \lambda_{0,n} \circ \tau'_0\ [Var(G'_0)]$.

Then the two derivations are identical up to variable renaming, that is

- $n = m$,
- $v_i = v'_i$, *for* $i = 1, \ldots, n$,
- $\pi_i = \pi'_i$, *for* $i = 1, \ldots, n$,
- *there exist substitutions* τ_i, τ'_i *such that*
  - $\tau_i(G_i{\downarrow}) = G'_i{\downarrow}$, $\tau'_i(G'_i{\downarrow}) = G_i{\downarrow}$,
  - $\lambda_{i,n} = \rho' \circ \lambda'_{i,m} \circ \tau_i\ [Var(G_i{\downarrow})]$, $\lambda'_{i,m} = \rho \circ \lambda_{i,n} \circ \tau'_i\ [Var(G'_i{\downarrow})]$,

  for $i = 1, \ldots, n$.

Proof. Without loss of generality we assume $n \leq m$. First we show by induction on n that the first n steps of the two derivations are identical up to variable renaming. For $n = 0$ nothing has to be shown. Assume therefore $n \geq 1$ and consider the associated $(\multimap \circ \xrightarrow{*})$-derivations

$$\lambda_{0,n}(G_0) \multimap_{[v_1,\pi_1]} \lambda_{1,n}(G_1) \xrightarrow{*} \lambda_{1,n}(G_1{\downarrow}) \multimap \ldots \xrightarrow{*} \lambda_{n,n}(G_n{\downarrow}) = Triv$$
$$\lambda'_{0,m}(G'_0) \multimap_{[v'_1,\pi'_1]} \lambda'_{1,m}(G'_1) \xrightarrow{*} \lambda'_{1,m}(G'_1{\downarrow}) \multimap \ldots \xrightarrow{*} \lambda'_{m,m}(G'_m{\downarrow}) = Triv'.$$

By Proposition 22, $\lambda_{0,n}(G_0) \multimap_{[v_1,\pi_1]} \lambda_{1,n}(G_1)$ and $\lambda'_{0,m}(G'_0) \multimap_{[v'_1,\pi'_1]} \lambda'_{1,m}(G'_1)$ are both left $\multimap$-reduction steps. From $\lambda'_{0,m}(G'_0) = (\rho \circ \lambda_{0,n} \circ \tau'_0)(G'_0) = \rho(\lambda_{0,n}(G_0))$ and $\lambda_{0,n}(G_0) = (\rho' \circ \lambda'_{0,m} \circ \tau_0)(G_0) = \rho'(\lambda'_{0,m}(G'_0))$ we deduce that $\lambda_{0,n}(G_0)$ and $\lambda'_{0,m}(G'_0)$ are identical up to variable renaming. By the uniqueness of left $\multimap$-reductions to trivial systems, which follows from Theorem 12, this implies $v_1 = v'_1$ and $\pi_1 = \pi'_1$.

From Lemma 8 and its proof we get the existence of substitutions τ_1 and τ'_1 with $\tau_1 \circ \delta_1 = \delta'_1 \circ \tau_0$ and $\tau'_1 \circ \delta'_1 = \delta_1 \circ \tau'_0$ such that $\tau_1(G_1) = G'_1$ and $\tau'_1(G'_1) = G_1$.

Since τ_1 and τ_1' are renaming substitutions, we get even $\tau_1(G_1\downarrow) = \tau_1(G_1)\downarrow = G_1'\downarrow$ and $\tau_1'(G_1'\downarrow) = \tau_1'(G_1')\downarrow = G_1\downarrow$.

From $\lambda_{1,n} \circ \delta_1 = \lambda_{0,n} = \rho' \circ \lambda_{0,m}' \circ \tau_0 = \rho' \circ \lambda_{1,m}' \circ \delta_1' \circ \tau_0 = \rho' \circ \lambda_{1,m}' \circ \tau_1 \circ \delta_1$ $[Var(G_0)]$ we deduce $\lambda_{1,n} = \rho' \circ \lambda_{1,m}' \circ \tau_1$ $[(Var(G_0) \setminus Dom(\delta_1)) \cup Im(\delta_1|_{Var(G_0)})]$. Since $(Var(G_0) \setminus Dom(\delta_1)) \cup Im(\delta_1|_{Var(G_0)}) = Var(\delta_1(G_0)) = Var(\delta_1(G_0[v_1 \leftarrow l_1])) \supseteq Var(\delta_1(P_1 \wedge G_0[v_1 \leftarrow r_1])) = Var(G_1) \supseteq Var(G_1\downarrow)$, this implies $\lambda_{1,n} = \rho' \circ \lambda_{1,m}' \circ \tau_1$ $[Var(G_1\downarrow)]$. In the same way, we can show that $\lambda_{1,m}' = \rho \circ \lambda_{1,n} \circ \tau_1'$ $[Var(G_1'\downarrow)]$.
Now we can apply the induction hypothesis and we get

- $v_i = v_i'$, for $i = 2, \ldots, n$
- $\pi_i = \pi_i'$, for $i = 2, \ldots, n$
- there exist substitutions τ_i, τ_i' such that
  - $\tau_i(G_i\downarrow) = G_i'\downarrow$, $\quad \tau_i'(G_i'\downarrow) = G_i\downarrow$,
  - $\lambda_{i,n} = \rho' \circ \lambda_{i,m}' \circ \tau_i$ $[Var(G_i\downarrow)]$, $\quad \lambda_{i,m}' = \rho \circ \lambda_{i,n} \circ \tau_i'$ $[Var(G_i'\downarrow)]$,

  for $i = 2, \ldots, n$.

From $G_n'\downarrow = \tau_n(G_n\downarrow) = Triv$ we finally conclude $n = m$. □

Assuming again that narrowing derivations starting from the same goal and using the same rules at the same occurrences produce the same narrowing substitution we get like before:

Corollary 25. *If normalizing conditional LSE narrowing enumerates two solutions σ and σ' which coincide up to variable renaming, then $\sigma = \sigma'$ holds and the two derivations coincide.*

Theorem 26. *Let R be a confluent and terminating conditional term rewrite system. Then for any normalizing conditional LSE narrowing derivation the narrowing substitution is normalized.*

Corollary 27. *Normalizing conditional LSE narrowing enumerates only normalized substitutions.*

6 Conditional Left-to-Right Basic Narrowing

For unconditional narrowing we proved in (Bockmayr, Krischer, and Werner 1993) that each LSE narrowing derivation is also a left-to-right basic narrowing derivation (cf. Hullot 1980; Herold 1986; Réty 1987). Based on this result we developed an efficient implementation of LSE narrowing (Werner, Bockmayr, and Krischer 1993, 1994). We show now that in the conditional case each *successful* LSE narrowing derivation is left-to-right basic. Therefore the ideas of (Werner, Bockmayr, and Krischer 1993) can also be used to get an efficient implementation of conditional LSE narrowing. For practical reasons we consider only normalizing conditional LSE narrowing. But the same results hold also in the non-normalizing case.

We start with some technical preliminaries needed for the definition of normalizing left-to-right basic narrowing. Occurrences in an equation e are given by sequences of numbers. For systems of equations, we assign to each equation a number to distinguish the different equations. An occurrence in a system of equations $e_1 \wedge \ldots \wedge e_n$ is therefore a pair (i, v) where i is the number of the equation e_i and v is an ordinary occurrence in e_i. Concatenation of occurrences is defined by $(i, v).\omega \stackrel{\text{def}}{=} (i, v.\omega)$ for $i \in \mathcal{N}$ and $v, \omega \in \mathcal{N}^*$. $FuOcc(T)$ denotes the set of non-variable occurrences in T. If we number the equations from left to right then the number of an equation may change during a narrowing step and consequently the definition of left-to-right basic narrowing becomes more complicated than necessary. To avoid this, we number equations from right to left.

Definition 28. An occurrence $v = (n, \omega)$ is *strictly left* of an occurrence $v' = (m, \omega')$, denoted $v \lhd v'$, iff $n > m$ or if $n = m$ and there exist occurrences o, u, u' and natural numbers i, i' such that $i < i', \omega = o.i.u$ and $\omega' = o.i'.u'$.

An occurrence $v = (n, \omega)$ is *strictly below* an occurrence $v' = (m, \omega')$, denoted $v \succ v'$ iff $n = m$ and there exists an occurrence $u \neq \epsilon$ such that $\omega = \omega'.u$.

Definition 29. Let $T \rightarrow_{[v, P \Rightarrow l \rightarrow r]} T'$ be a rewriting step. We say that the occurrence ω in T is an *antecedent* of the occurrence ω' in T' iff

- $\omega = \omega'$ and neither $\omega \succeq v$ nor $v \preceq \omega$ or
- there exists an occurrence ρ' of a variable x in r such that $\omega' = v.\rho'.o$ and $\omega = v.\rho.o$ where ρ is an occurrence of the same variable x in l (see Fig. 4 in (Bockmayr, Krischer, and Werner 1993) for an illustration).

Compared to unconditional narrowing the formulas for left-to-right basic occurrences are slightly more complicated. After a narrowing step we have to remove all occurrences left of or below the narrowing occurrence and we have to add not only the non-variable occurrences of the non-instantiated right-hand side but also those in the premise of the rule. After a $\rightarrow$-rewriting step, we have to replace the occurrences below the rewriting occurrence by the occurrences that are non-variable in the uninstantiated right-hand side or whose antecedents are left-to-right basic.

Definition 30. Let $G = g_m \wedge \ldots \wedge g_1$ be a system of equations, U a set of occurrences of G, and $\pi : e_n \wedge \ldots \wedge e_1 \Rightarrow l \rightarrow r$ a rewrite rule. For a conditional narrowing or rewriting step we define

$$\begin{aligned} LB(U, G \rightsquigarrow_{[v,\pi,\delta]} G') = LB(U, G \multimap_{[v,\pi,\delta]} G') &\stackrel{\text{def}}{=} (U \setminus \{u \mid u \lhd v \text{ or } u \succeq v\}) \\ &\cup \{v.o \mid o \in FuOcc(r)\} \cup \{(m+i, u) \mid 1 \leq i \leq n, u \in FuOcc(e_i)\} \\ LB(U, G \rightarrow_{[v,\pi,\delta]} G') \stackrel{\text{def}}{=} (U \setminus \{u \mid u \succeq v\}) &\cup \{v.o \mid o \in FuOcc(r)\} \\ &\cup \{v.o \mid o \in FuOcc(\delta(r)) \setminus FuOcc(r) \text{ and all antecedents of } v.o \text{ are in } U\} \end{aligned}$$

respectively. Given the normalizing conditional narrowing derivation

$$\begin{array}{llll} G_0{\downarrow} & \rightsquigarrow_{[v_1, P_1 \Rightarrow l_1 \rightarrow r_1, \delta_1]} & G_1 = G_{10} \rightarrow \ldots \rightarrow G_{1k_1} = G_1{\downarrow} \\ \vdots & & \\ G_{n-1}{\downarrow} & \rightsquigarrow_{[v_n, P_n \Rightarrow l_n \rightarrow r_n, \delta_n]} & G_n = G_{n0} \rightarrow \ldots \rightarrow G_{nk_n} = G_n{\downarrow} \end{array}$$

the sets of *left-to-right basic occurrences* are inductively defined by

$$LB'_0 \stackrel{\text{def}}{=} FuOcc(G_0\!\downarrow)$$
$$LB_i \stackrel{\text{def}}{=} LB(LB'_{i-1}, G_{i-1}\!\downarrow\,\rightsquigarrow_{[v_i,\pi_i,\delta_i]} G_i)$$
$$LB_{ij} \stackrel{\text{def}}{=} LB(LB_{i,j-1}, G_{i,j-1} \rightarrow_{[v_{ij},\pi_{ij},\delta_{ij}]} G_{i,j})$$

with $LB_{i0} = LB_i$ and $LB'_i = LB_{i,k_i}$, for $i = 1, \ldots n$ and $j = 1, \ldots, k_i$. For a *normalizing conditional left-to-right basic narrowing derivation* we require that $v_i \in LB'_{i-1}$, for all $i = 1, \ldots, n$, and $v_{ij} \in LB_{i,j-1}$, for all $i = 1, \ldots, n$ and $j = 1, \ldots, k_i$.

In contrast to the unconditional case, a conditional LSE narrowing derivation is not necessarily left-to-right basic.

Example 31. Given the rules $\pi_1 : s(0) + y \rightarrow s(y)$, and $\pi_2 : x \doteq p(0) \Rightarrow s(x) \rightarrow 0$, the derivation

$$z + 0 \doteq z \rightsquigarrow_{[1,\pi_1,\{z \mapsto s(0)\}]} s(0) \doteq s(0) \rightsquigarrow_{[2,\pi_2,\{\}]} 0 \doteq p(0) \wedge s(0) \doteq 0$$

is LSE but it is not left-to-right basic, since the last step takes place at a non-left-to-right-basic occurrence.

However, each successful LSE derivation is left-to-right basic. We prove this using the notion of sufficient largeness (Réty 1987).

Definition 32. A set U of occurrences in a system T is said to be *sufficiently large* on T, iff T/ω is $\rightarrow$-irreducible for all $\omega \in Occ(T) \setminus U$.

Lemma 33. *Let $\mathcal{R}$ be a conditional term rewrite system without extravariables such that no left-hand side of a rule is a variable. Let $H \multimap_{[v,\pi,\tau]} H'$ be a left $\multimap$-reduction step such that $H' \overset{*}{\multimap} Triv$ to some trivial system $Triv$. If U is sufficiently large on H and $U' \stackrel{\text{def}}{=} LB(U, H \multimap_{[v,\pi,\tau]} H')$, then $v \in U$ and U' is sufficiently large on H'.*

Proof. Since U is sufficiently large on H and since $H \multimap_{[v,\pi,\tau]} H' = \tau(P) \wedge H'' \overset{*}{\multimap} Triv$ implies $H \rightarrow_{[v,\pi,\tau]} H''$, the step $H \multimap_{[v,\pi,\tau]} H'$ satisfies $v \in U$. Since the strategy is innermost, $\tau(x)$ is $\rightarrow$-irreducible for all $x \in Var(l)$. This holds because l is not a variable and therefore $\tau(x)$ is a proper subterm of $\tau(l)$. Since the strategy is leftmost, the part of H strictly left of v is $\rightarrow$-irreducible. This shows that U' is sufficiently large on H'. □

Lemma 34. *Let $\mathcal{R}$ be a conditional term rewrite system without extravariables and let $H_0 \rightarrow_{[v_1,\pi_1,\delta_1]} \cdots \rightarrow_{[v_n,\pi_n,\delta_n]} H_n$ be a $\rightarrow$-derivation. If U_0 is sufficiently large on H_0 and $U_i \stackrel{\text{def}}{=} LB(U_{i-1}, H_{i-1} \rightarrow_{[v_i,\pi_i,\delta_i]} H_i)$ for $i = 1, \ldots, n$, then $v_i \in U_{i-1}$ for $i = 1, \ldots, n$, and U_n is sufficiently large on H_n.*

Proof. By induction on the length of the derivation. In the case $n = 0$ nothing has to be shown. Assume therefore that $H_0 \xrightarrow{*} H_{n-1}, n > 0$, with $v_i \in U_{i-1}$ $i = 1, \ldots, n-1$, and that U_{n-1} is sufficiently large on H_{n-1}. Then the occurrence v_{n-1} in $H_{n-1} \rightarrow_{[v_{n-1},\pi_{n-1}]} H_n$ must belong to U_{n-1}. If $\omega_n \in Occ(H_n) \setminus U_n$, then at least one antecedent ω_{n-1} of ω_n in G_{n-1} does not belong to U_{n-1}. Since U_{n-1} is sufficiently large on H_{n-1}, we deduce that $H_n/\omega_n = H_{n-1}/\omega_{n-1}$ is $\rightarrow$-irreducible. This shows that U_n is sufficiently large on H_n. □

Theorem 35. *Let $\mathcal{R}$ be a confluent and terminating conditional term rewrite system without extravariables such that no left-hand side of a rule is a variable. Any normalizing conditional LSE narrowing derivation $G \rightsquigarrow^{\downarrow *} Triv$ to a trivial system Triv is also a normalizing conditional left-to-right basic narrowing derivation.*

Proof. Consider a normalizing conditional LSE narrowing derivation

$$G_0{\downarrow} \rightsquigarrow_{[v_1,\pi_1,\delta_1]} G_1 \xrightarrow{*} G_1{\downarrow} \rightsquigarrow_{[v_2,\pi_2,\delta_2]} \cdots \rightsquigarrow_{[v_n,\pi_n,\delta_n]} G_n \xrightarrow{*} G_n{\downarrow} = Triv$$

and the corresponding rewriting derivation

$$H = H_0' \multimap_{[v_1,\pi_1]} H_1 \xrightarrow{*} H_1' \multimap \ldots \multimap_{[v_n,\pi_n]} H_n \xrightarrow{*} H_n' = H{\downarrow} = Triv,$$

with $H_i \stackrel{\text{def}}{=} \lambda_{i,n}(G_i)$, for $i = 1, \ldots, n$, and $H_i' \stackrel{\text{def}}{=} \lambda_{i,n}(G_i{\downarrow})$, for $i = 0, \ldots, n$ and let the sets of left-to-right basic narrowing occurrences be defined according to Definition 30. By induction on $m = 0, \ldots, n$, we prove that

- $v_i \in LB_{i-1}'$ for $i = 1, \ldots, m$,
- $v_{ij} \in LB_{i,j-1}$, for all $i = 1, \ldots, m$ and $j = 1, \ldots, k_i$, and
- LB_i' is sufficiently large on $G_i{\downarrow}$ and H_i' for $i = 0, \ldots, m$.

Since $\lambda_{0,n}|_{Var(G_0\downarrow)}$ is normalized by Theorem 26, $LB_0' = FuOcc(G_0{\downarrow})$ is sufficiently large on both $G_0{\downarrow}$ and $H_0' = \lambda_{0,n}(G_0{\downarrow})$.

Suppose the statement is true for $0 \leq m-1 < n$. By the induction hypothesis the statement holds for all $i \leq m-1$. Hence, LB_{m-1}' is sufficiently large on H_{m-1}'. Consider the rewriting step $H_{m-1}' \multimap_{[v_m,\pi_m]} H_m$. By Proposition 22, $H_{m-1}' \multimap_{[v_m,\pi_m]} H_m$ is a left $\multimap$-reduction step. Using Lemma 33, we can conclude that $v_m \in LB_{m-1}'$ and that the set $LB_m = LB(LB_{m-1}', H_{m-1}' \multimap_{[v_m,\pi_m]} H_m)$ is sufficiently large on H_m. Since $H_m = \lambda_{m,n}(G_m)$, this shows that LB_m is also sufficiently large on G_m. By Lemma 34, this implies that for the rewriting derivations $G_m = G_{m,0} \xrightarrow{*} G_{m,k_m} = G_m{\downarrow}$ and $H_m = H_{m,0} \xrightarrow{*} H_{m,k_m} = H_m'$, $v_{m,j} \in LB_{m,j-1}$ for $j = 1, \ldots, k_m$ holds and that LB_m' is sufficiently large on $G_m{\downarrow}$ and H_m'. Therefore, the statement is true for m. □

Corollary 36. *Normalizing conditional left-to-right basic narrowing is complete for confluent and decreasing conditional term rewrite systems.*

Example 37. We compare LSE and left-to-right basic narrowing on a conditional rewrite system specifying the integers with the $\leq$-predicate given in (Kaplan 1984a):

$$
\begin{array}{llll}
R = \{\ s(p(x)) & \rightarrow x, & p(s(x)) & \rightarrow x, \\
s(x) \leq y & \rightarrow x \leq p(y), & p(x) \leq y & \rightarrow x \leq s(y), \\
0 \leq 0 & \rightarrow \mathtt{true}, & 0 \leq p(0) & \rightarrow \mathtt{false}, \\
0 \leq x \doteq \mathtt{true} \Rightarrow 0 \leq s(x) & \rightarrow \mathtt{true}, & 0 \leq x \doteq \mathtt{false} \Rightarrow 0 \leq p(x) & \rightarrow \mathtt{false}\}
\end{array}
$$

The normalizing left-to-right basic narrowing tree for the goal $x \leq s(0) \doteq \mathtt{true}$ looks as follows:

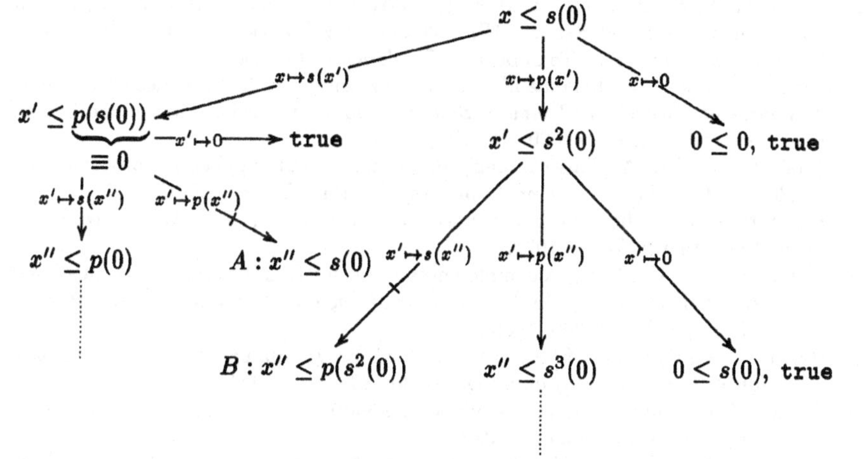

The tree can be pruned at node A due to the Sub-Test, and at node B due to the Sub-Test or the Epsilon-Test. Note that each solution is computed only once.

7 Conclusion

We have extended our results on LSE narrowing to the case of confluent and decreasing conditional term rewrite systems. Using the calculus of conditional rewriting without evaluation of the premise we were able to generalize most of our earlier results. This calculus, which establishes the connection between ordinary conditional rewriting and conditional narrowing, also makes clear where the differences to the unconditional case come from. Although, in the conditional case, some of the main results hold only for successful derivations this is not a restriction in practice. The methods developed in (Werner, Bockmayr, and Krischer 1993, 1994) for an efficient realization of LSE narrowing can also be used in the conditional case. Conditional narrowing is not complete for confluent and terminating conditional term rewrite systems with extravariables (Giovannetti and Moiso 1986). Extending LSE narrowing to systems with extravariables using additional hypotheses and techniques like those developed in (Middeldorp and Hamoen 1994) is currently being investigated.

References

S. Antoy, R. Echahed, and M. Hanus. A needed narrowing strategy. In *21st ACM Symposium on Principles of Programming Languages, POPL'94, Portland*, pages 268 – 279, 1994.

A. Bockmayr. *Contributions to the Theory of Logic-Functional Programming.* PhD thesis, Fakultät für Informatik, Univ. Karlsruhe, 1990. (in German).

A. Bockmayr. Conditional narrowing modulo a set of equations. *Applicable Algebra in Engineering, Communication and Computing*, 4(3):147 – 168, 1993.

A. Bockmayr, S. Krischer, and A. Werner. An optimal narrowing strategy for general canonical systems. In *Conditional Term Rewriting Systems, CTRS'92, Pont-à-Mousson, France.* Springer, LNCS 656, 1992.

A. Bockmayr, S. Krischer, and A. Werner. Narrowing strategies for arbitrary canonical systems. Technical Report MPI-I-93-233, Max-Planck-Institut für Informatik, Saarbrücken, July 1993. To appear in *Fundamenta Informaticae.*

N. Dershowitz and J.-P. Jouannaud. Rewrite systems. In Jan van Leeuwen, editor, *Handbook of Theoretical Computer Science*, volume B: Formal Models and Semantics, chapter 6, pages 244–320. Elsevier, 1990.

N. Dershowitz and D. A. Plaisted. Logic programming cum applicative programming. In *Proc. Intern. Symposium on Logic Programming, Boston.* IEEE, 1985.

M. Fay. First-order unification in an equational theory. In *4th Workshop on Automated Deduction, Austin, Texas*, 1979.

E. Giovannetti and C. Moiso. A completeness result for E-unification algorithms based on conditional narrowing. In *Foundations of Logic and Functional Programming, Trento.* Springer LNCS 306, 1986.

M. Hanus. The integration of functions into logic programming: From theory to practice. *Journal of Logic Programming*, 19&20:583 – 628, 1994.

A. Herold. Narrowing techniques applied to idempotent unification. SEKI-Report SR-86-16, Univ. Kaiserslautern, 1986.

G. Huet and D. C. Oppen. Equations and rewrite rules, A survey. In R. V. Book, editor, *Formal Language Theory.* Academic Press, 1980.

J. M. Hullot. Canonical forms and unification. In *Proc. 5th Conference on Automated Deduction, Les Arcs.* Springer, LNCS 87, 1980.

H. Hußmann. Unification in conditional-equational theories. Technical Report MIP-8502, Univ. Passau, Jan. 1985. Short version: EUROCAL 85, Linz, Springer, LNCS 204.

S. Kaplan. Conditional rewrite rules. *Theoretical Computer Science*, 33:175 – 193, 1984.

S. Kaplan. Fair conditional term rewriting systems: Unification, termination and confluence. Technical Report 194, L. R. I., Univ. Paris-Sud, 1984.

J. W. Klop. Term rewriting systems. In S. Abramski, D. M. Gabbay, and T. S. Maibaum, editors, *Handbook of Logic in Computer Science*, volume 2 - Background: Computational Structures, pages 1 – 116. Oxford Univ. Press, 1992.

A. Middeldorp and E. Hamoen. Completeness results for basic narrowing. *Applicable Algebra in Engineering, Communication and Computing*, 5:213–253, 1994.

P. Réty. Improving basic narrowing techniques. In *Proc. Rewriting Techniques and Applications RTA'87, Bordeaux.* Springer, LNCS 256, 1987.

A. Werner, A. Bockmayr, and S. Krischer. How to realize LSE narrowing. Technical Report 6/93, Fakultät für Informatik, Univ. Karlsruhe, December 1993.

A. Werner, A. Bockmayr, and S. Krischer. How to realize LSE narrowing. In *Algebraic and Logic Programming, ALP'94, Madrid*, pages 59 – 76. Springer, LNCS 850, 1994.

Preserving Confluence for Rewrite Systems with Built-in Operations

Reinhard Bündgen

Wilhelm-Schickard-Institut, Universität Tübingen
Sand 13, D-72076 Tübingen, Germany
phone: x7071/295459 — fax: x7071/295958
e-mail: ⟨buendgen@informatik.uni-tuebingen.de⟩

Abstract. We consider combinations of term rewriting with built-in operations which compute objects of some external domain from ground terms. Combining a canonical term rewriting system with a computation relation does in general not lead to a canonical simplification relation. We show how a canonical simplification system can be constructed for such rewrite systems with built-in operations. Decomposition free simplification systems never need to look at the internal structure of external objects and thus a previous computation never needs to be undone. For some interesting equational specifications the construction of decomposition free canonical simplification systems succeeds. Our construction of canonical simplification systems relies on simulating combined rewriting and computing by pure term rewriting. This allows us to use the wealth of all rewrite techniques and tools to investigate and manipulate these structures.

1 Introduction

Canonical term rewriting systems are an elegant and powerful tool to prove and compute in abstract domains. However for concrete domains they sometimes form an extremely inefficient framework for computation.

Example 1. Consider the Peano axiomatization of the natural numbers and +

$$\mathcal{R} = \{\, 0 + x \to x,\; s(y) + x \to s(y + x) \,\}.$$

The computation of $10^{10} + 10^{11}$ requires more than 10^{10} rewrites and storing the result takes $O(10^{11})$ space in memory. □

The goal of this paper is to combine both the proof-power of term rewriting systems and the algorithmic efficiency of functions designed for a particular domain. Such a particular domain may either be a data structure which is built-in into the implementation programming language of the term rewriting system or any other programming language (e. g., Boolean type, integers, ...) or it may be a user defined abstract data type implemented in any programming language

(e. g., infinite precision integers, polynomials, ...) or it may be a mixture of both (e. g., lists of integers).

A straightforward approach is to use *mixed terms* which may be built from the term signature, the term variables and *external objects*. The latter are data objects of a particular built-in domain which we will call an *external domain* from now on. Technically, external objects are incorporated into a term using some kind of *coercion operator* from the 'sort external domain' to the 'sort terms'. We want to allow for two *simplification operations* on mixed terms: (1) *reductions* by a canonical term rewriting system and (2) *computations*. A computation converts a mixed term m without variables into an external object provided that for each operator f in m a function f_D on external objects that implements f is known. The simplification relation is the union of the reduction and the computation relation. Provided all operators of the signature have an interpretation and the external domain is a model of the canonical term rewriting system the simplification relation is confluent and terminating on terms without variables. Unfortunately general confluence (including ground confluence on mixed terms) does not hold for the simplification relation. Thus combining canonical term rewriting systems and built-in operations may lead to an effective tool for computations but in general the combined system is not be suitable for theorem proving.

Example 2. Let $\mathcal{R}_{CR1}$ be the AC-canonical term rewriting system for commutative rings with ones ($+$ and $\cdot$ are both associative and commutative)

$$\begin{aligned}\mathcal{R}_{CR1} = \{\, & x+0 \to x, & x+-x &\to 0,\\ & x\cdot 0 \to 0, & y+x+-x &\to y,\\ & x\cdot 1 \to x, & -(x+y) &\to -x+-y,\\ & -0 \to 0, & x\cdot -y &\to -(x\cdot y),\\ & -(-x) \to x, & x\cdot(y+z) &\to (x\cdot y)+(x\cdot z)\}\end{aligned}$$

and the external domain be an implementation of $(\mathbb{Z}, 0, 1, -, +, \cdot)$. Then the mixed terms "30" $\cdot x +$ "70" $\cdot x$ and "100" $\cdot x$ are equal but do not have a common normal form w. r. t. the simplification relation. □

Some computer algebra systems (like Reduce [Hearn, 1991] or Mathematica [Wolfram, 1991]) provide for the combination of rewrite rules and built-in computations on an ad-hoc basis as parts of their programming language. In these systems it is left to the user that his rewrite specification together with the (built-in) evaluation strategy for applying rewrites and computations makes sense. Hence such rules may only be valid in certain computational contexts.

A canonical simplification relation provides a well-defined operational semantics for mixing rewrites and computations without giving up the declarative semantics of the equational specification (cf. [Avenhaus and Becker, 1994]). In this paper, we are concerned with the problem how the confluence of a canonical term rewriting system can be preserved in presence of built-in operations. In order to investigate the simplification relation we will simulate term rewriting combined with built-in computations in the initial model of the rewrite specification by pure term rewriting. This allows us to use well-known rewrite techniques and tools to study the simplification relation. We show how the Knuth-Bendix completion procedure [Knuth and Bendix, 1970] or its extension for rewriting modulo equational theories (e. g., [Peterson and Stickel, 1981]) can be used to find canonical simplification systems. We present some techniques to prove the termination of such combined systems too.

A second goal of our work is to maintain an *economy of computation*, i. e. we do not want to be forced to undo a previous computation. This leads to the discovery of decomposition free simplification systems. Yet it turns out that not every canonical term rewriting system can be extended to a decomposition free canonical simplification system.

2 Preliminaries

We assume the reader is familiar with the theory of term rewriting systems. For surveys on this topic see [Dershowitz and Jouannaud, 1990],[Klop, 1992] or [Plaisted, 1993].

Abstract Reduction Relations A *reduction relation* $\rightarrow_{\mathcal{A}} \subseteq \mathcal{D} \times \mathcal{D}$ is an asymmetric binary relation. Then $\leftarrow_{\mathcal{A}}$ is the inverse relation of $\rightarrow_{\mathcal{A}}$. $\leftrightarrow_{\mathcal{A}}$, $\rightarrow^*_{\mathcal{A}}$ and $\leftrightarrow^*_{\mathcal{A}}$ are the symmetric -, transitive and reflexive -, and the symmetric, transitive and reflexive closures of $\rightarrow_{\mathcal{A}}$ respectively. The relation $\rightarrow_{\mathcal{A}}$ is *terminating* if there is no infinite chain $a_1 \rightarrow_{\mathcal{A}} a_2 \rightarrow_{\mathcal{A}} \ldots$. An object which is irreducible w. r. t. $\rightarrow_{\mathcal{A}}$ is called a *normal form*. The relation $\rightarrow_{\mathcal{A}}$ is *confluent* if for all $a, b, c \in \mathcal{D}$ such that $b \leftarrow^*_{\mathcal{A}} a \rightarrow^*_{\mathcal{A}} c$ there is a $d \in \mathcal{D}$ with $b \rightarrow^*_{\mathcal{A}} d \leftarrow^*_{\mathcal{A}} c$. Confluent and terminating reduction relations are called *canonical*. They compute a unique normal form for each object.

Terms Throughout this paper, we denote by $\mathcal{F} = \bigcup_i \mathcal{F}_i$ the finite set of ranked *function symbols* (the $\mathcal{F}_n$ are the n-ary function symbols) and by $\mathcal{X}$ the set of *variables*. Each function symbol is implicitly accompanied by a *sort description* $f : s_1 \times \cdots \times s_n \rightarrow s$ for $f \in \mathcal{F}_n$ and the s, s_i are elements of a fixed set of sorts. In the same way each variable is assigned a fixed sort. *Terms* are defined recursively as follows: Each constant or variable that is assigned the sort s is a term of sort s. If $t_1, \ldots, t_n$ are terms of sorts $s_1, \ldots, s_n$ respectively and $f : s_1 \times \cdots \times s_n \rightarrow s$ then $f(t_1, \ldots, t_n)$ is a term of sort s. Nothing else is a (sort-correct) term. $T(\mathcal{F}, \mathcal{X})$ denotes the set of all sort-correct terms freely generated by $\mathcal{F}$ and $\mathcal{X}$. Terms without variables are called *ground terms*. For a term t, $\mathcal{X}(t)$ is the set of variables occurring in t. Let p be a *position* (or *occurrence*) in a term t. By λ we denote the empty (or top) position. Then $t|_p$ denotes the subterm of t at position p, $t(p)$ is the symbol labeling position p and for $s \in T(\mathcal{F}, \mathcal{X})$, $t[s]_p$ is the result of replacing in t the subterm at position p by s. Two positions p_1 and p_2 are *disjoint* if none of them is a prefix of the other. Thus subterms at disjoint positions of a term do not overlap. Let $p_1, \ldots, p_n$ be pairwise disjoint positions in t. Then the subterms at these positions can be replaced simultaneously and we write $t[s_1]_{p_1} \dots [s_n]_{p_n}$ omitting redundant parentheses.

A *substitution* $\sigma : \mathcal{X} \to T(\mathcal{F}, \mathcal{X})$ is a mapping from variables to terms. If we extend the application of a substitution σ to a term t, we write $t\sigma$ meaning that all variables x in t are simultaneously replaced by $\sigma(x)$. $t\sigma$ is an *instance* of t and we say that t is *more general than* s if s is an instance of t. If there is a substitution μ such that $s\mu = t\mu$ then s and t *unify* and μ is a *unifier* of s and t.

Term Rewriting Systems A *term rewriting system* $\mathcal{R}$ is a set of *rewrite rules* $l \to r$, where l and r are terms. l is called the *left-hand side* of the rule and r is its *right-hand side.* A rule is *left-linear* if in its left-hand side every variable occurs at most once. The term s reduces in one step to t, $s \to_{\mathcal{R}} t$, if $s = s[l\sigma]_p$, $l \to r \in \mathcal{R}$ and $t = s[r\sigma]_p$. $s|_p$ is then called a *redex.* A term is *ground reducible* if all its ground instances are reducible. A term rewriting system is *ground confluent* if it is confluent for ground terms.

A set $\mathcal{E}$ of equations $s \leftrightarrow t$ where s and t are terms induces a relation $\leftrightarrow_{\mathcal{E}} = \to_{\overline{\mathcal{E}}}$ where $\overline{\mathcal{E}} = \{s \to t, t \to s \mid s \leftrightarrow t \in \mathcal{E}\}$. A term rewriting system $\mathcal{R}$ is *equivalent* to a set of equations $\mathcal{E}$ if $\leftrightarrow^*_{\mathcal{R}} = \leftrightarrow^*_{\mathcal{E}}$. The Knuth-Bendix completion procedure [Knuth and Bendix, 1970] transforms on success a set of equations into an equivalent canonical term rewriting system. Let $l \to r$ and $l' \to r'$ be two rules where l contains a subterm s which unifies with l' such that the most general unifier of s and l' is μ. Then $l\mu$ can be reduced by each of the two rules and the two terms resulting from the two different one-step reductions are called a *critical pair.* Knuth and Bendix [Knuth and Bendix, 1970] showed that a terminating term rewriting system is confluent iff all critical pairs are confluent.

By the *initial model* of a rewrite specification $(\mathcal{F}, \mathcal{R})$, we understand the initial algebra of $\mathcal{F}$ modulo the equalities induced by interpreting the rules of $\mathcal{R}$ as equations. If $\mathcal{R}$ is canonical the initial model of $(\mathcal{F}, \mathcal{R})$ is isomorphic to the set of $\mathcal{R}$-irreducible ground terms.

Associativity and Commutativity In many applications in computer algebra associative and commutative operations play an important rôle. Therefore we also consider rewriting modulo AC. I. e., certain operators in $\mathcal{F}_{AC} \subseteq \mathcal{F}_2$ are known to be both associative and commutative. We assume that the basic algorithms (equality-test, match, unification) do have built-in knowledge of the associativity and commutativity of the operators in $\mathcal{F}_{AC}$ [Hullot, 1979, Stickel, 1981]. Also the Knuth-Bendix completion procedure can be extended to term rewriting systems with built-in AC-operators (e. g., [Peterson and Stickel, 1981], and [Jouannaud and Kirchner, 1986]).

Mixed Terms By D we will denote a set of *external objects* disjoint from the signature and variables. A *mixed term* is a structure obtained by replacing some leaf nodes of a term (variable or constant nodes) by external objects. In a many sorted term algebra D should be partitioned into $\biguplus_{s \in S} D_s$ where S is the set of sorts occurring in the term algebra. Then elements of D_s are considered as mixed terms of sort s.

Example 3. Let $x \in \mathcal{X}$ be of sort *Nat*, $+ : Nat \times Nat \rightarrow Nat \in \mathcal{F}$ and $57 \in \mathcal{D}_{Nat} = \mathbb{N}$ then $x + 57$ is a well-formed mixed term. □

The set of mixed terms generated by a signature $\mathcal{F}$, the variables in $\mathcal{X}$ and the external objects in D will be denoted by $T(\mathcal{F}, \mathcal{X}, D)$.

Computations Computations map a ground term to an external object. This can be described by an $\mathcal{F}$-algebra $\mathcal{D} = (D, \mathcal{F}_{\mathcal{D}})$ where

$$\mathcal{F}_{\mathcal{D}} = \{f_{\mathcal{D}} : D_{s_1} \times \cdots \times D_{s_n} \rightarrow D_s \mid f : s_1 \times \cdots \times s_n \rightarrow s \in \mathcal{F}\}$$

is a set of interpretation functions of the operators in $\mathcal{F}$. $\mathcal{D}$ will be called the *external domain.* The *computation relation* $\rightarrow_{\mathcal{D}} \subseteq T(\mathcal{F}, \mathcal{X}, D) \times T(\mathcal{F}, \mathcal{X}, D)$ can now be defined as $s \rightarrow_{\mathcal{D}} t$ if $s|_p = f(d_1, \ldots, d_n)$ for $d_1, \ldots, d_n \in D$ and $t = s[f_{\mathcal{D}}(d_1, \ldots, d_n)]_p$. Since the $f_{\mathcal{D}} \in \mathcal{F}_{\mathcal{D}}$ are total functions, $\rightarrow_{\mathcal{D}}$ is a canonical relation. Let $\phi : T(\mathcal{F}, \mathcal{X}, D) \rightarrow T(\mathcal{F}, \mathcal{X}, D)$ be the function that computes the $\rightarrow_{\mathcal{D}}$-normal form for each mixed term.

The Simplification Relation Given a term rewriting system $\mathcal{R}$ and an external domain $\mathcal{D}$, $\rightarrow_{\mathcal{R}} \cup \rightarrow_{\mathcal{D}}$ is a *simplification relation.* If $\mathcal{D}$ is the initial model of $\mathcal{R}$ the simplification relation is confluent on ground terms. But as demonstrated by Example 2, general confluence does not hold. In the remaining sections, we will investigate how we can recover general confluence for the simplification relation.

3 Simulating Mixed Terms

In this and the following sections, we will simulate mixed terms and simplification relations by pure terms and term rewriting systems. We do not suggest to use this simulation to incorporate computations into term rewriting systems. The purpose of this simulation is only to investigate simplification relations over mixed terms.

We want to simulate the external domain $\mathcal{D}$ by (ground) terms of a new sort. Therefore we assume that the external domain is isomorphic to the initial model of the original rewrite specification $\mathcal{R}$. Let $\mathcal{R}$ be a canonical term rewriting system over terms in $T(\mathcal{F}, \mathcal{X})$ built from operators in a finite set $\mathcal{F}$ and variables in $\mathcal{X}$. Both operators and variables are accompanied with sort information which assigns to each term in $T(\mathcal{F}, \mathcal{X})$ a unique sort $s \in \Sigma$. Then let $\overline{\Sigma} = \{\bar{s} \mid s \in \Sigma\}$ be a set of new sorts, $\overline{\mathcal{F}} = \{\bar{f} \mid f \in \mathcal{F}\}$ be a new signature such that the sort description of $\bar{f}$ is $\bar{f} : \bar{s}_1 \times \cdots \times \bar{s}_n \rightarrow \bar{s}$ iff that of f is $f : s_1 \times \cdots \times s_n \rightarrow s$ and $\overline{\mathcal{X}} = \{\bar{x} \mid x \in \mathcal{X}\}$ be a set of new variables. Terms in $T(\overline{\mathcal{F}}, \overline{\mathcal{X}})$ will be called *external terms* and variables in $\overline{\mathcal{X}}$ will be called *barred.* For $t \in T(\mathcal{F}, \mathcal{X})$ we define the *corresponding external term*

$$\bar{t} = \begin{cases} \bar{x} & \text{if } t = x \in \mathcal{X}, \\ \bar{f}(\bar{t}_1, \ldots, \bar{t}_n) & \text{if } t = f(t_1, \ldots, t_n). \end{cases}$$

Now the external domain can be simulated by external ground terms modulo the canonical term rewriting system $\overline{\mathcal{R}} = \{\bar{l} \to \bar{r} \mid l \to r \in \mathcal{R}\}$, i.e., each external object corresponds to a $\overline{\mathcal{R}}$-equivalence class of external ground terms. Since $\overline{\mathcal{R}}$ is canonical, each external object corresponds to an external ground term in $\overline{\mathcal{R}}$-normal form. Note that simulated mixed terms must not contain barred variables. To build well-formed 'mixed' terms, we need *coercion operators* $c_s : \bar{s} \to s$ which map terms of a new sort to terms of the original sort. Let $\mathcal{C} = \{c_s \mid s \in \Sigma\}$. Mixed terms can be simulated by elements of $T(\mathcal{F} \cup \overline{\mathcal{F}} \cup \mathcal{C}, \mathcal{X})$ which belong to the original sorts in Σ. For ease of notation we will use quotations to denote the coercions (e.g., "$\bar{t}$" instead of $c_s(t)$).[1]

The computation relation can now be simulated with the help of the term rewriting system

$$\widehat{\mathcal{R}} = \{f(\text{“}\bar{x}_1\text{”}, \ldots, \text{“}\bar{x}_n\text{”}) \to \text{“}\bar{f}(\bar{x}_1, \ldots, \bar{x}_n)\text{”} \mid f \in \mathcal{F}, \bar{x}_1, \ldots, \bar{x}_n \in \overline{\mathcal{X}}\}$$

where the quotation operators stand for the appropriate coercion operators. Note that $\widehat{\mathcal{R}} \cup \overline{\mathcal{R}}$ is confluent. A one-step computation

$$t[f(d_1, \ldots, d_n)]_p \to_{\mathcal{D}} t[f_{\mathcal{D}}(d_1, \ldots, d_n)]_p$$

then corresponds to first applying a rule from $\widehat{\mathcal{R}}$ to $t[f(\text{“}t_1\text{”}, \ldots, \text{“}t_n\text{”})]_p$ (where "t_i" simulates d_i) resulting in $t' = t[\text{“}f(t, \ldots, t_n)\text{”}]_p$ and then normalizing $t'|_p$ w. r. t. $\overline{\mathcal{R}}$. Note that the $\overline{\mathcal{R}}$-normalization has no direct analogy in the computation relation. We may say that a $\widehat{\mathcal{R}}$-reduction prepares a function call whereas the $\overline{\mathcal{R}}$-normalization simulates the computation of the function.

Computing the $\to_{\mathcal{D}}$-normal form $\phi(t)$ of a mixed term t is simulated by computing the $\widehat{\mathcal{R}} \cup \overline{\mathcal{R}}$-normal form of a term in $T(\mathcal{F} \cup \overline{\mathcal{F}}, \mathcal{X})$. The whole simplification relation can now be simulated using $\mathcal{S} = \mathcal{R} \cup \overline{\mathcal{R}} \cup \widehat{\mathcal{R}}$. Note that not every one-step reduction w. r. t. $\mathcal{S}$ corresponds to a one-step simplification.

4 Canonical Simplification

The simplification relation can be simulated by $\mathcal{S} = \mathcal{R} \cup \overline{\mathcal{R}} \cup \widehat{\mathcal{R}}$. Clearly $\mathcal{S}$ is terminating and ground confluent and two *simulated mixed terms* m and n are equal iff $m \leftrightarrow_{\mathcal{S}}^* n$. To prove such an equality we can use an (unfailing) completion procedure. Note that both $\mathcal{R} \cup \overline{\mathcal{R}}$ and $\widehat{\mathcal{R}} \cup \overline{\mathcal{R}}$ are confluent. Therefore the critical pair computation can be strongly restricted. The next theorem states that we can always find a canonical term rewriting system for $\mathcal{S}$.

Theorem 1. *Let $\mathcal{R}$ be a canonical term rewriting system and $\mathcal{S}$ is constructed from $\mathcal{R}$ as described above. Then there exists a canonical term rewriting system $\mathcal{S}^*$ with $\leftrightarrow_{\mathcal{S}}^* = \leftrightarrow_{\mathcal{S}^*}^*$.*

[1] Note that the overloading of the coercion operators can always correctly be resolved.

Proof: Let $\bar{\sigma}(t_1,\ldots,t_k)$ be a substitution which maps each variable x occurring in one of the terms $t_1,\ldots,t_k$ to the corresponding external variable "$\bar{x}$". Then let

$$\begin{aligned}\mathcal{S}^* = \{\, & l[\text{“}\overline{t_1}\text{”}]_{p_1}\ldots[\text{“}\overline{t_k}\text{”}]_{p_k} \to r\bar{\sigma}(t_1,\ldots,t_k) \mid \\ & l \to r \in \mathcal{R}, p_i \neq \lambda \text{ pairwise disjoint positions in } l, l|_{p_i} = t_i \notin \mathcal{X}\,\}.\end{aligned}$$

We define an operation ψ on terms that erases coercion operators and 'bars'. It is easy to see that

$$\forall s,t:\ s \to_{\mathcal{S}^*} t \ \Rightarrow\ \psi(s) = \psi(t) \text{ or } \psi(s) \to_{\mathcal{R}} \psi(t) \tag{1}$$

and

$$\forall s,t:\ s \to_{\mathcal{S}^*} t \Rightarrow (\psi(s) = \psi(t) \ \Leftrightarrow\ s \to_{\widehat{\mathcal{R}}} t). \tag{2}$$

From (1) and (2) follows that $\to_{\mathcal{S}^*}$ is terminating because $\to_{\mathcal{R}}$ and $\to_{\widehat{\mathcal{R}}}$ terminate. By construction of $\mathcal{S}^*$ we have also

$$\forall s,t:\ s \leftrightarrow^*_{\mathcal{S}} t \Rightarrow (\exists t':\ \psi(s) \to_{\mathcal{R}} \psi(t) \ \Rightarrow\ s \to_{\mathcal{S}^*} t' \leftrightarrow^*_{\widehat{\mathcal{R}}} t). \tag{3}$$

Therefore $\to_{\mathcal{S}^*}$ is confluent. □

Example 4. Given the Peano specification of + and · as in

$$\begin{aligned}\mathcal{R}_P = \{\, & 0 + x \to x, && s(x) + y \to s(x+y), \\ & 0 \cdot x \ \to 0, && s(x)\cdot y \ \to (x \cdot y) + y\,\},\end{aligned}$$

we get the following canonical rewrite system for the simplification relation

$$\begin{aligned}\mathcal{S}^*_P = \mathcal{R}_P \cup \overline{\mathcal{R}}_P \cup \widehat{\mathcal{R}}_P \cup \{\, & \text{“}\bar{0}\text{”} + x \to x, && \text{“}\overline{s(x)}\text{”} + y \to s(\text{“}\bar{x}\text{”} + y), \\ & \text{“}\bar{0}\text{”} \cdot x \ \to 0, && \text{“}\overline{s(x)}\text{”} \cdot y \ \to (\text{“}\bar{x}\text{”} \cdot y) + y\,\}.\end{aligned}$$

Note that $\mathcal{S}^*$ is not interreduced: 0 may be totally eliminated and rules 1 and 3 of $\mathcal{R}_P$ may be collapsed. □

Alternative canonical term rewriting systems can be found using term completion procedures provided appropriate term orderings can be found and the completion terminates successfully. A few such systems are presented in the Appendix.

5 Extended Term Rewriting Systems

The canonical term rewriting systems derived for the simulated simplification relation typically contain rules that are neither pure rewrite rule rules (in $T(\mathcal{F},\mathcal{X})$) nor describe the computation relation. Their meaning is rather to describe a combined rewriting and computation step or to restrict rule applications to terms containing particular external objects.

In this section, we want to extend term rewriting systems for mixed terms to include new rule types that describe the additional reductions needed to make the simplification relation canonical. We will derive these new rule types from term rewriting systems over $T(\mathcal{F} \cup \overline{\mathcal{F}} \cup \mathcal{C}, \mathcal{X} \cup \overline{\mathcal{X}})$. Remember that subterms underneath coercion operators correspond to external objects. Thus terms simulating mixed terms must not contain variables of $\overline{\mathcal{X}}$. Nevertheless it may be sensible to allow for barred variables in rewrite rules. Let us now characterize the intended meaning of external subterms occurring in a rewrite rule over $T(\mathcal{F} \cup \overline{\mathcal{F}} \cup \mathcal{C}, \mathcal{X} \cup \overline{\mathcal{X}})$.

1. An external ground subterm "$\bar{t}$" denotes the external object $\phi(t)$ which can be computed from the corresponding ground term t.
2. A barred variable "$\bar{x}$" of sort $\bar{s}$ should match only external objects of D_s.
3. An external non-ground subterm "$\bar{t}$" occurring in a left-hand side of a rule describes any external object that can be decomposed according to t. An external object can be *decomposed according to* t if it corresponds to an $\mathcal{R}$-irreducible ground term $t' \in T(\mathcal{F}, \emptyset)$ and t matches t'.
4. An external non-ground subterm "$\bar{t}$" occurring in a right-hand side of a rule denotes the external object that can be computed from the respective instance of t.

Extending term rewriting systems to deal with the aforementioned situations leads to the following extension of pure term rewriting systems.

Definition 2. Let $\mathcal{R}$ be a rewrite specification for terms over $T(\mathcal{F}, \mathcal{X})$ and $\mathcal{D}$ be the initial model of $\mathcal{R}$. Let ? and ! be new unary operators not occurring in $\mathcal{F}$.[2] A term rewriting system over the mixed terms in $T(\mathcal{F} \cup \{?, !\}, \mathcal{X}, D)$ is an *extended term rewriting system* for mixed terms in $T(\mathcal{F}, \mathcal{X}, D)$ if

- subterms of the form $?(t)$ occur only in left-hand sides and $t \in T(\mathcal{F}, \mathcal{X}, D)$ and
- subterms of the form $!(t)$ occur only in right-hand sides and $t \in T(\mathcal{F}, \mathcal{X}, D)$.

The operator ? is called a *mark for decomposition* and $?(t)$ denotes external objects which correspond to irreducible ground instances of t. The operator ! is called a *mark for immediate computation* and $!(t)$ denotes external objects which can be computed from the corresponding ground instances of t.

To simplify the notation we may write $\bar{x}$ instead of $?(x)$ or $!(x)$ to denote variables that match only external objects. If the arguments of ? or ! are external objects the two operators may be omitted. It it now easy to translate a term rewriting system over $T(\mathcal{F} \cup \overline{\mathcal{F}} \cup \mathcal{C}, \mathcal{X} \cup \overline{\mathcal{X}})$ to an extended term rewriting system and vice versa: a ground term "$\bar{t}$" can be translated to $\phi(t)$, "$\bar{x}$" translates to $\bar{x}$ and "$\bar{t}$" for t non-ground translates to $?(t)$ if it occurs in a left-hand side and to $!(t)$ otherwise.

[2] To be precise, there should be two such operators for each sort, but again overloading is harmless.

Example 5. The following are examples of simulated rules and their translations to extended rewrite rules:

$$
\begin{array}{llll}
(\text{“}\bar{x}\text{”} \cdot y) + y & \to \text{“}\overline{x+1}\text{”} \cdot y & \Rightarrow (\bar{x} \cdot y) + y & \to !(x+1) \cdot y, \\
\text{“}\overline{1+1}\text{”} \cdot x & \to x + x & \Rightarrow 2 \cdot x & \to x + x, \\
pop(\text{“}\overline{push(e,s)}\text{”}) \to \text{“}\bar{s}\text{”} & & \Rightarrow pop(?push(e,s)) \to \bar{s}. &
\end{array}
$$

□

Extended term rewriting systems can be viewed as a schematization of an (infinite) set of rewrite rules over a signature that has been extended by one constant for each object in the external domain $\mathcal{D}$. This point of view has been adapted in [Bündgen, 1994].

6 Decomposition Free Systems

Unfortunately most of the canonical term rewriting systems derived from $\mathcal{S}$ — in particular $\mathcal{S}^*$ from the proof of Theorem 1 — have an undesired property. Namely there may be rules $l \to r$ in the canonical system such that l contains an external non-variable subterm (e. g. $\overline{s(x)}$). As we have seen in the last section these subterms translate to terms marked for decomposition. Such rules are unwanted because they force us to look into the internal structure of an external object and to 'decompose' the external object. Thus they may even force us to undo a previous 'computation'.

Example 6. Consider $\mathcal{R}_P$ and $\mathcal{S}_P^*$ from Example 4. The mixed term rewriting system associated with $\mathcal{S}_P^*$ is

$$\mathcal{R}_P \cup \{?(s(x)) + y \to \bar{x} + y,\ \ ?(s(x)) \cdot y \to (\bar{x} \cdot y) + y\}$$

if we assume that the constant 0 and the external object 0 can be overloaded. When simplifying $s(5) \cdot y$ w. r. t. the external rewrite system we may compute $6 \cdot y$ as computations are considered to be built-in. But in order to prove that $6 \cdot y = (5 \cdot y) \cdot y$ we have to undo the previous computation by applying the rule $?(s(x)) \cdot y \to (\bar{x} \cdot y) + y$. □

For an external object this means, we need to decide its *atomicity* (i. e., if it corresponds to an irreducible constant) and for a non-atomic object we must further be able to *decompose* it. Hereby only decomposition into *mixed constructor terms* is necessary. By a *mixed constructor term* we mean a mixed term that becomes a $\mathcal{R}$-irreducible ground term if all its external subterms are replaced by corresponding $\mathcal{R}$-irreducible ground terms from $T(\mathcal{F}, \mathcal{X})$. Therefore we may disregard rules that contain ground reducible subterms marked for decomposition.

For term rewriting systems modulo equational theories like associativity and commutativity (AC), the mapping between irreducible ground terms and external objects is not necessarily one-to-one. Therefore the decomposition operation is not uniquely defined. In these cases, we are free to choose any decomposition function. This is a strong restriction on the matches on mixed terms needed to decide the applicability of an extended rewrite rule. It makes our reduction relation on mixed terms much simpler than the one proposed in [Avenhaus and Becker, 1994].

Example 7. Given the free Abelian group with one generator specified by an AC-canonical term rewriting system ($\mathcal{F} = \{0, 1, -, +\}$, $+ \in \mathcal{F}_{AC}$)

$$
\begin{aligned}
\mathcal{R}_{AG} = \{\, & 0 + x \to x, & -0 & \to 0, \\
& -x + x \to 0, & -x + x + y & \to y, \\
& -(-x) \to x, & -(x+y) & \to -x + -y \,\}
\end{aligned}
$$

and let the integers with addition and negation be the external domain. Then the external object 5 may be decomposed into 1+4 or into 2+3. A decomposition into 0+5 is not allowed because $0+(1+1+1+1+1)$ is reducible by $\mathcal{R}_{AG}$ and $-2+7$ is not a mixed constructor term either because $(-1+-1)+(1+1+1+1+1+1+1)$ is also reducible by $\mathcal{R}_{AG}$. □

Alternative canonical term rewriting systems for S can be searched using the Knuth-Bendix procedure or its extensions for completion modulo AC. In particular term completion procedures may be employed to search for term rewriting systems which keep the need for decomposing terms as low as possible by orienting critical pairs accordingly. In the optimal case we may even end up with a canonical term rewriting system that never forces us to decompose external objects.

Definition 3. Let $\mathcal{R}$ be a canonical term rewriting system. A term rewriting system $\mathcal{R}^*$ which is derived from $\mathcal{R} \cup \overline{\mathcal{R}} \cup \widehat{\mathcal{R}}$ is *decomposition free* (for $\mathcal{R}$) if for all $l \to r \in \mathcal{R}^* \setminus \overline{\mathcal{R}}$, $l|_p \in T(\bar{\mathcal{F}}, \bar{\mathcal{X}})$ implies $l|_p \in \bar{\mathcal{X}} \cup \bar{\mathcal{F}}_0$.

Theorem 4. *Let $\mathcal{R}$ be a canonical term rewriting system and $\mathcal{D}$ be an external domain which is isomorphic to the initial model of $\mathcal{R}$. Let $\overline{\mathcal{R}}$ and $\widehat{\mathcal{R}}$ be as described above. If equivalence in $\mathcal{D}$ is decidable and $\mathcal{R} \cup \overline{\mathcal{R}} \cup \widehat{\mathcal{R}}$ can be compiled into a decomposition free canonical term rewriting system then there is a canonical simplification relation which never needs to decompose external objects.*

Proof: follows directly from the translation scheme of Section 5. □

The requirement of Theorem 4 to decide the equality of external objects is only needed if there are rules with left-hand sides that contain constants or that are not linear w.r.t. external variables. Given that we have such a decision procedure, we may also allow for external ground terms in the left-hand sides of the new rules without being forced to ever decompose a mixed term to be simplified. In the Appendix the two kinds of decomposition free systems for the commutative ring with one are presented.

When we want to derive a decomposition free system for $\mathcal{R}$, we may use the Knuth-Bendix completion procedure trying to orient critical pairs so that the conditions of Definition 3 (or the weakened conditions discussed above) are not violated. Simulated mixed terms behave like the corresponding terms in $\mathcal{T}(\mathcal{F},\mathcal{X})$ with the exception that certain redexes are blocked. Thus it may happen that the confluence of a mixed critical pair can only be shown using an inductive theorem of $\mathcal{R}$ or $\overline{\mathcal{R}}$. Therefore it is advantageous if the specification $\mathcal{R}$ allows rewrite proofs for as many theorems of its initial model as possible.

Example 8. Given a specification

$$\mathcal{R}_a = \{\, app(nil,l) \to l,\;\; app(cons(x,l_1),l_2) \to cons(x,app(l_1,l_2)) \,\},$$

the derivation of a decomposition free system will compute the critical pair

$$(\, app(app(\text{“}\overline{cons(x,nil)}\text{”},l_1),l_2),\; app(\text{“}\overline{cons(x,nil)}\text{”},app(l_1,l_2)) \,).$$

This pair could be resolved if *app* were declared to be associative. Without this declaration our trial to find a decomposition free system for $\mathcal{R}_a$ fails. For

$$\begin{aligned}\mathcal{R}_A = \{\, & app(nil,l) \to l, \;\; app(cons(x,l_1),l_2) \to cons(x,app(l_1,l_2)),\\ & app(l,nil) \to l, \;\; app(app(l_1,l_2),l_3) \to app(l_1,app(l_2,l_3)) \,\}\end{aligned}$$

we obtain the decomposition free canonical term rewriting system

$$\begin{aligned}\overline{\mathcal{R}}_A \cup \widehat{\mathcal{R}}_A \cup \{\, & app(\text{“}\overline{nil}\text{”},l) && \to l,\\ & app(l,\text{“}\overline{nil}\text{”}) && \to l,\\ & app(cons(x,l_1),l_2) && \to cons(x,app(l_1,l_2)),\\ & app(app(l_1,l_2),l_3) && \to app(l_1,app(l_2,l_3)),\\ & cons(\text{“}\overline{x}\text{”},l) && \to app(\text{“}\overline{cons(x,nil)}\text{”},l),\\ & app(\text{“}\overline{l}_1\text{”},app(\text{“}\overline{l}_2\text{”},l_3)) && \to app(\text{“}\overline{app(l_1,l_2)}\text{”},l_3) \,\}.\end{aligned}$$

□

So far, we have found decomposition free canonical systems for a variety of specifications like Abelian (idempotent) monoids, (Abelian) groups, commutative rings with 1, modular rings, Boolean rings, lists. But for non-commutative rings or Peano axiomatizations of the natural numbers we could not find decomposition free systems.

An equational specification for which an (equational) theorem is valid in the variety iff it is valid in the initial model is called *inductively complete* [Paul, 1984]. As a matter of fact, so far we could find decomposition free canonical term rewriting systems for all the inductively complete canonical term rewriting systems we looked at. But inductive completeness of $\mathcal{R}$ is not a necessary precondition for the existence of decomposition free canonical systems.

7 Termination of Simplification Systems

In a canonical decomposition free system derived from $\mathcal{R} \cup \overline{\mathcal{R}} \cup \widehat{\mathcal{R}}$, the left-hand sides of the new rules contain external variables and the right-hand sides typically contain non-atomic external terms. Such new rules denote a rewrite immediately followed by computations and will be translated to extended rewrite rules containing marks for immediate computation. This combination of basic simplification steps may be necessary to guarantee the termination property of the system.

Example 9. The following decomposition free rule can be generated for commutative rings:

$$(x \cdot \text{“}\overline{y}\text{”}) + (x \cdot \text{“}\overline{z}\text{”}) \rightarrow x \cdot \text{“}\overline{y+z}\text{”}.$$

Splitting this rule into

$$(x \cdot \text{“}\overline{y}\text{”}) + (x \cdot \text{“}\overline{z}\text{”}) \rightarrow x \cdot (\text{“}\overline{y}\text{”} + \text{“}\overline{z}\text{”})$$

and the computation of + conflicts with the standard orientation of the distributivity law:

$$x \cdot (y + z) \rightarrow (x \cdot y) + (x \cdot z).$$

□

Trying to construct a decomposition free canonical rewrite system for $\mathcal{S}$ by Knuth-Bendix completion (modulo AC), a major difficulty is to find an appropriate term ordering which is compatible with $\mathcal{S}$ and the orientations of critical pairs needed to guarantee decomposition freedom. Here we want to present a technique to construct term orderings which proved helpful in many cases. Examples like the one dealing with the distributivity law (Example 9) suggest that we should take sort information into account.

The sort structure of terms in $T(\mathcal{F} \cup \overline{\mathcal{F}} \cup \mathcal{C}, \mathcal{X} \cup \overline{\mathcal{X}})$ is very special. Only coercion operators link barred sorts and non-barred sorts. Thus terms of barred sorts may be embedded in terms of unbarred sorts but not vice versa. Therefore it is not possible that a coercion operator occurs below another coercion operator.

Definition 5. A many sorted term algebra $T(\mathcal{F}, \mathcal{X})$ has a *strict 2-level sort hierarchy* if the set of sorts can be partitioned into two disjoint subsets $\Sigma = \Sigma_1 \uplus \Sigma_0$ and $\mathcal{F} = \mathcal{F}_1 \uplus \mathcal{F}_0 \uplus C$ such that all terms in $T(\mathcal{F}_1, \emptyset)$ have a sort in Σ_1, all terms in $T(\mathcal{F}_0, \emptyset)$ have a sort in Σ_0, and C contains only coercion operators from a sort in Σ_0 to a sort in Σ_1.

The set Σ_0 is called the set of *base sorts* and Σ_1 is called set of *(first) extension sorts.*

A term rewriting system is terminating if for all its rules the left-hand sides are greater than the corresponding right-hand sides w.r.t. a term ordering.

Definition 6. A *(quasi) term ordering* $\succsim$ is a terminating (quasi) ordering over terms that is stable under substitution ($\forall s, t, \sigma : s \succsim t \Rightarrow s\sigma \succsim t\sigma$) and compatible ($\forall s, t, u, p : s \succsim t \Rightarrow u[s]_p \succsim u[t]_p$).

Let us construct a quasi term ordering for term algebras with a strict 2-level sort hierarchy. Σ and $\mathcal{F}$ be as described in Definition 5 and C be a set of constants of sorts in Σ_1 disjoint from $\mathcal{F}$. $\succ$ be a term ordering over terms of the extension sorts augmented by C. Then we can extend $\succ$ to a relation $\succsim$ over $T(\mathcal{F} \cup C, \mathcal{X})$ with associated equivalence relation $\approx$:

1. $\succsim$ restricted to terms of the extension sort is equivalent to $\succ$.
2. For each coercion operator c_s in C there is a constant c in C such that for all terms $c_s(t)$, $c_s(t) \approx c$.
3. For each base sort s there is a constant c in C such that for all terms t of sort s, $t \approx c$.

It is straightforward to show that $\succsim$ is both stable under substitution and compatible.

Lemma 7. *Let $T(\mathcal{F}, \mathcal{X})$ be a term algebra with strict 2-level sort hierarchy as described above. Let $\succsim_1$ be a quasi term ordering such that all terms of the same base sort are equivalent and all terms with the same coercion operator at their tops are equivalent. Let $\succ_2$ be a term ordering. Then the lexicographic combination $(\succsim_1, \succ_2)$ is a term ordering.* □

Definition 5 and Lemma 7 can be extended to n-level sort hierarchies.

8 Variations and Extensions

In the last sections, we have imposed a strong restriction on the external domains we admit. Namely we demanded that the external domains be isomorphic to the initial model of the rewrite specification $\mathcal{R}$. This restriction can be relaxed to allow for external domains that are an initial model of a consistent enrichment of $\mathcal{R}$. This corresponds to augmenting $\overline{\mathcal{F}}$ by additional constants and function symbols and to adding rules to $\overline{\mathcal{R}}$ that do not reduce formerly irreducible ground terms. Since $\overline{\mathcal{F}}$ and $\overline{\mathcal{R}}$ may even become infinite, we can also admit external domains which cannot be specified equationally (like fields). The relation between canonical simplification relations and various external domains has been investigated in [Bündgen, 1994]. Again decomposition free canonical systems seem to be of great importance because they turn out to be canonical for all consistent models of $\mathcal{R}$.

In a many sorted specification there may be built-in operations only for some of the sorts. Consider for example the polynomial specification in [Bündgen, 1991] which distinguishes sorts for coefficients, indeterminates and polynomials. Building-in an external coefficient domain allows for specifying polynomials over infinite fields. We have found a decomposition free canonical system for polynomials with built-in coefficient domain. This can be used to relate term completion procedures and Buchberger's algorithm [Buchberger, 1965]. In [Bündgen, 1992] the author showed that Buchberger's algorithm for polynomials over the rational numbers can be simulated using a term rewriting system that needs infinitely many rules to specify the rational numbers and in the constraint based description of Buchberger's algorithm in [Bachmair and Ganzinger, 1994] the coefficient domain is considered the set of all ground equations needed to specify the coefficient field. In both cases the coefficient fields may be considered built-in.

It is also possible to consider models of non-consistent enrichments of $\mathcal{R}$ as external domains. However this will lead to simplification systems different from those described in the last section because the completion will yield new equalities on terms of the original sort.

Defined operators in $\mathcal{F}$ need not be mapped to a corresponding defined operator in $\overline{\mathcal{F}}$. This introduces two hierarchies of computing [Kaplan and Choppy, 1989]. Base computations can be considered as built-in operations on external objects and higher-level functions can be realized by rewrite rules. In this case the decomposition of the recursion argument can in general not be avoided.

Example 10. The definition of the sum operator

$$\mathcal{R} = \{\Sigma(0) \to 0,\ \Sigma(s(n)) \to s(n) + \Sigma(n)\}$$

will be transformed to the extended term rewriting system

$$\mathcal{R} \cup \{\Sigma(?(s(n))) \to !(s(n)) + \Sigma(\bar{n})\}.$$

If we know how to decompose external objects and the computation of the successor function, the addition and the equality of external objects is built-in, an innermost reduction strategy with priority to computation will compute the sum as efficiently as a corresponding functional program. □

Starting with a non-canonical equational specification, our simulation technique need not discover a finite (decomposition free) canonical simplification system. Yet, the equality of two mixed terms may be proved using the (unfailing) Knuth-Bendix completion on simulated mixed terms which in the worst case may involve the complete decomposition of external objects. On the other hand, we may also employ inductive completion procedures [Huet and Hullot, 1980],[Paul, 1984], [Jouannaud and Kounalis, 1989],[Küchlin, 1989] to that end if we are sure that the external domain forms an initial model of the equational specification.

If we are only concerned with ground confluence, it is possible to become even more ambitious. Avenhaus and Becker [Avenhaus and Becker, 1994] have elaborated a semantical framework for conditional term rewriting systems with built-in domains allowing for partially defined functions.

9 Conclusion

We have shown that simplification presented by a canonical term rewriting system and a computation relation for its initial model can always be extended to a canonical simplification system. We propose extended term rewriting systems to describe such simplification systems. The simulation technique we used to study the simplification relation allows us to investigate simplification systems w. r. t. all standard rewriting techniques (e. g., term orderings, completion, narrowing, ...). For some important equational specifications, we can construct decomposition free canonical simplification systems which never force us to undo a previous computation. So far we do not know syntactical criteria which ensure the existence of decomposition free canonical systems.

A primary intension of building external domains into term rewriting system was the need for improved efficiency. The potential to compute operators acting on ground terms leads a first step towards this direction. Yet, as we have seen, for deciding equality of mixed terms, we may be forced to undo previous simplifications. Decomposition free systems avoid redundant computations. They may however introduce redundant rewrites. It remains an important and challenging task to look for good simplification strategies to obtain the full benefit of combined rewrites and computations.

10 Acknowledgments

I like to thank anonymous referees for some helpful comments.

References

[Avenhaus and Becker, 1994] J. Avenhaus and K. Becker. Operational specifications with built-ins. In P. Enlbert, E. W. Mayr, and K. W. Wagner, editors, *STACS 94 (LNCS 775)*, pages 263–274. Springer-Verlag, 1994. (Proc. STACS'94, Caen, France, February 1994).

[Bachmair and Ganzinger, 1994] Leo Bachmair and Harald Ganzinger. Buchberger's algorithm: A constraind-based completion procedure. In *Constraints in Computational Logics*, 1994. (Proc. of CCL'94 in Munich, Germany, September 1994).

[Buchberger, 1965] Bruno Buchberger. *Ein Algorithmus zum Auffinden der Basiselemente des Restklassenringes nach einem nulldimensionalen Polynomideal.* PhD thesis, Universität Innsbruck, 1965.

[Bündgen, 1991] Reinhard Bündgen. Completion of integral polynomials by AC-term completion. In Stephen M. Watt, editor, *International Symposium on Symbolic and Algebraic Computation*, pages 70 – 78. ACM Press, 1991. (Proc. ISSAC'91, Bonn, Germany, July 1991).

[Bündgen, 1992] Reinhard Bündgen. Buchberger's algorithm: The term rewriter's point of view. In G. Kuich, editor, *Automata, Languages and Programming (LNCS 623)*, pages 380 -391, 1992. (Proc. ICALP'92, Vienna, Austria, July 1992).

[Bündgen, 1993] Reinhard Bündgen. Reduce the redex → ReDuX. In Claude Kirchner, editor, *Rewriting Techniques and Applications (LNCS 690)*, pages 446–450. Springer-Verlag, 1993. (Proc. RTA'93, Montreal, Canada, June 1993).

[Bündgen, 1994] Reinhard Bündgen. Combining computer algebra and rule based reasoning. In *AISMC-2 Conference/Workshop on Artificial Intelligence and Symbolic Mathematical Computing*, 1994. (to appear in LNCS).

[Dershowitz and Jouannaud, 1990] Nachum Dershowitz and Jean-Pierre Jouannaud. Rewrite systems. In Jan van Leeuven, editor, *Formal Models and Semantics*, volume B of *Handbook of Theoretical Computer Science*, chapter 6. Elsevier, 1990.

[Hearn, 1991] A. C. Hearn. *REDUCE User's Manual. Version 3.4*. Rand Publication, 1991.

[Hsiang, 1985] Jieh Hsiang. Refutational theorem proving using term-rewriting systems. *Artificial Intelligence*, 25:255–300, 1985.

[Huet and Hullot, 1980] Gérard Huet and Jean-Marie Hullot. Proofs by induction in equational theories with constructors. In *Proc. 21st FoCS*, pages 96–107, Los Angeles, CA, 1980.

[Hullot, 1979] Jean-Marie Hullot. Associative-commutative pattern matching. In *Fifth IJCAI*, Tokyo, Japan, 1979.

[Jouannaud and Kirchner, 1986] Jean-Pierre Jouannaud and Hélène Kirchner. Completion of a set of rules modulo a set of equations. *SIAM J. on Computing*, 14(4):1155–1194, 1986.

[Jouannaud and Kounalis, 1989] Jean-Pierre Jouannaud and Emmanuel Kounalis. Proofs by induction in equational theories without constructors. *Information and Computation*, 82:1–33, 1989.

[Kaplan and Choppy, 1989] Stéphane Kaplan and Christine Choppy. Abstract rewriting with concrete operators. In Nachum Dershowitz, editor, *Rewriting Techniques and Applications (LNCS 355)*, pages 178–186. Springer-Verlag, 1989. (Proc. RTA'89, Chapel Hill, NC, USA, April 1989).

[Klop, 1992] Jan Willem Klop. Term rewriting systems. In S. Abramsky, D. M. Gabbay, and T. S. E. Maibaum, editors, *Background: Computational Strcutures*, volume 2 of *Handbook of Logic in Computer Science*, chapter 1. Oxford University Press, 1992.

[Knuth and Bendix, 1970] Donald E. Knuth and Peter B. Bendix. Simple word problems in universal algebra. In J. Leech, editor, *Computational Problems in Abstract Algebra*. Pergamon Press, 1970. (Proc. of a conference held in Oxford, England, 1967).

[Küchlin, 1989] Wolfgang Küchlin. Inductive completion by ground proof transformation. In H. Aït-Kaci and M. Nivat, editors, *Resolution of Equations in Algebraic Structures*, volume 2 of *Rewriting Techniques*, chapter 7. Academic Press, 1989.

[Paul, 1984] Etienne Paul. Proof by induction in equational theories with relations between constructors. In *Ninth Colloquium on Trees in Algebra and Programming*, pages 211–225. Springer-Verlag, 1984.

[Peterson and Stickel, 1981] G. Peterson and M. Stickel. Complete sets of reductions for some equational theories. *Journal of the ACM*, 28:223–264, 1981.

[Plaisted, 1993] David Plaisted. Equational reasoning and term rewriting systems. In D. M. Gabbay, C. J. Hogger, and J. A. Robinson, editors, *Logical Foundations*, volume 1 of *Handbook of Logic in Artificial Intelligence and Logic Programming*, chapter 5. Oxford University Press, 1993.

[Stickel, 1981] Mark E. Stickel. A unification algorithm for associative-commutative functions. *JACM*, 28(3):423–434, July 1981.

[Wolfram, 1991] Stephen Wolfram. *Mathematica: a system for doing mathematics by computer.* Addison-Wesley, Redwood City, CA, 1991.

Appendix

Here we want to present some of the decomposition free term rewriting systems simulating canonical simplification systems. We only present the original rewrite specification $\mathcal{R}$ and the new rules $\mathcal{R}^x$ derived during the completion. $\mathcal{F}$ is always the set of constants and operators occurring in $\mathcal{R}$. Thus $\overline{\mathcal{R}}$ and $\widehat{\mathcal{R}}$ can easily be derived. Then the complete system is $\mathcal{R} \cup \overline{\mathcal{R}} \cup \widehat{\mathcal{R}} \cup \mathcal{R}^x$ (minus the collapsed rules of $\mathcal{R}$). All systems have been found using the ReDuX rewrite laboratory [Bündgen, 1993]. The termination proofs were performed as described in Section 7. The orderings described below refer to the quasi termination ordering $\succsim_1$.

Groups For the canonical term rewriting system for groups

$$\begin{array}{llllll}\mathcal{R}_G = \{ & x \cdot 1 & \to x, & 1 \cdot x & \to x, \\ & x \cdot x^{-1} & \to 1, & x^{-1} \cdot x & \to 1, \\ & (x \cdot y) \cdot z & \to x \cdot (y \cdot z), & 1^{-1} & \to 1, \\ & (x^{-1})^{-1} & \to x, & x^{-1} \cdot (x \cdot y) & \to y, \\ & x \cdot (x^{-1} \cdot y) & \to y, & (x \cdot y)^{-1} & \to y^{-1} \cdot x^{-1} \}\end{array}$$

we find

$$\begin{array}{lllll}\mathcal{R}_G^x = \{ & x \cdot \text{“}\bar{1}\text{”} \to x, & \text{“}\bar{1}\text{”} \cdot x & \to x, \\ & \text{“}\bar{1}\text{”}^{-1} \to \text{“}\bar{1}\text{”}, & \text{“}\bar{x}\text{”} \cdot (\text{“}\bar{y}\text{”} \cdot z) & \to \text{“}\overline{x \cdot y}\text{”} \cdot z \}.\end{array}$$

A Knuth-Bendix ordering can be used to prove the termination of the whole system.

Commutative Ring with 1 For commutative rings with one (see $\mathcal{R}_{CR1}$ from Example 2, $+, \cdot \in \mathcal{F}_{AC}$) there are two decomposition free canonical systems. The first one is decomposition free according to Definition 3.

$$\begin{array}{lllll}\mathcal{R}_{CR1a}^x = \{ & \text{“}\bar{0}\text{”} + x & \to x, & \text{“}\bar{0}\text{”} \cdot x & \to \text{“}\bar{0}\text{”}, \\ & \text{“}\bar{1}\text{”} \cdot x & \to x, & -(x) & \to x \cdot \text{“}\overline{-1}\text{”}, \\ & (x \cdot \text{“}\bar{y}\text{”}) + x & \to x \cdot \text{“}\overline{y+1}\text{”}, & (x \cdot \text{“}\bar{y}\text{”}) + (x \cdot \text{“}\bar{z}\text{”}) & \to x \cdot \text{“}\overline{y+z}\text{”}, \\ & x + x & \to x \cdot \text{“}\overline{1+1}\text{”} \ \}. & & \end{array}$$

To prove the termination we used the polynomial interpretation ordering

$$\{[+](x,y) \mapsto x+y+5, [\cdot](x,y) \mapsto xy, [-](x,y) \mapsto 2x+1, [\text{“.”}](x) \mapsto 2\}.$$

The second canonical system is decomposition free according to the relaxed version of this definition which allows barred ground terms in the left-hand sides. $\mathcal{R}^{x}_{CR1b} =$

$$\begin{array}{llll} \{\text{“}\bar{0}\text{”} + x & \to x, & \text{“}\bar{0}\text{”} \cdot x & \to \text{“}\bar{0}\text{”}, \\ \text{“}\bar{1}\text{”} \cdot x & \to x, & x + x & \to x \cdot \text{“}\overline{1+1}\text{”}, \\ -(x \cdot \text{“}\bar{y}\text{”}) & \to x \cdot \text{“}\overline{-y}\text{”}, & x \cdot \text{“}\overline{-1}\text{”} & \to -(x), \\ (x \cdot \text{“}\bar{y}\text{”}) + x & \to x \cdot \text{“}\overline{y+1}\text{”}, & (x \cdot \text{“}\bar{y}\text{”}) + -(x) & \to x \cdot \text{“}\overline{y+-1}\text{”}, \\ (x \cdot \text{“}\bar{y}\text{”}) + (x \cdot \text{“}\bar{z}\text{”}) & \to x \cdot \text{“}\overline{y+z}\text{”} \ \}. & & \end{array}$$

To prove the termination we used the polynomial interpretation ordering

$$\{[+](x,y) \mapsto x + y + 100, [\cdot](x,y) \mapsto xy, [-](x,y) \mapsto 1.1x + 0.5, [\text{“}.\text{”}](x) \mapsto 2\}.$$

The complete decomposition free systems for Abelian groups can be extracted from the system for commutative rings.

Boolean Rings The canonical term rewriting system for Boolean rings presented in [Hsiang, 1985] is ($\oplus, \wedge \in \mathcal{F}_{AC}$)

$$\begin{array}{llll} \mathcal{R}_{BR} = \{ x \oplus 0 \to x, & x \wedge 0 & \to 0, \\ \qquad x \wedge 1 \to x, & x \oplus x & \to 0, \\ \qquad x \wedge x \to x, & x \wedge (y \oplus z) & \to (x \wedge y) \oplus (x \wedge z) \ \}. \end{array}$$

It can be extended to a canonical decomposition free term rewriting system with

$$\begin{array}{lll} \mathcal{R}^{x}_{BR} = \{ x \oplus \text{“}\bar{0}\text{”} \to x, & x \wedge \text{“}\bar{0}\text{”} & \to \text{“}\bar{0}\text{”}, \\ \qquad x \wedge \text{“}\bar{1}\text{”} \to x, & (x \wedge \text{“}\bar{y}\text{”}) \oplus (x \wedge \text{“}\bar{z}\text{”}) & \to x \wedge \text{“}\overline{y \oplus z}\text{”} \ \}. \end{array}$$

The termination proof is analogous to that for commutative rings.

Hierarchical Termination*

Nachum Dershowitz

Department of Computer Science
University of Illinois at Urbana-Champaign
1304 West Springfield Avenue
Urbana, IL 61801, U.S.A.
nachum@cs.uiuc.edu

Abstract. From a practical perspective, it is important for programs to have modular correctness properties. Some (largely syntactic) sufficient conditions are given here for the union of terminating rewrite systems to be terminating, particularly in the hierarchical case, when one of the systems makes no reference to functions defined by the other.

1 Introduction

A *rewrite rule* is an equation between first-order terms used to replace equals-by-equals in one direction only. A rewrite system, that is a set of rewrite rules, is a form of applicative program that computes by reducing (that is, repeatedly rewriting) a variable-free term to its normal form (an unrewritable term), where the order in which rules are applied and the choice of subterm to which to apply a rule is arbitrary. Rewrite systems have other important applications in programming language semantics and automated deduction. For recent surveys of rewriting, see [Dershowitz and Jouannaud, 1990; Avenhaus and Madlener, 1990; Klop, 1992; Plaisted, 1993].

When no infinite sequences of rewrites are possible, a rewrite system is said to have the (*strong*, or *uniform*) *termination* property. In practice, one usually guarantees termination by devising a well-founded partial ordering for which a rewritten term is always smaller than the original. For a survey of methods of proving termination, see [Dershowitz, 1987]; for examples of these methods, see [Dershowitz, 1995].

Rewrite systems provide a simple, intuitive, nondeterministic functional language. As such, it would be quite valuable to be able to combine systems possessing desirable properties. In particular, we look for sufficient conditions under which the union of two terminating systems would be terminating. The conditions given here are based on syntactic restrictions of the systems in question. The first to consider modularity issues in rewriting was Bidoit [1981] with his "gracious" conditions.

A rule $l \rightarrow r$ is used to rewrite a term s containing an instance $l\sigma$ of its left-hand side l at some position p to $s[r\sigma]_p$, the same term, except that the subterm at p has

* This research was supported in part by the U. S. National Science Foundation under Grants CCR-90-07195 and CCR-90-24271, by a Lady Davis fellowship at the Hebrew University of Jerusalem, Israel, and by a Meyerhoff Visiting Professorship at the Weizmann Institute of Science, Rehovot, Israel.

been replaced by the corresponding instance $r\sigma$ of the rule's right-hand side r. We will have recourse to the notation $s|_p$ for the subterm at position p in term s.

Various modularity properties (such as termination and uniqueness of normal forms) have been intensely studied since the appearance of [Toyama, 1987b], particularly for combinations of systems that have no function symbols (or constants) in common.[1] Toyama [1987a] gave the following example, showing that even in this simplest case the combination of two terminating systems is not necessarily terminating:

$$\boxed{f(0,1,x) \to f(x,x,x) \quad \Big| \quad \begin{array}{c} g(x,y) \to x \\ g(x,y) \to y \end{array}} \tag{A}$$

In the next section, we review what is known about termination in this disjoint case, and present the major syntactic restrictions of interest in this regard. (For other modular properties, see [Middeldorp, 1990].) Then, in Section 3, the case where "constructor" symbols are shared by the two systems is considered.

Section 4 considers the "hierarchical" case where one system is allowed to refer to symbols appearing in the other, but not vice-versa. For example, we want to be able to use terminating systems for addition and multiplication in conjunction with a terminating program for factorial:

$$\begin{array}{r|c} fact(x) \to f(x, s(0)) & x \cdot 0 \to 0 \\ f(s(x), y) \to f(x, s(x) \cdot y) & x \cdot s(y) \to (x \cdot y) + x \\ \hline f(0, x) \to x\,. & x + 0 \to x \\ & x + s(y) \to s(x + y) \end{array} \tag{1}$$

The individual systems can be shown to terminate by showing that the terms in any derivation decrease in some well-founded measure (such as the natural numbers). For the right systems, one can, for example, let $[\![0]\!] = 2$, $[\![s(x)]\!] = [\![x]\!] + 1$, $[\![x \cdot y]\!] = [\![x]\!]^{2[\![y]\!]}$, and $[\![x + y]\!] = [\![x]\!][\![y]\!]$. For the "higher" system, let $[\![0]\!] = 1$, $[\![s(x)]\!] = [\![x]\!] + 1$, $[\![x \cdot y]\!] = [\![x]\!][\![y]\!]$, $[\![f(x,y)]\!] = [\![y]\!]([\![x]\!] + 1)!$, and $[\![fact(x)]\!] = ([\![x]\!] + 2)!$. With either system, whenever $s \to t$, we have $[\![s]\!] > [\![t]\!]$. The question is how to ensure that the union of the two systems terminates, without having to find an independent proof for the combination. The measures used above for the individual systems cannot be combined. (Termination of the combined system could be proved instead using the methods in [Dershowitz, 1982].) This eminently practical case has received virtually no attention.[2]

Section 5 mentions some results for the fully general case, when both systems can refer to all symbols, and is followed by a brief discussion of some remaining questions.

---

[1] Some authors reserve the designation "modular" for this disjoint case; we prefer, however, to use the term generically, specifying "disjoint," "shared constructors," or "hierarchical," as the case may be.

[2] A draft of this paper [Dershowitz, 1992] was distributed in December 1992.

2 Disjoint termination

Let A and B be disjoint sets of function symbols (including constants) and X be a set of variables. Let a *red* rewrite system contain terms built from A and X only (*red* terms), while a blue system has terms from B and X only (*blue* terms).[3] Supposing the red and blue systems are both terminating for all terms, that is, there are no infinite sequences of red rewrites, nor of blue rewrites, for terms constructed from A and B, then termination is said to be *modular* when the union is also terminating. The notion of modularity of properties for the disjoint vocabulary case was first studied by Toyama [1987b].

It is worth repeating the following bit of folk wisdom:

Proposition 1. *If a system is terminating for all terms constructed from symbols appearing in it (plus one new constant if the rules display none), then it also terminates for terms constructed from any richer set of symbols.*

Thus, to show that a red (blue) system terminates for "all" terms, it suffices to show termination for red (blue) terms.

Proof. Suppose a red system terminates for all red terms. One way to prove that it also terminates for mixed terms, containing "foreign" (blue) symbols, is to decompose mixed terms into pure red subterms, with some red constant replacing subterms of the components headed by blue symbols. The nesting depth of these pure-red components in a term cannot increase by rewriting. Terms are compared by looking lexicographically at a tuple of multisets, the most significant element of the tuple containing the uppermost red components, and so on. Multisets are compared in the multiset ordering [Dershowitz and Manna, 1979] and components in the red rewrite relation. For example, if f, a, b, c, and d are red, then $g(f(g(f(a,g(b))),f(a,f(g(d),g(d)))))$ would have components $\langle\{f(c,f(a,f(c,c)))\},\{f(a,c),d,d\},\{b\}\rangle$. □

Various sufficient conditions for modularity of termination (and related properties in subsequent sections) make use of the following decidable notions, all but the last of which are syntactic:[4]

- A *non-erasing* system has no rule with a variable on the left not also appearing on its right.
- A *non-collapsing* system has no rule with a variable as its right-hand side.
- A *right-linear* system has no rule with more than one occurrence of a variable on its right-hand side.
- A *non-duplicating* system[5] has no rule with more occurrences of a variable on its right-hand side than on its left. Of course, right-linear systems are non-duplicating, since we normally disallow rules that introduce a variable on the right not already on the left.

[3] With apologies to some previous authors, the color scheme has been changed here for added mnemonic value.

[4] These restrictions are ordered from the more severe to the less, taking a programmer's point of view.

[5] Called "conservative" in [Fernandez and Jouannaud, 1995].

- A *left-linear* system has no rule with more than one occurrence of a variable on its left-hand side.
- A *non-overlapping* system has no left-hand side that unifies with a non-variable subterm of another left-hand side or with a proper non-variable subterm of itself, after renaming variables in the terms so that they are disjoint. (This means there are no non-trivial "critical pairs" in the terminology of Knuth and Bendix [1970].) Clearly, rules with different colors at the top of their left sides cannot overlap.
- A *constructor-based* system is one in which no left-hand side has a symbol below the top that appears at the top of any left-hand side.
- An *overlaying* system is one in which no left-hand side unifies with a non-variable proper subterm of any left-hand side (including itself), after "standardizing apart" (renaming variables in the terms so that they are disjoint). Of course, non-overlapping and constructor-based systems are also overlaying.
- A *locally confluent* system is one for which any two terms that can be obtained each by one step of rewriting from the same term can both be rewritten in zero or more steps to the identical term. (Local confluence is decidable for finite terminating systems [Knuth and Bendix, 1970].) In particular, non-overlapping systems are locally confluent [Huet, 1980].

The following results are known:

Theorem 2 [Rusinowitch, 1987]. *The union of non-collapsing red and blue terminating systems is terminating.*

Theorem 3 [Rusinowitch, 1987]. *The union of non-duplicating red and blue terminating systems is terminating.*

Theorem 4 [Middeldorp, 1989]. *The union of a non-collapsing non-duplicating red terminating system with a blue terminating system is terminating.*

Theorem 5 [Toyama *et al.*, 1989]. *The union of left-linear locally-confluent red and blue terminating systems is terminating.*

Theorem 6 [Middeldorp and Toyama, 1991]. *The union of constructor-based locally-confluent red and blue terminating systems is terminating.*

Let us call overlaying locally-confluent systems *overlay-confluent*. The conditions of the previous theorem have been weakened to include this class of systems:

Theorem 7 [Gramlich, 1995]. *The union of overlay-confluent red and blue terminating systems is terminating.*

All the above results apply to the union of the following non-erasing, non-collapsing, non-duplicating, left-linear, non-overlapping systems:

$$x + s(y) \rightarrow s(x + y) \quad \Big| \quad p(x) - p(y) \rightarrow x - y\,, \tag{2}$$

where $+$ is red and $-$ is blue.

Generalizations of Theorems 2, 3, 4, and 7 will be proved in the sequel.

A more semantic approach was developed in [Gramlich, 1994; Ohlebusch, 1993] (based on the syntactic ideas in [Kurihara and Ohuchi, 1990]):

Theorem 8 **[Ohlebusch, 1993].** *The union of red and blue systems that are each terminating when joined with the system* $\{h(x,y) \rightarrow x, h(x,y) \rightarrow y\}$*, for new function symbol h not appearing in either system, is terminating. (This is an undecidable property.)*

Theorem 9 **[Ohlebusch, 1993].** *The union of a non-duplicating red system that is terminating when joined with* $\{h(x,y) \rightarrow x, h(x,y) \rightarrow y\}$*, for new function symbol h not appearing in either system, with a terminating blue system is terminating.*

3 Shared termination

Requiring that two systems have no symbols in common is much too restrictive in practice. In this section, we investigate the case where constructors are shared by the two systems (as also considered in [Middeldorp and Toyama, 1991; Gramlich, 1994; Ohlebusch, 1993]).

For our purposes, a *constructor* is any function symbol (in the given vocabulary) that never appears as the outermost symbol of a left-hand side (of either system), while a *defined symbol* is one that does. Let C be a set of *yellow* constructors, disjoint from A and B. Terms built from $A \cup C \cup X$ are *orange*; those over $B \cup C \cup X$ are *green*. An orange (green) term is deemed *bright* when its top symbol is red (blue).

An *orange* system has only orange terms; a *green* system, only green. Note that an orange (green) system must have a red (blue) symbol on the top of the left-hand side (since constructors never appear on the top left), and that red and blue symbols may be nested on either side of a rule. We will call an orange (green) system *bright* if the top symbol on the right is always red (blue). A bright system cannot have just a variable for right-hand side (that is, it is non-collapsing), nor can its right-hand side be headed by a constructor.[6] Defined (red or blue) symbols may be nested on either side of a rule, unless otherwise stated.

We count the number of alternations of red and blue symbols (ignoring yellow ones) along the path from the root leading to each symbol f in a term (in its tree representation) and assign a *level* to f accordingly. Yellow symbols in a term are assigned to the level of the nearest blue or red symbol preceding it along the path from the root. Orange and green rewrites can never increase the number of homogeneously colored layers in the terms of a derivation.

Were a shared system non-terminating, there would be a minimum number of layers for non-termination. An infinite derivation with that number of layers would have to have an infinite number of rewrites in the top layer (or else fewer layers would suffice for non-termination), as well as an infinite number below (or else the top system alone would be non-terminating).

The next theorem extends Theorem 2 (due to Rusinowitch [1987]) to systems with shared constructors.

Theorem 10 **[Gramlich, 1994].** *The union of bright-orange and bright-green terminating systems is terminating.*

[6] In the terminology of [Gramlich, 1994], it is not "constructor-lifting."

Thus,

$$x + s(y) \to s(x) + y \quad | \quad s(x) - s(y) \to x - y \tag{3}$$

is terminating (red +, blue −, yellow s), since each rule by itself is.

Proof. We use a well-founded ordering for which a rewrite within the top layer decreases the term in the ordering and a rewrite below the top layer does not increase it, precluding more than finitely many top rewrites. Since the systems are "brightly colored," layers never collapse, that is, the top layer never grows on account of a step below the top. Thus, to compare two terms, we simply compare their top layers—with one arbitrary term replacing all lower-level subterms—in the terminating rewrite relation of the top system. □

Similarly, the following theorem extends Theorem 3 [Rusinowitch, 1987] for shared constructors:[7]

Theorem 11. *The union of orange and green non-duplicating terminating systems is terminating.*

Proof. Rewriting does not increase the number of levels (except to add constructors at the top, which we can safely ignore). Consider the multiset of subterms below the top layer (the "aliens") in an infinite derivation in the combined system having no more levels than necessary for non-termination. If the top level is red (say), these subterms are headed by the highest blue symbol. Terms are compared in the union of the combined rewrite relation (which may be presumed terminating for terms of fewer layers) and the proper subterm relation. (This combined rewrite and subterm relation is terminating since the two commute, that is, any rewritten subterm is the subterm of the whole term rewritten.) Since the systems are non-duplicating, a top rewrite can only remove elements from the multiset (or leave them all intact). Furthermore, each of the ostensibly infinitely many rewrites below the top decreases the multiset in the rewrite relation. If a rewrite in the second, blue layer creates more than one (disjoint) blue subterm (connected by constructors), they are each smaller in the composition of the rewrite and subterm relation. Similarly, when a segment of the second layer collapses (which it can do in the non-bright case), some elements of the multiset may be replaced by some of their subterms. □

The following extends Theorem 4:

Theorem 12. *The union of a non-duplicating bright-orange terminating system with a green terminating system is terminating.*

Proof. Consider any derivation with a fixed number of layers. If the top layer is red, then the argument is just as in the previous proof. If the second layer is red, then we can use the same ordering as for Theorem 10. □

[7] A similar proof was given independently in [Ohlebusch, 1993]. In [Fernandez and Jouannaud, 1995], the result is extended to allow sharing of symbols other than constructors, provided the same proof method still applies.

The following non-terminating example [Dershowitz, 1981], with red f, blue 2, and yellow 0 and 1, shows the necessity of brightness (as in Theorem 10) or non-duplication (as in Theorem 11):

$$\begin{array}{l|c} f(0,1,x) \rightarrow f(x,x,x) & \begin{array}{c} 2 \rightarrow 0 \\ 2 \rightarrow 1 \end{array} \end{array} \quad \text{(B)}$$

Proposition 13 [Gramlich, 1995]. *An overlay-confluent system is terminating for a given term if, and only if, it is by innermost rewriting.*

This is analogous to the well-known fact that termination of call-by-value implies termination of call-by-name [Cadiou, 1972]. It includes, as a common special case, non-overlapping systems, proved in [Geupel, 1989].

Since, as it is easy to ascertain, innermost termination is preserved by unions of orange and green systems (cf. [Kurihara and Kaji, 1990; Gramlich, 1995]), it follows that[8]

Theorem 14. *The union of overlay-confluent orange and green terminating systems is terminating.*

In particular, non-overlapping systems can be combined. This extends Theorem 7 to systems with shared constructors, like:

$$\begin{array}{r|l} x + 0 \rightarrow x & \\ 0 + x \rightarrow x & s(x) \uparrow 0 \rightarrow s(0) \\ s(x) + y \rightarrow s(x+y) & x \uparrow s(y) \rightarrow (x \uparrow y) \cdot x\ . \\ x + s(y) \rightarrow s(x+y) & \end{array} \quad (4)$$

It also extends the result in [Middeldorp and Toyama, 1991] for constructor-sharing constructor-based systems.[9]

The overlaying requirement is necessary, as seen in this locally confluent example [Drosten, 1989]:

$$\begin{array}{l|c} f(0,1,x) \rightarrow f(x,x,x) & \\ f(x,y,z) \rightarrow 2 & g(x,y,y) \rightarrow x \\ \quad 0 \rightarrow 2 & g(x,x,y) \rightarrow y \\ \quad 1 \rightarrow 2 & \end{array} \quad \text{(C)}$$

Local-confluence is likewise essential (cf. Example (B)).

[8] This result also appears in [Gramlich, 1995].

[9] Middeldorp and Toyama [1991] also consider the case where certain rules are shared by both systems, also easily handled by our method.

4 Hierarchical termination

Suppose one has defined some blue functions, recursively, using green rules. Typically, these rules would reduce any green term to a yellow (constructor) normal form. Then, one goes ahead and defines red functions, also recursively, but using blue functions in an auxiliary manner. We are thinking of a system like the right half of System (1), where $\cdot$ is red, $+$ is blue, 0 and s are yellow. We'll call such systems, with bright-orange left-hand sides and arbitrary right-hand sides, *purple*. This common situation also arises in applicative programs; semantics (that is, knowing the values computed by the functions) are usually needed for termination proofs.

At least two approaches are possible. We can endeavor to show that any infinite derivation in the combined system could be rearranged to provide an infinite monochrome derivation, which is impossible. Or we can try to extend the results of the previous section which require that the number of layers not increase in a derivation (which is not in general true for the hierarchical case).

An easy result using the first approach is:

Theorem 15. *The union of a left-linear purple terminating system with a right-linear bright-green terminating system is terminating.*

This theorem applies, for example, to:

$$\begin{array}{r|l} x\cdot 0 \rightarrow 0 & \\ x\cdot s(y) \rightarrow (x\cdot y)+x & x+s(y)\rightarrow s(x)+y\,. \end{array} \tag{5}$$

It does not apply to (1) with its "dull" rule $x+0\rightarrow x$.

The necessity of brightness and right-linearity can both be seen from Example (B); left-linearity is needed to exclude:

$$\boxed{f(x,x)\rightarrow f(g(a),g(b)) \;\mid\; g(a)\rightarrow g(b)} \tag{D}$$

Here f is red, g is blue, and a and b are constructors. With blue symbols on both sides of purple rules, we invite non-termination, as in:

$$\boxed{\begin{array}{c|r} f(b)\rightarrow f(g(a)) & g(a)\rightarrow b \\ & b\rightarrow g(c) \end{array}} \tag{E}$$

(g and b are blue).

A more general version of this theorem will be proved at the end of the next section.

Right-linearity is not a very natural requirement. To get a better handle on the hierarchical non-right-linear case, we further restrict the form of purple rules.

Theorem 16. *The union of a left-linear overlay-confluent purple terminating system with an overlay-confluent bright-green terminating system is terminating.*

This theorem applies to System (5) and corrects the result in [Bidoit, 1981] for "gracious" systems by requiring that the green system be bright. Without brightness, we could be fooled by

$$\boxed{f(a) \rightarrow f(b) \quad | \quad b \rightarrow a} \tag{F}$$

where a is the only constructor.[10] We've already seen the need for left-linearity in Example (D). Example (B) shows the need for confluence (of the purple system at least).

Proof. The union is overlaying (since the purple left sides cannot unify with non-variable green subterms, nor green left sides with the orange subterms of purple left sides) and locally-confluent (by the Critical Pair Lemma [Knuth and Bendix, 1970; Huet, 1980], since the union cannot introduce any new overlapping left-hand sides). Hence, by Proposition 13, we need only show innermost termination. In any innermost derivation, there cannot be a purple step taking place in the variable part of a preceding green step, since that would mean that the purple step could have been applied to a proper subterm of the green redex. So, if a derivation has a green step immediately preceding a purple step, the two either occur at disjoint positions (neither at a subterm of the other redex), in which case they can be interchanged, or else the green step occurs in the variable part of the purple step (on account of brightness), in which case the left-linear purple step can be applied first, followed by some number of green steps (one for each occurrence of that variable on the right-hand side of the purple rule). Thus, from any innermost derivation with infinitely many purple steps, an infinite purple derivation could be constructed. □

A purple system is *red-increasing* if the maximum number of red symbols along a path from the root of a term to a leaf can increase in a derivation. For example,

$$\begin{array}{c} f(x, g(y)) \rightarrow f(y, g(x)) \\ g(x) \rightarrow x\,, \end{array} \tag{6}$$

where f and g are red, is red-increasing. If a purple system is not red-increasing, then neither is its union with a green system.

Theorem 10 has the following analogue in the hierarchical case:

Theorem 17. *The union of a left-linear non-red-increasing bright-purple terminating system with a bright-green terminating system is terminating.*

System (1), sans its three dull rules, is an example. Example (B) shows the need for brightness of both systems; (D) demonstrates the need for left-linearity.

Proof. Since the nesting of reds does not increase, were the union non-terminating, there would be a non-terminating derivation of minimal depth, with infinitely many rewrites at topmost red symbols. (Were the top red symbols to all become inactive, fewer levels would suffice for non-termination, since the green steps above the top red layer could not go on in perpetuity.) Transform this derivation by replacing

[10] It wouldn't help—nor would it make sense—to insist that all terms reduce to constructor terms via the purple system, since we could just add $f(x) \rightarrow a$ to this counter-example.

all subterms headed by a non-topmost red with some green constant, since bright purple steps at the second level cannot impinge on rewritability of higher purple or green redexes, and lower steps can certainly have no effect. That leaves an infinite derivation of terms having green above and below a single layer of red. If there are green symbols above the red, they cannot sustain more than finitely many green steps, since bright purple steps do not contribute to that green layer. Thus, any green step preceding a purple step must either be disjoint from the latter, or else it is in the variable part of the purple rule, since the green right-hand side is headed by a blue symbol, which can appear nowhere in the purple left side. As in Theorem 16, since the purple system is left-linear, that green step can be delayed until after the purple step, and then performed once for each occurrence of the variable in question on the right side. Hence, a derivation containing infinitely many purple steps in succession can be constructed, contradicting the presumed termination of the purple system on its own. □

A *flat* system is one in which red symbols are not nested on the left or right. That is, no path from the root symbol has more than one red symbol along it. Flat systems, in addition to being constructor-based, cannot invoke nested recursion.

Lemma 18. *Flat purple systems are not red-increasing.*

Even for flat systems, hierarchical termination is by no means ensured (Example (F)). In the remainder of this section, we develop sufficient conditions for termination in the flat case.

Violet (a bluish purple) systems have no blue symbols below a red on the right side, as in the right half of System (1). When innermost termination suffices, we can show:[11]

Theorem 19. *The union of a locally-confluent flat violet (only yellow below red) terminating system with an overlay-confluent green terminating system is terminating.*

The right half of System (1) is an example. More generally, this theorem applies to hierarchies of primitive-recursive definitions. It is a corollary of the one that follows.

If we desire to allow blue symbols to also appear below the red ones, we need to be able to ignore the effects of green rewriting. We will say that a terminating purple system is *oblivious (of green)* if it remains terminating even when the rules are replicated so that each bright-green subterm on a right-hand side (headed by a blue symbol) is replaced by all possible green (and yellow) variable-free terms. (We assume that there is at least one blue or yellow constant—or else we must add one so that the set of green terms is not void.) That is, a purple system R is oblivious of green terms if $R_g = \{l \rightarrow r[g]_p : l \rightarrow r \in R;\ r|_p \text{ is bright green};\ g \text{ is green}\}$ is also terminating. (Actually, we need only replace maximal green subterms.)

[11] This result also appeared in [Krishna Rao, 1992]. Instead of requiring flatness, Krishna Rao [1993] forbids those nestings that *seem* able to lead eventually to a blue symbol that can cause an increase in the depth of red. See also [Gramlich, 1995].

For System (1), the recursive rule on the right adds $f(s(x), y) \rightarrow f(x, ?)$, where ? is an arbitrary green term (containing any combination of $\cdot$, $+$, s, and 0). The extended system is still terminating.

Flat violet systems are oblivious by definition. Systems for which there is one green argument position that decreases (taking subterms, say) with each "recursive call" are also oblivious.

Obliviousness compensates for the appearance of blue symbols below reds on the right, and allows us to generalize the previous theorem:

Theorem 20. *The union of an oblivious locally-confluent flat purple terminating system with an overlay-confluent green terminating system is terminating.*

Note that green levels can grow deeper; hence, more than two levels of hierarchically-defined functions are possible. This result applies, for example, to the three parts of System (1), as well as to the following "tail recursive" program:

$$\begin{array}{rcl|l} sum(x) & \rightarrow & f(0, x) & x + 0 \rightarrow x \\ f(x, \epsilon) & \rightarrow & x & x + s(y) \rightarrow s(x + y)\ , \\ f(x, y \cdot z) & \rightarrow & f(x + y, z) & \end{array} \tag{7}$$

Flatness of purple right-hand sides is necessary as can be seen from the following non-terminating union:

$$\boxed{\begin{array}{rcl|l} f(x, x) & \rightarrow & f(a, g(x)) & g(x) \rightarrow x \\ a & \rightarrow & f(c, d) & \end{array}} \tag{G}$$

(f and a are red; g is blue). That non-increasing red depth is insufficient can be seen from the following variant:

$$\boxed{\begin{array}{rcl|l} f(x, x, a) & \rightarrow & f(a, g(x), a) & g(x) \rightarrow x \\ a & \rightarrow & f(c, d, d) & \end{array}} \tag{H}$$

These two systems are oblivious, since no *green* replacement for $g(x)$ can match the red a needed for the first purple rule to reapply. One part of (B) is not locally confluent, which explains its non-termination; (D) is not oblivious; the green left half of (C) is not overlaying.

Proof. By flatness, there is a bound on the depth of red symbols in any derivation. As was the case for Theorem 16, the union is locally confluent and overlaying, so we need only consider innermost rewriting (Proposition 13). Purple steps at the lowest red level may be followed by some green steps lower down. In an innermost derivation those green steps cannot be in the variable part of the purple right-hand side, since those are already in normal form. Thus, the purple system is oblivious of those green steps, and the net effect (reordering the green steps, as necessary, to follow immediately upon the red step that created them) is just a sequence of "oblivious purple steps," guaranteed to terminate with the lowest red level and everything below in normal form. Those subterms of red normal forms that have a red symbol at the top can play no further role in the derivation, since green rules cannot "see" them at all, nor can the applicability of constructor-based purple rules depend on red symbols below the redex. The only impact they can have is in allowing or disallowing a non-left-linear rule to fire. They can all, therefore, be replaced with one non-red constant, giving a term with fewer levels of red. Thus, any derivation with the original red normal forms can be mimicked by shallower terms. □

This proof only requires that the purple system be oblivious of green subterms that are below a red symbol.

By combining the commutation-based approach with the layer-based approach, we get the following modification of Theorem 11:

Theorem 21. *The union of an oblivious right-linear flat purple terminating system with a non-duplicating green terminating system is terminating.*

System (7) falls in this category. Without flatness we have non-termination, as before (G); non-duplication by green rewrites rules out systems like (B); right-linearity of purple cannot be weakened to non-duplication, witness:

$f(0,1,x,x) \rightarrow f(x,x,g(0,1),g(0,1))$	$g(x,y) \rightarrow x$ $g(x,y) \rightarrow y$

(I)

Proof. By flatness, the nesting of red does not increase, so we may consider an infinite derivation of minimal red depth. The multiset of subterms headed by second-layer red symbols (call them the "aliens") decreases with each rewrite at or below the second red layer and does not increase with a (purple or green) rewrite above—in the union of the subterm relation and the combined rewrite relation for terms with shallower nesting of red. Hence, from some point on, the minimal infinite derivation only has top red steps and green steps above the second layer. We can, therefore, replace all the aliens by a green constant without affecting any of those steps. This yields an infinite derivation of terms with green symbols above and below a single layer of red. (Flatness comes into play here; without it, new aliens would be produced by purple steps.) On account of right-linearity, any purple rewrite at a topmost red symbol followed immediately by a green rewrite within the position of a particular variable of the purple right-hand side can be rearranged to first apply the green rule as many times as necessary to rewrite the (one or more) occurrences of that variable on the left-hand side of the purple rule, followed by the same purple step. That can only transpire finitely many times (since green terminates), leaving an infinite sequence of oblivious purple steps, plus green steps in the top, non-duplicating, green layer. But there can be only finitely many of either, since green is non-duplicating. □

5 Combined termination

We mention here a few results for the non-hierarchical case in which either terminating system can refer to symbols appearing also in the other. We will refer to the systems as black and white. In particular, we will generalize the first two theorems of the previous section.

Proposition 22 **[Dershowitz and Hoot, 1995; Gramlich, 1995].** *A non-erasing, non-overlapping system terminates if it is normalizing (that is, if there is always some derivation leading to a normal form).*

This improves the result in [O'Donnell, 1977] which requires that the system be left-linear, and which, consequently, has the same behavior as Church's [1941] λ-I calculus.

We say that a white system *preserves normal forms* of a black system if the former always rewrites black normal forms to black normal forms.

Theorem 23. *The union of black and white non-erasing terminating systems, the union of which is non-overlapping, and such that the white system preserves normal forms of the black, is terminating.*

For example,

$$\begin{array}{c|c} x + s(y) \rightarrow s(x) + y & s(x) - s(y) \rightarrow x - y \\ x + 0 \rightarrow x & s(x) \uparrow 0 \rightarrow 1\,. \end{array} \tag{8}$$

Proof. Use the preceding proposition and the fact that the union is normalizing under the stated conditions, taking white normal forms of black normal forms. □

The necessity of preservation is demonstrated by Example (F); the need for non-overlapping, by (B).[12]

We say that terms s and t are *separate* if s does not unify with a renamed non-variable subterm of t, nor vice-versa.

Proposition 24. *A white system preserves normal forms of a left-linear black system whenever white right-hand sides and black left-hand sides are separate. (In particular, the white system must be non-collapsing.)*

Proof. White cannot create an occurrence of a black left-hand side. Since black is left linear, white cannot create a new black redex by making making the latter's variable parts equal. □

We have:

Theorem 25. *The union of a left-linear overlay-confluent black terminating system with an overlay-confluent white terminating system, such that white right-hand sides are separate from black left-hand sides, is terminating.*

Theorem 16 is a corollary.[13] The proof is unchanged. An example is

$$x \cdot (y + z) \rightarrow (x \cdot y) + (x \cdot z) \quad | \quad x \cdot x \rightarrow 0\,. \tag{9}$$

We need the following:

[12] Though the non-erasing requirement is needed for the above proposition, an example of non-termination for non-overlapping preserving systems is lacking.

[13] This idea of decomposing proofs of termination by looking at overlappings between rules, but ignoring the difficulties engendered by non-left-linear rules, appeared in [Pettorossi, 1981].

Lemma 26 [Raoult and Vuillemin, 1980]. *If u rewrites to v using a right-linear rule $l \rightarrow r$, and then to w using a left-linear rule $s \rightarrow t$, and r and s are separate, then w can also be derived from u using at least one application of $s \rightarrow t$ followed by some number of applications of $l \rightarrow r$.*

Theorem 27 [Bachmair and Dershowitz, 1986]. *The union of a left-linear black terminating system with a right-linear white terminating system, such that white right-hand sides are separate from black left-hand sides, is terminating.*

Theorem 15 is a corollary.[14] System (9) is again an example.

Proof. Since the white right-hand sides are separate from black left sides and both the white right sides and black left sides do not have repeated variables, by the preceding lemma, any white step followed by a black step can always be replaced by at least one black step followed by some number of white steps. By induction, any number of white steps followed by one black can be replaced by at least one black step followed by some number of white steps. Hence, from any derivation with infinitely many black steps, an infinite purely black derivation could be constructed, contradicting the assumption of black termination. □

6 Discussion

It appears that Theorem 20 is the most useful result we have obtained for hierarchical systems, since it does not require brightness, right-linearity, or non-duplication. Both systems must be overlaying and locally-confluent (which implies that innermost rewriting will lead to non-termination if any strategy can), but that is normal in a functional programming style. In the absence of prescience as to the semantics (normal-form computations) of the green system, the purple system must be oblivious of green rewrites taking place in arguments of red functions (or have no blue symbols below red recursive calls, as in Theorem 19), but that is similar to the situation with ordinary functional languages. The purple system must be flat, but—in future work—we hope to use more general notions of obliviousness (such as obliviousness of terms built from green symbols *and* subterms of the purple left side) to perhaps weaken some of our restrictions. In any case, we need to develop additional sufficient conditions for obliviousness.

There still seems to be room for improving the various results we have given here, though we have provided counter-examples to most (but not quite all) ways of relaxing the conditions for termination. Some extensions are obvious: Bright-green rules, as in Theorem 16 for example, were only needed to preclude a green rule "creating" a purple redex; a constructor on the top right of a green rule that does not appear below the defined function of a purple left-hand side poses no problem (they are "separate" in the terminology of Section 5). Transformation methods of [Bachmair and Dershowitz, 1986; Bellegarde and Lescanne, 1990] can perhaps be used to handle certain collapsing cases.

[14] This theorem was claimed in [Dershowitz, 1981], but an overly weak condition of separateness was implied. (The examples in [Bachmair and Dershowitz, 1986] were also wrong on this account.) This direction was pursued further in [Geser, 1989].

This paper has only considered modularity of termination. The preservation of other properties, such as existence and uniqueness of normal forms, is also worth exploring for hierarchical systems.

To conclude with one more example, consider the fact that none of the theorems we have given apply to the union of

$$\begin{aligned} mapf(\epsilon) &\rightarrow \epsilon \\ mapf(x \cdot y) &\rightarrow f(x) \cdot mapf(y) \end{aligned} \tag{10}$$

with an *arbitrary* (green) system for computing f, not containing *mapf*. If the latter is non-duplicating or overlay-confluent, then it's okay, but it is highly unlikely (in this case, at least) that *any* terminating green system could cause problems.

Acknowledgements

I thank Bernhard Gramlich, M. R. K. Krishna-Rao and Aart Middeldorp for their helpful comments.

References

[Avenhaus and Madlener, 1990] Jürgen Avenhaus and Klaus Madlener. Term rewriting and equational reasoning. In R. B. Banerji, editor, *Formal Techniques in Artificial Intelligence: A Sourcebook*, pages 1–41. Elsevier, Amsterdam, 1990.

[Bachmair and Dershowitz, 1986] Leo Bachmair and Nachum Dershowitz. Commutation, transformation, and termination. In J. H. Siekmann, editor, *Proceedings of the Eighth International Conference on Automated Deduction (Oxford, England)*, volume 230 of *Lecture Notes in Computer Science*, pages 5–20, Berlin, July 1986. Springer-Verlag.

[Bellegarde and Lescanne, 1990] François Bellegarde and Pierre Lescanne. Termination by completion. *Applicable Algebra in Engineering, Communication and Computing*, 1990.

[Bidoit, 1981] Michel Bidoit. *Une méthode de présentation de types abstraits: Applications.* PhD thesis, Université de Paris-Sud, Orsay, France, June 1981. Rapport 3045.

[Cadiou, 1972] J. M. Cadiou. *Recursive Definitions of Partial Functions and their Computations.* PhD thesis, Stanford University, Stanford, CA, March 1972.

[Church, 1941] Alonzo Church. *The Calculi of Lambda Conversion*, volume 6 of *Ann. Mathematics Studies.* Princeton University Press, Princeton, NJ, 1941.

[Dershowitz, 1981] Nachum Dershowitz. Termination of linear rewriting systems. In *Proceedings of the Eighth International Colloquium on Automata, Languages and Programming (Acre, Israel)*, volume 115 of *Lecture Notes in Computer Science*, pages 448–458, Berlin, July 1981. European Association of Theoretical Computer Science, Springer-Verlag.

[Dershowitz, 1982] Nachum Dershowitz. Orderings for term-rewriting systems. *Theoretical Computer Science*, 17(3):279–301, March 1982.

[Dershowitz, 1987] Nachum Dershowitz. Termination of rewriting. *J. Symbolic Computation*, 3(1&2):69–115, February/April 1987. Corrigendum: *4*, 3 (December 1987), 409–410; reprinted in *Rewriting Techniques and Applications*, J.-P. Jouannaud, ed., pp. 69—115, Academic Press, 1987.

[Dershowitz, 1992] Nachum Dershowitz. Hierarchical termination. Unpublished report, Leibnitz Center for Research in Computer Science, Hebrew University, Jerusalem, Israel, December 1992.

[Dershowitz, 1995] Nachum Dershowitz. 33 examples of termination. In H. Comon and J.-P. Jouannaud, editors, *French Spring School of Theoretical Computer Science Advanced Course on Term Rewriting (Font Romeux, France, May 1993)*, volume 909 of *Lecture Notes in Computer Science*, pages 16–26, Berlin, 1995. Springer-Verlag.

[Dershowitz and Hoot, 1995] Nachum Dershowitz and Charles Hoot. Natural termination. *Theoretical Computer Science*, 142(2):179–207, May 1995.

[Dershowitz and Jouannaud, 1990] Nachum Dershowitz and Jean-Pierre Jouannaud. Rewrite systems. In J. van Leeuwen, editor, *Handbook of Theoretical Computer Science*, volume B: Formal Methods and Semantics, chapter 6, pages 243–320. North-Holland, Amsterdam, 1990.

[Dershowitz and Manna, 1979] Nachum Dershowitz and Zohar Manna. Proving termination with multiset orderings. *Communications of the ACM*, 22(8):465–476, August 1979.

[Drosten, 1989] K. Drosten. *Termersetzungssysteme*. PhD thesis, Universitat Passau, Berlin, Germany, 1989. Informatik Fachberichte 210, Springer-Verlag.

[Fernandez and Jouannaud, 1995] Maribel Fernandez and Jean-Pierre Jouannaud. Modular termination of term rewriting systems revisited. In *Proceedings of the Eleventh Workshop on Specification of Abstract Data Types (Santa Margherita de Ligura, Italy)*, Lecture Notes in Computer Science, Berlin, 1995. Springer-Verlag.

[Geser, 1989] Alfons Geser. *Termination Relative*. PhD thesis, Universität Passau, Passau, West Germany, 1989.

[Geupel, 1989] Oliver Geupel. Overlap closures and termination of term rewriting systems. Report MIP-8922, Universität Passau, Passau, West Germany, July 1989.

[Gramlich, 1994] Bernhard Gramlich. Generalized sufficient conditions for modular termination of rewriting. *Applicable Algebra in Engineering, Communication and Computing*, 5:131–158, 1994. A preliminary version appeared in the Proceedings of the Third Internationnal Conference on Algebraic and Logic Programming, Lecture Notes in Computer Science 632, Springer-Verlag, Berlin, pp. 53–68, 1992.

[Gramlich, 1995] Bernhard Gramlich. Abstract relations between restricted termination and confluence properties of rewrite systems. *Fundamenta Informaticae*, September 1995. Preliminary versions appeared as "Relating Innermost, Weak, Uniform and Modular Termination of Term Rewriting Systems" in Proceedings of the Conference on Logic Programming and Automated Reasoning (St. Petersburg, Russia), A. Voronkov, ed., Lecture Notes in Artificial Intelligence 624, Springer-Verlag, Berlin, pp. 285–296 and as SEKI-Report SR-93-09, Fachbereich Informatik, Universität Kaiserslautern, Kaiserslautern, Germany, 1993.

[Huet, 1980] Gérard Huet. Confluent reductions: Abstract properties and applications to term rewriting systems. *J. of the Association for Computing Machinery*, 27(4):797–821, October 1980.

[Klop, 1992] Jan Willem Klop. Term rewriting systems. In S. Abramsky, D. M. Gabbay, and T. S. E. Maibaum, editors, *Handbook of Logic in Computer Science*, volume 2, chapter 1, pages 1–117. Oxford University Press, Oxford, 1992.

[Knuth and Bendix, 1970] Donald E. Knuth and P. B. Bendix. Simple word problems in universal algebras. In J. Leech, editor, *Computational Problems in Abstract Algebra*, pages 263–297. Pergamon Press, Oxford, U. K., 1970. Reprinted in *Automation of Reasoning 2*, Springer-Verlag, Berlin, pp. 342–376 (1983).

[Krishna Rao, 1992] M. R. K. Krishna Rao. Modular proofs for completeness of hierarchical systems. Unpublished report, December 1992.

[Krishna Rao, 1993] M. R. K. Krishna Rao. Completeness of hierarchical combinatins of term rewriting systems. In *Proceedings of the Thirteenth Conference on Foundations of Software Technology and Theoretical Computer Science (Bombay, India)*, volume 761 of *Lecture Notes in Computer Science*, pages 125–138, Berlin, 1993. Springer-Verlag.

[Kurihara and Kaji, 1990] Masahito Kurihara and Ikuo Kaji. Modular term rewriting systems and the termination. *Information Processing Letters*, 34:1–4, February 1990.

[Kurihara and Ohuchi, 1990] Masahito Kurihara and Azuma Ohuchi. Modularity of simple termination of term rewriting systems. *Journal of Information Processing Society*, 34:632–642, 1990.

[Middeldorp, 1989] Aart Middeldorp. A sufficient condition for the termination of the direct sum of term rewriting systems. In *Proceedings of the Fourth Symposium on Logic in Computer Science*, pages 396–401, Pacific Grove, CA, 1989. IEEE.

[Middeldorp, 1990] Aart Middeldorp. *Modular Properties of Term Rewriting Systems.* PhD thesis, Vrije Universiteit, Amsterdam, The Netherlands, 1990.

[Middeldorp and Toyama, 1991] Aart Middeldorp and Yoshihito Toyama. Completeness of combinations of constructor systems. In R. Book, editor, *Proceedings of the Fourth International Conference on Rewriting Techniques and Applications (Como, Italy)*, volume 488 of *Lecture Notes in Computer Science*, pages 174–187, Berlin, April 1991. Springer-Verlag.

[O'Donnell, 1977] Michael J. O'Donnell. *Computing in systems described by equations*, volume 58 of *Lecture Notes in Computer Science*. Springer-Verlag, Berlin, 1977.

[Ohlebusch, 1993] Enno Ohlebusch. On the modularity of termination of term rewriting systems. Report 11, Abteilung Informationstechnik, Universität Bielefeld, Bielefeld, Germany, 1993.

[Pettorossi, 1981] Alberto Pettorossi. Comparing and putting together recursive path orderings, simplification orderings and non-ascending property for termination proofs of term rewriting systems. In *Proceedings of the Eighth EATCS International Colloquium on Automata, Languages and Programming (Acre, Israel)*, volume 115 of *Lecture Notes in Computer Science*, pages 432–447, Berlin, July 1981. Springer-Verlag.

[Plaisted, 1993] David A. Plaisted. Equational reasoning and term rewriting systems. In D. Gabbay, C. Hogger, J. A. Robinson, and J. Siekmann, editors, *Handbook of Logic in Artificial Intelligence and Logic Programming*, volume 1, chapter 5, pages 273–364. Oxford University Press, Oxford, 1993.

[Raoult and Vuillemin, 1980] Jean-Claude Raoult and Jean Vuillemin. Operational and semantic equivalence between recursive programs. *J. of the Association for Computing Machinery*, 27(4):772–796, October 1980.

[Rusinowitch, 1987] Michael Rusinowitch. On termination of the direct sum of term-rewriting systems. *Information Processing Letters*, 26:65–70, 1987.

[Toyama, 1987a] Yoshihito Toyama. Counterexamples to termination for the direct sum for the direct sum of term rewriting systems. *Information Processing Letters*, 25:141–143, 1987.

[Toyama, 1987b] Yoshihito Toyama. On the Church-Rosser property for the direct sum of term rewriting systems. *J. of the Association for Computing Machinery*, 34(1):128–143, January 1987.

[Toyama *et al.*, 1989] Yoshihito Toyama, Jan Willem Klop, and Hendrik Pieter Barendregt. Termination for the direct sum of left-linear term rewriting systems. In Nachum Dershowitz, editor, *Proceedings of the Third International Conference on Rewriting Techniques and Applications (Chapel Hill, NC)*, volume 355 of *Lecture Notes in Computer Science*, pages 477–491, Berlin, April 1989. Springer-Verlag.

Well-foundedness of Term Orderings

M. C. F. Ferreira* and H. Zantema

Utrecht University, Department of Computer Science
P.O. box 80.089, 3508 TB Utrecht, The Netherlands
e-mail: {maria, hansz}@cs.ruu.nl

Abstract. Well-foundedness is the essential property of orderings for proving termination. We introduce a simple criterion on term orderings such that any term ordering possessing the subterm property and satisfying this criterion is well-founded. The usual path orders fulfil this criterion, yielding a much simpler proof of well-foundedness than the classical proof depending on Kruskal's theorem. Even more, our approach covers non-simplification orders like *spo* and *gpo* which can not be dealt with by Kruskal's theorem.
For finite alphabets we present completeness results, i. e., a term rewriting system terminates if and only if it is compatible with an order satisfying the criterion. For infinite alphabets the same completeness results hold for a slightly different criterion.

1 Introduction

The usual way of proving termination of a term rewriting system (TRS) is by finding a well-founded order such that every rewrite step causes a decrease according to this ordering. Proving well-foundedness is often difficult, in particular for recursively defined syntactic orderings. It is therefore desirable to have criteria that help decide whether a particular order is well-founded. A standard criterion of this type is implied by Kruskal's theorem: if a monotonic term ordering over a finite signature satisfies the subterm property then it is well-founded. However, this theorem does not apply for all terminating TRS's: there are terminating TRS's like $f(f(x)) \to f(g(f(x)))$ that are not compatible with any monotonic term ordering satisfying the subterm property. Even *recursive path order (rpo)* with lexicographic status over a varyadic alphabet, is not covered directly by Kruskal's theorem ([5]). This motivated us to look for other conditions ensuring well-foundedness. In this paper we remove the monotonicity condition and replace it by some decomposability condition. For orderings satisfying the subterm property and this decomposability condition we prove well-foundedness in a way that is inspired by Nash-Williams' proof of Kruskal's theorem ([10]; as it appears in [6]), but which is much simpler. A similar technique, for a particular order, has already been used by Kamin and Lévy ([9]). Standard orderings like

* Supported by NWO, the Dutch Organization for Scientific Research, under grant 612-316-041.

recursive path order ([1, 12]) and *semantic path order (spo)* ([9, 2]) trivially satisfy our conditions, yielding a simple proof of well-foundedness for these orders. Moreover, our conditions cover all terminating TRS's: a TRS terminates if and only if it is compatible with an order satisfying our conditions.

We are concerned essentially with term rewrite systems over finite signatures. In the case of an infinite signature the same conditions yield well-foundedness if the signature is provided with a partial well-order satisfying some natural compatibility with the given term ordering.

The rest of the paper is organized as follows. In section 2 we give some well-known notions on term rewriting and partial orders. On section 3, we introduce the notion of lifting of an order, which plays an essential role in the theory presented. On section 4 we present our well-foundedness criterion for orders on terms built over a finite signature and give some surprising completeness results involving orders closed under substitutions and orders that are total.

In section 5, we present a well-foundedness criterion for orders on terms built over infinite signatures. First we follow an approach similar to the one used in section 4. For that we need the existence of *well-quasi-orders* on the set of function symbols. This requirement is quite strong and to overcome it we introduce a different notion of lifting of orders on terms. Using this new notion we can present a very general and simple result on well-foundedness and show that in this case the completeness results of section 4 also hold. The criteria presented are used on section 6 to derive well-foundedness of *semantic path order* and *general path order*.

Finally we make some concluding remarks, including some comparison between our results and Kruskal's theorem.

2 Preliminaries

For the sake of self-containment we give some notions over term rewriting systems and orders. For more information the reader is referred to [4].

Let $\mathcal{F}$ be a signature (a set of function symbols) and $\mathcal{X}$ a set of variables with $\mathcal{F} \cap \mathcal{X} = \emptyset$. To each function symbol of $\mathcal{F}$ we associate a set of possible arities given by the function *arity*: $\mathcal{F} \rightarrow \mathcal{P}(\mathbb{N}) \setminus \emptyset$, where $\mathcal{P}(\mathbb{N})$ is the power set of $\mathbb{N}$. In the case that $arity(f)$ contains only one element for all $f \in \mathcal{F}$, we speak of a fixed-arity signature, otherwise we speak of a varyadic signature.

The set of all terms over $\mathcal{F}$ and $\mathcal{X}$ is defined inductively as usual and denoted by $\mathcal{T}(\mathcal{F}, \mathcal{X})$, the set of ground terms is denoted by $\mathcal{T}(\mathcal{F})$. In the sequel we will consider terms over different kinds of signature, for example finite or infinite signatures and finite or infinite sets of variables. We will make clear which restrictions apply at any point.

Given any term t, s is a *subterm* of t if we can write $t = C[s]$ for some context C. If $C[s] = f(\ldots, s, \ldots)$ and C is not the empty context, we say that s is a *principal subterm* of t. We define $|t|$ to be the depth of a term t. Recall that depth strictly decreases by taking (principal) subterms.

A term rewriting system (TRS) is a tuple $(\mathcal{F}, \mathcal{X}, R)$, where R is a subset of $\mathcal{T}(\mathcal{F}, \mathcal{X}) \times \mathcal{T}(\mathcal{F}, \mathcal{X})$. The elements of R are the so called rules of the TRS and are usually denoted by $l \to r$, with l a non-variable term and such that all the variables occurring in r also occur in l.

A TRS R induces a *rewrite relation* over $\mathcal{T}(\mathcal{F}, \mathcal{X})$, denoted by $\to_R$, as follows: $s \to_R t$ iff $s = C[l\sigma]$ and $t = C[r\sigma]$, for some context C, substitution σ and rule $l \to r \in R$. The transitive closure of $\to_R$ is denoted by $\to_R^+$ and its reflexive-transitive closure by $\to_R^*$. A TRS is called *terminating* (strongly normalizing or noetherian) if there exists no infinite sequence of the form $t_0 \to_R t_1 \to_R \ldots$.

We use the terminology *partial order* on a set S meaning an irreflexive and transitive relation on S, that we usually denote by $>$. By *quasi-order* we mean a reflexive and transitive relation, usually denoted by $\geq$. Any quasi-order contains a strict partial order, namely $\geq \setminus \leq$, and an equivalence relation $\geq \cap \leq$, that we usually denote by $\sim$.

A partial order or quasi-order over a set S is said to be *well-founded* if it doesn't admit infinite descending chains of the form

$$x_0 > x_1 > x_2 > \ldots$$

We extend the terminology well-founded to the elements of S: we say that $x \in S$ is well-founded if x does not occur in an infinite descending chain as above. Obviously an order $>$ on a set S is well-founded if and only if all elements $s \in S$ are well-founded.

We are interested on orders on the set of terms $\mathcal{T}(\mathcal{F}, \mathcal{X})$. An order $>$ on $\mathcal{T}(\mathcal{F}, \mathcal{X})$ is said to be *monotonic* if $s > t$ implies $C[s] > C[t]$, for any context C. Given a TRS R and a order $>$ on $\mathcal{T}(\mathcal{F}, \mathcal{X})$, we say that $>$ is *compatible* with R if $s > t$ whenever $s \to_R t$.

An order on $\mathcal{T}(\mathcal{F}, \mathcal{X})$ is said to have the *subterm property* if $f(t_1, \ldots, t_n) > t_i$, for any $f \in \mathcal{F}$ and terms $t_1, \ldots, t_n \in \mathcal{T}(\mathcal{F}, \mathcal{X})$, where $n \in arity(f)$.

3 Liftings and Status

As mentioned before, we replace monotonicity by another condition. This condition relates the comparison between $f(s_1, \ldots, s_m)$ and $f(t_1, \ldots, t_n)$ to the comparison of the sequences $\langle s_1, \ldots, s_m \rangle$ and $\langle t_1, \ldots, t_n \rangle$. Here we need to describe how an ordering on terms is lifted to an ordering on sequences of terms. To be able to conclude well-foundedness it is essential that this lifting preserves well-foundedness.

Definition 1. Let $(S, >)$ be a partial ordered set and $S^* = \cup_{n \in \mathbb{N}} S^n$. We define a *lifting* to be a partial order $>^\lambda$ on S^* for which the following holds: for every $A \subseteq S$, if $>$ restricted to A is well-founded, then $>^\lambda$ restricted to A^* is also well-founded. We use the notation $\lambda(S)$ to denote all possible liftings of $>$ on S^*.

A typical example of a lifting is the *multiset extension* of an order. The usual *lexicographic extension* on unbounded sequences is not a lifting. Just take $S = \{0, 1\}$ with $1 > 0$, then

$$\langle 1 \rangle >^{\lambda} \langle 01 \rangle >^{\lambda} \langle 001 \rangle >^{\lambda} \langle 0001 \rangle >^{\lambda} \dots$$

If the lexicographic comparison is restricted to sequences whose size is bounded by some fixed natural N, then this is indeed a lifting.

Another type of lifting is a constant lifting, i. e., any fixed well-founded partial order on S^*. Clearly other liftings can be defined, for example as combinations of the ones mentioned. In particular, combinations of multiset and lexicographic order can be very useful. In a partial order $(S, >)$ where $a > b$ and c is incomparable with a and b, one cannot conclude $\langle a, c, c \rangle >^{\lambda} \langle c, b, a \rangle$, for the multiset lifting nor for any lexicographic lifting. If we define $>^{\lambda}$ by

$$\langle s_1, \dots, s_m \rangle >^{\lambda} \langle t_1, \dots, t_n \rangle \iff \begin{cases} (m = n = 3) & \text{and} \\ \langle s_1, s_2 \rangle >^{mul} \langle t_1, t_2 \rangle & \text{or} \\ (\langle s_1, s_2 \rangle =^{mul} \langle t_1, t_2 \rangle) & \text{and } s_3 > t_3 \end{cases}$$

it is not difficult to see that $>^{\lambda}$ satisfies the definition of lifting and also satisfies $\langle a, c, c \rangle >^{\lambda} \langle c, b, a \rangle$. This lifting will be used to obtain

$$f(s(x), y, y) >_{rpo} f(y, x, s(x))$$

Classical $>_{rpo}$ cannot be used to compare these two terms.

Definition 1 is intended to be applied to terms over varyadic function symbols. If we consider signatures with fixed arity function symbols we can simplify the notion of lifting: instead of taking liftings of any order we need only take liftings of fixed order, i. e., the lifting is going to be a partial order over S^n, for a fixed natural number n. This is a special case of a lifting to S^*: $>^{\lambda}$ is defined on S^* to be the order one has in mind for S^n on sequences of length n, while all other pairs of sequences are defined to be incomparable with respect to $>^{\lambda}$.

Again typical examples of liftings are the *lexicographic extension* of $>$ on sequences and the *multiset extension* of $>$ restricted to multisets of a fixed size.

We are interested in orders on terms so from now on we choose $S = \mathcal{T}(\mathcal{F}, \mathcal{X})$, with $\mathcal{F}$ containing varyadic function symbols, and we fix a partial order $>$ on $\mathcal{T}(\mathcal{F}, \mathcal{X})$.

Definition 2. Given $(\mathcal{T}(\mathcal{F}, \mathcal{X}), >)$, a *status function* (with respect to $>$) is a function $\tau : \mathcal{F} \to \lambda(\mathcal{T}(\mathcal{F}, \mathcal{X}))$, mapping every $f \in \mathcal{F}$ to a lifting $>^{\tau(f)}$.

Again for the case of fixed-arity signatures, a status function will associate to each function symbol $f \in \mathcal{F}$ a order n lifting $>^{\lambda}$ on $\mathcal{T}(\mathcal{F}, \mathcal{X})^n$, where n is the arity of f.

The following status will be used later in connection with the semantic path order. Let $>$ be a partial order and $\succeq$ a well-founded quasi-order, both defined on $\mathcal{T}(\mathcal{F}, \mathcal{X})$. Write $\succ$ for the strict part of $\succeq$ (i. e., $\succ = \succeq \setminus \preceq$) and $\sim$ for the equivalence relation induced by $\succeq$ (i. e., $\sim = \succeq \cap \preceq$). For each $f \in \mathcal{F}$ the lifting $\tau(f)$ is given by

$$\langle s_1, \ldots, s_k \rangle >^{\tau(f)} \langle t_1, \ldots, t_m \rangle \iff \begin{cases} s \succ t, \text{ or} \\ s \sim t \text{ and } \langle s_1, \ldots, s_k \rangle >^{mul} \langle t_1, \ldots, t_m \rangle \end{cases}$$

for any $k, m \in arity(f)$ and where $>^{mul}$ is the multiset extension of $>$, $s = f(s_1, \ldots, s_k)$ and $t = f(t_1, \ldots, t_m)$. It is not difficult to see that $>^{\tau(f)}$ is indeed a partial order on $\mathcal{T}(\mathcal{F}, \mathcal{X})^*$ and that $>^{\tau(f)}$ respects well-foundedness, being therefore a lifting.

4 Finite signatures

In this section we present one of the main results of this paper. For the sake of simplicity we restrict ourselves to finite signatures. Surprisingly we do not need to fix the arities of the function symbols. Infinite signatures will be treated separately.

4.1 Main result

In the following we consider the set of terms $\mathcal{T}(\mathcal{F}, \mathcal{X})$, over the set of varyadic function symbols $\mathcal{F}$ and such that $\mathcal{F} \cup \mathcal{X}$ is finite.

Recall that a term $t \in \mathcal{T}(\mathcal{F}, \mathcal{X})$ is *well-founded* (with respect to a certain order $>$ on $\mathcal{T}(\mathcal{F}, \mathcal{X})$) if there are no infinite descending chains starting with t.

We introduce some notation.

Definition 3. Let $>$ be a partial order over $\mathcal{T}(\mathcal{F}, \mathcal{X})$ and τ a status function with respect to $>$. We say that $>$ is *decomposable* with respect to τ if $>$ satisfies

- if $f(s_1, \ldots, s_k) > f(t_1, \ldots, t_m)$ then either
  - $\exists 1 \leq i \leq k : s_i \geq f(t_1, \ldots, t_m)$, or
  - $\langle s_1, \ldots, s_k \rangle >^{\tau(f)} \langle t_1, \ldots, t_m \rangle$.

for all $f \in \mathcal{F}$, $k, m \in arity(f)$ and terms $s_1, \ldots, s_k, t_1, \ldots, t_m \in \mathcal{T}(\mathcal{F}, \mathcal{X})$.

We can now present the main result of this section.

Theorem 4. *Let $>$ be a partial order over $\mathcal{T}(\mathcal{F}, \mathcal{X})$ and τ a status function with respect to $>$. Suppose $>$ has the subterm property and is decomposable with respect to τ, then $>$ is well-founded.*

Proof. Suppose that $>$ is not well-founded and take an infinite descending chain $t_0 > t_1 > \cdots > t_n > \cdots$, minimal in the following sense

- $|t_0| \leq |s|$, for all non-well-founded terms s;
- $|t_{i+1}| \leq |s|$, for all non-well-founded terms s such that $t_i > s$.

Note that from the first minimality condition follows that any principal subterm of t_0 is well-founded. Assume that $t_{i+1} = f(u_1, \ldots, u_k)$ and some u_j, with $1 \leq j \leq k$, is not well-founded. From the subterm property and transitivity of $>$, we obtain $t_i > t_{i+1} > u_j$, hence the second minimality condition yields $|t_{i+1}| \leq |u_j|$ which is a contradiction. We conclude that all principal subterms of any term t_i, $i \geq 0$, are well-founded.

Since $\mathcal{F} \cup \mathcal{X}$ is finite, the (infinite) sequence $(t_i)_{i \geq 0}$ must contain a subsequence $(t_{\phi(i)})_{i \geq 0}$ with $t_{\phi(i)} = f(u_{i,1}, \ldots u_{i,n_i})$, for a fixed $f \in \mathcal{F}$. By hypothesis, for each $i \geq 0$, either

- $\exists 1 \leq j \leq n_i : \; u_{i,j} \geq t_{\phi(i+1)}$; or
- $\langle u_{i,1}, \ldots, u_{i,n_i} \rangle >^{\tau(f)} \langle u_{i+1,1}, \ldots, u_{i+1,n_{i+1}} \rangle$.

Since all terms $u_{i,j}$ are well-founded, the first case never occurs. Consequently we have an infinite descending chain

$$\langle u_{0,1}, \ldots, u_{0,n_0} \rangle >^{\tau(f)} \langle u_{1,1}, \ldots, u_{1,n_1} \rangle >^{\tau(f)} \langle u_{2,1}, \ldots, u_{2,n_2} \rangle >^{\tau(f)} \ldots$$

Since $>$ is well-founded over the set $\bigcup_{i \geq 0} (\bigcup_{j=1}^{n_i} \{u_{i,j}\})$, this contradicts the assumption that $\tau(f)$ preserves well-foundedness. □

Theorem 4 provides a way of proving well-foundedness of orders on terms, including orders which are not closed under contexts nor closed under substitutions.

Consider the *recursive path order* with status ([1, 12]) whose definition we present below.

Definition 5. (RPO with status) Let $\rhd$ be a partial order on $\mathcal{F}$ and τ a status function with respect to $>_{rpo}$. Given two terms s, t we say that $s >_{rpo} t$ iff $s = f(s_1, \ldots, s_m)$ and either

1. $t = g(t_1, \ldots, t_n)$ and
   (a) $f \rhd g$ and $s >_{rpo} t_i$, for all $1 \leq i \leq n$, or
   (b) $f = g$, $\langle s_1, \ldots, s_m \rangle >_{rpo}^{\tau(f)} \langle t_1, \ldots, t_n \rangle$ and $s >_{rpo} t_i$, for all $1 \leq i \leq n$; or
2. $\exists 1 \leq i \leq m : \; s_i >_{rpo} t$ or $s_i = t$.

Irreflexivity and transitivity of $>_{rpo}$ are cumbersome but not difficult to check. Well-foundedness of $>_{rpo}$, as defined in definition 5, follows from theorem 4. If we take the definition of $>_{rpo}$ over a precedence that is a quasi-order with the additional condition that each equivalence class of function symbols has one status associated, well-foundedness is still a direct consequence of theorem 4. We remark that by using our definition of lifting and status, definition 4 is a generalization of $>_{rpo}$ orders as found in the literature. With this definition we are able to prove termination of the following TRS (originally from [7]):

$$f(s(x), y, y) \rightarrow f(y, x, s(x))$$

For that we use a lifting given earlier, namely

$$\langle s_1,\ldots,s_m\rangle >^{\lambda} \langle t_1,\ldots,t_n\rangle \iff \begin{cases} (m=n=3) & \text{and} \\ \langle s_1,s_2\rangle >^{mul} \langle t_1,t_2\rangle & \text{or} \\ (\langle s_1,s_2\rangle =^{mul} \langle t_1,t_2\rangle) & \text{and } s_3 > t_3 \end{cases}$$

and then take $>_{rpo}^{\tau(f)} = >_{rpo}^{\lambda}$. Termination of this system cannot be handled by earlier versions of $>_{rpo}$.

In section 6 we shall see that well-foundedness of both *semantic path order* and *general path order* also follow from theorem 4.

4.2 Completeness results

The next result states that the type of term orderings described in theorem 4 covers all terminating TRS's.

Theorem 6. *A TRS R is terminating if and only if there is an order > over $\mathcal{T}(\mathcal{F},\mathcal{X})$ and a status function τ satisfying the following conditions:*

- *> has the subterm property*
- *> is decomposable with respect to τ*
- *if $s \to_R t$ then $s > t$.*

Proof. The "if" part follows from theorem 4: the order > is well-founded and the assumption $\to_R \subseteq >$ implies that R is terminating.

For the "only-if" part we define the relation > on $\mathcal{T}(\mathcal{F},\mathcal{X})$ by:

$$s > t \iff s \neq t \text{ and } \exists C[\,] : s \to_R^* C[t]$$

By definition, the relation > is irreflexive and has the subterm property. Transitivity is checked straightforwardly using termination of R.

We check that > is well-founded. Suppose it is not and let $s_0 > s_1 > \cdots$ be an infinite descending chain. By definition of >, for each $i \geq 0$, we have $s_i \to_R^* C_i[s_{i+1}]$, for some context $C_i[\,]$, so we obtain the infinite chain

$$s_0 \to_R^* C_0[s_1] \to_R^* C_0[C_1[s_2]] \to_R^* \cdots$$

From termination of R, we conclude that there is an index $j \geq 0$ such that

$$s_j = C_j[s_{j+1}] = C_j[C_{j+1}[s_{j+2}]] = \ldots$$

Since the sequence is infinite and $C_k[\,] \neq \Box$ (since $s_k \neq s_{k+1}$), for all $k \geq j$, this is a contradiction.

For each function symbol $f \in \mathcal{F}$ we define $>^{\tau(f)}$ by:

$$\langle u_1,\ldots u_k\rangle >^{\tau(f)} \langle v_1,\ldots,v_m\rangle \iff f(u_1,\ldots,u_k) > f(v_1,\ldots,v_m)$$

for any $k,m \in arity(f)$. Since > is well-founded, we see that $>^{\tau(f)}$ is indeed a lifting.

From the above reasoning follows that all the conditions of theorem 4 are satisfied. Finally if $s \to_R t$, we obviously have $s \to_R^* C[t]$, with C the empty context. Since R is terminating we must have $s \neq t$ and consequently $s > t$. □

An alternative proof of theorem 6 can be given using the fact that a TRS R is terminating if and only if it is compatible with a semantic path order; in the proof of this fact the same order as above is used. Since *spo* fulfils the conditions of theorem 4, as we shall see in section 6, this provides an alternative proof for theorem 6.

The order defined in the proof of theorem 6 has the additional property of being closed under substitutions (but not under contexts). Consequently we also have the following stronger result.

Theorem 7. *A TRS R is terminating if and only if there is an order $>$ over $\mathcal{T}(\mathcal{F},\mathcal{X})$ and a status function τ satisfying the following conditions:*

- *$>$ has the subterm property*
- *$>$ is decomposable with respect to τ*
- *$>$ is closed under substitutions*
- *if $s \rightarrow_R t$ then $s > t$.*

An interesting question raised by J.-P. Jouannaud is what can be said about totality of orders satisfying the conditions of theorem 4. It turns out that totality can very easily be achieved as we now show. However totality is not compatible with closedness under substitutions. First we present a well-known lemma.

Lemma 8. *Any partial well-founded order $>$ on a set A can be extended to a total well-founded order on A.*

Proof. (Sketch). A possible way of proving this result is via Zorn's lemma. Consider K, the set of partial orders $(S, >_S)$ satisfying the following conditions:

1. $S \subseteq A$.
2. if $s, t \in S$ and $s > t$ then $s >_S t$.
3. $>_S$ is total and well-founded in S.
4. if $s > t$ and $t \in S$ then $s \in S$ and $s >_S t$.

We turn K to a partially ordered set itself by defining the order $\sqsubset$ as follows[2]:

$$(S, >_S) \sqsubset (T, >_T) \iff \begin{cases} S \subset T \text{ (as sets)}, >_S \subset >_T, \text{ and} \\ \text{if } s >_T t \text{ and } s \in S \text{ then } t \in S \text{ and } s >_S t \end{cases}$$

It is easy to check that $\sqsubset$ is a partial order on K and that the conditions of Zorn's lemma are satisfied. We therefore establish the existence of a maximal element in K which turns out to be a total well-founded order extending the original one. □

Theorem 9. *A TRS R is terminating if and only if there is an order $>$ over $\mathcal{T}(\mathcal{F},\mathcal{X})$, and a status function τ satisfying the following conditions:*

- *$>$ has the subterm property*

[2] Note that Zorn's lemma cannot be applied with the usual subset ordering since well-foundedness is not preserved under infinite unions.

– $>$ *is decomposable with respect to* τ
– $>$ *is total*
– *if* $s \rightarrow_R t$ *then* $s > t$.

Proof. Again the "if" part follows from theorem 4: the order $>$ is well-founded and the assumption $\rightarrow_R \subseteq >$ implies that R is terminating.

For the "only-if" part we use theorems 4 and 6. Since R is terminating, by theorem 6 there is an order $\gg$ on $\mathcal{T}(\mathcal{F}, \mathcal{X})$ and a status function τ satisfying the conditions of theorem 4 and such that $s \rightarrow_R t \Rightarrow s \gg t$. By theorem 4 the order $\gg$ is well-founded, but not necessarily total. By lemma 8, let $>$ be a total well-founded order extending $\gg$. Since $\gg$ has the subterm property, so does $>$. Furthermore $>$ is also compatible with $\rightarrow_R$, for if $s \rightarrow_R t$ then $s \gg t$ and so $s > t$. In order to apply theorem 4 we still have to define a status function τ for which $>$ is decomposable. For each function symbol $f \in \mathcal{F}$ we define: $\langle u_1, \ldots u_k \rangle >^{\tau(f)} \langle v_1, \ldots, v_m \rangle \iff f(u_1, \ldots, u_k) > f(v_1, \ldots, v_m)$, for any $k, m \in arity(f)$. Since $>$ is well-founded, $>^{\tau(f)}$ is indeed a lifting. Theorem 4 now gives the result. □

The previous result may seem a bit strange since it tells us that we can achieve totality on all terms and not only ground terms. This is so because we do not impose any closure conditions on the order. Note that a total order on $\mathcal{T}(\mathcal{F}, \mathcal{X})$ is never closed under substitutions as long as $\mathcal{X}$ contains more than one element. As for closure under contexts, this property is usually not maintained by naive extensions of the order, it may even make the existence of certain extensions impossible. In our case the conditions imposed are subterm property and compatibility with the reduction relation and so any extension will comply with those conditions whenever the original order does.

5 Infinite Signatures

In the previous section we presented some results which are applicable to orders and TRS's over finite signatures. Here we turn to the infinite case, i. e., we consider the set of terms over an infinite alphabet $\mathcal{F}$, with varyadic function symbols, and an infinite set of variables $\mathcal{X}$. As usual we require that $\mathcal{F} \cap \mathcal{X} = \emptyset$.

We first discuss orders which are based on a precedence on the set of function symbols. Afterwards we will present another simplified approach in which we can dispense with the precedence. This approach is based on a generalization of the notion of lifting.

5.1 Precedence-based orders

It turns out that theorem 4 can also be extended to infinite signatures. We do however need to impose some extra conditions.

We introduce some more notation. Let $\trianglerighteq$ be a quasi-order over $\mathcal{F}$, called a *precedence*. We denote the strict partial order $\trianglerighteq \setminus \trianglelefteq$ by $\triangleright$ and the equivalence relation $\trianglerighteq \cap \trianglelefteq$ by $\sim$.

Definition 10. Given an order $>$ on $\mathcal{T}(\mathcal{F},\mathcal{X})$ and a precedence $\trianglerighteq$ on $\mathcal{F}$, we say that $>$ is *compatible* with $\trianglerighteq$ if whenever $f(s_1,\ldots,s_m) > g(t_1,\ldots,t_n)$ and $g \triangleright f$ then $s_i \geq g(t_1,\ldots,t_n)$, for some $1 \leq i \leq m$.

In theorem 4 we only needed to take into account comparisons between terms with the same head function symbol, but now we also need to consider the comparisons between terms whose head function symbols are equivalent under the precedence considered. As a consequence we need to impose some constraint on the status associated with a function symbol.

Definition 11. Given a precedence $\trianglerighteq$ on $\mathcal{F}$, an order $>$ on $\mathcal{T}(\mathcal{F},\mathcal{X})$ and a status function τ, with respect to $>$, we say that τ and $\trianglerighteq$ are *compatible* if whenever $f \sim g$ then $\tau(f) = \tau(g)$.

As usual a *well quasi-order*, abbreviated to *wqo*, is a quasi-order $\succeq$ such that any extension of it is well-founded. We can now formulate theorem 4 for infinite signatures:

Theorem 12. *Let $\trianglerighteq$ be a precedence on $\mathcal{F}$, $>$ a partial order over $\mathcal{T}(\mathcal{F},\mathcal{X})$, and τ a status function with respect to $>$, such that that both $>$ and $\trianglerighteq$ and τ and $\trianglerighteq$ are compatible. Suppose $>$ has the subterm property and satisfies the following condition:*

- $\forall f, g \in \mathcal{F},\ m \in \text{arity}(f), n \in \text{arity}(g), s_1,\ldots,s_m,t_1,\ldots,t_n \in \mathcal{T}(\mathcal{F},\mathcal{X})$:
  if $f(s_1,\ldots,s_m) > g(t_1,\ldots,t_n)$ *with* $f \sim g$, *then either*
  - $\exists 1 \leq i \leq m : s_i \geq g(t_1,\ldots,t_n)$, *or*
  - $\langle s_1,\ldots,s_m\rangle >^{\tau(f)} \langle t_1,\ldots,t_n\rangle$.

Suppose also that $\trianglerighteq$ is a wqo on $\mathcal{F}\setminus\mathcal{F}_0$ and $>$ is well-founded on $\mathcal{X}\cup\mathcal{F}_0$, where $\mathcal{F}_0 = \{f \in \mathcal{F} : \text{arity}(f) = \{0\}\}$. Then $>$ is well-founded on $\mathcal{T}(\mathcal{F},\mathcal{X})$.

Proof. We proceed, as in proof of theorem 4, by contradiction. First we remark that any infinite descending sequence $(t_i)_{i\geq 0}$ contains an infinite subsequence $(t_{\phi(i)})_{i\geq 0}$ such that $arity(t_{\phi(i)}) \neq \{0\}$, for if that would not be the case, the sequence would contain infinitely many variables or constants, contradicting the fact that $>$ is well-founded on $\mathcal{X}\cup\mathcal{F}_0$.

We take a minimal infinite descending sequence $(t_i)_{i\geq 0}$, in the same sense as in theorem 4. Again, as remarked in the proof of theorem 4, from the minimality of $(t_i)_{i\geq 0}$, the subterm property and transitivity of $>$, it follows that all (principal) subterms of any term t_i, $i \geq 0$, are well-founded.

Let $root(t)$ be the head symbol of the term t. Consider the infinite sequence $(root(t_i))_{i\geq 0}$. From the first observation above it follows that this sequence contains infinitely many terms such that the root function symbol of those terms has arities greater than 0. Consequently and since $\trianglerighteq$ is a *wqo* on $\mathcal{F}\setminus\mathcal{F}_0$, we can conclude that this sequence contains an infinite subsequence $(root(t_{\phi(i)}))_{i\geq 0}$ such that $root(t_{\phi(i+1)}) \trianglerighteq root(t_{\phi(i)})$ and $arity(root(t_{\phi(i)})) \neq \{0\}$, for all i.

Furthermore the infinite sequence $(root(t_i))_{i\geq 0}$ contains no infinite subsequence $(root(t_{\psi(i)}))_{i\geq 0}$ such that $root(t_{\psi(i+1)}) \sim root(t_{\psi(i)})$, for all i. Suppose it is not so and let $(root(t_{\psi(i)}))_{i\geq 0}$ be such a sequence. Since $t_{\psi(i)} > t_{\psi(i+1)}$, by hypothesis we must have

1. $s_{i,k} \geq t_{\psi(i+1)}$, with $s_{i,k}$ a principal subterm of $t_{\psi(i)}$, or
2. $\langle s_{i,1}, \ldots, s_{i,k_{\psi(i)}} \rangle >^{\lambda} \langle s_{i+1,1}, \ldots, s_{i+1,k_{\psi(i+1)}} \rangle$, where $>^{\lambda}$ is the lifting given by the status of $root(t_{\psi(0)})$[3], and $s_{i,1}, \ldots, s_{i,k_{\psi(i)}}$ and $s_{i+1,1}, \ldots, s_{i+1,k_{\psi(i+1)}}$ are the principal subterms of respectively $t_{\psi(i)}$ and $t_{\psi(i+1)}$, for all i.

Due to the minimality of $(t_i)_{i\geq 0}$ and the subterm property, case 1 above can never occur. Therefore we have an infinite descending sequence

$$\langle s_{0,1}, \ldots, s_{0,k_{\psi(0)}} \rangle >^{\lambda} \langle s_{1,1}, \ldots, s_{1,k_{\psi(1)}} \rangle >^{\lambda} \langle s_{2,1}, \ldots, s_{2,k_{\psi(2)}} \rangle >^{\lambda} \ldots$$

Since $>$ is well-founded on $\bigcup_{i\geq 0} \bigcup_{j=1}^{k_{\psi(i)}} \{s_{i,j}\}$, this contradicts the definition of lifting.

Therefore, and without loss of generality, we can state that the infinite subsequence $(root(t_{\phi(i)}))_{i\geq 0}$ has the additional property $root(t_{\phi(i+1)}) \rhd root(t_{\phi(i)})$, for all i.[4] Since $t_{\phi(i)} > t_{\phi(i+1)}$ and $>$ is compatible with $\unrhd$, we must have $u \geq t_{\phi(i+1)}$, for some principal subterm u of $t_{\phi(i)}$, contradicting the minimality of $(t_i)_{i\geq 0}$. □

Some remarks are in order. Since there are no substitutions involved, there is no essential difference between elements of $\mathcal{X}$ and $\mathcal{F}_0$. The condition stating that $>$ is well-founded on $\mathcal{X}$ is imposed to disallow the bizarre case where we can have an infinite descending sequence constituted solely by variables. Usually (e. g. in Kruskal's theorem) it is required that the precedence $\unrhd$ be a *wqo* over $\mathcal{F}$, we can however relax that condition to $\unrhd$ being a *wqo* over $\mathcal{F} \setminus \mathcal{F}_0$ provided $>$ is also well-founded on $\mathcal{F}_0$. This is weaker than requiring that $\unrhd$ be a *wqo* on $\mathcal{F}$. The *wqo* requirement cannot be weakened to well-foundedness as the following example shows. Consider $\mathcal{F} = \{f_i | i \geq 0\}$ with $arity(f_i) = \{1\}$, for all $i \geq 0$. Let $>$ be an order on $\mathcal{T}(\mathcal{F}, \mathcal{X})$ with the subterm property and such that

$$f_0(x) > f_1(x) > f_2(x) > \ldots$$

Take $\unrhd$ to be the empty precedence. Obviously $\unrhd$ is well-founded and all the other conditions of theorem 12 are satisfied, however the order $>$ is not well-founded.

If we remove the condition "$>$ is well-founded on $\mathcal{X} \cup \mathcal{F}_0$", and strengthen the condition on $\unrhd$ to "$\unrhd$ is a *wqo* on $\mathcal{F} \cup \mathcal{X}$", then the same statement as above can be proved (and the proof is very similar). In this case and for finite signatures, theorem 4 is a direct consequence of theorem 12, since the discrete order is a *wqo* and the compatibility conditions are trivially fulfilled.

Theorem 12 holds in particular for precedences that are *partial well-orders (pwo's)*. In this case we only need to compare terms with the same root function symbol and the compatibility condition of definition 11 is trivially verified.

---

[3] Recall that for equivalent function symbols, their status coincides.

[4] Strictly speaking, an infinite subsequence of this sequence has that property.

As in the finite case, well-foundedness of orders as *rpo* over infinite signatures, is a consequence of theorem 12. For that we only need to extend the well-founded precedence to a total well-founded one, maintaining the equivalence part the same, which is then a *wqo*. All the other conditions also hold, so the theorem can be applied.

If it is the case that $\mathcal{F}$ is finite but we allow $\mathcal{X}$ to be an infinite set, the conditions imposed on the order on theorem 4 are not enough to guarantee that the order is well-founded: any non-well-founded order defined only in $\mathcal{X}$ is a counter-example. However, in the presence of an infinite set of variables, well-foundedness of $>$ on $\mathcal{T}(\mathcal{F},\mathcal{X})$ is equivalent to well-foundedness of $>$ on $\mathcal{X}$, i. e., theorem 4 can be rewritten as:

Theorem 13. *Let $>$ be a partial order over $\mathcal{T}(\mathcal{F},\mathcal{X})$ and τ a status function with respect to $>$. Suppose $>$ has the subterm property and is decomposable with respect to τ. Then $>$ is well-founded on $\mathcal{T}(\mathcal{F},\mathcal{X})$ if and only if $>$ is well-founded on $\mathcal{X}$.*

One direction is trivial, the other is a consequence of theorem 12, by taking an empty precedence. Note that for $\mathcal{F}$ finite and $\mathcal{X}$ infinite, theorems 6, 7 and 9, hold under the additional assumption that the order considered is well-founded when restricted to $\mathcal{X}$.

Another interesting result arises if we relax the requirements on the precedence and strengthen the ones on the order.

Theorem 14. *Let $\trianglerighteq$ be a well-founded precedence on $\mathcal{F}$, $>$ a partial order over $\mathcal{T}(\mathcal{F},\mathcal{X})$, and τ a status function with respect to $>$, such that that τ and $\trianglerighteq$ are compatible. Suppose $>$ has the subterm property and satisfies the following condition:*

- $\forall f,g \in \mathcal{F},\ m \in \text{arity}(f), n \in \text{arity}(g), s_1,\ldots,s_m,t_1,\ldots,t_n \in \mathcal{T}(\mathcal{F},\mathcal{X})$: *if* $f(s_1,\ldots,s_m) > g(t_1,\ldots,t_n)$ *then either*
  - $\exists 1 \le i \le m : s_i \ge g(t_1,\ldots,t_n)$, *or*
  - $f \triangleright g$, *or*
  - $f \sim g$ *and* $\langle s_1,\ldots,s_m\rangle >^{\tau(f)} \langle t_1,\ldots,t_n\rangle$.

Then $>$ is well-founded on $\mathcal{T}(\mathcal{F},\mathcal{X})$.

The proof is very similar to the proof of theorem 12, therefore we omit it. Note that well-foundedness of *rpo*, for an arbitrary well-founded precedence, is a direct consequence of this result. In the "classical" approach, first the precedence has to be extended via lemma 8 to a well-founded total precedence, maintaining the equivalence part, before Kruskal's theorem yields the desired result.

It would also be interesting to have a theorem similar to theorem 6 for the case of infinite signatures. However for infinite signatures the empty relation is not a *wqo* any longer and it is not clear how to choose an appropriate *wqo*. A possibility is to take $\trianglerighteq$ defined by $f \sim g$ for any $f,g \in \mathcal{F}$, which is trivially a *wqo*, however this choice will not always work as the following example shows. Consider the infinite terminating TRS given by

$$a_i \rightarrow a_j$$

for any $i \geq 0$ and any $0 \leq j < i$ and where each a_i is a constant. Then any order compatible with R will never be compatible with a precedence in which $a_i \sim a_j$, for all $i, j \geq 0$.

Another alternative is to take a total well-founded order on $\mathcal{F}$, again by definition a *wqo*, but then other compatibility problems arise. Just consider the rule

$$a \rightarrow f(0)$$

If we choose the precedence as an arbitrary total well-founded order on $\mathcal{F}$, we may have $f \rhd a$, and the conditions of theorem 12 will never hold.

5.2 Generalizing liftings on orders

The decomposability restriction $\langle s_1, \ldots, s_m \rangle >^{\tau(f)} \langle t_1, \ldots, t_n \rangle$ has the inconvenience of forgetting about the root symbols of the terms compared. In the case of finite signatures, that is irrelevant since we only need to compare terms with the same head symbol and the symbol can be encoded in the status τ. For infinite signatures however, that information is essential, since given an infinite sequence of terms we no longer have the guarantee that it contains an infinite subsequence of terms having the same root symbol. As a consequence we need to impose some strong conditions both on the set of function symbols and on the status and order used. A way of relaxing these conditions is by remembering the information lost with the decomposition and this can be achieved by changing the definition of lifting.

In this section we present another condition for well-foundedness on term orderings. Now we do not require the existence of an order or quasi-order on the set of function symbols $\mathcal{F}$. Instead we will use a different definition of lifting for orderings on terms.

Definition 15. Let $(\mathcal{T}(\mathcal{F}, \mathcal{X}), >)$ be a partial ordered set of terms. We define a *term lifting* to be a partial order $>^{\Lambda}$ on $\mathcal{T}(\mathcal{F}, \mathcal{X})$ for which the following holds: for every $A \subseteq \mathcal{T}(\mathcal{F}, \mathcal{X})$, if $>$ restricted to A is well-founded, then $>^{\Lambda}$ restricted to $\bar{A}$ is also well-founded, where

$$\bar{A} = \{f(t_1, \ldots, t_n) : f \in \mathcal{F},\ n \in \mathit{arity}(f),\ \text{and } t_i \in A, \text{ for all } i, 0 \leq i \leq n\}$$

We use the notation $\Lambda(>)$ to denote all possible term liftings of $>$ on $\mathcal{T}(\mathcal{F}, \mathcal{X})$.

We remark that term liftings can make use of liftings and status functions since the well-foundedness requirement is preserved. Given an order $>$ on $\mathcal{T}(\mathcal{F}, \mathcal{X})$, every lifting in the sense of definition 1 induces a term lifting of the same order as follows:

$$f(s_1, \ldots, s_m) >^{\Lambda} g(t_1, \ldots, t_n) \iff \langle s_1, \ldots, s_m \rangle >^{\lambda} \langle t_1, \ldots, t_n \rangle$$

We present a new well-foundedness criterion.

Theorem 16. *Let $>$ be a partial order on $\mathcal{T}(\mathcal{F},\mathcal{X})$ and let $>^{\Lambda}$ be a term lifting of $>$. Suppose $>$ has the subterm property and satisfies the following condition:*

- $\forall f,g \in \mathcal{F},\ m \in \text{arity}(f), n \in \text{arity}(g), s_1,\dots,s_m,t_1,\dots,t_n \in \mathcal{T}(\mathcal{F},\mathcal{X})$: *if* $s = f(s_1,\dots,s_m) > g(t_1,\dots,t_n) = t$ *then either*
  - $\exists 1 \leq i \leq m : s_i \geq g(t_1,\dots,t_n)$, *or*
  - $s >^{\Lambda} t$

Then $>$ is well-founded on $\mathcal{T}(\mathcal{F},\mathcal{X})$.

Proof. Suppose that $>$ is not well-founded and take an infinite descending chain $t_0 > t_1 > \cdots > t_n > \cdots$, minimal in the same sense as in the proof of theorem 4, i. e.,

- $|t_0| \leq |s|$, for all non-well-founded terms s;
- $|t_{i+1}| \leq |s|$, for all non-well-founded terms s such that $t_i > s$.

As remarked in the proof of theorem 4, from the minimality of $(t_i)_{i\geq 0}$, the subterm property and transitivity of $>$, it follows that all principal subterms of any term t_i, $i \geq 0$, are well-founded.

Since $t_i > t_{i+1}$, for all $i \geq 0$, we must have

1. $u_i \geq t_{i+1}$, for some principal subterm u_i of t_i, or
2. $t_i >^{\Lambda} t_{i+1}$

Due to the minimality of the sequence, the first case above can never occur. Therefore we have an infinite descending chain

$$t_0 >^{\Lambda} t_1 >^{\Lambda} t_2 >^{\Lambda} \dots$$

But due also to minimality, the order $>$ is well-founded over the set of terms $A = \{u : u \text{ is a principal subterm of } t_i, \text{ for some } i \geq 0\}$. By definition of term lifting we have that $>^{\Lambda}$ is well-founded over

$$\bar{A} = \{f(u_1,\dots,u_k) : f \in \mathcal{F}, k \in arity(f) \text{ and } u_i \in A, \text{ for all } 1 \leq i \leq k\}$$

and since $\{t_i : i \geq 0\} \subseteq \bar{A}$, we get a contradiction. □

It is interesting to remark that theorem 4 is a consequence of theorem 16. To see that we define the following order $\gg$:

$$s \gg t \iff (root(s) = root(t)) \text{ and } (s > t)$$

Now we define the following term lifting

$$f(s_1,\dots,s_m) \gg^{\Lambda} g(t_1,\dots,t_n) \iff (f = g) \text{ and } \langle s_1,\dots,s_m\rangle >^{\tau(f)} \langle t_1,\dots,t_n\rangle$$

where $>^{\tau(f)}$ is the lifting associated by the status function τ to the function symbol f. It is not difficult to see that since the lifting $>^{\tau(f)}$ respects well-foundedness of $>$, $\gg^{\Lambda}$ is a well-defined term lifting. Now theorem 16 gives well-foundedness of $\gg$. But since non-well-foundedness of $>$ would imply non-well-foundedness of $\gg$ (by an argument similar to the proof of theorem 4), we are done.

Furthermore when $\mathcal{F}$ is finite, theorem 16 is also a consequence of theorem 4 (i. e., they are equivalent). For that we define the status

$$\langle s_1, \ldots, s_m \rangle >^{\tau(f)} \langle t_1, \ldots, t_n \rangle \iff f(s_1, \ldots, s_m) >^{A} f(t_1, \ldots, t_n)$$

It is now not difficult to check that the other implication holds.

Due to the required existence of a partial order on the set of function symbols, the relation of this theorem with theorems 12 and 14 is not yet clear.

An important consequence of the use of term liftings is that we manage to recover the completeness results stated in section 4.2 and that we could not state for precedence-based orders.

Theorem 17. *Let R be a TRS over an infinite varyadic signature. Then R is terminating if and only if there is an order $>$ over $\mathcal{T}(\mathcal{F}, \mathcal{X})$ and a term lifting $>^{A}$ satisfying the following conditions:*

- *$>$ has the subterm property (and $>$ is closed under substitutions)*
- *$\forall f, g \in \mathcal{F},\ m \in \text{arity}(f), n \in \text{arity}(g), s_1, \ldots, s_m, t_1, \ldots, t_n \in \mathcal{T}(\mathcal{F}, \mathcal{X})$:* *if $s = f(s_1, \ldots, s_m) > g(t_1, \ldots, t_n) = t$ then either*
  - *$\exists 1 \leq i \leq m : s_i \geq g(t_1, \ldots, t_n)$, or*
  - *$s >^{A} t$*
- *if $s \rightarrow_R t$ then $s > t$.*

Proof. Sketch. The "if" part follows from theorem 16: the order $>$ is well-founded and the assumption $\rightarrow_R \subseteq >$ implies that R is terminating.

For the "only-if" part the proof is similar to the proof of theorem 6. We define again the relation $>$ on $\mathcal{T}(\mathcal{F}, \mathcal{X})$: $s > t \iff s \neq t$ and $\exists C[\,] : s \rightarrow^{*}_{R} C[t]$. The only different part is the definition of term lifting. Since the order $>$ is well-founded we can use it as the term lifting itself. □

As for the finite case the completeness result concerning totality also holds and the proof is very similar, so we omit it.

Theorem 18. *Let R be a TRS over an infinite varyadic signature. Then R is terminating if and only if there is an order $>$ over $\mathcal{T}(\mathcal{F}, \mathcal{X})$ and a term lifting $>^{A}$ satisfying the following conditions:*

- *$>$ has the subterm property*
- *$\forall f, g \in \mathcal{F},\ m \in \text{arity}(f), n \in \text{arity}(g), s_1, \ldots, s_m, t_1, \ldots, t_n \in \mathcal{T}(\mathcal{F}, \mathcal{X})$:* *if $s = f(s_1, \ldots, s_m) > g(t_1, \ldots, t_n) = t$ then either*
  - *$\exists 1 \leq i \leq m : s_i \geq g(t_1, \ldots, t_n)$, or*
  - *$s >^{A} t$*
- *$>$ is total*
- *if $s \rightarrow_R t$ then $s > t$.*

6 Semantic Path Order and General Path Order

In this section we show how well-foundedness of *semantic path order* [9] and *general path order* [3] can be derived using either theorem 4 or theorem 16.

Definition 19. **(Semantic Path Order)** Let $\geq$ be a well-founded quasi-order on $\mathcal{T}(\mathcal{F})$. The *semantic path order* $\succeq_{spo}$ is defined on $\mathcal{T}(\mathcal{F})$ as follows:
$s = f(s_1, \ldots, s_m) \succeq_{spo} g(t_1, \ldots, t_n) = t$ if either

1. $s > t$ and $s \succ_{spo} t_i$, for all $1 \leq i \leq n$, or
2. $s \sim t$ and $s \succ_{spo} t_i$, for all $1 \leq i \leq n$ and $\langle s_1, \ldots, s_m \rangle \succeq_{spo}^{mul} \langle t_1, \ldots, t_n \rangle$, where $\succeq_{spo}^{mul}$ is the multiset extension of $\succeq_{spo}$, or
3. $\exists i \in \{1, \ldots, m\} : s_i \succeq_{spo} t$.

It can be seen that the $\succ_{spo}$ has the subterm property and is in general not monotonic.

In the case the alphabet we consider is finite, define the following status. Let $\succeq$ be the well-founded quasi-order used in the definition of $\succ_{spo}$. For each $f \in \mathcal{F}$ the lifting $\tau(f)$ is given by

$$\langle s_1, \ldots, s_k \rangle \succ_{spo}^{\tau(f)} \langle t_1, \ldots, t_m \rangle \iff \begin{cases} s \succ t, \text{ or} \\ s \sim t \text{ and } \langle s_1, \ldots, s_k \rangle \succ_{spo}^{mul} \langle t_1, \ldots, t_m \rangle \end{cases}$$

for any $k, m \in arity(f)$ and where $\succ_{spo}^{mul}$ is the multiset extension of $\succ_{spo}$, $s = f(s_1, \ldots, s_k)$ and $t = f(t_1, \ldots, t_m)$. It is not difficult to see that $\succ_{spo}^{\tau(f)}$ is indeed a partial order on $\mathcal{T}(\mathcal{F}, \mathcal{X})^*$ and that $\succ_{spo}^{\tau(f)}$ respects well-foundedness, being therefore a lifting. Since $\succ_{spo}$ has the subterm property and satisfies the other conditions of theorem 4, its well-foundedness follows from application of the theorem.

For the case we consider an infinite signature, we define the following term lifting: for $s = f(s_1, \ldots, s_m)$ and $t = g(t_1, \ldots, t_n)$

$$s \succ_{spo}^{\Lambda} t \iff \begin{cases} (s \succ t) & \text{or} \\ (s \sim t) & \text{and } \langle s_1, \ldots, s_m \rangle \succ_{spo}^{mul} \langle t_1, \ldots, t_n \rangle \end{cases}$$

where again $\succeq$ is the well-founded quasi-order used in the definition of $\succ_{spo}$. Since $\succ$ is well-founded and the multiset extension respects well-foundedness, $\succ_{spo}^{\Lambda}$ is indeed a term lifting. Using this term lifting, we can apply theorem 16 to conclude that $\succ_{spo}$ is well-founded.

The *general path order*, that we denote by $\succeq_{gpo}$, was introduced in [3]. We present the definition and show how well-foundedness of this order can be derived from theorem 4 or theorem 16.

Definition 20. A *termination function* θ is a function defined on the set of terms $\mathcal{T}(\mathcal{F}, \mathcal{X})$ and is either

1. a homomorphism from terms to a set S such that

$$\theta(f(s_1, \ldots, s_n)) = f_\theta(\theta(s_1), \ldots, \theta(s_n))$$

2. an extraction function that given a term associates to it a multiset of principal subterms, i. e.,

$$\theta(f(s_1,\ldots,s_n)) = [s_{i_1},\ldots,s_{i_k}]$$

where $i_1,\ldots,i_k \in \{1,\ldots,n\}$.

Definition 21. A *component order* $\phi = \langle\theta,\geq\rangle$ consists of a termination function defined on the set $\mathcal{T}(\mathcal{F})$ of ground terms, along with an associated well-founded quasi-order $\geq$ (defined on the codomain of θ).

Definition 22. (General Path Order) Let $\phi_i = \langle\theta_i,\geq_i\rangle$, with $0 \leq i \leq k$, be component orders, such that if θ_j is an extraction function then $\geq_j$ is the multiset extension of the general path order itself. The induced *general path order* $\succeq_{gpo}$ is defined on $\mathcal{T}(\mathcal{F})$ as follows: $s = f(s_1,\ldots,s_m) \succ_{gpo} g(t_1,\ldots,t_n) = t$ if either

1. $\exists i \in \{1,\ldots,m\} : s_i \succeq_{gpo} t$ or
2. $s \succ_{gpo} t_j$, for all $1 \leq j \leq n$, and $\Theta(s) >_{lex} \Theta(t)$, where $\Theta = \langle\theta_0,\ldots,\theta_k\rangle$ and $>_{lex}$ is the lexicographic combination of the component orderings θ_i with $0 \leq i \leq k$.

The equivalence part is defined as: $s = f(s_1,\ldots,s_m) \sim_{gpo} g(t_1,\ldots,t_n) = t$ if $s \succ_{gpo} t_j$, for all $1 \leq j \leq n$, and $t \succ_{gpo} s_j$, for all $1 \leq j \leq m$, and $\theta_i(s) \sim_i \theta_i(t)$, for all $0 \leq i \leq k$, and where $\sim_i$ is the equivalence contained in $\geq_i$.

It is known ([3]) that $\succeq_{gpo}$ is a quasi-order with the subterm property.

Well-foundedness of $\succeq_{gpo}$ is a consequence of the results previously presented. For the case of finite signatures we define the following status

$$\langle s_1,\ldots,s_m\rangle \succ_{gpo}^{\tau(f)} \langle t_1,\ldots,t_n\rangle \iff \Theta(f(s_1,\ldots,s_m)) >_{lex} \Theta(f(t_1,\ldots,t_k))$$

where as in definition 22, $\Theta(v) = \langle\theta_0(v),\ldots,\theta_k(v)\rangle$ and $>_{lex}$ is the lexicographic combination of the component orderings θ_i with $0 \leq i \leq k$. If θ_i is an homomorphism to a well-founded set, then θ_i is obviously a lifting, and if θ_i is a multiset extracting function, since the multiset construction preserves well-foundedness, we also have that θ_i is a lifting. Finally the finite lexicographic composition of liftings is still a lifting. As a consequence $\succ_{gpo}^{\tau(f)}$ is a well-defined status, and since $\succ_{gpo}$ has the subterm property and satisfies the other conditions of theorem 4, we can apply this result to conclude $\succ_{gpo}$ is well-founded.

For infinite signatures, well-foundedness of $\succ_{gpo}$ is a consequence of theorem 16. If we define the term lifting $\succ_{gpo}^{A}$ as Θ, we see that $\succ_{gpo}^{A}$ is indeed well-defined. Since the other conditions of theorem 16 are satisfied, we can apply it to conclude well-foundedness of $\succ_{gpo}$. Finally it is interesting to remark that if we allow the termination function to be not only a multiset extraction function but an arbitrary lifting, we obtain a generalization of $\succ_{gpo}$ whose well-foundedness can still be derived from the results presented.[5]

[5] For other similar generalization of $\succ_{gpo}$ see [8].

7 Conclusions

We presented some criteria for proving well-foundedness of orders on terms. Our approach was inspired by Kruskal's theorem but is simpler. Kruskal's theorem (and extensions as the one in [11]) is a stronger result in the sense that it establishes that a certain order is a *well-quasi-order* (or *partial-well-order*). Our result allows to conclude well-foundedness directly. However the essential difference is the domain of application: Kruskal's theorem implies well-foundedness of orders extending any monotonic order with the subterm property, hence only covers simplification orders and it is well-known that those orders do not cover all terminating TRS's. Our criteria do not require monotonicity and as a consequence, cover all terminating TRS's.

For infinite signatures we managed to present a well-foundedness criterion even simpler and the completeness results still hold.

References

1. DERSHOWITZ, N. Orderings for term rewriting systems. *Theoretical Computer Science 17*, 3 (1982), 279–301.
2. DERSHOWITZ, N. Termination of rewriting. *Journal of Symbolic Computation 3*, 1 and 2 (1987), 69–116.
3. DERSHOWITZ, N., AND HOOT, C. Topics in termination. In *Proceedings of the 5th Conference on Rewriting Techniques and Applications* (1993), C. Kirchner, Ed., vol. 690 of *Lecture Notes in Computer Science*, Springer, pp. 198–212.
4. DERSHOWITZ, N., AND JOUANNAUD, J.-P. Rewrite systems. In *Handbook of Theoretical Computer Science*, J. van Leeuwen, Ed., vol. B. Elsevier, 1990, ch. 6, pp. 243–320.
5. FERREIRA, M. C. F., AND ZANTEMA, H. Syntactical analysis of total termination. In *Proceedings of the 4th International Conference on Algebraic and Logic Programming* (1994), G. Levi and M. Rodríguez Artalejo, Eds., vol. 850 of *Lecture Notes in Computer Science*, Springer, pp. 204–222.
6. GALLIER, J. H. What's so special about Kruskal's theorem and the ordinal Γ_0? A survey of some results in proof theory. *Annals of Pure and Applied Logic 53* (1991), 199–260.
7. GEERLING, M. Termination of term rewriting systems. Master's thesis, Utrecht University, 1991.
8. GESER, A. An improved general path order. Tech. Rep. MIP-9407, University of Passau, 1994.
9. KAMIN, S., AND LÉVY, J. J. Two generalizations of the recursive path ordering. University of Illinois, 1980.
10. NASH-WILLIAMS, C. S. J. A. On well-quasi ordering finite trees. *Proc. Cambridge Phil. Soc. 59* (1963), 833–835.
11. PUEL, L. Using unavoidable sets of trees to generalize Kruskal's theorem. *Journal of Symbolic Computation 8* (1989), 335–382.
12. STEINBACH, J. Extensions and comparison of simplification orderings. In *Proceedings of the 3rd Conference on Rewriting Techniques an Applications* (1989), N. Dershowitz, Ed., vol. 355 of *Lecture Notes in Computer Science*, Springer, pp. 434–448.

A New Characterisation of AC-Termination and Application

Jean-Michel Gélis

LRI, bâtiment 490, CNRS URA 410
Université de Paris 11
F91405 Orsay cedex, France
mail : gelis@lri.fr

1 Introduction

The work we report on in this paper originates from the APLUSIX project [Nicaud and al. 93]. In this project, we have developed an Intelligent Tutoring System that aims at teaching high school students polynomial factorization. This system implements rewriting modulo an equivalence relation ≡ with factorization, grouping, collapsing and development rules. For example, rewriting modulo ≡ of $x+4x(90x^2-40)$ with the factorization rule $A^2-B^2 \rightarrow (A+B)(A-B)$:

$$x+4x(90x^2-40) \equiv x+4x*10[(3x)^2-2^2]$$
$$\rightarrow \quad x+4x*10[(3x+2)(3x-2)] \equiv x+40x(3x+2)(3x-2),$$

is implemented by a production rule first determinating the relevant equivalent expression $x+4x*10[(3x)^2-2^2]$ (by means of a concept *square*) and then applying the rewriting rule.

We think intuitively that factorization, grouping and collapsing rules are terminating modulo ≡. To prove this theoretical result, we first reduce rewriting modulo ≡ to another rewriting modulo AC in a set of *interpreted terms*. For instance, the above rewriting is reduced to:

$$x+x(xx+k) =_{AC} x+x(xx+k) \rightarrow x+x[(x+k)(x+k)] =_{AC} x+x(x+k)(x+k),$$

using the factorization rule: $xx+k \rightarrow (x+k)(x+k)$.

To prove AC-termination of factorization, grouping and collapsing rules in the set of interpreted terms, we first try, unsuccessfully, polynomial interpretations [Dershowitz 87] and associative path' orderings [Bachmair and Plaisted 85]. Other methods are required. We try to use Dershowitz' theorem [87] and establish another form well-suited to our application.

We think that such an *actual* example is likely to interest the rewriting community. New termination orderings can perhaps be inferred from the deep understanding of this existing system and from the impossibility of applying some well-known termination orderings.

In this paper, we define our rewriting system R including an infinite number of rules. We define a sub-system R_6 including only 6 rules of the R system. We establish that AC-termination of R_6 cannot be proved with a polynomial interpretation nor with an associative path ordering nor with the reduction ordering constructed by Narendran and Rusinowitch [91]. Finally, we prove the AC-termination of the R system by means of another form of Dershowitz' theorem [87].

2 Two Forms of an AC-Termination Theorem

2.1. Theorem 1

Dershowitz [87] proved the following theorem. We denote by $\bar{t}$ the flattened form of t.

Theorem 1. *Let $\mathcal{R}$ be a rewrite system over some set $\mathcal{T}$ of terms and $\mathcal{F}$ a set of associative-commutative symbols. The rewrite relation $\mathcal{R}/AC$ is terminating if, and only if, it fulfils the following conditions (H_1):*

There exists a well-founded ordering > on flattened terms in $\bar{\mathcal{T}}$ such that:

(1) $\overline{l\sigma} > \overline{r\sigma}$ for each rule $l \to r$ of $\mathcal{R}$ and for any substitution σ in $\mathcal{T}$ for the variables of the rule;

(2) $\overline{f(l\sigma, N\sigma)} > \overline{f(r\sigma, N\sigma)}$
for each rule $l \to r$ of $\mathcal{R}$ whose left-hand-side l or right-hand side r has associative-commutative root symbol $f \in \mathcal{F}$ or for which r is just a variable and
for any substitution σ in $\mathcal{T}$ for the variables of the rule and for N (a new variable, otherwise not occurring in the rule);

(3) $\bar{u} \overset{\mathcal{R}/AC}{\longrightarrow} \bar{v}$ and $\bar{u} > \bar{v}$ imply $\overline{h(\ldots u \ldots)} > \overline{h(\ldots v \ldots)}$
for all terms u and v in $\mathcal{T}$ and $\overline{h(\ldots u \ldots)}$ and $\overline{h(\ldots v \ldots)}$ in $\bar{\mathcal{T}}$.

2.2. Theorem 2

Theorem 2 is obtained from Theorem 1 by replacing the third strict monotonicity condition by a larger one and a sub-term property.

Theorem 2. *Let $\mathcal{R}$ be a rewrite system over some set $\mathcal{T}$ of terms and $\mathcal{F}$ a set of associative-commutative symbols. The rewrite relation $\mathcal{R}/AC$ is terminating if, and only if, it fulfils the following conditions (H_2):*

There exists a well-founded ordering > on flattened terms in $\bar{\mathcal{T}}$ such that:

(1) $\overline{l\sigma} > \overline{r\sigma}$ for each rule $l \to r$ of $\mathcal{R}$ and for any substitution σ in $\mathcal{T}$ for the variables of the rule;

(2) $\overline{f(l\sigma, N\sigma)} > \overline{f(r\sigma, N\sigma)}$
for each rule $l \to r$ of $\mathcal{R}$ whose left-hand-side l or right-hand side r has associative-commutative root symbol $f \in \mathcal{F}$ or for which r is just a variable and
for any substitution σ in $\mathcal{T}$ for the variables of the rule and for N (a new variable, otherwise not occurring in the rule);

(3) $\bar{u} \overset{\mathcal{R}/AC}{\longrightarrow} \bar{v}$ and $\bar{u} \geq \bar{v}$ imply $\overline{h(\ldots u \ldots)} \geq \overline{h(\ldots v \ldots)}$
for all terms u and v in $\mathcal{T}$ and $\overline{h(\ldots u \ldots)}$ and $\overline{h(\ldots v \ldots)}$ in $\bar{\mathcal{T}}$.

(4) $\overline{h(\ldots u \ldots)} \geq \bar{u}$ for all terms u in $\mathcal{T}$ and $\overline{h(\ldots u \ldots)}$ in $\bar{\mathcal{T}}$.

The proof of Theorem 2 relies on the following result:

Property 1. *The conditions (H_1) and (H_2) are equivalent.*

PROOF SKETCH: We denote by $>$ and $>$ the ordering relations involved respectively in (H_1) and (H_2). The terms below u, v and w are supposed to be flattened.

(1) Let us suppose that (H_1) is true. If u is AC-rewritten to v, the three (H_1) conditions imply that $u > v$ (we call this implication (I)). We denote $>_{st}$ the strict subterm relation, i.e. $u >_{st} v$ if and only if there exists a context w=h(...) such that u=h(...v...). Let us consider the relation $>_{st} \cup >$. The $>_{st} \cup >$ relation is irreflexive and transitive. Moreover, $>$ commutes over $>_{st}$ because if we suppose that we have $u >_{st} v > w$, then there exists a context h(...) such that u=h(...v...). Therefore, as $v > w$, part (3) of (H_1) implies that h(...v...) $>$ h(...w..). Consequently, we have u=h(...v...) $>$ h(...w..) $>_{st}$ w and it follows that $>$ commutes over $>_{st}$. As $>$ commutes over $>_{st}$ and as both $>$ and $>_{st}$ are terminating, it follows [Dershowitz 87] that $>_{st} \cup >$ is also terminating. Let us denote by $>$ the $>_{st} \cup >$ ordering. As $>$ fulfils the (H_1) conditions, it follows that $>$ fulfils the (H_2) conditions. For instance, the third (H_2) condition is verified because since u is AC-rewritten to v, $u > v$ (because of the implication (I)), and therefore h(...u...) $>$ h(...v...) (because of the third (H_1) condition). Hence h(...u...) $\geq$ h(...v...).

(2) Let us suppose that (H_2) is true. For any term t, we define M(t) as the multiset of all strict subterms of the flattened form $\overline{t}$ of t. We define the $>$ ordering by $u > v$ if and only if M(u) >> M(v) where >> is the multiset ordering induced from $>$. The $>$ ordering fulfils the (H_1) conditions.

Theorem 2 is another form of Theorem 1 that is useful when strict monotonicity is hard to establish.

3 A Rewriting Rules System and a Sub-System

In this paragraph, we define our rewrite system R whose AC-termination can be proved by Theorem 2 and cannot be proved with an associative path ordering nor with a polynomial interpretation nor with the reduction ordering constructed by Narendran and Rusinowitch [91].

More precisely, terms are defined over a set including constants (rewriting variables are considered as constants) and two associative and commutative functions denoted + and *. We define the infinite set R of rewriting rules over the set of terms and a subset of 6 rules we name R_6. We prove that:

(i) R is AC-terminating, using Theorem 2. *It implies that R_6 is also AC-terminating.*

(ii) AC-termination of R_6 cannot be proved using any associative path ordering or any polynomial interpretation or the Narendran and Rusinowitch's reduction ordering [91].

3.1. Terms

Let V be a countable set of rewriting variables named x, y, z... and C a countable set of other constants denoted a, b, c... We use F for the function set comprising any constant of V and C, and the two associative and commutative symbols + and *. The corresponding set of terms is denoted T.

3.2. The Rewrite System R_6

The system R_6 includes six rules. Multiplication is implicit in this formulation, e.g. we denote xx+yy by (x*x)+(y*y).

Definition 1. *We call the set of the following six rules R_6:*

(1)	$xx+yy$	$\rightarrow$	$(x+y)(x+y)$,	(4)	xyz	$\rightarrow$	$xy+xz+yz$,
(2)	$x+xxx+yyy$	$\rightarrow$	$(x+x+y)(x+x+y)(x+x+y)$,	(5)	xy	$\rightarrow$	x,
(3)	$xy+xz$	$\rightarrow$	$x(y+z)$,	(6)	xy	$\rightarrow$	$x+y$.

3.3. Degree and Multisets of Degrees

The definition of the rule system R requires the definition of degree and of two "degree" multisets.

We define the notion of *degree* on terms in a similar way one defines the degree of polynomials.

Definition 2. *Let t be a term. The degree of t is the non-negative integer defined by:*

$$deg(\sum_{i=1}^{n} a_i) = \max_{i\in\{1...n\}} (a_i); \quad deg(\prod_{i=1}^{n} a_i) = \sum_{i=1}^{n} deg(a_i); \quad deg(a)=1 \text{ if } a \text{ is a constant.}$$

For instance : deg((ab+c)(a+b))=3 and deg(a(a+b)+c)=2.
Note that the given definition implies that for any term t: deg(t)≠0.

We define also two multisets of degrees on terms.

Definition 3. *Let t be a term. The* additive multiset *of degrees of t is the multiset defined by:*

$$m_s(\sum_{i=1}^{p} a_i) = \bigcup_{i=1}^{p} m_s(a_i); \quad m_s(u)=\{deg(u)\} \text{ if } u \text{ is not a sum.}$$

For instance : m_s(ab+(a+b)(b+c)+a(b+ac)+a)={2,2,3,1}, m_s(a+(a+ab))={1,1,2} and m_s(a(a+b)(ac+b))={4}.

Definition 4. *Let t be a term. The* multiplicative multiset *of degrees of t is the multiset defined by:*

$$m_p(\prod_{i=1}^{n} a_i) = \bigcup_{i=1}^{p} m_s(a_i); \quad m_p(u)=\{deg(u)\} \text{ if } u \text{ is not a product.}$$

For instance : $m_p(a(ac+b)(abcd+a))=\{1,2,4\}$, $m_p(a(a(a+b)))=\{1,1,1\}$ and $m_p(a+ab)=\{2\}$.

3.4. The Rewriting Rule System R

We first define factorization, additive grouping and collapsing rules.

Definition 5. *A rule $l \to r$ is a* factorization *if and only if, for any substitution σ in T:*

$$deg(l\sigma)=deg(r\sigma) \text{ and } m_p(l\sigma)>>m_p(r\sigma).$$

EXAMPLE: $(xx+yy)(x+y) \to (x+y)(x+y)(x+y)$ and $xx+yy+xy \to (x+y)(x+y)$ are factorizations.

Definition 6. *A rule $l \to r$ is an* additive grouping *if and only if, for any substitution σ in T:*

$$deg(l\sigma)=deg(r\sigma) \text{ and } m_s(l\sigma)>>m_s(r\sigma).$$

EXAMPLE: The rules $x+x+y \to x+y$, $xy+xy \to xy$ and $x+x+x \to x$ are additive groupings.

Definition 7. *A rule $l \to r$ is a* collapsing rule *if and only if, for any substitution σ in T:*

$$deg(l\sigma)>deg(r\sigma).$$

EXAMPLE: The rules $xy \to x+y$ and $x(x+y+z) \to x+y+z$ are collapsing rules.

We now define our rewrite system R comprising an *infinite number* of rules.

Definition 8. *We call R the rewrite system over T including all factorization, additive grouping and collapsing rules.*

3.5. R_6 is a Sub-System of R

Property 2. *The rule system R_6 is a sub-system of R.*

The three first rules of R_6 are factorization rules and the three last ones are collapsing rules.

4 Some Impossibilities of Proving AC-Termination of Sub-System R_6

4.1. The Sub-System R_6 and an Associative Path Ordering

Property 3. *AC-termination of the rule sub-system R_6 cannot be proved using any associative path ordering.*

PROOF : Three cases may occur. We show that there exists in each case a rule whose AC-termination can't be proved with the help of any associative path ordering. Let's recall that the associative path ordering requires a precedence on the function symbol satisfying the associative path condition.

This is not the required inequality. It's therefore impossible to prove AC-termination using the associative path ordering in this case.

There are no more cases left. Proving that the relation R_6/AC is terminating using an associative path ordering is therefore impossible. ♦

Note that only two R_6 rules:

$$xx+yy \rightarrow (x+y)(x+y)$$

and

$$xy \rightarrow x+y$$

are enough to prove this result.

4.2. The Sub-System R_6 and a Polynomial Interpretation

Property 4. *AC-termination of the rules sub-system R_6 cannot be proved using any polynomial interpretation.*

Let p be a given polynomial interpretation. From now on, we write for the polynomial interpretation p and for the AC-symbols + and *:

$$p(x*y) = ap(x)p(y)+b[p(x)+p(y)]+c$$

and

$$p(x+y) = mp(x)p(y)+n[p(x)+p(y)]+q,$$

where a, b, c, m, n and q are non negative integers such that $ac+b=b^2$ and $mq+n=n^2$.

We recall that, for proving termination with a polynomial interpretation p, we have to establish for any rule $l \rightarrow r$, that: $p(l)>p(r)$,

for any positive value of p(x) greater than a given positive integer x_0 where x is any variable of the rule $l \rightarrow r$.

To prove the previous property, we need first some lemmas.

Lemma 1. *If p is a polynomial interpretation and * denotes an AC-symbol function, we have:*

$$p(x*y*z)=a^2p(x)p(y)p(z)+ab[p(x)p(z)+p(y)p(z)+p(x)p(y)]+b^2[p(x)+p(y)+p(z)]+bc+c.$$

PROOF: We just have to calculate.

$$p(x*y*z) = p\big((x*y)*z\big) = ap(x*y)p(z)+b[p(x*y)+p(z)]+c$$

$$= a\big[ap(x)p(y)+b[p(x)+p(y)]+c\big]p(z)+b\big[ap(x)p(y)+b[p(x)+p(y)]+c+p(z)\big]+c.$$

The equality $b+ac=b^2$ implies that:

$$p(x*y*z) = a^2p(x)p(y)p(z)+ab[p(x)p(z)+p(y)p(z)+p(x)p(y)]+b^2[p(x)+p(y)+p(z)]+bc+c.$$

♦

Lemma 2. *We suppose we can prove that the relation R_6/AC is terminating with the polynomial interpretation p. Then, it's impossible that m=n=a=b=0 (in other words, p can't be constant for + and *).*

PROOF: Let's suppose that m=n=a=b=0. Hence: p(x+y)=q and p(xy)=c.

Let's consider the factorization rule:

$$xx+yy \rightarrow (x+y)(x+y).$$

As termination of the R_6/AC relation can be proved with the polynomial interpretation p, we infer that:

$p(xx+yy) \quad > \quad p((x+y)(x+y))$,

or: (1) $q \quad > \quad c$

Let's now consider the collapsing rule:

$xy \quad \rightarrow \quad x+y.$

As termination of the R_6/AC relation can be proved with the polynomial interpretation p, we infer that:

$p(xy) \quad > \quad p(x+y)$

or: (2) $c \quad > \quad q.$

We get a contradiction between the inequalities (1) and (2). So, it's impossible that m=n=a=b=0. ♦

Lemma 3. *We suppose we can prove that the relation R_6/AC is terminating with the polynomial interpretation p. Then, it's impossible that m=n=0.*

PROOF: Let's suppose that m=n=0. Hence: $p(x+y)=q$ and $p(xy)=ap(x)p(y)+b[p(x)+p(y)]+c$.

Let's consider the factorization rule:

$xy+xz \rightarrow x(y+z)$

We have:

$p(xy+xz)=q$ and $p(x(y+z))=aqp(x)+b[p(x)+q]+c.$

As AC-termination can be proved with the polynomial interpretation p, we infer that:

$p(xy+xz) \quad > \quad p(x(y+z))$,

or $q \quad > \quad aqp(x)+b[p(x)+q]+c.$

But, we can't have a=b=0 (Lemma 2). So, this previous inequality is wrong, because q is a integer and p(x) may be any positive integer.

So, we get a contradiction and it's impossible that m=n=0. ♦

Lemma 4. *We suppose we can prove that the relation R_6/AC is terminating with the polynomial interpretation p. Then, it's impossible that a=b=0.*

PROOF: Let's suppose that a=b=0. Hence: $p(xy)=c$.

Let's consider the collapsing rule:

$xy \quad \rightarrow \quad x$

As AC-termination can be proved with the polynomial interpretation p, we infer that:

$p(xy) \quad > \quad p(x)$,

or $c \quad > \quad p(x).$

This inequality is wrong, because p(x) is likely to take any positive value greater than a given positive integer x_0.

So, we get a contradiction and it's impossible that a=b=0. ♦

Lemma 5. *We suppose we can prove that the relation R_6/AC is terminating with the polynomial interpretation p. Then, it's impossible that m=a=0 (in other words, the interpretations of + and * cannot be linear at the same time).*

PROOF: Let's suppose that m=a=0. Hence: $p(x+y)=n(x+y)+q$ and $p(xy)=b(x+y)+c$.

Let's consider the factorization rule:

$x+xxx+yyyy \quad \rightarrow \quad (x+x+y)(x+x+y)(x+x+y)$

*First case: + and * can't be compared in the precedence.*

In this case, there is not any distributive law. All terms are therefore in the required normal form and we have: A(xx+yy)=xx+yy and A((x+y)(x+y))=(x+y)(x+y). Let's now consider the R_6 rule:

$$xx+yy \rightarrow (x+y)(x+y).$$

The term xx is embedded in (x+y)(x+y), therefore xx $<_{rpo}$ (x+y)(x+y). In the same way:

$$yy <_{rpo} (x+y)(x+y).$$

Hence we have: xx $\not>_{rpo}$ (x+y)(x+y) and yy $\not>_{rpo}$ (x+y)(x+y). Since we suppose that the two functions * and + can't be compared in the precedence, this implies that:

$$xx+yy \not>_{rpo} (x+y)(x+y).$$

Therefore: $A(xx+yy) \not>_{rpo} A((x+y)(x+y)).$

Note that A(xx+yy) $\neq$ A((x+y)(x+y)). So, using Bachmair and Plaisted's notation, we conclude that there exists a substitution σ_L (equal to identity here) of Λ(xx+yy,(x+y)(x+y)) such that:

$$A((xx+yy)\sigma_L) \not>_{apo} A(((x+y)(x+y))\sigma_L).$$

Therefore AC-termination is impossible to prove in this case using any associative path ordering.

*Second case: + < *.*

The distributive laws are: (x+y)z → xz+yz and x(y+z) → xy+xz. Let's now consider the R_6 factorization rule:

$$xx+yy \rightarrow (x+y)(x+y).$$

The respective normal forms of both rule sides are: A(xx+yy)=xx+yy and A((x+y)(x+y))=xx+xy+yx+yy. Further, xx+yy is embedded in xx+xy+yx+yy and then:

$$xx+yy <_{rpo} xx+xy+yx+yy.$$

It results in:

$$A(xx+yy) <_{rpo} A((x+y)(x+y)).$$

Note that A(xx+yy) $\neq$ A((x+y)(x+y)). Therefore this implies that there exists a substitution σ_L (equal to identity) of Λ(xx+yy,(x+y)(x+y)) such that:

$$A((xx+yy)\sigma_L) <_{apo} A(((x+y)(x+y))\sigma_L).$$

This is not the required inequality. It's therefore impossible to prove AC-termination using the associative path ordering in this case.

*Third case: * < +.*

In this case, the distributive laws are: (xy)+z → (x+z)(y+z) and x+(yz) → (x+y)(x+z). Let's now consider the R_6 collapsing rule:

$$xy \rightarrow x+y.$$

Both terms involved in the rule are in normal form. Hence: A(xy)=xy and A(x+y)=x+y. As x $<_{rpo}$ x+y, y $<_{rpo}$ x+y and *<+ in the precedence, we conclude that:

$$xy <_{rpo} x+y,$$

or $A(xy) <_{rpo} A(x+y).$

Note that A(xy) $\neq$ A(x+y). Therefore it implies that there exists a substitution σ_L (equal to identity) of Λ(xy,(x+y)) such that:

$$A((xy)\sigma_L) <_{apo} A((x+y)\sigma_L).$$

Using Lemma 1, we have:
$p(xxx)=3b^2p(x)+bc+c$ and $p(yyy)=3b^2p(y)+bc+c$.
So: $p(x+xxx+yyy) = n^2[p(x)+3b^2p(x)+bc+c+3b^2p(y)+bc+c]+nq+q$
$= n^2p(x)+3b^2n^2p(x)+2bcn^2+2cn^2+3b^2n^2p(y)+nq+q.$
Futher, we have: $p(x+x+y)=n^2[p(x)+p(x)+p(y)]+nq+q$.
So: $p((x+x+y)(x+x+y)(x+x+y)) = 3b^2p(x+x+y)+bc+c$
$= 3b^2[n^2[p(x)+p(x)+p(y)]+nq+q]+bc+c$
$= 6b^2n^2p(x)+3b^2n^2p(y)+3b^2nq+3b^2q+bc+c.$

As AC-termination can be proved with the polynomial interpretation p, we infer that: $p(x+xxx+yyy) > p((x+x+y)(x+x+y)(x+x+y))$,
$n^2p(x)+3b^2n^2p(x)+2bcn^2+2cn^2+3b^2n^2p(y)+nq+q >$
$6b^2n^2p(x)+3b^2n^2p(y)+3b^2nq+3b^2q+bc+c$ or:
$n^2p(x)-3b^2n^2p(x) > k$ (where k is a constant, not depending on p(x) nor p(y))
$(1-3b^2)p(x) > k.$

This inequality has to be true for any p(x) such that $p(x) > x_0$ (where x_0 is an non-negative integer). Hence, it's necessary that:

$$1-3b^2 \geq 0, \text{ or } 0 \leq b \leq \frac{1}{\sqrt{3}} \text{ because b is non-negative.}$$

As b is an integer, the only possibility is: b=0.
So, we obtain that a=b=0, which is impossible because of Lemma 4.

Lemma 6. *We suppose we can prove termination of the R_6/AC relation with the polynomial interpretation p. Then, it's impossible that m=0.*

PROOF: Let's suppose that m=0.
Hence: $p(x+y)=n[p(x)+p(y)]+q$ and $p(xy)=ap(x)p(y)+b[p(x)+p(y)]+c$.
Let's consider the factorization rule:

$$xx+yy \rightarrow (x+y)(x+y)$$

We have:
$p(xx+yy) = n[a[p(x)]^2+2bp(x)+c+a[p(y)]^2+2bp(y)+c]+q$
$= an[p(x)]^2+2bnp(x)+cn+an[p(y)]^2+2bnp(y)+cn+q$
$p((x+y)(x+y)) = a[n[p(x)+p(y)]+q]^2+2b[n[p(x)+p(y)]+q]+c$
$=$
$a[n^2[p(x)]^2+n^2[p(y)]^2+2n^2p(x)p(y)+2nqp(x)+2nqp(y)+q^2]+2bnp(x)+2bnp(y)+2bq+c$
$=an^2[p(x)]^2+an^2[p(y)]^2+2an^2p(x)p(y)+2anqp(x)+2anqp(y)+aq^2+2bnp(x)+2bnp(y)+2bq$
$+c$. As AC-termination can be proved with the polynomial interpretation p, we infer that: $p(xx+yy) > p((x+y)(x+y))$.
So, after simplifying:
$an[p(x)]^2+an[p(y)]^2+2cn+q >$
$an^2[p(x)]^2+an^2[p(y)]^2+2an^2p(x)p(y)+2anqp(x)+2anqp(y)+aq^2+2bq+c$
$2cn+q > an(n-1)[p(x)]^2+an(n-1)[p(y)]^2+2an^2p(x)p(y)+2anqp(x)+2anqp(y)+aq^2+2bq+c$
$k > an(n-1)[p(x)]^2+an(n-1)[p(y)]^2+2an^2p(x)p(y)+2anqp(x)+2anqp(y)$,
where k is a constant.

All the coefficients occuring in the previous inequality $an(n-1)$, $2an^2$ and $2an$ are non-negative, because a and n are non-negative integers and $n(n-1)$ is equal to 0 when n is equal to 0).

So, as the previous inequality has to be true for any value of $p(x)$ and $p(y)$ respectively greater than the positive integers x_0 and y_0, we conclude that $an=0$, which implies $a=0$ or $n=0$.

But we supposed that $m=0$. Hence only two cases may occur:

(1) $m=0$ and $a=0$: this case can't occur because of Lemma 5.

(2) $m=0$ and $n=0$: this case can't occur too because of Lemma 3.

The two only cases are impossible, therefore m can't be equal to 0. ♦

PROOF OF THE PROPERTY: *The termination of the R_6/AC relation can't be proved using any polynomial interpretation.*

We suppose that such an interpretation exists denoted p.

Let's now consider the collapsing rule:

$$xyz \quad \rightarrow \quad xy+xz+yz.$$

We have:

$$p(xyz) = a^2p(x)p(y)p(z)+ab[p(x)p(z)+p(y)p(z)+p(x)p(y)]+b^2[p(x)+p(y)+p(z)]+bc+c$$

$$p(xy+xz+yz)=m^2\big[ap(x)p(y)+b[p(x)+p(y)]+c\big]\big[ap(x)p(z)+b[p(x)+p(z)]+c\big]\big[ap(y)p(z))+b[p(y)+p(z)]+c\big]+...$$

We know that all constants a, b, c, m, n, p are non-negative integers, either zero or greater than 1. Futher, all variables $p(x)$, $p(y)$ and $p(z)$ are non-negative integers, greater than the given constants x_0, y_0 and z_0 we can suppose greater than 1.

We now prove that $p(xyz) \leq p(xy+xz+yz)$. We consider any term of $p(xyz)$ and find a greater term in $p(xy+xz+yz)$. We have establish that $m \neq 0$ (Lemma 6). But m is a non-negaive integer, hence:

$$m \geq 1.$$

We have:

$$a^2p(x)p(y)p(z) \quad \leq \quad m^2a^3[p(x)]^2[p(y)]^2[p(z)]^2$$

(because a is either 0 or greater than 1, and m, $p(x)$, $p(y)$ and $p(z)$ are greater than 1)

$$abp(x)p(y) \quad \leq \quad m^2ab^2[p(x)]^2[p(y)]^2$$

(and similar inequalities for $abp(y)p(z)$ and $abp(x)p(z)$)

$$b^2p(x) \quad \leq \quad m^2b^3[p(x)]^2p(y)$$

(and similar inequalities for $b^2p(y)$ and $b^2p(z)$)

$$bc \quad \leq \quad m^2bc^2[p(x)]$$

$$c \quad \leq \quad m^2c^3.$$

Therefore, we conclude that:

(1) $\qquad p(xyz) \leq p(xy+xz+yz).$

But we supposed that p is a polynomial interpretation able to prove the termination of the R_6/AC relation, and consequently of the rule $xyz \rightarrow xy+xz+yz$. Hence:

(2) $\qquad p(xyz) > p(xy+xz+yz).$

The inequalities (1) and (2) can't be true at the same time and so we get a contradiction which prove that no polynomial interpretation may exist to prove that the relation R_6/AC is terminating. ♦

4.3. The Sub-System R_6 and the Ordering given by Narendran and Rusinowitch

Property 5. *AC-termination of the rules sub-system R_6 cannot be proved using the reduction ordering constructed by Narendran and Rusinowitch.*

PROOF: Let us consider the ground instance abc → ab+ac+bc of the fourth R_6 rule. For a term u, the interpretation I(u) is a polynomial. An indeterminate is associated with each function symbol. We denote respectively by A, B, C, X and Y the indeterminates associated with a, b, c, * and +. We have:

$$I(abc)=(X+1)[X^2+2X)(A+1)(B+1)(C+1)+(X+1)(A+B+C+3)+1];$$
$$I(ab)=(X+1)[X^2+2X)(A+1)(B+1)+(X+1)(A+B+2)+1];$$
$$I(ab+ac+bc)=(Y+1)[Y^2+2Y)I(ab)I(ac)I(bc)+(Y+1)(I(ab)+I(ac)+I(ac))+1].$$

Note that the polynomial degrees are:

$$deg[I(abc)]=6;$$
$$deg[I(ab)]=deg[I(ac)]=deg[I(bc)]=5;$$
$$deg[I(ab+ac+bc)]=18.$$

Polynomials are considered as multisets of monomials. Two monomials are compared first by their degrees and second, when these degrees are equal, by a lexicographic ordering. We denote by $>_N$ the multiset extension of the ordering on monomials. We have: I(abc) $\not>_N$ I(ab+ac+bc) because I(ab+ac+bc) includes a monomial having a strictly greater degree (18) than the degree of any monomial of I(abc). Therefore, this ordering cannot be used for proving AC-termination. ♦

5 Proof of the AC-Termination of R

The proof we propose relies on the m_{ps} and m_{sp} multisets of degrees we define now.

Definition 9. *Let s be a term. The* multiplicative-additive multiset of degrees *of s is the multiset defined by, if n is an integer greater than 2:*

- *if* $s=\prod_{i=1}^{n} s_i$ *then* $m_{ps}(s)=\bigcup_{i=1}^{n} m_s(s_i)$;
- *if s is not a product, then* $m_{ps}(s)=m_s(s)$.

EXAMPLE : We have:
m_{ps}([abcd+a+ab(a+b)][a+b(c+d)+a])={4, 1, 3, 1, 2, 1} ; m_{ps}(aa+b)={2, 1}.

Definition 10. *Let s be a term. The* additive-mulitplicative multiset of degrees *of s is the multiset defined by, if n is an integer greater than 2:*

- *if* $s=\sum_{i=1}^{n} s_i$ *then* $m_{sp}(s)=\bigcup_{i=1}^{n} m_p(s_i)$;
- *if s is not a sum, then* $m_{sp}(s)=m_p(s)$.

EXAMPLE: We have: $m_{sp}((aaa+b)(ab+c)(aa+b)+(a+d)(abcd+a)+a)=\{3,2,2,1,4,1\}$ and

$m_{sp}(a(ab+c))=\{1,2\}$.

Property 6. *The rule system R is AC-terminating.*

PROOF SKETCH: We use Theorem 2. We denote by **N** the set of non-negative integers and M(**N**) the set of finite multisets over **N**. We define the following mapping τ from the set T of terms to $\mathbf{N} \times M(\mathbf{N}) \times M(\mathbf{N}) \times M(\mathbf{N})$ by:

$$\tau(u) = (\deg(\bar{u}), m_p(\bar{u}), m_{ps}(\bar{u}), m_{sp}(u)),$$

where m_{ps} and m_{sp} are the previous multisets of degrees. We use the usual ordering on **N**, the induced ordering >> on finite multisets M(**N**) and the lexicographic ordering › in $\mathbf{N} \times M(\mathbf{N}) \times M(\mathbf{N}) \times M(\mathbf{N})$. We define the > ordering in T by: u > v if and only if $\tau(u) \succ \tau(v)$. The > relation fulfils the conditions (H_2) of Theorem 2 and consequently R is AC- terminating. We provide the whole proof of AC-termination in the appendix.♦

6 Conclusion

In this paper, we have defined a rewriting system originating from an implemented system and proved AC-termination of a precise rule set using a new characterisation. We have proved that some classical methods are useless to show AC-termination. A deep understanding of these failures may perhaps lead to new AC-orderings. A precedence ordering on the function symbols seems to be an important constraint for the system we have. Moreover, second terms of some rewriting rules have a more *complex* form than the first ones (as for the rule xyz → xy+xz+yz).

A promising direction of research is perhaps to design termination orderings preferably based on orderings not involving term root symbols (as precedence orderings) but rather involving characteristics of terms not depending on the term complexity (as the multisets m_p...).

Bibliography

Bachmair L. and Plaisted D.A. [85] : *Termination Orderings for Associative-Commutative Rewriting Systems*, Journal of Symbolic Computation, n° 1, pp 329-349.

Dershowitz N. [87] : *Termination of Rewriting*, Journal of Symbolic Computation, n° 3, pp 69-116.

Narendran P. and Rusinowitch M. : *Any ground associative commutative theory has a finite canonical system.* In Fourth International Conference on Rewriting Techniques and Applications, LNCS 488, p 423-434, Como, Italy, April 1991. Springer-Verlag.

Nicaud J.F., Gélis J.M., Saïdi M. [93] : A Framework for Learning Polynomial Factoring with New Technologies, ICCE'93 . Taïwan 1993.

Appendix

We make precise the proof of the property:

Property 6. *The rule system R is AC-terminating.*

•• We now propose some lemmas useful to the termination proof.

Lemma 1. *If u, v and w are terms, we have:*

- *if u is not a product, then* $m_{ps}(u)=m_s(u)$ *;*
- *if u is not a sum, then* $m_{sp}(u)=m_p(u)$ *;*
- *if u and v are not sums, then:*
  $m_{sp}(u+w)>>m_{sp}(v+w)$ *is equivalent to* $m_p(u)>>m_p(v)$ *;*
- *if u and v are not products, then:*
  $m_{ps}(u*w)>>m_{ps}(v*w)$ *is equivalent to* $m_s(u)>>m_s(v)$.

In a similar way, we have with the $\geq\geq$ *ordering:*

- *if u and v are not sums, then:*
  $m_{sp}(u+w)\geq\geq m_{sp}(v+w)$ *is equivalent to* $m_p(u)\geq\geq m_p(v)$ *;*
- *if u and v are not products, then:*
  $m_{ps}(u*w)\geq\geq m_{ps}(v*w)$ *is equivalent to* $m_s(u)\geq\geq m_s(v)$.

Lemma 2. *If* $u \to v$ *is an instance of an additive grouping, then u is a sum..*

Lemma 3. *If* $u \to v$ *is an instance of a factorization, then one of the two following assumptions is true:*

- *u is a sum and v is a product;*
- *u and v are products.*

Lemma 4. *If s AC-rewrites to t with any rule of the R system, then* $deg(s) \geq deg(t)$.

•• We now provide the proof of Property 6.

PROOF: Our proof uses Theorem 2.

We define the set $\mathbb{W}$ as $\mathbf{N} \times M(\mathbf{N}) \times M(\mathbf{N}) \times M(\mathbf{N})$ where **N** is the integer set and M(**N**) the set of finite mutisets over **N**.

We consider in $\mathbb{W}$ the lexicographic ordering (>, >>, >>, >>) where > is the usual ordering in the integer set and >> its induced ordering in M(**N**).

We define the mapping τ from $\overline{T}$ to $\mathbb{W}$ such that:

$$\tau(u) = (deg(u), m_p(u), m_{ps}(u), m_{sp}(u)).$$

We consider the lexicographic ordering in $\mathbb{W}$ induced by the respetive ordering > in **N** and >> in M(**N**). As > in **N** is a well-founded ordering, we can infer [Dershowitz 87] that >> on M(**N**) and the lexicographic ordering in $\mathbb{W}$ are also well-founded.

We denote by > the lexicographic ordering in $\mathbb{W}$. We have $\tau(u) > \tau(v)$

iff: deg(u) > deg(v),

or deg(u) = deg(v) and $m_p(u) >> m_p(v)$,

or deg(u) = deg(v), $m_p(u) = m_p(v)$ and $m_{ps}(u) >> m_{ps}(v)$,

or deg(u) = deg(v), $m_p(u) = m_p(v)$, $m_{ps}(u) = m_{ps}(v)$ and $m_{sp}(u) >> m_{sp}(v)$.

In the same way, if:

$deg(u) \geq deg(v)$, $m_p(u) \geq\geq m_p(v)$, $m_{ps}(u) \geq\geq m_{ps}(v)$ and $m_{sp}(u) \geq\geq m_{sp}(v)$,

we can conclude that:

$\tau(u) \geq \tau(v)$.

We now have to prove the different conditions given in Theorem 2. For clarity, we use the following conventions:

CONVENTIONS: All terms occurring in Theorem 2 and which are arguments of the τ function are flattened and from now on, we don't overline them and consider they are implicitely flattened. Moreover, to simplify notations, we write:

- $\tau(\overline{l\sigma}) > \tau(\overline{r\sigma})$ where $l\sigma \rightarrow r\sigma$ is any R rule instance

  as $\tau(d) > \tau(e)$;

- $\tau[\ \overline{f(g\sigma,N\sigma)}\ ] > \tau[\ \overline{f(d\sigma,N\sigma)}\ ]$

  where $l\sigma \rightarrow r\sigma$ is any rule instance and N a new variable not occurring in the rule

  as $\tau(f(d,s)) > \tau(f(e,s))$,

  where s may have f as root symbol

- $\bar{u} \xrightarrow{R_P/ac} \bar{v}$ and $\tau(\bar{u}) \geq \tau(\bar{v}) \Rightarrow \tau[h(...\bar{u}...)] \geq \tau[h(...\bar{v}...)]$

  where $h(...\bar{u}...)$ and $h(...\bar{v}...)$ are flattened terms

  as $u \rightarrow v$, $\tau(u) \geq \tau(v) \Rightarrow \tau(h(u,w)) \geq \tau(h(v,w))$

  where w may have h as root symbol and neither u nor v don't have h as root symbol.

- $\tau[h(..,\bar{u},..)] \geq \tau[\bar{u}]$ *as* $\tau[h(...u...)] \geq \tau(u)$

OVERVIEW: We summarize below the non-obvious proofs of the different conditions that the > ordering and the τ function have to verify. We adapt these conditions to T (the only function symbols of T terms are "+" and "*") and to the rewriting system R including either factorizations, groupings or collapsings. The follwing table make precise why we can conclude in each case that $\tau(s_1)>\tau(s_2)$ or $\tau(s_1)\geq\tau(s_2)$. The partial proofs are developed thereafter.

CONDITION	deg	m_p	m_{ps}	m_{sp}
$\tau(d) > \tau(e)$				
where $d \rightarrow e$ is a factorization	=	>>		
where $d \rightarrow e$ is a grouping	=	*proof 2.2.1.* e product: >> else: =	else: >>	
where $d \rightarrow e$ is a collapsing	>			
$\tau(d+s) > \tau(e+s)$ d or e being sums				
where $d \rightarrow e$ is a factorization	=	=	d sum: >> else: =	else: >>
where $d \rightarrow e$ is a grouping	=	=	>>	
where $d \rightarrow e$ is a collapsing	≥	≥≥	>>	
$\tau(d*s) > \tau(e*s)$ d or e being products				
where $d \rightarrow e$ is a factorization	=	>>		
where $d \rightarrow e$ is a grouping	=	e product: >> else: =	else: >>	
where $d \rightarrow e$ is a collapsing	>			
$u \rightarrow v$, $\tau(u) \geq \tau(v)$, u and v not sums $\Rightarrow$ $\tau(u+w) \geq \tau(v+w)$	≥	≥≥	collapsing at level 1 >> else: ≥≥	*proof 9.4* else: ≥≥
$u \rightarrow v$, $\tau(u) \geq \tau(v)$, u and v not products $\Rightarrow$ $\tau(u*w) \geq \tau(v*w)$	≥	≥≥	*proof 10.3* ≥≥	≥≥
$\tau(u+w) \geq \tau(w)$ u not being a sum	≥	≥≥	≥≥	≥≥
$\tau(u*w) \geq \tau(w)$ u not being a product	>			

We provide some partial proofs we mentioned above.

PROOF 2.2.1: *$m_p(d) >> m_p(e)$ where $d \rightarrow e$ is a grouping. We assume that e is a product.*

Lemma 2 implies that d is a sum. Consequently:

(1) $m_p(d)=\{deg(d)\}$.

As e is a product, we have:

(2) $m_p(e)=\{deg(e_1),...,deg(e_n)\}$,

where e_i are the arguments of the product e. By definition of degree concept:

(3) $deg(e_i)>0$, for all i = 1 ... n.

Moreover:

(4) $deg(e_1)+...+deg(e_n)=deg(e)$.

The (3) and (4) assertions imply that:

(5) $deg(e)>deg(e_i)$, for all i = 1 ... n.

A factorization verifies deg(d)=deg(e), hence:

(6) $deg(d)>deg(e_i)$, for all i = 1 ... n.

Consequently:

(7) $\{deg(d)\}>>\{deg(e_1),...,deg(e_n)\}$,

or, because of (1) and (2) :

(8) $m_p(d)>>m_p(e)$.

We conclude that $\tau(d) > \tau(e)$ in this case.

PROOF 9.4: *$u \rightarrow v$, $\tau(u) \geq \tau(v)$, u and v not sums $\Rightarrow$ $m_{sp}(u+w) \geq\geq m_{sp}(v+w)$.*

We suppose that u is not rewritten to v by a collapsing applied at level 1.

We distinguish three sub-cases, according as u is rewritten to v by a factorization applied at the level 1, or by a grouping applied at the level 1 or a rule applied at a level equal or greater than 2.

case 1 : u is rewritten to v by a factorization applied at the level 1.

We recall that neither u nor v are sums. Let $d \rightarrow e$ be the factorization applied at the level 1 allowing the rewriting of u to v. The following table summarizes all the possible cases that may appear, assuming that neither u nor v are sums. We first prove for each case that $m_p(u)>>m_p(v)$.

u	v	$m_p(u)\geq\geq m_p(v)$ is true because:
d product	e product	Lemma 5.
d*s	e*s	Lemma 5 implies that $m_p(d)\geq\geq m_p(e)$. Therefore, we have $m_p(d*s)\geq\geq m_p(e*s)$.

According to the point 3 of Lemma 1, we infer that if $m_p(u)\geq\geq m_p(v)$ then $m_{sp}(u+w)\geq\geq m_{sp}(v+w)$.

case 2 : u is rewritten to v by a grouping applied at the level 1.

We recall that neither u nor v are sums. Let $d \rightarrow e$ be the grouping applied at the level 1 allowing the rewriting of u to v. The following table summarizes all the possible cases that may appear, assuming that neither u nor v are sums. We first prove for each case that $m_p(u) >> m_p(v)$. Lemma 2 makes precise that d is a sum.

u	v	$m_p(u) \geq\geq m_p(v)$ is true because:
d*s d: sum	e*s e: product	Let the product e be $e_1 * ... e_n$. For all i=1...n, we have: $deg(e) > deg(e_i)$. It follows that: $\{deg(e)\} >> \{deg(e_1),...,deg(e_n)\}$. As deg(d)=deg(e) (by definition of groupings), then: $\{deg(d)\} >> \{deg(e_1),...,deg(e_n)\}$. Besides, d is a sum (Lemma 2), therefore $m_p(d)=\{deg(d)\}$. Therefore, we have: $m_p(d) >> \{deg(e_1),...,deg(e_n)\}$. Futhermore e is the product $e_1 * ... e_n$, hence: $m_p(e)=\{deg(e_1),...,deg(e_n)\}$. It follows that $m_p(d) >> m_p(e)$ and $m_p(d*s) >> m_p(e*s)$.
d*s d: sum	e*s e: sum or constant	d is a sum (Lemma 2), e is not a product. Hence: $m_p(d)=\{deg(d)\}$ and $m_p(e)=\{deg(e)\}$. As deg(d)=deg(e) (definition of groupings), we have $m_p(d)=m_p(e)$ and $m_p(d*s)=m_p(e*s)$.

According to the point 3 of Lemma 1, we infer that if $m_p(u) \geq\geq m_p(v)$ then $m_{sp}(u+w) \geq\geq m_{sp}(v+w)$.

case 3 : u is rewritten to v by a rule applied at a level equal or greater than 2. As the terms u and v are not sums, they can only be products. If we denote as $=_c$ the equality modulo the multiplication commutativity, the term $u =_c \prod_{i=1}^{p} w_i * w$ is rewritten to $v =_c \prod_{i=1}^{p} w_i * w'$ where the rewriting rule is applied in w rewritten to w' (neither w nor w' are products because all terms we consider are flattened).

Furthermore $m_p(u)=\{deg(w_1),...,deg(w_p),deg(w)\}$ and $m_p(v)=\{deg(w_1),...,deg(w_p),deg(w')\}$ (by definition of m_p). As $deg(w) \geq deg(w')$ (Lemma 4) then $m_p(u) \geq\geq m_p(v)$. According to the point 3 of Lemma 1, we infer that if $m_p(u) \geq\geq m_p(v)$ then $m_{sp}(u+w) \geq\geq m_{sp}(v+w)$.

PROOF 10.3: *$u \rightarrow v$, $\tau(u) \geq \tau(v)$, u and v not products $\Rightarrow$ $m_{ps}(u*w) \geq\geq m_{ps}(v*w)$.*

We distinguish 4 cases.

case 1: u is rewritten to v with a factorization at the first level.

Let $e \rightarrow f$ be the factorization applied at the first level 1 such that u is rewritten to v. The following table summarize all the possible cases that may appear, assuming that neither u nor v are products. We first prove for each case that $m_s(u) \gg m_s(v)$ (Lemma 3 proves that f is a product).

u	v	$m_s(u) \geq\geq m_s(v)$ is true because:
e+s e : sum	f+s f : product	Si $e=e_1+...+e_n$, on a : $m_s(e)=\{deg(e_1),...,deg(e_n)\}$. As v is a product (Lemma 3), we have: $m_s(f)=\{deg(f)\}$. As deg(e)=deg(f) (definition of factorizations) and as e is a sum, there exists k such that: $deg(e_k)=deg(f)$. By definition of the >> ordering, and as $m_s(e)=\{deg(e_1),..,deg(e_k),..,deg(e_n)\}$, $m_s(f)=\{deg(f)\}$ and $deg(e_k)=deg(f)$, we can conclude that $m_s(e) \gg m_s(f)$. It results that: $m_s(e+s) \gg m_s(f+s)$.
e+s e: product	f+s f :product	As e and f are products: $m_s(e)=\{deg(e)\}$ and $m_s(f)=\{deg(f)\}$. As: deg(e)=deg(f) (definition of factorizations), it results that $m_s(e)=m_s(f)$ and $m_s(e+s)=m_s(f+s)$.

We have proved that $m_s(u) \geq\geq m_s(v)$ in all cases. Therefore, it results because of the point 4 of Lemma 1 that $m_{ps}(u*w) \geq\geq m_{ps}(v*w)$.

case 2: u is rewritten to v with a grouping at the first level.

Let e → f be the grouping applied at the first level 1 such that u is rewritten to v. The following table summarize all the possible cases that may appear, assuming that neither u nor v are products. We first prove for each case that $m_s(u) \gg m_s(v)$ (Lemma 2 proves that u is a sum).

u	v	$m_s(u) \geq\geq m_s(v)$ is true because:
e sum	f sum, x or constant	Because of the definition of groupings, we have: $m_s(u) \gg m_s(v)$.
e+s	f+s	Because of the definition of groupings, we have: $m_s(e) \gg m_s(f)$. It follows that $m_s(e+s) \gg m_s(f+s)$.

We have proved that $m_s(u) \geq\geq m_s(v)$ in all cases. Therefore, it results because of the point 4 of Lemma 1 that $m_{ps}(u*w) \geq\geq m_{ps}(v*w)$.

case 3: u is rewritten to v with a collapsing rule at the first level.

Let e → f be the collapsing transformation applied at the first level 1 such that u is rewritten to v. The following table summarize all the possible cases that may appear, assuming that neither u nor v are products. We first prove for each case that $m_s(u) \gg m_s(v)$.

u	v	$m_s(u) \geq\geq m_s(v)$ is true because:
e sum	f sum, x or constant	As deg(u)>deg(v) (definition of collapsing rules) and as $e=e_1+...+e_n$, then there exists u_k argument of u such that $deg(u_k)=deg(u)$ and consequently $deg(u_k)>deg(v)$. Hence $deg(u_k)$ is strictly greater than the degree of f and, if f is a sum, than the degree of any additive argument. Hence: $\{deg(e_k)\}>>m_s(f)$. As $m_s(e)=\{deg(e_1),...,deg(e_k),...,deg(e_n)\}$, it results that: $m_s(u)>>m_s(v)$.
e constant	f	We have: deg(e)=1. As deg(e)>deg(f), it results that: deg(f)=0. This case is therefore impossible because the degree of any term is different from 0.
e+s e: product	f+s	We have: deg(e)>deg(f) and, if f is a sum, deg(e) is strictly greater than the degree of any additive argument. Hence: $\{deg(e_k)\}>>m_s(f)$. As e is a product: $m_s(e)=\{deg(e)\}$. It results that: $m_s(e)>>m_s(f)$. Hence: $m_s(e+s) \geq\geq m_s(f+s)$.
e+s e: sum	f+s	As deg(e)>deg(f) (definition of collapsing rules) and as $e=e_1+...+e_n$, then there exists u_k argument of u such that $deg(u_k)=deg(u)$ and consequently $deg(u_k)>deg(v)$. Hence $deg(u_k)$ is strictly greater than the degree of f and, if f is a sum, than the degree of any additive argument. Hence: $\{deg(e_k)\}>>m_s(f)$. As $m_s(e)=\{deg(e_1),...,deg(e_k),...,deg(e_n)\}$, it results that: $m_s(e)>>m_s(f)$. Hence: $m_s(e+s) \geq\geq m_s(f+s)$.
e+s constant	f+s	We have: deg(e)=1. As deg(e)>deg(f), it results that: deg(f)=0. This case is therefore impossible because the degree of any term is different from 0.

We have proved that $m_s(u) \geq\geq m_s(v)$ in all cases. Therefore, it results because of the point 4 of Lemma 1 that $m_{ps}(u*w) \geq\geq m_{ps}(v*w)$.

case 4: u is rewritten to v by a rule applied at a level equal or greater than 2.
The terms u and v are not products, therefore they are sums. If $=_{ac}$ is the equality modulo associativity and commutativity, the term $u =_{ac} \sum_{i=1}^{p} w_i + w$ is rewritten to $v =_{ac} \sum_{i=1}^{p} w_i + w'$, the rule being applied in w rewritten to w' (w and w' are not sums).

Further: $m_s(u)=\{deg(w_1),\ldots,deg(w_p),deg(w)\}$ and $m_s(v)=\{deg(w_1),\ldots,deg(w_p),deg(w')\}$. As $deg(w)\geq deg(w')$ (Lemma 4) then $m_s(u)\geq\geq m_s(v)$. Therefore, it results because of the point 4 of Lemma 1 that $m_{ps}(u*w)\geq\geq m_{ps}(v*w)$.♦

Relative Normalization in Orthogonal Expression Reduction Systems

John Glauert and Zurab Khasidashvili

School of Information Systems, UEA
Norwich NR4 7TJ England
jrwg@sys.uea.ac.uk, zurab@sys.uea.ac.uk *

Abstract. We study reductions in orthogonal (left-linear and non-ambiguous) Expression Reduction Systems, a formalism for Term Rewriting Systems with bound variables and substitutions. To generalise the normalization theory of Huet and Lévy, we introduce the notion of *neededness* with respect to a set of reductions Π or a set of terms $\mathcal{S}$ so that each existing notion of neededness can be given by specifying Π or $\mathcal{S}$. We imposed natural conditions on $\mathcal{S}$, called *stability*, that are sufficient and necessary for each term not in $\mathcal{S}$-normal form (i.e., not in $\mathcal{S}$) to have at least one $\mathcal{S}$-needed redex, and repeated contraction of $\mathcal{S}$-needed redexes in a term t to lead to an $\mathcal{S}$-normal form of t whenever there is one. Our relative neededness notion is based on tracing *(open) components*, which are occurrences of contexts not containing *any* bound variable, rather than tracing redexes or subterms.

1 Introduction

Since a normalizable term, in a rewriting system, may have an infinite reduction, it is important to have a *normalizing* strategy which enables one to construct reductions to normal form. It is well known that the leftmost-outermost strategy is normalizing in the λ-calculus. For Orthogonal Term Rewriting Systems (OTRSs), a general normalizing strategy, called the *needed* strategy, was found by Huet and Lévy in [HuLé91]. The needed strategy always contracts a *needed* redex – a redex whose residual is to be contracted in any reduction to normal form. Huet and Lévy showed that any term t not in normal form has a needed redex, and that repeated contraction of needed redexes in t leads to its normal form whenever there is one; we refer to it as the *Normalization Theorem*. They also defined the class of *strongly sequential* OTRSs where a needed redex can efficiently be found in any term.

Barendregt et al. [BKKS87] generalized the concept of neededness to the λ-calculus. They studied neededness not only w.r.t. normal forms, but also w.r.t. head-normal forms – a redex is *head-needed* if its residuals are contracted in each reduction to a head-normal form. The authors proved correctness of the two

* This work was supported by the Engineering and Physical Sciences Research Council of Great Britain under grant GR/H 41300.

needed strategies for computing normal forms and head-normal forms, respectively. Khasidashvili defined a similar normalizing strategy, called the *essential* strategy, for the λ-calculus [Kha88] and OTRSs [Kha93]. The strategy contracts *essential* redexes – the redexes that have *descendants* under any reduction. The notion of descendant is a refinement of that of *residual* – the descendant of a contracted redex is its contractum, while it does not have residuals. This refined notion allows for much simpler proofs of correctness of the essential strategy in OTRSs and the λ-calculus, which generalize straightforwardly to all Orthogonal Expression Reduction Systems (OERSs). Kennaway and Sleep [KeSl89] used a generalization of Lévy's labelling for the λ-calculus [Lév78] to adapt the proof from [BKKS87] to the case of Klop's OCRSs [Klo80], which can also be applied to OERSs. Khasidashvili [Kha94] showed that in *Persistent* OERSs, where redex-creation is limited, one can find *all* needed redexes in any term. Gardner [Gar94] described a *complete* way of encoding neededness information using a type assignment system in the sense that using the principal type of a term one can find all the needed redexes in it (the principal type cannot be found efficiently, as one might expect). Antoy et al. [AEH94] designed a needed narrowing strategy.

In [Mar92], Maranget introduced a different notion of neededness, where a redex u is needed if it has a residual under any reduction that does not contract the residuals of u. This neededness notion makes sense also for terms that do not have a normal form, and coincides with the notion of essentiality [Kha93] (essentiality makes sense for all subterms, not only for redexes). In [Mar92], Maranget studied also a strategy that computes a (in fact, the 'minimal' in some sense) weak head-normal form of a term in an OTRS. Normalization w.r.t. another interesting set of 'normal forms', that of constructor head-normal forms in constructor OTRSs, is studied by Nöcker [Nök94].

A question arises naturally: what are the properties that a set of terms must possess in order for the neededness theory of Huet and Lévy still to make sense? The main contribution of this paper is to provide a solution to that question. We introduce the notion of *neededness* w.r.t. a set of reductions Π or a set of terms $\mathcal{S}$ so that each existing notion of neededness can be given by specifying Π or $\mathcal{S}$. For example, *Huet&Lévy-neededness* is neededness w.r.t. the set of normal forms, *Maranget-neededness* is neededness w.r.t. all fair reductions, *head-neededness* is neededness w.r.t. the set of head-normal forms, etc. We impose natural conditions on $\mathcal{S}$, called *stability*, that are sufficient and necessary for each term not in $\mathcal{S}$-normal form (i.e., not in $\mathcal{S}$) to have at least one $\mathcal{S}$-needed redex, and repeated contraction of $\mathcal{S}$-needed redexes in a term t to lead to an $\mathcal{S}$-normal form of t whenever there is one.

A set $\mathcal{S}$ of terms is stable if it is *closed under parallel moves*: for any $t \notin \mathcal{S}$, any $P : t \twoheadrightarrow o \in \mathcal{S}$, and any $Q : t \twoheadrightarrow e$, the final term of P/Q, the residual of P under Q, is in $\mathcal{S}$; and is *closed under unneeded expansion*: for any $e \xrightarrow{u} o$ such that $e \notin \mathcal{S}$ and $o \in \mathcal{S}$, a residual of u is contracted in any reduction from e to a term in $\mathcal{S}$. We present a counterexample to show that the $\mathcal{S}$-needed strategy is not *hypernormalizing* for every stable $\mathcal{S}$, i.e., an $\mathcal{S}$-normalizable term may possess a reduction which never reaches a term in $\mathcal{S}$ even though $\mathcal{S}$-needed redexes are contracted infinitely many times, but $\mathcal{S}$-unneeded redexes are contracted as well. Therefore, *multistep* $\mathcal{S}$-needed reductions need not be $\mathcal{S}$-normalizing. This is because a 'non-standard' situation – when a component under an $\mathcal{S}$-unneeded component is $\mathcal{S}$-needed – may occur for some, call it *irregular* stable sets $\mathcal{S}$. However, if $\mathcal{S}$ is a regular stable set, then the $\mathcal{S}$-needed strategy is again hypernormalizing, and the multistep $\mathcal{S}$-needed strategy is normalizing.

Our relative neededness notion is based on tracing *(open) components*, which are occurrences of contexts not containing *any* bound variable, rather than tracing redexes or subterms. We therefore introduce notions of *descendant* and *residual* for components that are invariant under Lévy-equivalence. A component of a term $t \notin \mathcal{S}$ is called *$\mathcal{S}$-needed* if at least one descendant of it is 'involved' in any $\mathcal{S}$-normalizing reduction; a redex is *$\mathcal{S}$-needed* if so is its pattern. Besides generality, this approach to defining the neededness notion via components is crucial from a technical point of view, because components of a term in an OERS enjoy the same 'disjointness' property that subterms of a term in an OTRS possesses: residuals of disjoint components of a term in an OERS remain disjoint, and this allows for simpler proofs.

The rest of the paper is organized as follows. In the next section, we review Expression Reduction Systems (ERS), a formalism for higher order rewriting that we use here [Kha90, Kha92]; define the descendant relation for components, and show that it is invariant under Lévy-equivalence. Section 3 establishes equivalence of Maranget's neededness and our essentiality for OERSs. In section 4, we introduce the relative notion of neededness. In section 5, we sketch some properties of the labelling system of Kennaway&Sleep [KeSl89] for OERSs needed to define a *family-relation* among redexes. We prove correctness of the $\mathcal{S}$-needed strategy for finding terms of $\mathcal{S}$, for all stable $\mathcal{S}$, in section 6. The conclusions appear in section 7.

2 Orthogonal Expression Reduction Systems

Klop introduced *Combinatory Reduction Systems* (CRSs) in [Klo80] to provide a uniform framework for reductions with substitutions (also referred to as higher-order rewriting) as in the λ-calculus [Bar84]. Restricted rewriting systems with substitutions were first studied in Pkhakadze [Pkh77] and Aczel [Acz78]. Several interesting formalisms have been introduced later [Nip93, Wol93, OR94]. We refer to Klop et al. [KOR93] and van Oostrom [Oos94] for a survey. Here we use a system of higher order rewriting, *Expression Reduction Systems* (ERSs), defined in Khasidashvili [Kha90, Kha92] (ERSs are called CRSs in [Kha92]); the present formulation is simpler.

Definition 2.1 Let Σ be an *alphabet*, comprising *variables*, denoted by x, y, z; *function symbols*, also called *simple operators*; and *operator signs* or *quantifier signs*. Each function symbol has an *arity* $k \in N$, and each operator sign σ has an *arity* (m, n) with $m, n \neq 0$ such that, for any sequence $x_1, \ldots, x_m$ of pairwise distinct variables, $\sigma x_1 \dots x_m$ is a *compound operator* or a *quantifier* with *arity* n. Occurrences of $x_1, \dots, x_m$ in $\sigma x_1 \dots x_m$ are called *binding variables*. Each quantifier $\sigma x_1 \dots x_m$, as well as the corresponding quantifier sign σ and binding variables $x_1 \dots x_m$, has a *scope indicator* $(k_1, \dots, k_l)$ to specify the arguments in which $\sigma x_1 \dots x_m$ binds all free occurrences of $x_1, \dots, x_m$. *Terms* are constructed from variables using functions and quantifiers in the usual way.

Metaterms are constructed similarly from *terms* and *metavariables* $A, B, \dots$, which range over terms. In addition, *metasubstitutions*, expressions of the form $(t_1/x_1, \dots, t_n/x_n)t_0$, with t_j arbitrary metaterms, are allowed, where the *scope* of each x_i is t_0. Metaterms without metasubstitutions are *simple metaterms*. An *assignment* maps each metavariable to a term over Σ. If t is a metaterm and θ is an assignment, then the *θ-instance* $t\theta$ of t is the term obtained from t by replacing metavariables with their values under θ, and by replacing metasubstitutions $(t_1/x_1, \dots, t_n/x_n)t_0$, in right to left order, with the result of substitution of terms $t_1, \dots, t_n$ for free occurrences of $x_1, \dots, x_n$ in t_0.

For example, a β-redex in the λ-calculus appears as $Ap(\lambda x\, t, s)$, where Ap is a function symbol of arity 2, and λ is an operator sign of arity (1,1) and scope indicator (1). Integrals such as $\int_s^t f(x)\,dx$ can be represented as $\int x\, s\, t\, f(x)$ using an operator sign $\int$ of arity (1,3) and scope indicator (3).

Definition 2.2 An *Expression Reduction System* (ERS) is a pair (Σ, R), where Σ is an *alphabet*, described in Definition 2.1, and R is a set of *rewrite rules* $r : t \to s$, where t and s are closed metaterms (i.e., no free variables) such that t is a simple metaterm and is not a metavariable, and each metavariable that occurs in s occurs also in t.

Further, each rule r has a set of *admissible assignments* $AA(r)$ which, in order to prevent undesirable confusion of variable bindings, must satisfy the condition that:

(a) for any assignment $\theta \in AA(r)$, any metavariable A occurring in t or s, and any variable $x \in FV(A\theta)$, either every occurrence of A in r is in the scope of some binding occurrence of x in r, or no occurrence is.

For any $\theta \in AA(r)$, $t\theta$ is an *r-redex* or an *R-redex*, and $s\theta$ is the *contractum* of $t\theta$. We call R *simple* if right-hand sides of R-rules are simple metaterms.

Our syntax is similar to that of Klop's CRSs [Klo80], but is closer to the syntax of the λ-calculus and of First Order Logic. For example, the β-rule is written as $Ap(\lambda x A, B) \to (B/x)A$, where A and B can be instantiated by any terms; the η-rule is written as $\lambda x(Ax) \to A$ which requires that an assignment θ is admissible iff $x \notin (A\theta)$, otherwise an x occurring in $A\theta$ and therefore bound in $\lambda x(A\theta x)$ would become free. A rule like $f(A) \to \exists x(A)$ is also allowed, but an assignment θ with $x \in A\theta$ is not. The recursor rule from [AsLa93] is written as $\mu(\lambda x A) \to (\mu(\lambda x A)/x)A$. $\exists x A \to (\tau x(A)/x)A$ and $\exists! x A \to \exists x A \wedge \forall x \forall y(A \wedge (y/x)A \Rightarrow x = y)$ are rules corresponding to familiar definitions.

Below we restrict ourselves to the case of non-conditional ERSs, i.e., ERSs where an assignment is admissible iff the condition (a) of Definition 2.2 is satisfied. We ignore questions relating to renaming of bound variables. As usual, a rewrite step consists of replacement of a redex by its contractum. Note that the use of metavariables in rewrite rules is not really necessary – free variables can be used instead, as in TRSs. We will indeed do so at least when giving TRS examples.

To express substitution, we use the S-reduction rules

$$S^{n+1}x_1 \dots x_n A_1 \dots A_n A_0 \to (A_1/x_1, \dots, A_n/x_n)A_0, \quad n = 1, 2, \dots,$$

where S^{n+1} is the *operator sign of substitution* with arity $(n, n+1)$ and scope indicator $(n+1)$, and $x_1, \dots, x_n$ and $A_1, \dots, A_n$, A_0 are pairwise distinct variables and metavariables. Thus S^{n+1} binds free variables only in the last argument. The difference with β-rules is that S-reductions can only perform β-developments of λ-terms [Kha92].

Notation 2.1 We use a, b, c, d for constants, t, s, e, o for terms and metaterms, u, v, w for redexes, and N, P, Q for reductions. We write $s \subseteq t$ if s is a subterm of t. A one-step reduction in which a redex $u \subseteq t$ is contracted is written as $t \xrightarrow{u} s$ or $t \to s$ or just u. We write $P : t \twoheadrightarrow s$ or $t \overset{P}{\twoheadrightarrow} s$ if P denotes a reduction of t to s. $|P|$ denotes the length of P. $P + Q$ denotes the concatenation of P and Q.

Let $r : t \to s$ be a rule in an ERS R and let $\theta \in AA(r)$. Subterms of a redex $v = t\theta$ that correspond to metavariables of t are the *arguments* of v, and the rest is the *pattern* of v. Subterms of v rooted in the pattern are called the *pattern-subterms* of v. If R is a simple ERS, then arguments, pattern, and pattern-subterms are defined analogously in the contractum $s\theta$ of v.

We now recall briefly the definition of *descendant* of subterms as introduced in [Kha88, Kha93, Kha92] for the λ-calculus, TRSs, and ERSs, respectively. First, we need to split an ERS R into a *TRS-part* R_f and the *substitution-part* S. For any ERS R, which we assume does not contain symbols S^{n+1}, R_f is the ERS obtained from R by adding symbols S^{n+1} in the alphabet and by replacing in right-hand sides of the rules all metasubstitutions of the form $(t_1/x_1, \dots, t_n/x_n)t_0$ by $S^{n+1}x_1 \dots x_n t_1 \dots t_n t_0$, respectively. For example, the β_f rule would be $Ap(\lambda x A, B) \to S^2 x B A$. If R is simple, then $R_{fS} =_{def} R_f =_{def} R$. Otherwise $R_{fS} =_{def} R_f \cup S$. For each step $C[t\theta] \xrightarrow{u} C[s\theta]$ in R there is a reduction $P : C[t\theta] \to_{R_f} C[s'\theta] \twoheadrightarrow_S C[s\theta]$ in R_{fS}, where $C[s'\theta] \twoheadrightarrow_S C[s\theta]$ is the rightmost innermost normalizing S-reduction. We call P the *refinement* of u. The notion of *refinement* generalizes to R-reductions with 0 or more steps.

Let $t \xrightarrow{u} s$ be an R_f-reduction step and let e be the contractum of u in s. For each argument o of u there are 0 or more arguments of e. We will call them *u-descendants* of o. We refer to the i-th (from the left, $i > 0$) descendant of o also as the *(u, i)-descendant* of o. Correspondingly, subterms of o have 0 or more *descendants*. By definition, the *descendant*, referred to also as the *$(u, *)$-descendant*, of each pattern-subterm of u is e. It is clear what is to be meant by the *descendant* of a subterm $s' \subseteq t$ that is not in u. We call it also the *$(u, *)$-descendant* of s'. In an S-reduction step $C[S^{n+1}x_1 \dots x_n t_1 \dots t_n t_0] \xrightarrow{u} C[(t_1/x_1, \dots, t_n/x_n)t_0]$, the argument t_i and its subterms have the same number of descendants as the number of free occurrences of x_i in t_0; the i-th descendant is referred to as the

(u,i)-*descendant.* Every subterm of t_0 has exactly one descendant, the $(u,0)$-*descendent* (in particular, the $(u,0)$-descendants of free occurrences of $x_1,\ldots,x_n$ in t_0 are the substituted subterms). The descendant or $(u,*)$-*descendent* of the contracted redex u itself is its contractum. The pairs (u,i) and $(u,*)$ are called the *indexes* of corresponding descendants. The descendants of all redexes except the contracted one are called *residuals.*

The notions of *descendant* and *residual* extend by transitivity to arbitrary R_{fS}-reductions; *indexes* of descendants and residuals are sequences of indexes of immediate (i.e., under one step) descendants and residuals in the chain leading from the initial to the final subterm. If P is an R-reduction, then P descendants are defined to be the descendants under the refinement of P. The *ancestor* relation is the converse of the descendant relation.

We call a *component* an occurrence of a context that does not contain *any* bound variables; that is, neither variables bound from above in the term, nor variables for which the binder is in the component, belong to the component. Since we also consider occurrences of the empty context $[\,]$, which has an arity 1, we will think of a component as a pair $(context, path)$, where the path characterizes the *position* of the component in the term (usually, a position is a chain of natural numbers). Thus an *empty component* or *empty occurrence* is a pair $([\,], path)$. If terms are represented by trees, then the empty occurrence $([\,], path)$ can be seen as the *connection* at the *top* of the symbol at the position *path*.

Obviously, a component C can be considered as its *corresponding subterm* (the subterm rooted at the position of C), with some subterms (the *arguments* of C), removed. In particular, a subterm s with itself removed becomes the empty occurrence at the position of s. We say that an empty occurrence $([\,], path)$ in a term t is *in* a subterm or component e in t if *path* is a *non-top* position of e. We use C and the letters s, t, e, o used for terms to denote components as well. We write $s \sqsubseteq t$ if s is a component of t.

The concept of *descendent* can be extended to components in the following way:

Definition 2.3 Let R be an ERS, C be a component of a term t in R_{fS}, let $s = C[s_1,\ldots,s_n]$ be the corresponding subterm of C in t, let $t \xrightarrow{u} t'$ in R_{fS}, and let $o' \subseteq t'$ be the contractum of u. The pattern of u will be denoted by $pat(u)$ and the pattern of the contractum by $cpat(u)$. We define u-*descendants* of C by considering all relative positions of u and C.

(1) $C \cap u = \emptyset$ (so if C is an empty component, C is not in u). Then the $(u,*)$-descendant of C is the $(u,*)$-descendant of s with the $(u,*)$-descendants of s_j removed. (If C is an empty occurrence $([\,], path)$, then its $(u,*)$-descendant is the same pair.)

(2) C is the empty occurrence at the top of an argument o of u. Then the (u,i)-descendant of C is the empty occurrence at the top-position of the (u,i)-descendant of o, if the latter exists; if o doesn't have u-descendants, then C doesn't have u-descendants either. (Thus, the (u,i)-descendant of C is the (u,i)-descendant of its corresponding subterm $s = o$ with itself removed.)

(3) u is an R_f-redex and C is in an argument of u. Then the (u, i)-descendant of C is the (u, i)-descendant of s with the (u, i)-descendants of s_j removed ($i \geq 1$).

(4) u is an S-redex and C is in an argument of u. Then the (u, i)-descendant of C is the (u, i)-descendant of s with the (u, i)-descendants of s_j removed ($i \geq 0$).

(5) u is an R_f-redex and $pat(u)$ is in C. Then the $(u, *)$-*descendant* of C is the $(u, *)$-descendant of s with the descendants of s_j removed (see Figure 1). In particular, if C is *collapsed*, i.e. $pat(u) = C$ and $cpat(u) = \emptyset$, then the $(u, *)$-descendant of C is the empty occurrence at the position of σ'.

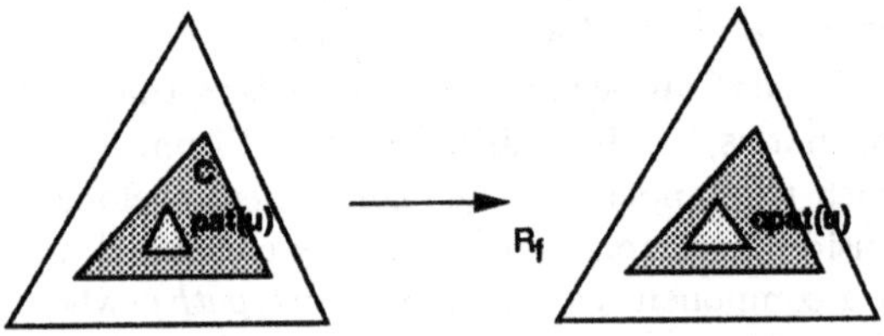

Fig. 1.

(6) u is an S-redex, say with two arguments (for simplicity) $u = S^2x\,e\,o$, with the top in C, and let $s = C'[s_1, \ldots, S^2x\,C_e[s_k, \ldots, s_l]\,C_o[s_{l+1}, \ldots, s_m], \ldots, s_n]$, with $C = C'[\ldots, S^2x\,C_e[\,]\,C_o[\,], \ldots]$, $e = C_e[s_k, \ldots, s_l]$, and $o = C_o[s_{l+1}, \ldots, s_m]$. Then, the $(u, *)$-*descendants* of C are the $(u, *)$-descendant of s with the descendants of the subterms s_j removed, and the descendants of C_e, as defined in (4) (see Figure 2, where $k = l = 1$ and $l+1 = m = 2$). Note that if $C = Pat(u) = S$, then C_e and C_o are empty components, C is *collapsed*, and its descendants are the empty occurrences at the top positions of the descendants of u and its arguments.

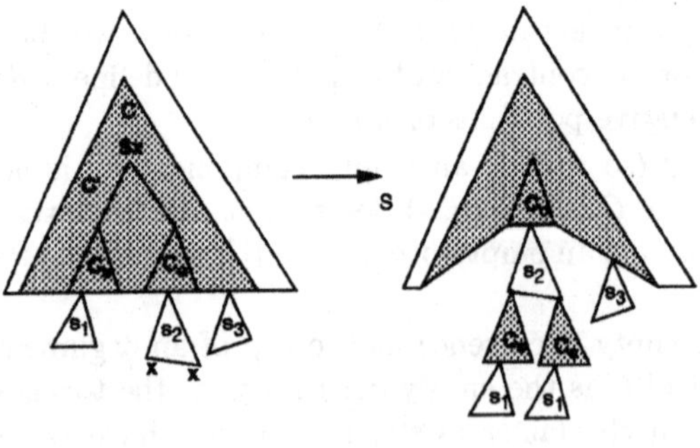

Fig. 2.

(7) u is an R_f-redex, $pat(u)$ and C partially overlap (i.e., neither contains another), and the top of u is (not necessarily strictly) below the top of C. Then the $(u,*)$-descendant of C is the $(u,*)$-descendant of s with the descendants of the arguments of u that do not overlap with C and the descendants of s_i that do not overlap with $pat(u)$ removed (see Figure 3). In addition, if the top symbol of σ' doesn't belong to the (above) $(u,*)$-descendant of C, then the empty component at the top of σ' is also a $(u,*)$-descendant of C.

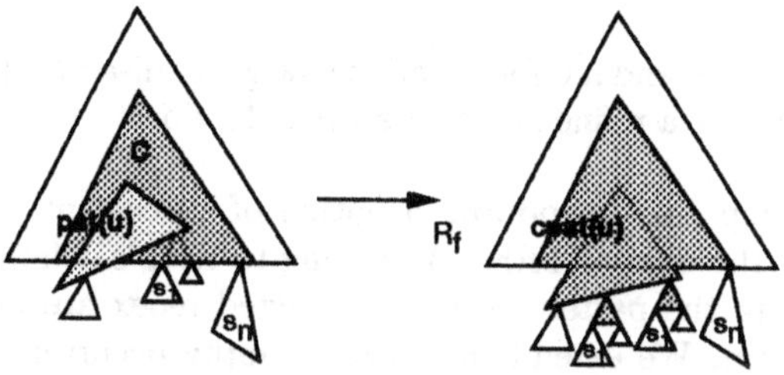

Fig. 3.

(8) u is an R_f-redex, $pat(u)$ and C partially overlap, and the top of C is below the top of u. Then the $(u,*)$-descendant of C is the $(u,*)$-descendant of s with the descendants of s_i that do not overlap with $pat(u)$ and the descendants of the arguments of u that do not overlap with C removed (see Figure 4). In addition, if the top symbol of σ' doesn't belong to the (above) $(u,*)$-descendant of C, then the empty component at the top of σ' is also a $(u,*)$-descendant of C.

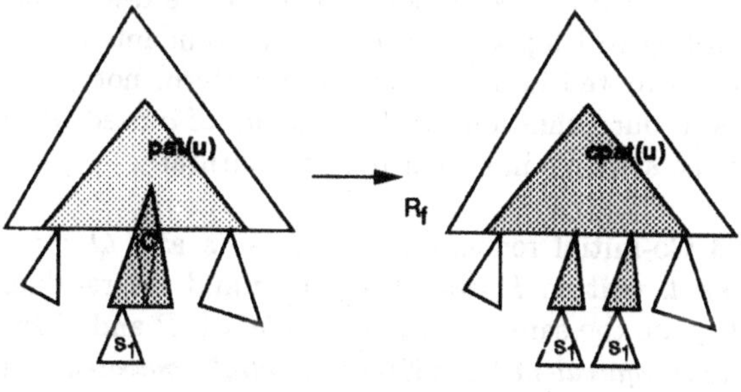

Fig. 4.

(9) u is an R_f-redex and $pat(u)$ contains C (C may be an empty component). Then the $(u,*)$-descendant of C is the contractum-pattern of u (the latter may also be empty) (see Figure 5).

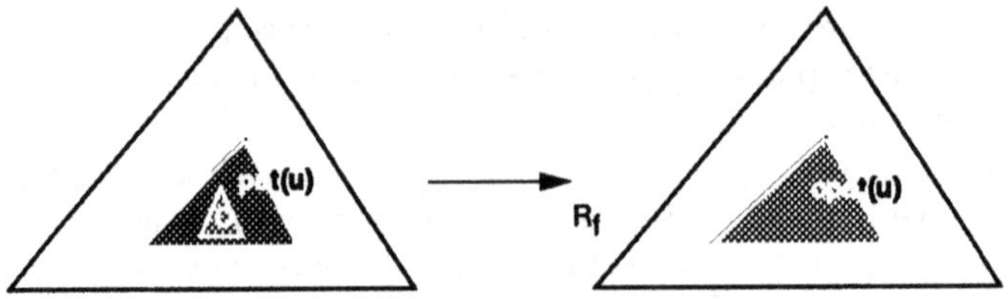

Fig. 5.

The notion of descendant for components generalises by transitivity to all R_{fS} reductions and via refinements to all R-reductions.

In [Klo92], Klop also introduced a notion of *descendant of (an occurrence of) a symbol* for the case of TRSs. According to Klop's definition, descendants of every symbol in the pattern of the contracted redex are *all* symbols in the contractum pattern. We extend this also to empty occurrences in the pattern of the contracted redex. If the rule is collapsing, i.e., the right-hand side is a variable, then Klop defines the descendant of pattern-symbols of the contracted redex to be the topmost symbol of the contractum. We define the descendant of the pattern-symbols in that case to be the empty occurrence at the position of the contractum. We can define the notion of descendant of a symbol in the same spirit for S-reduction steps by declaring that the descendants of the top S-symbol of a contracted S-redex u are the empty occurrences at the positions of the descendant of u and the descendants of its arguments. We define the descendants of other symbols, in particular of bound variables, to be the top-symbols of the descendants of the corresponding subterms. This gives us a definition of descendant for symbols for all ERSs. Now, it is not difficult to check that descendants of a component are composed of the 'corresponding' descendants of its symbols and occurrences, and similarly for the descendants of subterms. Note that, according to Klop's definition, the descendants of a pattern-subterm o are all subterms rooted in the contractum-pattern, not just the contractum of the redex, as in our definition; so descendants of o need not be composed of descendants of the symbols in o, which is less natural.

Definition 2.4 Co-initial reductions $P : t \twoheadrightarrow s$ and $Q : t \twoheadrightarrow e$ are called *Hindley-equivalent*, written $P \approx_H Q$, if $s = e$ and the residuals of a redex of t under P and Q are the same occurrences. We call P and Q respectively *Klop-equivalent, strictly equivalent* [Kha92], or *strictly* equivalent*, written $P \approx_K Q$, $P \approx_{st} Q$, or $P \approx_{st}^* Q$, if $s = e$ and P-descendants and Q-descendants of respectively any symbol, subterm, or component of t are the same occurrences in s and e.

Definition 2.5 A rewrite rule $t \to s$ in an ERS R is *left-linear* if t is linear, i.e., no metavariable occurs more than once in t. R is *left-linear* if each rule in R is so. $R = \{r_i \mid i \in I\}$ is *non-ambiguous* or *non-overlapping* if in no term redex-patterns can overlap, i.e., if r_i-redex u contains an r_j-redex u' and $i \neq j$, then u' is in an argument of u, and the same holds if $i = j$ and u' is a proper subterm of u. R is *orthogonal* (OERS) if it is left-linear and non-overlapping.

As in the the λ-calculus [Bar84], for any co-initial reductions P and Q, one can define in OERSs the notion of *residual of P under Q*, written P/Q, due to Lévy [Lév80], via the notion of *development* of a set of redexes in a term. We write $P \trianglelefteq Q$ if $P/Q = \emptyset$ ($\trianglelefteq$ is the *Lévy-embedding* relation); P and Q are called *Lévy-equivalent* or *permutation-equivalent* (written $P \approx_L Q$) if $P \trianglelefteq Q$ and $Q \trianglelefteq P$. It follows immediately from the definition of / that if P and Q are co-initial reductions in an OERS, then $(P + P')/Q \approx_L P/Q + P'/(Q/P)$ and $P/(Q + Q') \approx_L (P/Q)/Q'$.

Theorem 2.1 Let P and Q be co-initial reductions in an OERS R. Then:
(1) ($CR(res)$: **Church-Rosser for residuals**) $P + Q/P \approx_H Q + P/Q$.
(2) ($CR(sub)$: **Church-Rosser for subterms**) $P + Q/P \approx_{st} Q + P/Q$.
(3) ($CR(sym)$: **Church-Rosser for symbols**) $P + Q/P \approx_K Q + P/Q$.
(4) ($CR(com)$: **Church-Rosser for components**) $P + Q/P \approx_{st}^* Q + P/Q$.

Proof. (1) is proved in [Klo80]. (2) is obtained in [Kha92]. The proof of (3) is routine (it is enough to consider the case when $|P| = |Q| = 1$). (4) follows both from (3) and (2), since descendants of a component can be defined both via descendants of symbols and via descendants of subterms. (Note that (2) can also be derived from (3) for the same reason.)

3 Neededness, Essentiality and Unabsorbedness

In this section, we recall Huet&Lévy and Maranget's notion of neededness, and relate them to the notion of essentiality, in OERSs. We also prove existence of an essential redex in any term, in an OERS, not in normal form.

Definition 3.1 A redex u in t is *Huet&Lévy-needed* [HuLé91, Lév80] if in each reduction of t to normal form (if any) at least one residual of u is contracted; u is *Maranget-needed* [Mar92] if u has at least one residual under any reduction starting from t that does not contract residuals of u.

Definition 3.2 A subterm s in t is *essential* (written $ES(s,t)$) if s has at least one descendant under any reduction starting from t and is *inessential* (written $IE(s,t)$) otherwise [Kha93, Kha88].

Definition 3.3 A subterm s of a term t is *unabsorbed in a reduction* $P : t \twoheadrightarrow e$ if none of the descendants of s appear in redex-arguments of terms in P, and is *absorbed in P* otherwise; s is *unabsorbed in t* if it is unabsorbed in any reduction starting from t and *absorbed in t* otherwise [Kha93].

Remark 3.1 It is easy to see that a redex $u \subseteq t$ is unabsorbed iff u is *external* [HuLé91] in t. Clearly, unabsorbedness implies essentiality, and Huet&Levy- and Maranget-neededness coincide for normalizable terms. □

Definition 3.4 Let $P : t \twoheadrightarrow s$ and o be a subterm or a component in t. Then we say that P *deletes* o if o doesn't have P-descendants.

Definition 3.5 (1) Let $P : t \rightarrow t_1 \rightarrow \ldots$ and $u \subseteq t_i$. Then u is called *erased* in P if there is $j > i$ such that u does not have residuals in t_j. P is *fair* if each redex in any t_i is erased in P [Klo92].

(2) We call a reduction P starting from t *strictly cofinal* if for any $Q : t \twoheadrightarrow e$ there is an initial part $P' : t \twoheadrightarrow s$ of P, and a reduction $Q' : e \twoheadrightarrow s$ such that $P' \approx_{st} Q + Q'$.

Lemma 3.1 Any strictly cofinal reduction P starting from t deletes all inessential subterms of t.

Proof. Immediate from Definitions 3.2 and 3.5.

For example, fair reductions are strictly cofinal: Klop's proof of cofinality of fair reductions (Theorem 12.3 in [Klo80]) can be modified to a proof of strict cofinality of fair reductions by using $CR(sub)$ (Theorem 2.1.(2)) instead of the CR theorem.

The following lemma from [Kha94] follows from $CR(sub)$ (Theorem 2.1.(2)); the proofs are same as for OTRS [Kha93].

Lemma 3.2 (1) Let $P : t \twoheadrightarrow t'$ and $s \subseteq t$. Then $IE(s,t)$ iff all P-descendants of s are inessential in t'.

(2) Let $t \xrightarrow{u} t'$ and $e \subseteq s \subseteq t$. Then any u-descendant of e is contained in some u-descendant of s.

(3) Let $e \subseteq s \subseteq t$ and $ES(e,t)$. Then $ES(s,t)$.

(4) Let s be a pattern-subterm of a redex $u \subseteq t$. Then $ES(u,t)$ iff $ES(s,t)$.

Notation We write $t = (t_1//s_1, \ldots, t_n//s_n)s$ if $s_1, \ldots, s_n$ are disjoint subterms in s and t is obtained from s by replacing them with $t_1, \ldots, t_n$, respectively.

Definition 3.6 Let $t = (t_1//s_1, \ldots, t_n//s_n)s$ in an OERS R and let $P : s = e_0 \xrightarrow{v_0} e_1 \xrightarrow{v_1} \ldots$ be an R_{fS}-reduction. We define the reduction $P\|(t) : t = o_0 \xrightarrow{u_0} o_1 \xrightarrow{u_1} \ldots$ as follows.

(1) If v_0 is an R_f-redex and its pattern does not overlap with $s_1, \ldots, s_n$, then u_0 is the corresponding subterm of v_0 in $t = o_0$.

(2) If v_0 is an R_f-redex that is not inside the subterms $s_1, \ldots, s_n$ and its pattern does overlap with some of $s_1, \ldots, s_n$, then $u_0 = \emptyset$.

(3) If v_0 is an S-redex that is outside the replaced subterms, then u_0 is the corresponding S-redex in t_0.

(4) If v_0 is in some of subterms $s_1, \ldots, s_n$, then $u_0 = \emptyset$.

In the first case, o_1 is obtained from e_1 by replacing the descendants of $s_1, \ldots, s_n$ with the corresponding descendants of $t_1, \ldots, t_n$, respectively; in the second case, o_1 is obtained from e_1 by replacing the descendant of v_0 by the descendant of its corresponding subterm in o_0, and by replacing the descendants of the subterms s_i that do not overlap with v_0 by the corresponding descendants of t_i; in the third case, o_1 is obtained from e_1 by replacing outermost descendants of $s_1, \ldots, s_n$ with the corresponding descendants of $t_1, \ldots, t_n$; in the fourth case, o_1 is obtained from e_1 by replacing the descendants of $s_1, \ldots, s_n$ with the corresponding descendants of $t_1, \ldots, t_n$, respectively. Thus, in o_1 we can choose the redex u_1 analogously, and so on. (Note that $P\|(t)$ depends not only on P and t, but also on the choice of $s_1, \ldots, s_k$, but the notation does not give rise to ambiguity.)

Lemma 3.3 Let $s_1, \ldots, s_n$ be inessential in s, in an OERS R, and let $t = (t_1//s_1, \ldots, t_n//s_n)s$. Then $t_1, \ldots, t_n$ are inessential in t.

Proof. We show by induction on $|Q|$ that if an R_{fS}-reduction $Q : s \twoheadrightarrow o$ deletes $s_1, \ldots, s_n$, then $Q\|(t)$ deletes $t_1, \ldots, t_n$; such a Q exists by Lemma 3.1. Let $Q = v + Q'$, $s \xrightarrow{v} e$, and $s'_1, \ldots, s'_m$ be all the v-descendants of $s_1, \ldots, s_n$. By Lemma 3.2.(1), s'_i are inessential in e. By Definition 3.6, if $v\|(t) : t \rightarrow o$, then o is obtained from e by replacing some inessential subterms that contain $s'_1, \ldots, s'_m$, and all the descendants of $t_1, \ldots, t_n$ also are in the replaced subterms of o. By the induction assumption, the replaced subterms in o are inessential. Hence, by Lemma 3.2.(3), all the descendants of $t_1, \ldots, t_n$ in o are inessential and, by Lemma 3.2.(1), $t_1, \ldots, t_n$ are inessential in t.

It follows immediately from Definition 3.6 and the proof of Lemma 3.3 that replacement of inessential subterms in a term does not effect its normal form.

Corollary 3.1 Any term not in normal form, in an OERS, contains an essential redex.

Proof. If all redexes in t were inessential, their replacement by fresh variables would yield a term in normal form containing inessential subterms, a contradiction.

Existence of an unabsorbed redex in any term not in normal form can be proved exactly as in OTRSs [Kha93] (the proof does not use the proof of Corollary 3.1).

Proposition 3.1 Let t be a term in an OERS R and let u be a redex in t. Then u is essential in t iff it is Maranget-needed in t.

Proof. ($\Leftarrow$) Let $IE(u, t)$. Further, let $FV(u) = \{x_1, \ldots, x_n\}$, let f be a fresh n-ary function symbol that does not occur in the left-hand sides of rewrite rules (we can safely add such a symbol to the alphabet, if necessary), and let $s = (f(x_1, \ldots, x_n)//u)t$. By Lemma 3.3, $IE(f(x_1, \ldots, x_n), s)$, i.e., there is some reduction P starting from s such that $f(x_1, \ldots, x_n)$ does not have P-descendants. Let $r : f(x_1, \ldots, x_n) \to u$ be a rule and $Q : s \to t$ be the r-reduction step. Obviously, $R \cup \{r\}$ is orthogonal. Hence $P + Q/P \approx_{st} Q + P/Q$ and therefore u does not have P/Q-descendants. But P/Q does not contract the residuals of u. Thus u is not Maranget-needed. ($\Rightarrow$) From Definitions 3.2 and 3.1.

4 Relative Notions of Neededness

In this section, we introduce notions of neededness relative to a set of reductions Π and to a set of terms $\mathcal{S}$. We show how all existing notions of neededness can be obtained by specifying Π or $\mathcal{S}$; $\mathcal{S}$-neededness is also a special case of Π-neededness. We introduce *stability* of a set of terms, in an OERS, and show that if $\mathcal{S}$ is not stable, contraction of $\mathcal{S}$-needed redexes in a term t need not terminate at a term in $\mathcal{S}$ even if t can be reduced to a term in $\mathcal{S}$. It is the aim of the last section to show that if $\mathcal{S}$ is stable, then a $\mathcal{S}$-needed strategy is $\mathcal{S}$-normalizing.

Definition 4.1 (1) We call a reduction P starting from a term t in an OERS R *external* to a component $e \sqsubseteq t$ if there is no redex executed in P whose pattern overlaps with a descendant of e (e can be empty). We call P *external* to a redex $u \subseteq t$ if P is external to $pat(u)$, i.e., if P doesn't contract the residuals of u.

(2) Let Π be a set of reductions. We call $e \sqsubseteq t$ *Π-needed* if there is no $P \in \Pi$ starting from t that is external to e, and call it *Π-unneeded* otherwise.

(3) We call $e \sqsubseteq t$ *P-(un)needed* if it is $\langle P \rangle_L$-(un)needed, where $\langle P \rangle_L$ is the set of all reductions Lévy-equivalent to P.

(4) Let $\mathcal{S}$ be a set of terms in R, and let $\Pi_{\mathcal{S}}$ be the set of $\mathcal{S}$-*normalizing* reductions, i.e., reductions that end at a term in $\mathcal{S}$. We call $e \sqsubseteq t$ *$\mathcal{S}$-(un)needed* if it is *$\Pi_{\mathcal{S}}$-(un)needed*.

(5) If $o \subseteq t$, then we call o *$(P, \Pi, \mathcal{S}$-$)$unneeded* (*needed*) if so is $Int(o)$, the component obtained from the subterm o by removing *all* bound variables. However, if not otherwise stated, we say that a redex $u \subseteq t$ (which is a subterm) is *$(P, \Pi, \mathcal{S}$-$)$unneeded* (*needed*) if so is its pattern.

(6) We say that $P : t \twoheadrightarrow o$ *$\mathcal{S}$-suppresses* $e \sqsubseteq t$ if P is $\mathcal{S}$-normalizing and is external to e. We say that P *$\mathcal{S}$-suppresses* $u \subseteq t$ if it $\mathcal{S}$-suppresses $pat(u)$, i.e., is $\mathcal{S}$-normalizing and is external to u. (Obviously, a redex or component is $\mathcal{S}$-unneeded iff it is $\mathcal{S}$-suppressed by some $\mathcal{S}$-normalizing reduction.)

We write $NE_{(P,\Pi,M)}(e, t)$ $(UN(e, t))$ if e is (P, Π, M)-needed (unneeded).

If $o \sqsubseteq t$ and $P : t \twoheadrightarrow s$ is external to o, then we call the descendants of o also the *residuals* of o. A component that does overlap with the pattern of a contracted redex does not have residuals. Note that if $u \subseteq t$, then $P : t \twoheadrightarrow s$ is external to $pat(u)$ iff P does not contract the residuals of u, because orthogonality of the system implies that if the pattern $pat(v)$ of a redex v contracted in P does overlap with a residual of $pat(u)$, then $pat(u) = pat(v)$. Hence u is $\mathcal{S}$-needed iff at least one residual of it is contracted in each reduction from t to a term in $\mathcal{S}$ (the intended notion of neededness). Obviously, any redex in a term that is not $\mathcal{S}$-normalizable is $\mathcal{S}$-needed; we call such redexes *trivially $\mathcal{S}$-needed.*

The following definition introduces the property of sets of terms for which it is possible to generalise the Normalization Theorem:

Definition 4.2 We call a set $\mathcal{S}$ of terms *stable* if:

(a) $\mathcal{S}$ is *closed under parallel moves*: for any $t \notin \mathcal{S}$, any $P : t \twoheadrightarrow o \in \mathcal{S}$, and any $Q : t \twoheadrightarrow e$, the final term of P/Q is in $\mathcal{S}$; and

(b) $\mathcal{S}$ is *closed under unneeded expansion*: for any $e \xrightarrow{u} o$ such that $e \notin \mathcal{S}$ and $o \in \mathcal{S}$, u is $\mathcal{S}$-needed.

Remark 4.1 Of course, a set closed under reduction is closed under parallel moves as well. But a set closed under parallel moves, even if closed under unneeded expansion, need not be closed under reduction. Indeed, consider $R = \{f(x) \to g(x,x),\ a \to b\}$, and take $\mathcal{S} = \{g(a,a), g(b,b)\}$. The only one-step $\mathcal{S}$-normalizing reductions are $g(a,b) \to g(b,b)$, $g(b,a) \to g(b,b)$, $f(a) \to g(a,a)$, and $f(b) \to g(b,b)$. Therefore, one can check that $\mathcal{S}$ is closed under unneeded expansion. Also, $\mathcal{S}$ is closed under parallel moves, since the right-bottom term $g(b,b)$ in the diagram below, which is the only non-trivial diagram to be checked, is in $\mathcal{S}$. However, $\mathcal{S}$ is not closed under reduction, since, e.g., $g(a,a) \to g(b,a)$, $g(a,a) \in \mathcal{S}$, but $g(b,a) \notin \mathcal{S}$. Note that the second occurrence of a in $g(a,a)$ is $\mathcal{S}$-unneeded, but its residual in $g(b,a)$ is $\mathcal{S}$-needed.

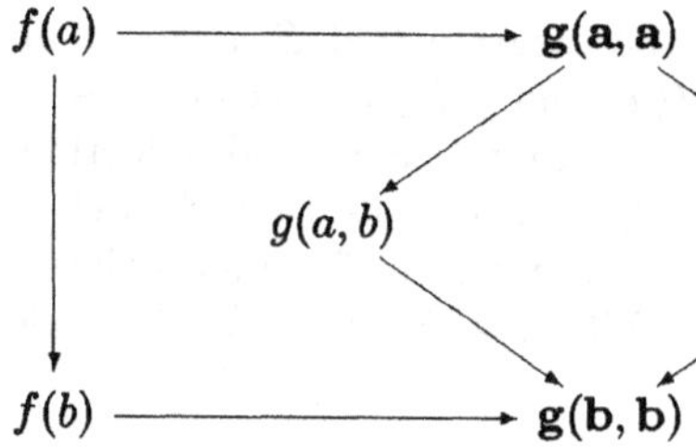

□

The most appealing examples of stable sets, for an OERS, are the set of normal forms [HuLé91], the set of head-normal forms [BKKS87], the set of weak-head-normal forms (a partial result is in [Mar92]), and the set of constructor-head-normal forms for constructor TRSs [Nök94]. The sets of terms having (resp. not having) (head-, constructor-head-) normal forms are stable as well. The graph G_s of a term s (which consists of terms to which s is reducible) is closed under reduction, but need not be closed under unneeded expansion. For example, the graph $G_{I(x)} = \{I(x), x\}$ of $I(x)$ is closed under reduction but is not closed under unneeded expansion: $I(I(x))$ can be reduced to $I(x)$ by reducing either I-redex (according to the rule $I(x) \to x$). Hence *none* of the redexes in $I(I(x))$ are $\mathcal{S}$-needed. Thus the closure of $\mathcal{S}$ under unneeded expansion is a necessary condition for the normalization theorem.

We say that a set $\mathcal{S}$ of terms is *closed under ($\mathcal{S}$)-normalization* if any reduct of every $\mathcal{S}$-normalizable term is still $\mathcal{S}$-normalizable. Obviously, sets closed under parallel moves are closed under normalization as well. Even if $\mathcal{S}$ is closed under unneeded expansion, closure of $\mathcal{S}$ under normalization is also necessary for the normalization theorem to be valid for $\mathcal{S}$. Indeed, consider $R = \{f(x) \to g(x,x),\ a \to b\}$, take $\mathcal{S} = \{g(a,b)\}$, and take $t = f(a)$. Then $t \to g(a,a) \to g(a,b)$ is an $\mathcal{S}$-needed $\mathcal{S}$-normalizing reduction, while after the $\mathcal{S}$-needed step $t \to f(b)$, the term $f(b)$ is not $\mathcal{S}$-normalizable any more (the only redex in $f(b)$ is only trivially $\mathcal{S}$-needed). However, the following example shows that closure of $\mathcal{S}$ under normalization (even in combination with closure of $\mathcal{S}$ under unneeded expansion) is not enough; closure of $\mathcal{S}$ under parallel moves is necessary.

Example 4.1 Let $R = \{f(x) \to g(x,x),\ a \to b,\ b \to a\}$ and $\mathcal{S} = \{g(a,b)\}$. Since the reduction preserves the height of a term and the property to be a ground term, only the terms in the following diagram are $\mathcal{S}$-normalizable.

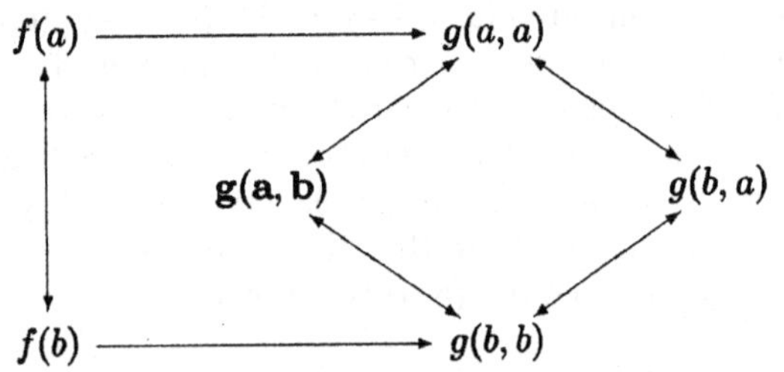

Therefore, it is clear from the diagram that $\mathcal{S}$ is closed under normalization. It is easy to see that, in $f(a)$ and $f(b)$, all the redexes are $\mathcal{S}$-needed; hence $f(a) \to f(b) \to f(a) \to \ldots$ is an infinite $\mathcal{S}$-needed reduction that never reaches $\mathcal{S}$ (there are many others). One can check that $\mathcal{S}$ is closed under unneeded expansion. Thus the reason for the failure of the normalization theorem is that, as it can be seen from the following diagram, $\mathcal{S}$ is not closed under parallel moves.

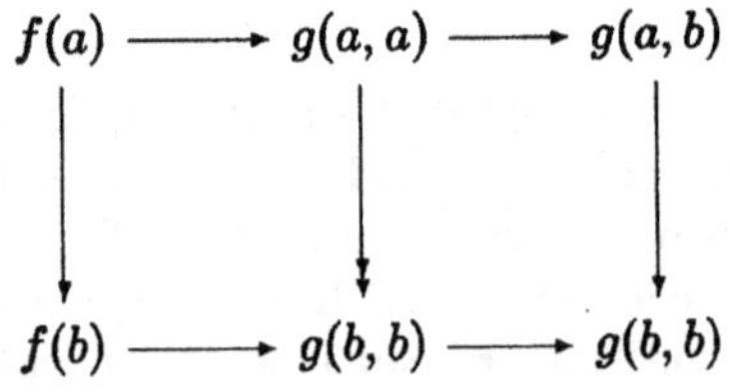

□

Proposition 4.1 A redex $u \subseteq t$ is Maranget-needed iff it is needed w.r.t. the set of all fair reductions starting from t.

Proof. ($\Rightarrow$) Let u be Maranget-needed in t. Then any fair P starting from t should contract a residual of u (to erase it). ($\Leftarrow$) If u is not Maranget-needed, i.e., there is a reduction $Q : t \twoheadrightarrow e$ in which u is erased and that does not contract the residuals of u, then there is a reduction Q' such that $Q^* = Q + Q'$ is fair and obviously u is not Q^*-needed, a contradiction.

Proposition 4.2 A redex $u \subseteq t$ is essential iff u (or $pat(u)$) is needed w.r.t. the set of all fair reduction starting from t.

Proof. An immediate corollary of Proposition 3.1 and Proposition 4.1.

Lemma 4.1 Let $P : t \twoheadrightarrow s$ be external to $e \sqsubseteq t$, and let $o \subseteq t$ be the subterm corresponding to e. Then any descendant of o along P is the subterm corresponding to some P-descendant of e.

Proof. Immediate from Definition 2.3, since P is external to e.

Proposition 4.3 A subterm $s \subseteq t$ is inessential iff there is a reduction that is external to $Int(s)$ and deletes it.

Proof. ($\Leftarrow$) Immediate from Lemma 4.1. ($\Rightarrow$) Let $x_1, \ldots, x_n$ be the list of occurrences of bound variables in s from left to right, let f be a fresh n-ary function symbol not occurring in left-hand sides of rewrite rules, and let $t^* = (f(x_1, \ldots, x_n)//s)t$. Since $IE(s,t)$, there is Q starting from t that deletes s. Therefore, it follows from Definition 3.6 that $P = ((Q\|t^*)\|t)$ is external to $Int(s)$ and deletes it.

5 A Labelling for OERSs

In Kennaway&Sleep [KeSl89] a labelling is introduced for OCRSs, based on the labelling system of Klop [Klo80], which is in turn a generalization of the labelling system for the λ-calculus introduced by Lévy [Lév78]. Each *label* of Kennaway&Sleep [KeSl89] is a tuple of labels, built up from a set of *base* labels. For any OERS R, terms in the corresponding labelled OERS R^L are those of R where each subterm has one or more labels, represented as a *string* of labels. A labelling of a term is *initial* if all its subterms are labelled by different base labels. The *signature* of a labelled term is the tuple of all its labels, from left to right. Rules of R^L are those of R where pattern-symbols in left-hand sides are labelled by a string of labels except for the head-symbol, which has just one label (a string of length one). Each subterm (including metavariables) in the right-hand side of a rule bears the signature of the corresponding left-hand side. Further, a *redex-index* of a redex is the maximal depth of nesting in the labels of the corresponding left-hand side of the rule. The *index* $Ind(P)$ of a reduction P is the maximal redex-index of redexes contracted in it.

The crucial properties of the labelling are given by the following propositions.

Proposition 5.1 [KeSl89] If a step $t \xrightarrow{u} s$ in an OERS R creates a redex $v \subseteq s$, then, for any labelling t^l of t, the corresponding step $t^l \xrightarrow{u^{l'}} s^{l''}$ in the corresponding labelled OERS R^L creates a redex v^{l^*} whose label l^* contains the label l' of u. Thus $Ind(u^{l'}) < Ind(v^{l^*})$. If $w \subseteq s^{l''}$ is a residual of a redex $w' \subseteq t^l$, then w and w' have the same labels, thus $Ind(w) = Ind(w')$.

Corollary 5.1 [KeSl89] Let P and Q be co-initial reductions such that P creates a redex u and Q does not contract residuals of any redex of t having a residual contracted in P. Then the redexes in $u/(Q/P)$ are created by P/Q and Q/P is external to u.

Proposition 5.2 [Klo80, Lév78] Any reduction in which only redexes with a bounded redex-index are contracted is terminating.

Remark 5.1 The above propositions are obtained for OCRS, but it is straightforward to carry them over OERSs. □

Definition 5.1 (1) For any co-initial reductions P and Q, the redex Qv in the final term of Q (read as v *with history* Q) is called a *copy* of a redex Pu if $P \trianglelefteq Q$, i.e., $P + Q/P \approx_L Q$, and v is a Q/P-residual of u; the *zig-zag* relation $\simeq_z$ is the symmetric and transitive closure of the copy relation [Lév80]. A *family* relation is an equivalence relation among redexes with histories containing the zig-zag relation.

(2) For any co-initial reductions P and Q, the redexes Qv and Pu are in the same *labelling-family* if for any initial labelling of the initial term of P and Q, they bear the same labels.

Proposition 5.1 implies that the labelling-family relation is indeed a family relation. As pointed out in [AsLa93], for OERSs in general the zig-zag and labelling family relations do not coincide. Below by family we always mean the labelling-family.

6 The Relative Normalization Theorem

In this section, we pre nt a uniform proof of correctness of the needed strategy that works for all stable sets of 'normal forms'. Our proof is different from all known proofs because properties of needed and unneeded components are different in the general case (the main difference is that a component under an unneeded component may be needed). However, the termination argument we use is the same as in [KeSl89] and in [Mar92], and is based on Proposition 5.2. The main idea and a proof in the same spirit is already in [Lév80].

Below in this section S always denotes a stable set of terms.

Lemma 6.1 Let $t \xrightarrow{w} s$, $v \subseteq t$, $o \sqsubseteq t$, and let $pat(v) \cap o = \emptyset$. Further, let $v' \subseteq s$ be a w-residual of v and $o' \sqsubseteq s$ be a w-descendant of o. Then $pat(v') \cap o' = \emptyset$.

Proof. Immediate from Definition 2.3.

Corollary 6.1 Let F be a set of redexes in t, and let every redex $v \in F$ be external to $s \sqsubseteq t$. Then any development of F is external to s.

Lemma 6.2 Let $t \xrightarrow{v} s$, $P : t \twoheadrightarrow o$, $e \sqsubseteq t$, and v be external to e. Then v/P is external to every P-descendant of e.

Proof. By Lemma 6.1, every P-residual of v is externa to each P-descendant of e, and the lemma follows from Corollary 6.1.

Lemma 6.3 (1) Let $s_1, \ldots, s_n \sqsubseteq s$ be disjoint, let $P : s \twoheadrightarrow e \neq \emptyset$ be external to $s_1, \ldots, s_n$, let $P^* : s \twoheadrightarrow s^*$, let $s_1^*, \ldots, s_m^*$ be all P^*-descendants of $s_1, \ldots, s_n$ in s^*, and let $Q = P/P^* : s^* \twoheadrightarrow e^*$. Then Q is external to $s_1^*, \ldots, s_m^*$.

(2) If P $\mathcal{S}$-suppresses $s_1, \ldots, s_n$, then Q $\mathcal{S}$-suppresses $s_1^*, \ldots, s_m^*$.

Proof. (1) By induction on $|P|$. Let $P = v + P'$, let $s'_1, \ldots, s'_l$ be v-descendants of $s_1, \ldots, s_n$, and let $s_1'^*, \ldots, s_k'^*$ be P^*/v-descendants of $s'_1, \ldots, s'_l$. By $CR(com)$ (Theorem 2.1.(4)), $s_1'^*, \ldots, s_k'^*$ are v/P^*-descendants of $s_1^*, \ldots, s_m^*$. By the induction assumption, $P'/(P^*/v)$ is external to $s_1'^*, \ldots, s_k'^*$. But by Lemma 6.2 v/P^* is external to $s_1^*, \ldots, s_m^*$; hence $P/P^* = v/P^* + P'/(P^*/v)$ is external to $s_1^*, \ldots, s_m^*$.

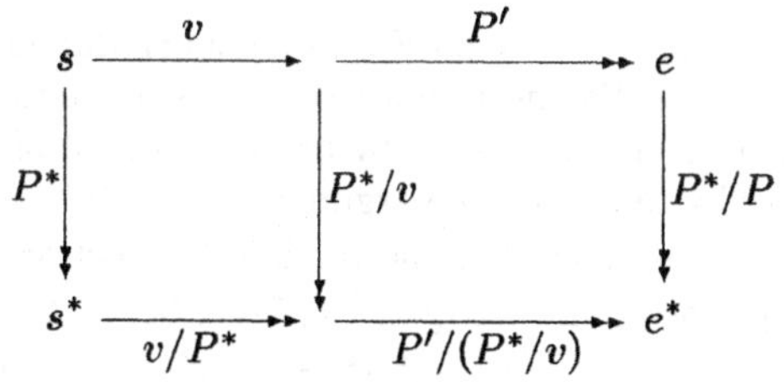

(2) By (1) and closure of $\mathcal{S}$ under parallel moves.

Corollary 6.2 (1) Descendants of $\mathcal{S}$-unneeded redexes of $t \notin \mathcal{S}$ remain $\mathcal{S}$-unneeded.

(2) Residuals of $\mathcal{S}$-unneeded redexes of $t \notin \mathcal{S}$ remain $\mathcal{S}$-unneeded.

Lemma 6.4 (1) Let $t \xrightarrow{u} t'$ and $e \sqsubseteq s \sqsubseteq t$. Then any u-descendant of e is contained in some u-descendant of s.

(2) Let $e \sqsubseteq s \sqsubseteq t$ and $NE_{\mathcal{S}}(e,t)$. Then $NE_{\mathcal{S}}(s,t)$.

(3) Let $u \subseteq t$ and let $s \sqsubseteq pat(u)$. Then $NE_{\mathcal{S}}(u,t)$ iff $NE_{\mathcal{S}}(s,t)$.

Proof. (1) By Definition 2.3.

(2) By (1) and Definition 4.1.

(3) From Definition 4.1, since a reduction $\mathcal{S}$-suppresses s iff it $\mathcal{S}$-suppresses u (orthogonality of the system implies that any redex whose pattern contains a symbol from a residual of $pat(u)$ coincides with $pat(u)$ and hence contains a symbol from a residual of s as well).

Note that if a component $s \sqsubseteq t$ is below $o \sqsubseteq t$, then $UN_{\mathcal{S}}(o,t)$ does not necessarily imply $UN_{\mathcal{S}}(s,t)$, although the inessentiality of the subterm corresponding to o implies that of the subterm corresponding to e (Lemma 3.2.(3)). Take for example $R = \{f(x) \to g(x),\ a \to b\}$, and take for $\mathcal{S}$ the set of terms not containing occurrences of a. Then $\mathcal{S}$ is stable, a is $\mathcal{S}$-needed in $f(a)$, but $f(a)$ is not.

Lemma 6.5 Let $t \notin \mathcal{S}$, $t \xrightarrow{u} t'$, $UN_{\mathcal{S}}(u,t)$, and let $u' \subseteq t'$ be a u-new redex. Then $UN_{\mathcal{S}}(u',t')$.

Proof. $UN_{\mathcal{S}}(u,t)$ implies existence of $P : t \twoheadrightarrow e$ that $\mathcal{S}$-suppresses $pat(u)$; thus P is external to u. By Corollary 5.1, P/u is external to u'. Also, P/u is $\mathcal{S}$-normalizing since $\mathcal{S}$ is closed under parallel moves. Hence u' is $\mathcal{S}$-unneeded.

We call $P : t_0 \to t_1 \to \ldots$ *$\mathcal{S}$-needed* if it contracts only $\mathcal{S}$-needed redexes.

Theorem 6.1 (Relative Normalization) Let $\mathcal{S}$ be a stable set of terms in an OERS R.

(1) Any $\mathcal{S}$-normalizable term $t \notin \mathcal{S}$ in R contains an $\mathcal{S}$-needed redex.

(2) If $t \notin \mathcal{S}$ is $\mathcal{S}$-normalizable, then any $\mathcal{S}$-needed reduction starting from t eventually ends at a term in $\mathcal{S}$.

Proof. (1) Let $P : t \twoheadrightarrow s \xrightarrow{u} e$ be an $\mathcal{S}$-normalizing reduction that doesn't contain terms in $\mathcal{S}$ except for e. By the stability of $\mathcal{S}$, u is $\mathcal{S}$-needed. By Corollary 6.2.(2) and Lemma 6.5, it is either created by or is a residual of an $\mathcal{S}$-needed redex in s, and (1) follows by repeating the argument.

(2) Let $P : t \twoheadrightarrow s$ be an $\mathcal{S}$-normalizing reduction that doesn't contain terms in $\mathcal{S}$ except for e, and let $Q : t \xrightarrow{u_0} t_1 \xrightarrow{u_1} \ldots$ be an $\mathcal{S}$-needed reduction. Further, let $Q_i : t \xrightarrow{u_0} t_1 \xrightarrow{u_1} \ldots \xrightarrow{u_{i-1}} t_i$ and $P_i = P/Q_i$ $(i \geq 1)$. By Proposition 5.1, $Ind(P_i) \leq Ind(P)$. Since Q is $\mathcal{S}$-needed and P_i is $\mathcal{S}$-normalizing (by the closure of $\mathcal{S}$ under parallel moves), at least one residual of u_i is contracted in P_i. Therefore, again by Proposition 5.1, $Ind(u_i) \leq Ind(P_i)$. Hence $Ind(Q) \leq Ind(P)$ and Q is terminating by Proposition 5.2.

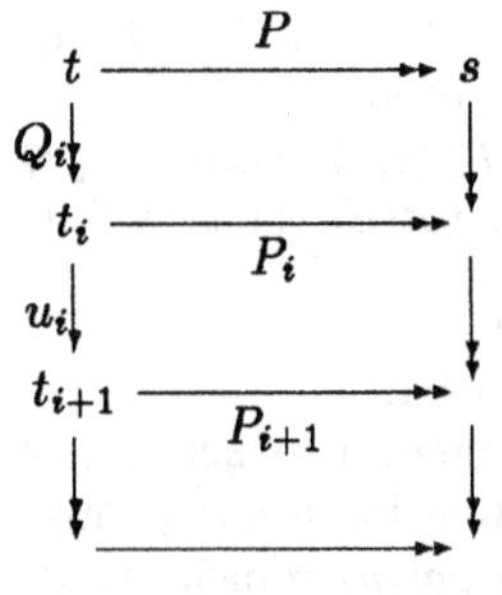

Lemma 6.6 Let t be $\mathcal{S}$-normalizable, let $t \xrightarrow{u} s$, $e \sqsubseteq t$, $NE_{\mathcal{S}}(e,t)$, and $pat(u) \cap e = \emptyset$. Then e has at least one $\mathcal{S}$-needed u-residual in s. In particular, any $\mathcal{S}$-needed redex $v \subseteq t$ different from u has an $\mathcal{S}$-needed residual.

Proof. Let $P : s \twoheadrightarrow o$ be an $\mathcal{S}$-needed $\mathcal{S}$-normalizing reduction; there is one by Theorem 6.1. Then if all u-residuals of e were $\mathcal{S}$-unneeded, P would $\mathcal{S}$-suppress them, and $u + P$ would $\mathcal{S}$-suppress e, a contradiction.

We call a stable set $\mathcal{S}$ *regular* if $\mathcal{S}$-unneeded redexes cannot duplicate $\mathcal{S}$-needed ones. One can show using Lemma 6.6 (e.g., as in [KeSl89] or in [Kha88]) that, for any regular stable $\mathcal{S}$, the $\mathcal{S}$-needed strategy is $\mathcal{S}$-*hypernormalizing*. That is, a term is $\mathcal{S}$-normalizing iff it does not have a reduction which contracts infinitely many $\mathcal{S}$-needed redexes. However, this is not the case for some irregular stable $\mathcal{S}$. Indeed, consider the OTRS $R = \{f(x) \to h(f(x), f(x)),\ a \to b\}$ and take for $\mathcal{S}$ the set of terms not containing occurrences of a. Then the reduction $f(a) \to h(f(a), f(a)) \to h(f(b), f(a)) \to h(f(b), h(f(a), f(a))) \to h(f(b), h(f(b), f(a))) \to \ldots$ contracts infinitely many $\mathcal{S}$-needed redexes, while $f(a) \to f(b)$ is $\mathcal{S}$-normalizing. This example shows also that multistep $\mathcal{S}$-needed reductions need not be $\mathcal{S}$-normalizing — just omit in the above reduction the initial step and group each pair of consecutive steps as a single multistep. Recall that multistep needed reductions are normalizing in the λ-calculus [Lév80]. The same holds for all regular stable $\mathcal{S}$; this follows immediately from hypernormalization of the $\mathcal{S}$-needed strategy for such $\mathcal{S}$.

7 Conclusions and Future Work

We have introduced a relative notion of neededness and proved the Relative Normalization Theorem in OERSs. We expect that this and other results of this paper can be proved for other higher-order rewriting systems too. Analogous questions arise for other strategies. For example, how can one construct reductions that avoid head-normal forms? Besides strong sequentiality for normal forms studied in Huet&Lévy [HuLé91], strong sequentiality is studied w.r.t. head-normal forms in Kennaway [Ken94]. Investigation of relative strong sequentiality and related to it strictness analysis (see e.g., [Nök94]) seems also as an interesting topic for future research. In forthcoming papers, we extend the theory of relative normalization in two directions: we study *minimal* and *optimal* relative normalization in OERSs, and study relative normalization in an abstract setting (in *Deterministic Residual Structures* and in *Family Structures*).

Acknowledgements

We thank J. R. Kennaway, F. van Raamsdonk, and M. R. Sleep for useful discussions, and J.-J. Lévy, L. Maranget, and P.-A. Melliès for help in overcoming difficulties in an early version of the paper. The use of empty descendants in the definition of descendants of components was suggested by L. Maranget. Some of the diagrams were drawn using P. Taylor's diagram package.

References

[Acz78] Aczel P. A general Church-Rosser theorem. Preprint, University of Manchester, 1978.

[AEH94] Antoy S., Echahed R., Hanus M. A needed narrowing strategy. In: Proc. of the 21st ACM Symposium on Principles of Programming Languages, POPL'94, Portland, Oregon, 1994.

[AsLa93] Asperti A., Laneve C. Interaction Systems I: The theory of optimal reductions. Mathematical Structures in Computer Science, vol. 11, Cambridge University Press, 1993, p. 1-48.

[Bar84] Barendregt H. P. The Lambda Calculus, its Syntax and Semantics. North-Holland, 1984.

[BKKS87] Barendregt H. P., Kennaway J. R., Klop J. W., Sleep M. R. Needed Reduction and spine strategies for the lambda calculus. Information and Computation, v. 75, no. 3, 1987, p. 191-231.

[DeJo90] Dershowitz N., Jouannaud J.-P. Rewrite Systems. In: J. van Leeuwen ed. Handbook of Theoretical Computer Science, Chapter 6, vol. B, 1990, p. 243-320.

[Gar94] Gardner P. Discovering needed reductions using type theory. In: Proc. of the 2nd International Symposium on Theoretical Aspects of Computer Software, TACS'94, Springer LNCS, v. 789, M. Hagiya, J. C. Mitchell, eds. Sendai, 1994, p. 555-574.

[HuLé91] Huet G., Lévy J.-J. Computations in Orthogonal Rewriting Systems. In: Computational Logic, Essays in Honor of Alan Robinson, J.-L. Lassez and G. Plotkin, eds. MIT Press, 1991.

[Ken94] Kennaway J. R. A conflict between call-by-need computation and parallelism. Workshop on conditional (and typed) term rewriting systems, Jerusalem, 1994.

[KeSl89] Kennaway J. R., Sleep M. R. Neededness is hypernormalizing in regular combinatory reduction systems. Preprint, School of Information Systems, University of East Anglia, Norwich, 1989.

[Kha88] Khasidashvili Z. β-reductions and β-developments of λ-terms with the least number of steps. In: Proc. of the International Conference on Computer Logic COLOG'88, Tallinn 1988, Springer LNCS, v. 417, P. Martin-Löf and G. Mints, eds. 1990, p. 105–111.

[Kha90] Khasidashvili Z. Expression Reduction Systems. Proceedings of I. Vekua Institute of Applied Mathematics of Tbilisi State University, vol. 36, 1990, p. 200-220.

[Kha92] Khasidashvili Z. The Church-Rosser theorem in Orthogonal Combinatory Reduction Systems. Report 1825, INRIA Rocquencourt, 1992.

[Kha93] Khasidashvili Z. Optimal normalization in orthogonal term rewriting systems. In: Proc. of the 5th International Conference on Rewriting Techniques and Applications, RTA'93, Springer LNCS, vol. 690, C. Kirchner, ed. Montreal, 1993, p. 243-258.

[Kha94] Khasidashvili Z. On higher order recursive program schemes. In: Proc. of the 19th International Colloquium on Trees in Algebra and Programming, CAAP'94, Springer LNCS, vol. 787, S. Tison, ed. Edinburgh, 1994, p. 172-186.

[Klo80] Klop J. W. Combinatory Reduction Systems. Mathematical Centre Tracts n. 127, CWI, Amsterdam, 1980.

[Klo92] Klop J. W. Term Rewriting Systems. In: S. Abramsky, D. Gabbay, and T. Maibaum eds. Handbook of Logic in Computer Science, vol. II, Oxford University Press, 1992, p. 1-116.

[KOR93] Klop J. W., van Oostrom V., van Raamsdonk F. Combinatory reduction systems: introduction and survey. In: To Corrado Böhm, J. of Theoretical Computer Science 121, 1993, p. 279-308.

[Lév78] Lévy J.-J. Réductions correctes et optimales dans le lambda-calcul, Thèse de l'Université de Paris VII, 1978.

[Lév80] Lévy J.-J. Optimal reductions in the Lambda-calculus. In: To H. B. Curry: Essays on Combinatory Logic, Lambda-calculus and Formalism, Hindley J. R., Seldin J. P. eds, Academic Press, 1980, p. 159-192.

[Mar92] Maranget L. La stratégie paresseuse. Thèse de l'Université de Paris VII, 1992.

[Nip93] Nipkow T. Orthogonal higher-order rewrite systems are confluent. In: Proc. of the 1st International Conference on Typed Lambda Calculus and Applications, TLCA'93, Springer LNCS, vol. 664, Bazem M., Groote J.F., eds. Utrecht, 1993, p. 306-317.

[Nök94] Nöcker E. Efficient Functional Programming. Compilation and Programming Techniques. Ph. D. Thesis, Katholic University of Nijmegen, 1994.

[Oos94] Van Oostrom V. Confluence for Abstract and Higher-Order Rewriting. Ph. D. Thesis, Free University of Amsterdam, 1994.

[OR94] Van Oostrom V., van Raamsdonk F. Weak orthogonality implies confluence: the higher-order case. In: Proc. of the 3rd International Conference on Logical Foundations of Computer Science, 'Logic at St. Petersburg', LFCS'94, Springer LNCS, vol. 813, Narode A., Matiyasevich Yu. V. eds. St. Petersburg, 1994. p. 379-392.

[Pkh77] Pkhakadze Sh. Some problems of the Notation Theory (in Russian). Proceedings of I. Vekua Institute of Applied Mathematics of Tbilisi State University, Tbilisi, 1977.

[Wol93] Wolfram D.A. The Clausal Theory of Types. Cambridge Tracts in Theoretical Computer Science, vol. 21, Cambridge University Press, 1993.

On Termination and Confluence of Conditional Rewrite Systems

Bernhard Gramlich*

Fachbereich Informatik, Universität Kaiserslautern
Postfach 3049, D-67653 Kaiserslautern, Germany
gramlich@informatik.uni-kl.de

Abstract. We consider the problem of verifying confluence and termination of conditional term rewriting systems. Recently we have obtained some interesting results for unconditional term rewriting systems (TRSs) which are non-overlapping or, more generally, locally confluent overlay systems. These results provide sufficient criteria for termination plus confluence in terms of restricted termination and confluence properties (Gramlich 1994a). Here we generalize our approach to the conditional case and show how to solve the additional complications due to the presence of conditions in the rules. Our main result can be stated as follows: Any conditional TRS (CTRS) which is an innermost terminating overlay system such that all (conditional) critical pairs are joinable is complete, i.e., terminating and confluent.

1 Introduction and Overview

Due to the fact that termination is a fundamental property of TRSs but undecidable in general (Huet & Lankford 1978), many sufficient criteria, techniques and methods for proving termination have been developed, see (Dershowitz 1987) for a survey. Most practically applicable approaches are based on reduction orderings, i.e. well-founded term orderings which are stable w.r.t. substitutions and monotonic w.r.t. the term structure.

Here we follow another line of reasoning and investigate under which conditions one can infer termination (plus confluence) from restricted termination (and confluence) properties. More precisely, we are interested in cases for which termination of rewriting under some strategy implies termination under arbitrary strategies. Particularly interesting from a practical point of view is the innermost reduction strategy since it corresponds closely to the evaluation mechanism employed in many rewriting based computation models, e.g. in functional programming languages or formalisms for recursive function definitions. Note moreover that for the latter case sophisticated techniques for verifying termination by means of recursion analysis are available, cf. e.g. (Boyer & Moore 1979; Walther 1988). The notion of termination in this functional framework

* This research was supported by the 'Deutsche Forschungsgemeinschaft, SFB 314 (D4-Projekt)'.

corresponds to innermost termination from a more abstract rewriting point of view. Since innermost termination may be much easier to verify than general termination, our results bear considerable practical relevance.

The first results about termination of rewriting under such restrictions and its relation to termination (of unrestricted rewriting) have been obtained for orthogonal, i.e. left-linear and non-overlapping (unconditional) TRSs (cf. e.g. (Klop 1992) for a survey). Recently we have shown (Gramlich 1994a) that most of these results can be generalized to non-overlapping TRSs. The main result of (Gramlich1994a) states that even for locally confluent (unconditional) overlay systems innermost termination suffices for guaranteeing general termination and hence also completeness.

Since it is well-known that conditional rewriting is much more difficult to handle in theory and practice, it was not clear a priori that our results carry over to the conditional case. But this is indeed possible (for the main results) as can be shown by a careful analysis and inspection of the proofs for the unconditional case and by taking into account the additional complications arising with CTRSs. From a technical point of view one of the main problems in generalizing our results is the well-known fact that *variable overlaps* may be critical for CTRSs and need to be explicitly handled. Another related problem with CTRSs is the fact that – in contrast to the unconditional case – joinability of critical pairs is not sufficient anymore for guaranteeing local confluence.

The rest of the paper is structured as follows. After introducing some basic definitions and notions we present and discuss in section 3 our main results on restricted termination and confluence properties of CTRSs. Missing details can be found in (Gramlich 1993, 1994a). Finally we briefly discuss some related work.

2 Preliminaries

We assume familiarity with the basic terminology needed for dealing with (C)TRSs, cf. e.g. (Dershowitz et al. 1990; Klop 1992; Middeldorp 1993).

A TRS $\mathcal{R}$ is *(strongly) terminating* or *strongly normalizing (SN)* if $\rightarrow$ is noetherian, i.e. if there is no infinite reduction sequence $s_1 \rightarrow s_2 \rightarrow s_3 \rightarrow \cdots$. It is said to be *weakly terminating* or *weakly normalizing (WN)* if every term can be reduced to some normal form. If $s \rightarrow t$ then, in order to make explicit the position p of the reduced subterm, the applied substitution σ and the applied rule $l \rightarrow r$, we shall sometimes use the notation $s \rightarrow_{p,\sigma,l\rightarrow r} t$ or $s \rightarrow_p t$. The topmost (or *root*) position in a term is denoted by λ. By $s \rightarrow^k t$ we mean that s reduces to t in k steps. A reduction step $s \rightarrow t$ by applying some rule of $\mathcal{R}$ at position p in s is *innermost* if every proper subterm of s/p is irreducible. In that case we also write $s \rightarrow t$. $\mathcal{R}$ is *(strongly) innermost terminating* or *(strongly) innermost normalizing (SIN)* if every sequence of innermost reduction steps terminates. It is *weakly innermost terminating* or *weakly innermost normalizing (WIN)* if every term can be reduced to some normal form by innermost reduction steps.

A TRS is *confluent* or has the *Church-Rosser property (CR)* if ${}^* \leftarrow \circ \rightarrow^* \subseteq \rightarrow^* \circ {}^* \leftarrow$ and *weakly Church-Rosser (WCR)* or *locally confluent* if $\leftarrow \circ \rightarrow$

$\subseteq \rightarrow^* \circ\ ^*\leftarrow$.[2] $\mathcal{R}$ has the property WCR^1 if $t \leftarrow s \rightarrow u$ and $t \neq u$ imply that there exists some v with $t \rightarrow v \leftarrow u$. A confluent and terminating TRS is said to be *convergent* or *complete* . If $l_1 \rightarrow r_1$, $l_2 \rightarrow r_2$ are two rules of $\mathcal{R}$ (w.l.o.g. we assume that they do not share common variables), with p some non-variable position of $l_2 \rightarrow r_2$ such that l_1 and l_2/p are unifiable with most general unifier σ then $\langle\sigma(l_2[p \leftarrow r_1]), \sigma(r_2)\rangle$ is said to be a *critical pair (CP)* of $\mathcal{R}$ (obtained by overlapping $l_1 \rightarrow r_1$ into $l_2 \rightarrow r_2$ at position p), provided that p is not the top position if the two rules are renamed versions of the same rule. Joinability of all critical pairs is abbreviated by JCP. A critical pair of the form $\langle s, s\rangle$ is said to be *trivial*.

A TRS $\mathcal{R}$ is said to be *non-overlapping (NO)* if there is no critical pair between rules of $\mathcal{R}$. *Weakly non-overlapping* TRSs have only trivial critical pairs. $\mathcal{R}$ is *left-linear (LL)* if every variable occurs at most once in every left hand side of an $\mathcal{R}$-rule. $\mathcal{R}$ is *orthogonal / weakly orthogonal* if it is left-linear and non-overlapping / weakly non-overlapping. It is *non-erasing (NE)* if r and l have the same set of variables for every rule $l \rightarrow r \in \mathcal{R}$. If every critical pair of a TRS $\mathcal{R}$ is obtained by an *overlay*, i.e. by overlapping left hand sides of rules at top position then $\mathcal{R}$ is said to be an *overlay system (OS)*.

By $P(\mathcal{R})$ we mean that the TRSs $\mathcal{R}$ has property P. Moreover we also ambiguously use the notation $P(t)$ for terms t provided there is a sensible local interpretation for $P(t)$. For instance, $CR(t)$ is to denote the property that whenever we have $t \rightarrow^* v$ and $t \rightarrow^* w$ then there exists a term s with $v \rightarrow^* s$ and $w \rightarrow^* s$.

Some basic properties of TRSs (or, more generally, *abstract reduction systems*, cf. (Newman 1942; Klop1992)) which are explicitly or implicitly used subsequently, are the following:

- WCR and SN imply CR (*Newman's Lemma*).
- WCR^1 and WN imply SN.
- WCR^1 implies CR.

For TRSs the following well-known results are fundamental.

- WCR is equivalent to JCP (*Critical Pair Lemma*, (Huet 1980)).
- JCP and SN imply CR (Knuth & Bendix 1970).

Moreover, we need some basic terminology about conditional term rewriting systems (CTRSs). A CTRS is set of *conditional rewrite rules* of the form

$$s_1 = t_1 \ \wedge \ldots \wedge \ s_n = t_n \implies l \rightarrow r$$

. Here we require $l \notin \mathcal{V}$ and $\mathcal{V}(r) \subseteq \mathcal{V}(l)$, i.e. no variable left hand sides and no extra variables on the right hand side. Extra variables in conditions are allowed if not stated otherwise. Depending on the interpretation of the equality sign in the conditions of rewrite rules, different reduction relations may be associated with a given CTRS as usual. In a *join CTRS* $\mathcal{R}$ the equality sign in the conditions of rewrite rules is interpreted as joinability, i.e. as $\downarrow$. In a *normal CTRS* $\mathcal{R}$

[2] Here, '$\circ$' denotes relation composition.

a condition $s = t$ is satisfied if $s \to^* t$ and t is a ground normal form (w.r.t. the unconditional version of $\mathcal{R}$). *Semi-equational* CTRSs are obtained by interpreting the equality sign in the conditions as convertibility, i.e. as $\overset{*}{\leftrightarrow}$. A *generalized* CTRS has rules of the form $P_1 \wedge \ldots, \wedge P_n \Longrightarrow l \to r$ where the conditions P_i, $1 \leq i \leq n$, are formulated in a general mathematical framework, e.g. in some first order language.

For the sake of readability we shall use in the following some compact notations for conditional rules and conjunctions of conditions. When writing $P \Longrightarrow l \to r$ for some conditional rewrite rule then P stands for the conjunction of all conditions. Similarly, $P\downarrow$ means joinability of all conditions in P, and σP means that all conditions in P are instantiated by σ.

Definition 1 (reduction relation, depth). The reduction relation corresponding to a given (join, semi-equational or normal) CTRS $\mathcal{R}$ is inductively defined as follows ($\Box$ denotes $\downarrow$ or $\overset{*}{\leftrightarrow}$, respectively):

$$\begin{aligned} \mathcal{R}_0 &= \emptyset\,, \\ \mathcal{R}_{i+1} &= \{\sigma l \to \sigma r \mid P \Longrightarrow l \to r \in \mathcal{R}, \sigma u \Box_{\mathcal{R}_i} \sigma v \text{ for all } u \Box v \text{ in } P\}\,,^3 \\ s \to_{\mathcal{R}} t :&\Longleftrightarrow s \to_{\mathcal{R}_i} t \text{ for some } i \geq 0, \text{ i.e. } \to_{\mathcal{R}} = \bigcup_{i \geq 0} \to_{\mathcal{R}_i}\,. \end{aligned}$$

If $s \to_{\mathcal{R}} t$ then the *depth* of $s \to_{\mathcal{R}} t$ is defined to be the minimal n with $s \to_{\mathcal{R}_n} t$. For $s \to^*_{\mathcal{R}} t$ and $s \downarrow_{\mathcal{R}} t$ depths are defined analogously. More precisely, if $s \to^*_{\mathcal{R}} t$ then the *depth* of $s \to^*_{\mathcal{R}} t$ is defined to be the minimal n with $s \to^*_{\mathcal{R}_n} t$. The *depth* of $s \downarrow_{\mathcal{R}} t$ is the minimal n with $s \downarrow_{\mathcal{R}_n} t$. If the depth of $s \to^*_{\mathcal{R}} t$ is at most n we denote this by $s \xrightarrow{n}{}^{*}_{\mathcal{R}} t$.

Remark. Note that instead of a CTRS $\mathcal{R}$ one may somehow equivalently consider the extended system $\mathcal{R}' := \mathcal{R} \uplus \{eq(x,x) \to true\}$. More precisely, taking – within a many-sorted framework – $\mathcal{R}' := \mathcal{R} \uplus \{eq(x,x) \to true\}$, with eq a fresh binary function symbol and $true$ a fresh constant of a new sort (with x a variable of the 'old' sort), it is easily shown that $\mathcal{R}'$ is a conservative extension of $\mathcal{R}$ in the following sense: for all 'old' terms s, t we have:[4] $s \xrightarrow{n}_{\mathcal{R}} t \Longleftrightarrow s \xrightarrow{n}_{\mathcal{R}'} t$, $s \to_{\mathcal{R}} t \Longleftrightarrow s \to_{\mathcal{R}'} t$, $s \Box^n_{\mathcal{R}} t \Longleftrightarrow eq(s,t) \Box^n_{\mathcal{R}'} true$ (for $n \geq 1$),[5] $eq(s,t) \xrightarrow{n}{}^{*}_{\mathcal{R}'} eq(u,v) \Longleftrightarrow s \xrightarrow{n}{}^{*}_{\mathcal{R}} u \wedge t \xrightarrow{n}{}^{*}_{\mathcal{R}} v$, $eq(s,t) \xrightarrow{n}{}^{*}_{\mathcal{R}'} true \Longleftrightarrow \exists w : eq(s,t) \xrightarrow{n}{}^{*}_{\mathcal{R}} eq(w,w) \to_{\mathcal{R}'} true$ (again for $n \geq 1$), for $\Box$ denoting $\downarrow$ or $\overset{*}{\leftrightarrow}$, respectively. From these properties it is straightforward to infer that properties like termination, confluence, local confluence and joinability of critical pairs are

[3] Note in particular that all unconditional rules of $\mathcal{R}$ are contained in $\mathcal{R}_1$ (because the empty conditions are vacuously satisfied) as well as all conditional rules with trivial conditions only, i.e. conditions of the form $s \Box s$. In fact, rules of the latter class can be considered to be essentially unconditional.

[4] Note that $s \Box^n_{\mathcal{R}} t$ is to denote that the depth of $s \Box_{\mathcal{R}} t$ is at most n.

[5] In order to obtain the equivalence $s \Box^n_{\mathcal{R}} t \Longleftrightarrow eq(s,t) \Box^n_{\mathcal{R}'} true$ for $n = 0$, too, one would have to include the rule $eq(x,x) \to true$ into $\mathcal{R}_0$ instead of $\mathcal{R}_1$ in Definition 1 as it is sometimes done in the literature.

not affected by considering $\mathcal{R}'$ instead of $\mathcal{R}$, or vice versa. Note in particular, that for join CTRSs the equivalence $s \downarrow_{\mathcal{R}}^{n} t \iff eq(s,t) \downarrow_{\mathcal{R}'}^{n} true$ (for $n \geq 1$) means: $s \downarrow_{\mathcal{R}}^{n} t \iff eq(s,t) \xrightarrow{n}{}^{*}_{\mathcal{R}'} true$, since $true$ is irreducible. This (depth preserving!) encoding of joinability into reducibility by means of an equality predicate is particularly useful for proof-technical reasons as we shall see later on (in the proof of Theorem 7).

Definition 2 (*conditional critical pairs*). Let $\mathcal{R}$ be a join CTRS, and let $P_1 \Longrightarrow l_1 \to r_1$ and $P_2 \Longrightarrow l_2 \to r_2$ be two rewrite rules of $\mathcal{R}$ which have no variables in common. Suppose[6] $l_1 = C[t]_p$ with $t \notin \mathcal{V}$ for some (possibly empty) context $C[]_p$ such that t and l_2 are unifiable with most general unifier σ, i.e. $\sigma(t) = \sigma(l_1/p) = \sigma(l_2)$. Then $\sigma(P_1) \wedge \sigma(P_2) \Longrightarrow \sigma(C[r_2]) = \sigma(r_1)$ is said to be a *(conditional) critical pair* of $\mathcal{R}$. If the two rules are renamed versions of the same rule of $\mathcal{R}$, we do not consider an overlap at top position (i.e. with $p = \lambda$). A (conditional) critical pair $P \Longrightarrow s = t$ is said to be *joinable* if $\sigma(s) \downarrow_{\mathcal{R}} \sigma(t)$ for every substitution σ with $\sigma(P) \downarrow$. A substitution σ which satisfies the conditions, i.e. for which $\sigma(P) \downarrow$ holds, is said to be *feasible*. Otherwise σ is *unfeasible*. Analogously, a (conditional) critical pair is said to be *feasible (unfeasible)* if there exists some (no) feasible substitution for it.

Note that testing joinability of conditional critical pairs is in general much more difficult than in the unconditional case since one has to consider all substitutions which satisfy the correspondingly instantiated conditions. Moreover, the Critical Pair Lemma does not hold for CTRSs in general as shown e.g. by the following example.

Example 1. (Bergstra & Klop 1986) Consider the join CTRS

$$\mathcal{R} = \left\{ \begin{array}{r} x \downarrow f(x) \Longrightarrow f(x) \to a \\ b \to f(b) \,. \end{array} \right.$$

Here we get $f(b) \to a$ due to $b \downarrow f(b)$ and hence $f(f(b)) \to f(a)$. We also have $f(f(b)) \to a$ because of $f(b) \downarrow f(f(b))$. But a and $f(a)$ do not have a common reduct which is easily shown. Thus $\mathcal{R}$ is not locally confluent despite the lack of critical pairs. Note moreover that $\mathcal{R}$ is even orthogonal when considered as unconditional TRS, i.e. when omitting the condition in the first rule.

Definition 3. (Bergstra & Klop 1986; Klop 1992) Let $\mathcal{R}$ be a CTRS and let $\mathcal{R}_u$ be its unconditional version, i.e. $\mathcal{R}_u := \{l \to r \mid P \Longrightarrow l \to r \in \mathcal{R}\}$. Then $\mathcal{R}$ is said to be *left-linear / non-overlapping / weakly non-overlapping / orthogonal / weakly orthogonal)* if $\mathcal{R}_u$ is left-linear / non-overlapping / weakly non-overlapping / orthogonal / weakly orthogonal.

---

[6] Here and subsequently we shall freely make use of *contexts* like $C[]$, $C[]_p$, $C[]_\Pi$, $C[,\ldots,]$ which denote terms with one or more 'holes' at some position p or, more generally, at some set Π of mutually disjoint positions. This is easily formalized by treating the 'holes' as fresh constant symbols or as fresh variables.

According to this definition Example 1 above shows that orthogonal CTRSs need not be confluent. But note that the CTRS $\mathcal{R}$ defined in Example 1 is not innermost terminating.

Definition 4. A CTRS $\mathcal{R}$ is said to be a *(conditional) overlay system (OS)* if $\mathcal{R}_u$ is an unconditional overlay system.

The careful reader may have observed that the definition of being non-overlapping above is somehow rather restrictive. Namely, the case that there exist conditional critical pairs all of which are infeasible (and hence should not be 'properly critical') is not covered. Analogously, it may be the case that for some CTRS $\mathcal{R}$ all feasible (conditional) critical pairs are overlays. This observation may be exploited to slightly generalize the notions of being an overlay system and being non-overlapping as it is done in (Gramlich 1993). In fact, all results presented here for non-overlapping and overlay CTRSs do also hold for these slightly more general 'semantic' versions of being an overlay system and of being non-overlapping.

The other basic notions for unconditional TRSs introduced above generalize in a straightforward manner to CTRSs.

In the following we shall tacitly assume that all CTRSs considered are join CTRSs (which is the most important case in practice), except for cases where another kind of CTRSs is explicitly mentioned.

3 Restricted Termination and Confluence Properties of Conditional Term Rewriting Systems

We shall study now under which conditions various restricted kinds of termination imply (strong) termination (and also confluence under some additional assumptions) of CTRSs. Firstly we summarize the most important known results on confluence and termination of unconditional orthogonal TRSs.

Theorem 5. *(Rosen 1973; O'Donnell 1977; Klop 1992) Any orthogonal TRS $\mathcal{R}$ is confluent and satisfies the following properties:*

(1a) $\forall t : [WIN(t) \iff SIN(t) \iff SN(t)]$.
(1b) $WIN(\mathcal{R}) \iff SIN(\mathcal{R}) \iff SN(\mathcal{R})$.
(2) If $s \underset{i}{\to} t$ *and* $SN(t)$ *then* $SN(s)$.
(3a) $NE(\mathcal{R}) \implies [\forall t : [WN(t) \iff WIN(t) \iff SIN(t) \iff SN(t)]]$.
(3b) $NE(\mathcal{R}) \implies [WN(\mathcal{R}) \iff WIN(\mathcal{R}) \iff SIN(\mathcal{R}) \iff SN(\mathcal{R})]$.

Our recently obtained results on confluence and termination of TRSs show that all normalization properties of Theorem 5 above (i.e. (1)-(3)) still hold for non-overlapping but not necessarily left-linear TRSs (Gramlich 1994a). Moreover we have shown there that any innermost terminating (unconditional) overlay system with joinable critical pairs is terminating and hence confluent and complete.

For CTRSs much less is known concerning similar criteria for termination and confluence (see Example 1 above which shows that orthogonal CTRSs need not be confluent). Next we summarize the most important known criteria for confluence of (possibly non-terminating) CTRSs.

Let $\mathcal{R}$ be a CTRS with rewrite relation $\rightarrow$ and let P be an n-ary predicate on the set of terms of $\mathcal{R}$. Then P is said to be *closed with respect to* $\rightarrow$ (Klop 1992) if for all terms t_i, t_i' such that $t_i \rightarrow^* t_i'$ $(i = 1, \ldots, n)$: $P(t_1, \ldots, t_n) \Longrightarrow P(t_1', \ldots, t_n')$. $\mathcal{R}$ is said to be *closed* if all conditions (appearing in some conditional rewrite rule of $\mathcal{R}$), viewed as predicates with the variables ranging over $\mathcal{R}$-terms, are closed with respect to $\rightarrow$.

Theorem 6.

(1) Any generalized, weakly orthogonal, closed CTRS is confluent (O'Donnell 1977; Klop1992).
(2) Any weakly orthogonal, semi-equational CTRS is confluent.[7]
(3) Any weakly orthogonal, normal CTRS is confluent (Bergstra & Klop 1986; Klop 1992).

Under the assumption of termination and joinability of all critical pairs some further confluence criteria for CTRSs are possible if additional requirements are fulfilled, cf. (Dershowitz et al. 1988) for details.

In the following we shall show that our results on restricted termination and confluence properties of non-overlapping and even of overlay systems can be generalized to the conditional case. This generalization has to take into account the additional complications arising with CTRSs. In particular, we need a kind of 'local completeness' property implying that variable overlaps are not critical for certain conditional overlay systems. More precisely, we have the following result which is a generalized local version of Lemma 2 in (Dershowitz et al. 1988)[8] which in turn is the main technical result for inferring confluence of terminating CTRSs, provided that all conditional critical pairs are joinable overlays (cf. Theorem 4 in (Dershowitz et al. 1988), cf. also Theorem 6.2 in (Wirth & Gramlich 1994) which handles the more general case of positive / negative conditional rewrite systems). Note that extra variables (in conditions) are allowed here.

Theorem 7. *Let $\mathcal{R}$ be a CTRS with $OS(\mathcal{R})$ and $JCP(\mathcal{R})$ and let s be a term with $SN(s)$. Furthermore let $C[]_\Pi$ be a context (with 'holes' at some set Π of mutually disjoint positions), and t, u, v be terms. Then we have the following implication:*

$$u = C[s]_\Pi \rightarrow^* v \wedge s \rightarrow^* t \quad \Longrightarrow \quad C[t]_\Pi \downarrow v\,.$$

Proof. A detailed proof is provided in the Appendix. ■

[7] This is a corollary of (1).

[8] In Lemma 2 of (Dershowitz et al. 1988) the proof (i.e. the induction ordering) makes use of the general termination assumption $SN(\mathcal{R})$ for the considered CTRS $\mathcal{R}$. Our proof of Theorem 7 has a similar structure but is based on a slightly different notion of *depth* and – more importantly – the induction ordering only needs the local termination assumption $SN(s)$.

Note that Theorem 7 is interesting in itself because it entails a non-trivial structural confluence property for (conditional) overlay systems with joinable critical pairs without a full termination assumption (it is applicable even in situations where the whole system need not terminate and may have other potential applications than the ones mentioned below), namely in the following sense.

Lemma 8. *Let $\mathcal{R}$ be a CTRS with $OS(\mathcal{R})$ and $JCP(\mathcal{R})$ and let u, v, w be terms with $v \;{}^*\!\!\leftarrow u \rightarrow^* w$. Then $v \downarrow w$ holds provided that v is obtained from u by performing only reductions in strongly normalizing (parallel) subterms of u, formally: $u = C[s_1, \ldots, s_n]$, $v = C[t_1, \ldots, t_n]$ for some context $C[, \ldots,]$, and $SN(s_i)$, $s_i \rightarrow^* t_i$ for $i = 1, \ldots, n$.*

Proof. Straightforward by repeated application of Theorem 7. ∎

One may ask now whether the assumption $OS(\mathcal{R})$ is really crucial in the above results. This is indeed the case, clearly, for CTRSs, since there exist terminating CTRSs with joinable critical pairs that are non-confluent, hence necessarily not even locally confluent (cf. (Dershowitz et al. 1988) for some illustrative counterexamples). A very simple counterexample (due to Aart Middeldorp) is the following:

Example 2.

$$\mathcal{R} = \begin{cases} g(x) \downarrow f(x) \Longrightarrow g(x) \rightarrow f(b) \\ \qquad\qquad\qquad\quad f(a) \rightarrow g(a) \\ \qquad\qquad\qquad\qquad\;\; a \rightarrow b \end{cases}$$

This system is easily shown to be terminating and the only critical peak (between the last two rules) $f(b) \leftarrow f(a) \rightarrow g(a)$ is joinable since $g(a) \rightarrow f(b)$ by the first rule (due to $g(a) \downarrow f(a)$). But we have $g(b) \leftarrow g(a) \rightarrow f(b)$ (again due to $g(a) \downarrow f(a)$) with both $g(b)$ and $f(b)$ irreducible.

For unconditional TRSs such a counterexample involving a terminating TRS with joinable critical pairs cannot exist since for TRSs – in contrast to CTRSs – the Critical Pair Lemma holds (Huet 1980), i.e. the property $WCR(\mathcal{R})$ is equivalent to $JCP(\mathcal{R})$. By Newman's Lemma (Newman 1942) any such TRS must then be confluent. Hence, in the unconditional case the above question may be rephrased as follows: Does there exist a locally confluent (non-terminating) TRS which is not an overlay system violating the confluence property in Lemma 8 (and Theorem 7) above? This is indeed the case as shown by the following simple example.

Example 3.

$$\mathcal{R} = \begin{cases} f(a) \rightarrow f(b) \\ f(b) \rightarrow f(a) \\ \quad a \rightarrow c \\ \quad b \rightarrow d \end{cases}$$

This system which is not an overlay system is clearly locally confluent, but non-confluent (and hence necessarily non-terminating). We have for instance $f(c) \leftarrow f(a) \rightarrow^* f(d)$ with both $f(c)$ and $f(d)$ irreducible, and moreover, in the step $f(a) \rightarrow f(c)$ the proper subterm a of $f(a)$ which is contracted clearly satisfies $SN(a)$.

Thus, even for unconditional TRSs, Theorem 7 and Lemma 8 capture indeed a non-trivial confluence property of overlay systems!

But let us return now to the conditional case. Similar to Lemma 8 above we obtain from Theorem 7 in particular the following sufficient criterion for a variable overlap in CTRSs to be non-critical.

Lemma 9. *Let $\mathcal{R}$ be a CTRS with $OS(\mathcal{R})$ and $JCP(\mathcal{R})$, and let s, t be terms with $s \rightarrow_{p,\sigma,P \Longrightarrow l \rightarrow r} t$. Furthermore let σ' be given with $\sigma \rightarrow^* \sigma'$, i.e. $\sigma(x) \rightarrow^* \sigma'(x)$ for all $x \in dom(\sigma)$, such that $SN(\sigma(x))$ holds for all $x \in dom(\sigma)$. Then we have: $s = C[\sigma(l)]_p \rightarrow^* C[\sigma'(l)]_p \rightarrow_{p,\sigma',P \Longrightarrow l \rightarrow r} C[\sigma'(r)]_p$ (due to $\sigma'(P) \downarrow$), and $t = C[\sigma(r)]_p \rightarrow^* C[\sigma'(r)]_p$, for some context $C[]_p$.*

Proof. Straightforward by verifying $\sigma'(P) \downarrow$ using the encoding of $s \downarrow_{\mathcal{R}} t$ into $eq(s,t) \rightarrow^*_{\mathcal{R}'} true$[9] and applying Lemma 8 together with the fact that *true* is irreducible. ■

Choosing $C[]_\Pi$ to be the empty context (and accordingly $\Pi = \{\lambda\}$) in Theorem 7 we obtain as corollary the following local version of a confluence criterion.

Corollary 10. *Let $\mathcal{R}$ be a CTRS with $OS(\mathcal{R})$ and $JCP(\mathcal{R})$, and let s be a term with $SN(s)$. Then we have $CR(s)$. In other words, for a conditional overlay system with joinable critical pairs, a term is strongly normalizing if and only if it is complete.*

The termination assumption concerning s in this result is crucial as demonstrated by the following example.

Example 4 (example 1 continued). Here

$$\mathcal{R} = \begin{cases} x \downarrow f(x) \Longrightarrow f(x) \rightarrow a \\ \qquad\qquad\qquad\quad b \rightarrow f(b)\,. \end{cases}$$

is clearly an overlay system with joinable critical pairs (it is even non-overlapping). Moreover we have $f(f(b)) \rightarrow a$ and $f(f(b)) \rightarrow f(a)$ but not $a \downarrow f(a)$. Obviouly, $SN(f(f(b)))$ does not hold due to the presence of the rule $b \rightarrow f(b)$ in $\mathcal{R}$ (note that we even do not have $SIN(f(f(b)))$).

Under the stronger assumption of global termination we get from Corollary 10 the following known critical pair criterion for confluence of conditional overlay systems.

Theorem 11. *(Dershowitz et al. 1988) A terminating CTRS which is an overlay system such that all its conditional critical pairs are joinable is confluent, hence complete.*

---

[9] Cf. the Remark after Definition 1.

3.1 Non-Overlapping CTRS

Now let us consider non-overlapping CTRSs. We shall show that all normalization properties of Theorem 5 also hold for non-overlapping CTRSs. Since the proofs are very similar to those of the corresponding results for unconditional non-overlapping TRSs in (Gramlich 1994a) we shall only mention additional problems arising with CTRSs. Throughout this subsection we assume that $\mathcal{R}$ is a non-overlapping CTRS, i.e. $NO(\mathcal{R})$ holds.

Let us start with an easy result about innermost reduction.

Lemma 12. *Innermost reduction in* $\mathcal{R}$ *is* WCR^1*, i.e.* $WCR^1(\underset{i}{\to})$ *holds.*

Proof. Cf. (Gramlich 1994a). ∎

Corollary 13. *Innermost reduction in* $\mathcal{R}$ *is confluent, i.e.* $CR(\ \to)$ *holds.*

The following result shows that for non-overlapping systems the existence of a terminating innermost derivation for some term t implies that any innermost derivation initiated by t is finite.

Lemma 14. $\forall t : [WIN(t) \iff SIN(t)]$.

Proof. Cf. (Gramlich 1994a). ∎

Furthermore, strong innermost normalization is equivalent to strong normalization.

Theorem 15. $\forall t : [SIN(t) \iff SN(t)]$.

Proof. For a proof we refer to the more general Theorem 21. ∎

Combined with Lemma 14 this yields the following

Corollary 16. $\forall t : [WIN(t) \iff SIN(t) \iff SN(t)]$.

The next result says that innermost reduction steps in non-overlapping CTRSs cannot be critical in the sense that they may destroy the possibility of infinite derivations.

Lemma 17. *If* $s \underset{i}{\to} t$ *and* $SN(t)$ *then* $SN(s)$.

Proof. This is an immediate consequence of Corollary 16. ∎

Furthermore, as in the unconditional case, the non-erasing property is crucial for the equivalence of weak and strong termination of non-overlapping CTRSs.

Lemma 18. *Suppose* $NE(\mathcal{R})$. *If* $s \to t$ *and* $SN(t)$ *then* $SN(s)$.

Proof. As in (Gramlich 1994a) by using Lemma 17 and Corollary 16. Here we additionally need the fact that variable overlaps are not critical provided the substitution part is strongly normalizing, i.e. Lemma 9. ∎

Lemma 19. *If* $NE(\mathcal{R})$ *then* : $\forall t : [WN(t) \iff WIN(t) \iff SIN(t) \iff SN(t)]$.

Proof. As in (Gramlich 1994a) using Lemma 18. ■

Finally let us summarize the results obtained for non-overlapping, but not necessarily left-linear CTRSs.

Theorem 20. *Any non-overlapping CTRS* $\mathcal{R}$ *satisfies the following properties:*

(1a) $\forall t : [WIN(t) \iff SIN(t) \iff SN(t)]$.
(1b) $WIN(\mathcal{R}) \iff SIN(\mathcal{R}) \iff SN(\mathcal{R})$.
(2) *If* $s \to t$ *and* $SN(t)$ *then* $SN(s)$.
(3a) $NE(\mathcal{R}) \implies [\forall t : [WN(t) \iff WIN(t) \iff SIN(t) \iff SN(t)]]$.
(3b) $NE(\mathcal{R}) \implies [WN(\mathcal{R}) \iff WIN(\mathcal{R}) \iff SIN(\mathcal{R}) \iff SN(\mathcal{R})]$.
(4) $WIN(\mathcal{R}) \implies CR(\mathcal{R})$.
(5) $NE(\mathcal{R}) \wedge WN(\mathcal{R}) \implies CR(\mathcal{R})$.

Proof. (1)-(3) have been shown above. To verify the confluence criteria (4) and (5) one has to combine (1) and (3), respectively, with Theorem 21 (below), observing that a non-overlapping CTRS is in particular an overlay system with joinable critical pairs. ■

In fact, all properties of Theorem 20 do also hold for the slightly more general class of weakly non-overlapping CTRSs (where trivial critical pairs are allowed), as can be verified by a careful inspection of the proofs for the non-overlapping (and the overlay) case.

3.2 Conditional Overlay Systems with Joinable Critical Pairs

Furthermore we can show that Theorem 15 above may be generalized by requiring only $OS(\mathcal{R}) \wedge JCP(\mathcal{R})$ instead of the more restrictive $NO(\mathcal{R})$.

Theorem 21. *For any CTRS* $\mathcal{R}$ *we have:*

(a) $OS(\mathcal{R}) \wedge JCP(\mathcal{R}) \wedge SIN(\mathcal{R}) \implies SN(\mathcal{R}) \wedge CR(\mathcal{R})$, *and*
(b) $OS(\mathcal{R}) \wedge JCP(\mathcal{R}) \implies [\forall s : [SIN(s) \implies SN(s) \wedge CR(s)]]$.

Proof. The proof (by minimal counterexample) is essentially the same as in (Gramlich 1994a) for the unconditional case, but taking into account that we have to ensure two additional properties which one obtains for free in the unconditional case: Namely, in an unconditional overlay system with joinable critical pairs a term is complete if and only if it strongly normalizing, and secondly, variable overlaps are not critical (if the substitution part is strongly normalizing).[10]

[10] More precisely, this property is needed in the following form: if $\sigma l \to \sigma r$ then $(\sigma\downarrow)l \to (\sigma\downarrow)r$ using the same rule $P \implies l \to r$ where σ is assumed to be strongly normalizing, and $(\sigma\downarrow)$ denotes the corresponding normalized substitution which is obtained by replacing σx by its (unique) normal form, for all $x \in dom(\sigma)$.

Now, for conditional overlay systems with joinable critical pairs the first property is ensured by Corollary 10, and the second one by Lemma 9. Furthermore, the minimal counterexample considered must be an infinite derivation

$$s = s_0 \rightarrow s_1 \rightarrow s_2 \rightarrow \dots$$

such that in each step the redex contracted is either terminating (hence complete) or, if not, all its proper subterms are terminating (hence complete). The latter property for those steps $s_i = C[\sigma l]_p \rightarrow_{\sigma,P\Longrightarrow l\rightarrow r} C[\sigma r]_p = s_{i+1}$ which contract a non-terminating redex $\sigma l = s_i/p$ ensures[11] that the substitution part is terminating, hence that Lemma 9 is applicable and yields $C[(\sigma\downarrow)l]_p \rightarrow_{\sigma\downarrow,P\Longrightarrow l\rightarrow r} C[(\sigma\downarrow)r]_p$. ■

Theorem 21 states that any (strongly) innermost terminating (conditional) overlay system with joinable critical pairs is (strongly) terminating and confluent (part (a)), hence complete, which even holds in the stronger local version (part (b)). In other words, for (conditional) overlay systems it suffices to verify innermost termination and joinability of all critical pairs in order to infer general termination and confluence, i.e. completeness. The non-triviality of this result is obvious taking into account the fact that for CTRSs the Critical Pair Lemma does not hold in general and almost all known sufficient criteria for confluence presume even stronger properties than termination plus joinability of (conditional) critical pairs.

Finally let us mention that it is possible to apply the above results to the analysis of properties of (disjoint and restricted forms of non-disjoint) combinations of CTRSs, as it is done in (Gramlich 1994a) for the case of unconditional TRSs. But this generalization requires a careful analysis and is not straightforward, since for instance the properties weak termination, weak innermost termination and (strong) innermost termination are not preserved in general under disjoint unions of CTRSs (Middeldorp 1993; Gramlich 1994b). Some very recently obtained positive and negative results in this direction are presented in (Gramlich 1994b). Related results for combinations of constructor systems can be found in (Middeldorp & Toyama 1993) for the unconditional case and (Middeldorp 1994) for the conditional case.

Acknowledgement: I am indebted to Aart Middeldorp for suggesting a couple of improvements in (Gramlich 1994a) on which the present paper is partially based.

References

J. A. Bergstra and J. W. Klop. Conditional rewrite rules: Confluence and termination. *Journal of Computer and System Sciences*, 32:323–362, 1986.

[11] This stronger minimality property is not necessary for the proof of the unconditional version of Theorem 21 since for unconditional TRSs $\sigma l \rightarrow_{\lambda,\sigma,l\rightarrow r} \sigma r$ and $\sigma \rightarrow^* \sigma'$ always imply $\sigma' l \rightarrow_{\lambda,\sigma',l\rightarrow r} \sigma' r$.

R.S. Boyer and J S. Moore. *A Computational Logic*. Academic Press, 1979.

N. Dershowitz. Termination of rewriting. *Journal of Symbolic Computation*, 3(1):69–116, 1987.

N. Dershowitz and J.-P. Jouannaud. Rewrite systems. In J. van Leeuwen, editor, *Formal models and semantics, Handbook of Theoretical Computer Science*, volume B, chapter 6, pages 243–320. Elsevier - The MIT Press, 1990.

N. Dershowitz, M. Okada, and G. Sivakumar. Confluence of conditional rewrite systems. In S. Kaplan and J.-P. Jouannaud, editors, *Proc. 1st Int. Workshop on Conditional Term Rewriting Systems*, volume 308 of *Lecture Notes in Computer Science*, pages 31–44. Springer-Verlag, 1988.

B. Gramlich. New abstract criteria for termination and confluence of conditional rewrite systems. SEKI-Report SR-93-17, Fachbereich Informatik, Universität Kaiserslautern, 1993.

B. Gramlich. Abstract relations between restricted termination and confluence properties of rewrite systems. *Fundamenta Informaticae*, 1994, special issue on term rewriting systems, to appear. A preliminary version appeared as "Relating Innermost, Weak, Uniform and Modular Termination of Term Rewriting Systems" in Proc. LPAR'92, LNAI 624, pp. 285–296, 1992, see also SEKI-Report SR-93-09, Fachbereich Informatik, Universität Kaiserslautern, 1993.

B. Gramlich. On modularity of termination and confluence properties of conditional rewrite systems. In G. Levi and M. Rodríguez-Artalejo, editors, *Proc. 4th Int. Conf. on Algebraic and Logic Programming, Madrid, Spain*, volume 850 of *Lecture Notes in Computer Science*, pages 186–203, Springer-Verlag, 1994.

G. Huet and D. Lankford. On the uniform halting problem for term rewriting systems. Technical Report 283, INRIA, 1978.

G. Huet. Confluent reductions: Abstract properties and applications to term rewriting systems. *Journal of the ACM*, 27(4):797–821, oct 1980.

D.E. Knuth and P.B. Bendix. Simple word problems in universal algebra. In J. Leech, editor, *Computational Problems in Abstract Algebra*, pages 263–297. Pergamon Press, Oxford, U. K., 1970. Reprinted 1983 in "Automation of Reasoning 2", Springer, Berlin, pp. 342-376.

J.W. Klop. Term rewriting systems. In S. Abramsky, D. Gabbay, and T. Maibaum, editors, *Handbook of Logic in Computer Science*, volume 2, chapter 1, pages 2–117. Clarendon Press, Oxford, 1992.

A. Middeldorp. Modular properties of conditional term rewriting systems. *Information and Computation*, 104(1):110–158, May 1993.

A. Middeldorp. Completeness of combinations of conditional constructor systems. *Journal of Symbolic Computation*, 17:3–21, 1994.

A. Middeldorp and Y. Toyama. Completeness of combinations of constructor systems. *Journal of Symbolic Computation*, 15:331–348, September 1993.

M.H.A. Newman. On theories with a combinatorial definition of equivalence. *Annals of Mathematics*, 43(2):223–242, 1942.

M.J. O'Donnell. *Computing in Systems Described by Equations*, volume 58 of *Lecture Notes in Computer Science*. Springer-Verlag, 1977.

B.K. Rosen. Tree-manipulating systems and Church-Rosser theorems. *Journal of the ACM*, 20:160–187, 1973.

C. Walther. Argument bounded algorithms as a basis for automated termination proofs. In E. Lusk and R. Overbeek, editors, *Proc. of the 9th Int. Conf. on Automated Deduction*, volume 310 of *Lecture Notes in Computer Science*, pages 601–622. Springer-Verlag, 1988.

C.-P. Wirth and B. Gramlich. A constructor-based approach for positive/negative conditional equational specifications. *Journal of Symbolic Computation*, 17:51–90, 1994.

Appendix (proof of Theorem 7)

Proof. Let $\mathcal{R}$ be given as above. W.l.o.g. let us assume that $\mathcal{R}$ contains the rule $eq(x,x) \to true$ (with eq a fresh binary function symbol and $true$ a fresh constant of new sort, and x some 'old' variable).[12] For a context $C[]_\Pi$ and terms s, t, u, v we define the predicate $P(s,t,u,v,C[]_\Pi)$ by the following implication:

$$SN(s) \wedge u = C[s]_\Pi \to^* v \wedge s \to^* t \quad \Longrightarrow \quad C[t]_\Pi \downarrow v\,.$$

We have to show $P(s,t,u,v,C[]_\Pi)$ for all s, t, u, v, $C[]_\Pi$. To this end it is sufficient that $Q(s,n,k)$ defined by

$$\forall C[]_\Pi,\, t,\, u,\, v,:\; SN(s) \wedge u = C[s]_\Pi \xrightarrow{n}{}^{k} v \wedge s \to^* t \quad \Longrightarrow \quad C[t]_\Pi \downarrow v$$

holds for all s and for all n, k. We will show this by contradiction as follows. Assume that there exists a counterexample, i.e. $\langle s,n,k\rangle$ with $Q(s,n,k)$ not holding for. That means we have

$$(*)\quad SN(s) \quad \wedge \quad u = C[s]_\Pi \xrightarrow{n}{}^{k} v \quad \wedge \quad s \to^* t$$

for some $C[]_\Pi$, t, u, v, but

$$(**)\quad \neg(C[t]_\Pi \downarrow v)\,.$$

Now we define a complexity measure for $Q(\bar{s},\bar{n},\bar{k})$ by the triple $\langle \bar{s},\bar{n},\bar{k}\rangle$ using the lexicographic combination $\succ := lex(>_1,>_2,>_3)$ with $>_1 := (\to \cup >_{st})^+|_{below_s}$, $>_2 := >_3 := > := >_{\mathbb{N}}$, where – for some binary (ordering) relation R – $R|_{below_s}$ is defined by $R \cap (\{t|sR^*t\}^2)$, for comparing these triples. Now, $>_1$ is well-founded — due to $SN(s)$ — and $>_2 = >_3 = >_{\mathbb{N}}$ is obviously well-founded, too. Hence, their lexicographic combination $\succ$ is also well-founded. Thus, we may assume w.l.o.g. that

$$(***)\quad \langle s,n,k\rangle \text{ is a minimal counterexample w.r.t. } \succ\,.$$

In order to obtain a contradiction we proceed by case analysis and induction[13] showing that $\langle s,n,k\rangle$ cannot be a (minimal) counterexample.

If $u = v$ (i.e. $n = k = 0$) or $s = t$ we are done since $(**)$ is violated. Otherwise, let $s \to s' \to^* t$. If we can show that $C[s']_\Pi \downarrow v$ holds then by induction (on the first component) we get $C[t]_\Pi \downarrow v$ because we have $s \to s'$, hence $s >_1 s'$. But this is a contradiction to $(***)$. We shall distinguish the following cases:

[12] Cf. the Remark after Definition 1.

[13] This means that we shall exploit the minimality assumption of the counterexample $\langle s,n,k\rangle$.

(1) Proper subterm case (see Figure 1): If the first step $s \to s'$ reduces a proper subterm of s, i.e. $s \to_p s'$ for some $p > \lambda$, then we have

$$C[s]_\Pi = C[C'[s/p]_p]_\Pi = (C[C'[]_p]_\Pi)[s/p]_{\Pi p} \to^+ C[s']_\Pi = (C[C'[]_p]_\Pi)[s'/p]_{\Pi p}$$

with $s = C'[s/p]_p$ for some context $C'[]_p$, hence $C[s']_\Pi \downarrow v$ as desired by induction on the first component because $s >_{st} s/p = \sigma(l_1)$ implies $s >_1 s/p$.

Figure 1 : Proper subterm case

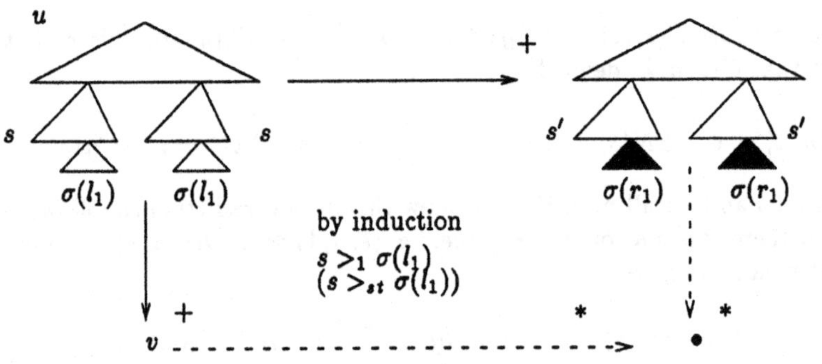

(2) Otherwise, we may suppose

$$s \to_{\lambda,\sigma,P_1 \Longrightarrow l_1 \to r_1} s', \text{ i.e. } s = \sigma(l_1)\ ,\ s' = \sigma(r_1) \text{ and } \sigma(P_1)\downarrow$$

for some rule $P_1 \Longrightarrow l_1 \to r_1 \in \mathcal{R}$ and some substitution σ. Moreover assume

$$u = C[s]_\Pi \xrightarrow{n}_{q,\tau,P \Longrightarrow l \to r} u' \xrightarrow{n}{}^{k-1} v\,,$$

i.e. $C[s]_\Pi/q = \tau(l)$, $u'/q = \tau(r)$ and $\tau(P)\downarrow$, for n minimal with $u \xrightarrow{n}{}^{*} v$ and $k \geq 1$ minimal with $u \xrightarrow{n}{}^{k} v$. Then we have to distinguish the following four subcases according to the relative positions of q and Π:

(2.1) $q \mid \Pi$ (disjoint peak, see Figure 2.1): Then we have $u = C[s]_\Pi \xrightarrow{n}_q C'[s]_\Pi = u' \xrightarrow{n}{}^{k-1} v$, $C'[s]_\Pi \to^* C'[s']_\Pi$ and $C[s]_\Pi \to^* C[s']_\Pi \to_q C'[s']_\Pi$ for some context $C'[,\ldots,]$, hence by induction on the second or third component ($n \geq n'$, $k > k-1$) $C'[s']_\Pi \downarrow v$ and thus $C[s']_\Pi \to C'[s']_\Pi \downarrow v$ as desired.

(2.2) $q \in \Pi$ (critical peak, see Figure 2.2): In this case we have a critical peak which is an instance of a critical overlay of $\mathcal{R}$, i.e. $s = \sigma(l_1) = \tau(l)$. Since all conditional critical pairs are joinable (overlays) we know that there exists some term w with $s = \sigma(l_1) \to \sigma(r_1) = s' \to^* w$ and $s = \tau(l) \to \tau(r) \to^* w$. Obviously, we have $u = C[s]_\Pi \to_{P \Longrightarrow l \to r} (C[\tau(l)]_\Pi)[q \leftarrow \tau(r)] =$

Figure 2.1 : Disjoint peak case

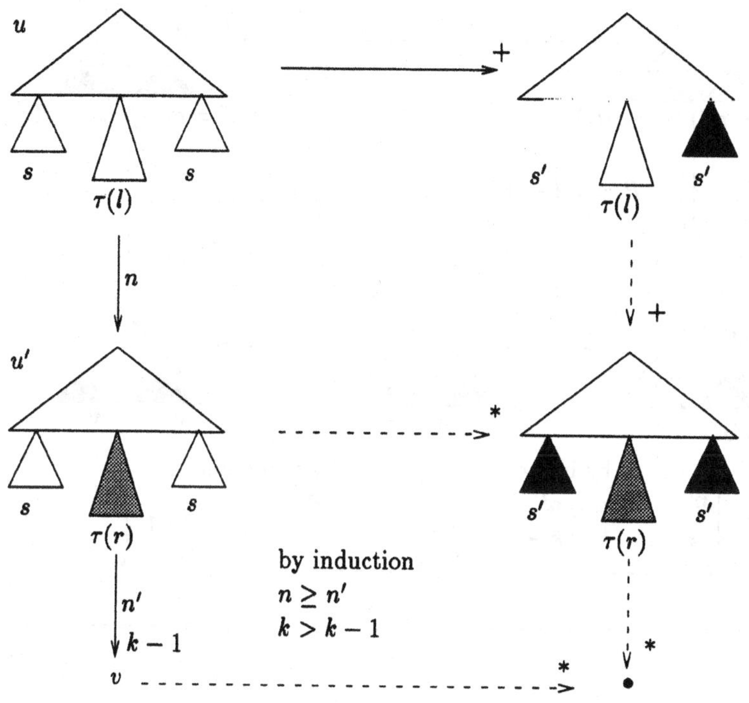

$u' \to^*_{P \Longrightarrow l \to r} C[\tau(r)]_\Pi$. For $|\Pi| = 1$ we obtain $\Pi = \{q\}$ and $(C[\tau(l)]_\Pi)[q \leftarrow \tau(r)] = u' = C[\tau(r)]_\Pi \to^* v$. Otherwise, we have $C[\tau(l)]_\Pi[q \leftarrow \tau(r)] \xrightarrow{n',\ k-1} v$ with $n' \leq n$. Hence, by induction on the second or third component we obtain $C[\tau(r)]_\Pi \downarrow v$ (due to $n' \leq n$, $k-1 < k$). Moreover, $\tau(r) \to^* w$ yields $C[\tau(r)]_\Pi \to^* C[w]_\Pi$ which by induction on the first component implies $C[w]_\Pi \downarrow v$ (due to $s = \tau(l) \to \tau(r)$, hence $s >_1 \tau(r)$). Thus, $C[s']_\Pi = C[\sigma(r_1)]_\Pi \to^* C[w]_\Pi \downarrow v$ because of $\sigma(r_1) \to^* w$. Hence we get $C[s']_\Pi \downarrow v$ as desired. The remaining case is that of a *variable overlap*, either above or in some subterm $C[s]_\Pi/\pi = s$ ($\pi \in \Pi$) of $C[s]_\Pi$. Note that a critical peak which is not an overlay cannot occur due to $OS(\mathcal{R})$.

(2.3) $q < \pi$ for some $\pi \in \Pi$ (variable overlap above, see Figure 2.3): Let Π' be the set of positions of those subterms $s = \sigma(l_1)$ of $u/q = \tau(l)$ which correspond to some $u/\pi = s$, $\pi \in \Pi$. Formally, $\Pi' := \{\pi' \mid q\pi' \in \Pi\}$. Moreover, for every $x \in dom(\tau)$, let $\Delta(x)$ be the set of positions of those subterms s in $\tau(x)$ which are rewritten into s' in the derivation $u = C[s]_\Pi \to^+ C[s']_\Pi$, i.e. $\Delta(x) := \{\rho' \mid \exists \rho : l/\rho = x \wedge \rho\rho' \in \Pi'\}$. Then τ' is defined by $\tau'(x) := \tau(x)[\rho' \leftarrow$

Figure 2.2 : Critical peak (overlay)

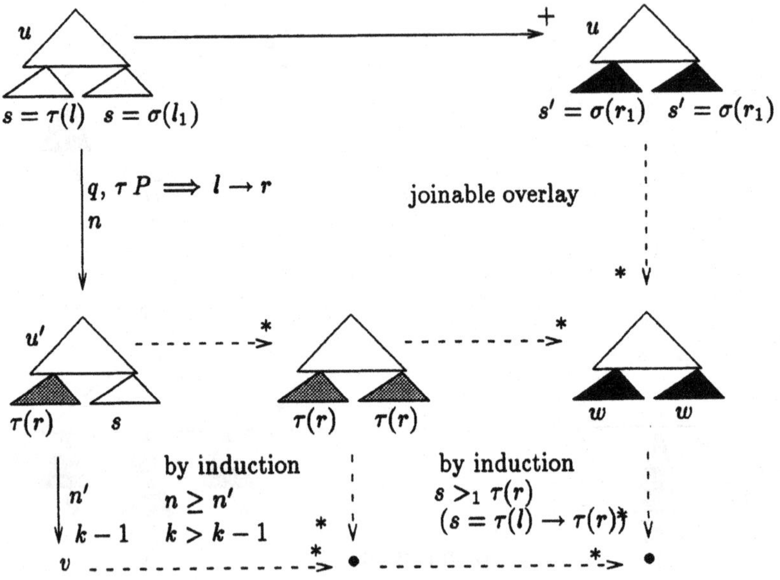

$s' \mid \rho' \in \Delta(x)]$ for all $x \in dom(\tau)$. Obviously, we have $\tau(x) \rightarrow^*_{P_1 \Rightarrow l_1 \rightarrow r_1} \tau'(x)$ for all $x \in dom(\tau)$. Thus we get

$$u = C[s]_\Pi = C'[\tau(l)]_q \xrightarrow{n}_{q,\tau,P \Rightarrow l \rightarrow r} C'[\tau(r)]_q = u' \xrightarrow{n',\, k-1} v$$

for some context $C'[]_q$ and some $n' \leq n$,

$$u = C[s]_\Pi \rightarrow^*_{P_1 \Rightarrow l_1 \rightarrow r_1} C[s']_\Pi \rightarrow^*_{P_1 \Rightarrow l_1 \rightarrow r_1} C''[\tau'(l)]_q$$

for some context $C''[]_q$, and

$$u' = C'[\tau(r)]_q \rightarrow^*_{P_1 \Rightarrow l_1 \rightarrow r_1} C''[\tau'(r)]_q .$$

Moreover we have

$$C''[\tau'(l)]_q \rightarrow_{q,\tau',P \Rightarrow l \rightarrow r} C''[\tau'(r)]_q$$

by induction (due to $\tau'(P)\downarrow$ as shown below) and finally

$$C[s']_\Pi \rightarrow^* C''[\tau'(r)]_q \downarrow v$$

as desired by induction (on the second or third component) due to $n \geq n'$, $k > k-1$.

It remains to prove the claim $\tau'(P)\downarrow$. This means that we have to show $\tau'(z_1)\downarrow\tau'(z_2)$ for all $z_1\downarrow z_2\in P$. If P is empty or trivially satisfied (i.e. $n\le 1$) we are done. Otherwise, we know by assumption that $\tau(z_1)\downarrow\tau(z_2)$ for all $z_1\downarrow z_2\in P$ in depth $n-1$. This means $eq(\tau(z_1),\tau(z_2))\xrightarrow{n-1^*} true$. By construction of τ' we know $\tau(z_1)\rightarrow^*_{P_1\Longrightarrow l_1\rightarrow r_1}\tau'(z_1)$, $\tau(z_2)\rightarrow^*_{P_1\Longrightarrow l_1\rightarrow r_1}\tau'(z_2)$. Moreover, $eq(\tau(z_1),\tau(z_2))$ is of the form $E[s]_Q$, for some context $E[]_Q$, such that $E[s]_Q\rightarrow^* E[s']_Q = eq(\tau'(z_1),\tau'(z_2))$. By induction on the second component (due to $n>n-1$) we obtain $E[s']_Q = eq(\tau'(z_1),\tau'(z_2))\downarrow true$, hence $eq(\tau'(z_1),\tau'(z_2))\rightarrow^* true$ (since *true* is irreducible) which means that $\tau'(z_1)$ and $\tau'(z_2)$ are joinable (without using the rule $eq(x,x)\rightarrow true$). This finishes the proof of the claim $\tau'(P)\downarrow$. Summarizing we have shown $C[s']_\Pi\downarrow v$ as desired.

(2.4) $\pi<q$ for some $\pi\in\Pi$ (variable overlap below, see Figure 2.4): Remember that we have $u/\pi=\sigma(l_1)=s$ and $u/q=\tau(l)$. Now let $q', q'', q''', \Pi', \Pi''$ and contexts $C'[]_q, D[]_{q'''}, D'[]_{\Pi'}, D''[]_{\Pi''}$ be (uniquely) defined by $u=C'[\tau(l)]_q$, $q=\pi q'$, $q'=q''q'''$, $l_1/q''=x\in\mathcal{V}$, $\sigma(x)=D[\tau(l)]_{q'''}$, $\Pi'=\{\pi'\mid l_1/\pi'=x\}$, $\Pi''=\{\pi''\mid r_1/\pi''=x\}$, $\sigma(l_1)=D'[D[\tau(l)]_{q'''}]_{\Pi'}$, $\sigma(r_1)=D''[D[\tau(l)]_{q'''}]_{\Pi''}$. Moreover let σ' be the substitution on $V(l_1)$ defined by

$$\sigma'(y)=\begin{cases}\sigma(y)\,,\ y\neq x\\ D[\tau(r)]_{q'''}\,,\ y=x,\sigma(x)=D[\tau(l)]_{q'''}\,.\end{cases}$$

Then we get

$$\begin{aligned}C[s]_\Pi &= C[\sigma(l_1)]_\Pi\\ &= C[D'[D[\tau(l)]_{q'''}]_{\Pi'}]_\Pi \rightarrow^+_{\sigma,P_1\Longrightarrow l_1\rightarrow r_1} C[s']_\Pi = C[\sigma(r_1)]_\Pi\\ &= C[D''[D[\tau(l)]_{q'''}]_{\Pi''}]_\Pi \rightarrow^*_{\tau,R\Longrightarrow l\rightarrow r} C[D''[D[\tau(r)]_{q'''}]_{\Pi''}]_\Pi\\ &= C[\sigma'(r_1)]_\Pi\,,\end{aligned}$$

and

$$\begin{aligned}C[s]_\Pi &= C[\sigma(l_1)]_\Pi = C[D'[D[\tau(l)]_{q'''}]_{\Pi'}]_\Pi\\ &= C'[\tau(l)]_q \rightarrow_{q,\tau,P\Longrightarrow l\rightarrow r} u'\\ &= C'[\tau(r)]_q \rightarrow^*_{\tau,P\Longrightarrow l\rightarrow r} C[D'[D[\tau(r)]_{q'''}]_{\Pi'}]_\Pi = C[\sigma'(l_1)]_\Pi\,.\end{aligned}$$

By induction on the first component we obtain

$$C[s]_\Pi = C[\sigma(l_1)]_\Pi \rightarrow^+ C[\sigma'(l_1)]_\Pi\downarrow v$$

(due to $s>_{st}\tau(l)$, hence $s>_1\tau(l)$). Moreover, we get

$$C[\sigma'(l_1)]_\Pi \rightarrow^*_{\sigma',P_1\Longrightarrow l_1\rightarrow r_1} C[\sigma'(r_1)]_\Pi$$

since $\sigma'(P_1)\downarrow$ is satisfied by induction on the first component ($s>_{st}\tau(l)$, hence $s>_1\tau(l)$). [14] Furthermore we have $C[\sigma'(r_1)]_\Pi\downarrow v$ by induction on the

[14] Note that – in contrast to case 2.3 – a 'proper *eq*-reasoning' can be avoided here (by applying the induction hypothesis twice), since the inductive argument needed does not involve the *depth* of rewriting.

first component (due to $s = \sigma(l) \to^+ \sigma'(l_1)$, hence $s >_1 \sigma'(l_1)$). Summarizing we have shown

$$C[s']_\Pi \to^* C[\sigma'(r_1)]_\Pi \downarrow v$$

as desired.

Thus, for all cases we have shown $C[s']_\Pi \downarrow v$ yielding a contradiction to $(***)$, hence we are done. ■

Figure 2.3 : Variable overlap above

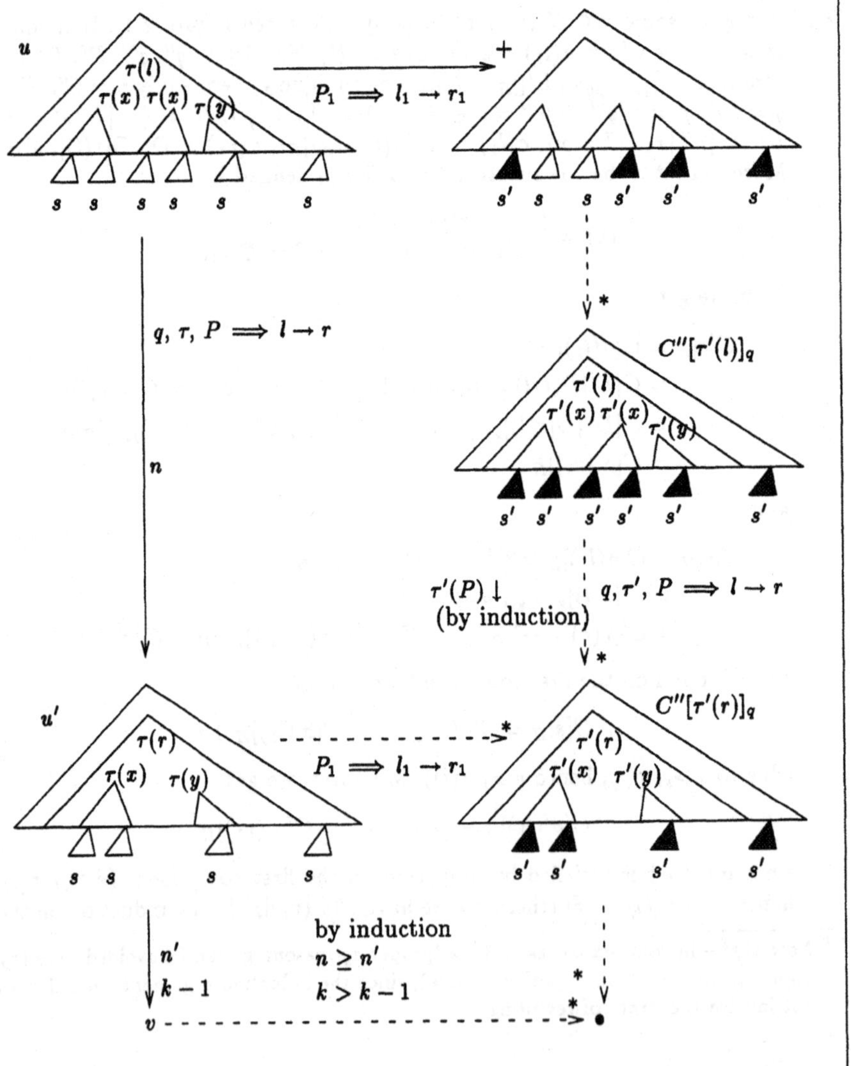

Figure 2.4 : Variable overlap below

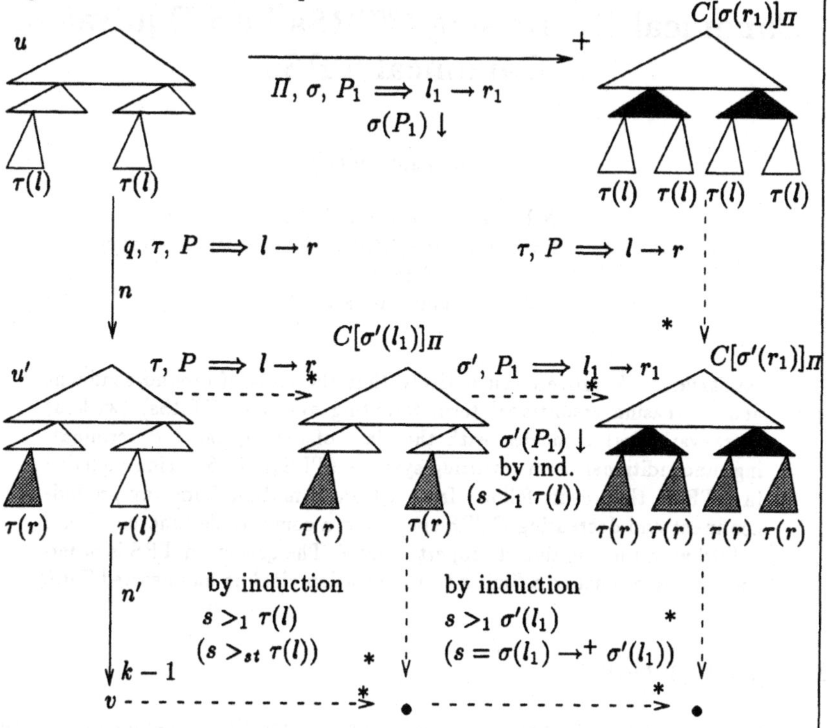

How to Transform Canonical Decreasing CTRSs into Equivalent Canonical TRSs

Claus Hintermeier

INRIA-Lorraine & CRIN-CNRS
615, rue du Jardin Botanique, BP101, 54600 Villers-lès-Nancy
France
E-mail: hinterme@loria.fr

Abstract. We prove constructively that the class of ground-confluent and decreasing conditional term rewriting systems (CTRSs) (without extra variables) coincides with the class of orthogonal and terminating, unconditional term rewriting systems (TRSs). TRSs being included in CTRSs, this result follows from a transformation from any ground-confluent and decreasing CTRS specifying a computable function f into a TRS with the mentioned properties for f. The generated TRS is order-sorted, but we outline a similar transformation yielding an unsorted TRS.

1 Introduction

Many formalisms for algebraic specifications use conditional equations. Several specific procedures (like narrowing, induction techniques etc.) extending the unconditional case, have been defined. However, it is well-known that finite unconditional algebraic specifications with hidden functions are sufficient to describe any computable algebra [Bergstra and Tucker, 1987], i.e. any algebra with decidable word problem. Therefore, finite conditional algebraic specifications with or without hidden functions cannot add any further computable algebras to the class of those being expressible by finite unconditional specifications with hidden functions.

Analogously, canonical TRSs allow to simulate any Turing machine (cf. [Huet and Lankford, 1978; Dauchet, 1989]). Consequently, adding conditions cannot result in more expressivity with respect to computable functions. This result was confirmed by the construction of a canonical TRS with hidden functions for any decidable number algebra by Bergstra and Tucker in [Bergstra and Tucker, 1980]. However, apart from the case of orthogonal [Bergstra and Klop, 1986] and *easily safely transformable* [Giovannetti and Moisi, 1987] CTRSs, no constructive transformation from CTRSs into unconditional TRSs was developed and proven to be sound yet up to our knowledge.

The general goal of this paper is thus to find a transformation of an arbitrary ground-confluent, decreasing CTRS into an equivalent TRS, close to the CTRS – in the sense of size, complexity and algebraic interpretation – and without explicit restriction of the evaluation strategy. The transformation can be used in the context of parallel implementations of conditional term rewriting, where it allows to reuse specific hardware/compiling techniques developed for the unconditional case [Goguen, 1987; Aida *et al.*, 1990; Kirchner and Viry, 1992].

Confluence and termination, as required here, is also a common restriction used to achieve completeness of ***narrowing***. We therefore discuss the use of our transformation for proving soundness and completeness of bottom-up strategies for conditional narrowing by reusing the proofs for corresponding strategies in the unconditional case.

Related Work

One way to get a transformation from CTRSs to corresponding TRSs is the construction of a Turing machine, which can be coded as a single left-linear unconditional rewrite rule [Dauchet, 1989]. However, working with such Turing machine encodings is of course not practical. Another way is the compilation of the CTRS, as proposed, e.g., by S. Kaplan [Kaplan, 1987], who transforms CTRSs into LISP code. Slight changings in the code generation, eliminating LISP operations involving static memory (*setq*) and other built-ins (like the equality test *eq*), result in a program that can be interpreted in combinatory logics. The latter having only left-linear, unconditional axioms, we get the wanted result.

Nevertheless, the correctness of the compiled code is only guaranteed under innermost evaluation [Kaplan, 1987]. This is an additional explicit restriction, which we would like to avoid, as well as the code size incrementation by a factor asymptotically proportional with $n * log(n)$ (supposing n be the size of the original code), which is inherent to the translation into combinatory logics. Note that the code size also depends on the signature, since the size of the replacement of the LISP built-in *eq* grows asymptotically like m^2, if the considered signature contains m function symbols.

Similarly, the target code for the CTRS compilation may be the one of an abstract machine, like in numerous implementations integrating functional and logic programming paradigms (e.g. [Bosco *et al.*, 1989]), often extending the Warren abstract machine principles. This is also impractical, since abstract machines typically use static memory and we do not know of any interpreter for the corresponding code in form of an unconditional term rewriting system.

For orthogonal, i.e. left-linear and non-overlapping (not necessarily terminating) CTRSs, an effective transformation was given in [Bergstra and Klop, 1986]. In the case of a ***easily safely transformable*** CTRS, a non-linear generalisation of orthogonal CTRSs for rules containing at most one condition, a similar transformation can be found in [Giovannetti and Moisi, 1987]. H. Aida, J. Goguen and J. Meseguer proposed a transformation of non-linear canonical CTRSs in [Aida *et al.*, 1990] using additional arguments and explicit strategy restrictions, but without detailed proofs.

In the functional-logic programming domain, [Dershowitz and Plaisted, 1988] and [Dershowitz and Okada, 1990] discuss the use of transformations of conditional into unconditional equational theories. In [Josephson and Dershowitz, 1989], a transformation of conditional equations into unconditional ones is given in the context of narrowing, i.e. it is possible to backtrack when a condition is not satisfied, in contrast to the work presented in this paper. [Sivakumar, 1989] contains a transformation for the purpose of handling non-decreasing equations and for the detection of unfeasible conditions during completion of CTRSs.

A transformation of first-order logics into discriminator varieties, i.e. algebras containing an *if-then-else* operation, is proposed in [Burris, 1992], based on work by McKenzie and Ackermann. Burris and Jeong (unpublished) also proved linear reducibilty of first-order logic to unconditional equational logics. Further (unpublished) transformations were used by Chtourou/Rusinowitch and Arts/Zantema, respectively, in the termination domain, however, without preserving equivalences.

Outline of the Transformation

Let $\mathcal{R}$ be a CTRS without extra variables. The rules in $\mathcal{R}$ are in general of the form $(g \to d \text{ if } \bigwedge_{i\in[1..n]} s_i \; R \; s_i')$, where $\mathcal{V}ar(d) \cup_{i\in[1..n]} (\mathcal{V}ar(s_i) \cup \mathcal{V}ar(s_i')) \subseteq \mathcal{V}ar(g)$ and R is one of the relations $\overset{*}{\longleftrightarrow}_{\mathcal{R}}$ (1), $\downarrow_{\mathcal{R}}$ (2) or $\overset{!}{\longrightarrow}_{\mathcal{R}}$ (3), defined as in [Dershowitz and Okada, 1990]. If all rules in $\mathcal{R}$ use conditions of type (1) only, then $\mathcal{R}$ is called *natural*. In case of type (2) conditions, it is called *join*-CTRS and when only type (3) conditions occur, then $\mathcal{R}$ is a *normal*-CTRS.

$\mathcal{R}$ is called *decreasing*, if there is a well-founded conservative extension $>$ of the rewrite relation $\to_{\mathcal{R}}$ which contains the proper subterm relation and satisfies $\sigma(l) > \sigma(s_i), \sigma(t_i)$ for all substitutions σ and all $i \in [1..n]$. Our general conditions for the CTRS to be transformed are decreasingness and confluence on ground terms. Furthermore, as usual for practical term rewriting systems, we assume the signature and the rule set to be finite. Under these conditions, natural and join CTRSs coincide and any join CTRS can be simulated by a normal CTRS [Dershowitz and Okada, 1990], with the help of auxiliary functions which semantically play the role of hidden operations. Hence, we can assume w.l.o.g. the CTRS to be normal.

Consider the following example of a confluent and terminating Σ-CTRS for concatenating lists using the constructors *nil* and *cons* from [Dershowitz and Okada, 1990]:

$$\mathcal{R} = \left\{ \begin{array}{ll} car(cons(x,y)) \to x, & null(nil) \to tt, \\ cdr(cons(x,y)) \to y, & null(cons(x,y)) \to ff, \\ \multicolumn{2}{l}{append(x,y) \to y \;\text{ if }\; null(x) \overset{*}{\longrightarrow}_R tt,} \\ \multicolumn{2}{l}{append(x,y) \to cons(car(x), append(cdr(x),y)) \;\text{ if }\; null(x) \overset{*}{\longrightarrow}_R ff} \end{array} \right\}$$

Now, assume we try to replace the two axioms for *append* by the following unconditional one:

$$append(x,y) \to if\ null(x)\ then\ y\ else\ cons(car(x), append(cdr(x),y))$$

Remark that the *if-then-else* used here is the unconditional term rewriting operation with the following two axioms:

$$\textit{if } tt \textit{ then } x \textit{ else } y \rightarrow x, \qquad \textit{if } ff \textit{ then } x \textit{ else } y \rightarrow y$$

However, the resulting TRS is now non-terminating (cf. [Dershowitz and Okada, 1990]), since the rule can always be applied to the occurrence of *append* in the right hand side. The basic idea of our transformation consists in describing equationally the application of a conditional rewrite rule. Evaluation of the conditions and their satisfiability test is performed by forcing implicitly an innermost, also called bottom-up, strategy for the evaluation of functions. This can easily be realized by the use of three ordered sorts. The first one, called *IrrTerms*, contains all irreducible terms, which have a top function symbol of the form f^{irr}. The second one, called *Terms*, contains additionally Σ-terms, which may have subterms of sort *IrrTerms*.

The last one, *FrozenTerms* is for so-called frozen terms, i.e. for those containing a function symbol of the form f^{fz}. All possible right hand sides r of rules from $\mathcal{R}$ are kept frozen, in order to inhibit their reduction, until the corresponding condition evaluates to true. Therefore, we replace r by a term where every function symbol g is replaced by g^{fz}, for which no axioms are given. When a matcher is found and the instantiated condition is satisfied, then and only then r will be transformed into its original form while applying the matcher.

All arguments of a function are required to be of sort *IrrTerms*. Functions will therefore only be applied if all arguments are irreducible, i.e. when we are sure to get either T (true) or F (false) when normalising the condition of a rule in $\mathcal{R}$. If none of these conditions is satisfied, then we are sure to get an irreducible term, i.e. we can replace the top function symbol f by f^{irr}. This technique requires matching to be done by rewriting. Hence, we need to distinguish variables in non-linear matching patterns, which is impossible without negation. The adopted solution to this problem replaces variables x in matching patterns, conditions and right hand sides of rules in $\mathcal{R}$ by corresponding constants d_x which can easily be distinguished.

To illustrate this, let d_x, d_y be the constants representing the variables x and y, respectively, in matching patterns. In the following we assume x, y to be of sort *IrrTerms*. Now, to come back to the example of $\mathcal{R}$, our transformation yields the following rule as replacement for *append*'s conditional axioms:

$$\begin{array}{lll}
append(x,y) \rightarrow & try & matchpb(append^{irr}(d_x,d_y), append^{irr}(x,y)) \\
 & cond & (null^{fz}(d_x) == tt^{irr}) \\
 & then & d_y \\
 & else\ try & matchpb(append^{irr}(d_x,d_y), append^{irr}(x,y)) \\
 & cond & (null^{fz}(d_x) == ff^{irr}) \\
 & then & cons^{fz}(car^{fz}(d_x), append^{fz}(cdr^{fz}(d_x), d_y)) \\
 & else & append^{irr}(x,y)
\end{array}$$

Remark that all used operations are defined by unconditional rewrite rules given in detail in the sequel of the paper. Informally described, the *try-cond-*

then-else operation takes four arguments and tries first to solve the matching problem *matchpb*($append^{irr}(d_x, d_y), append^{irr}(x, y)$) in the first argument. If successful, it instantiates the condition in the second argument with the found matcher. Let us focus on the lower *try-cond-then-else* term, i.e. the condition is $null^{fz}(d_x) == ff^{irr}$. Recall that instantiating a term results in removing all *fz*-marks. Hence, the terms in the condition get reducible and normalising them results in irreducible terms, i.e. marked overall by *irr*. This allows us to decide the satisfaction of the condition through simple syntactic equality as done by evaluating the operation $==$ via rewriting, giving either T or F.

Now, if the condition evaluates to T, the term *append*(x, y) is replaced by the subterm corresponding with the right hand side of *append*'s second axiom in $\mathcal{R}$, namely $cons^{fz}(car^{fz}(d_x), append^{fz}(cdr^{fz}(d_x), d_y))$, instantiated by the matcher. Remark that only then this subterm corresponding with the right hand side gets reducible since the *fz* marks are again removed. This freezing of the former right hand side until the corresponding condition evaluates to true prevents us from non-termination as in the naive transformation presented first. In case that both subterms representing the conditions of *append*'s rules in $\mathcal{R}$ evaluate to F, *append*(x, y) is replaced by $append^{irr}(x, y)$, i.e. the *irr*-mark is added on top.

Let $\xrightarrow{!}_{\mathcal{R}_0}$ stand for the normalisation relation for any TRS $\mathcal{R}_0$. Analogous considerations for the other rules in $\mathcal{R}$ and operation symbols without axioms in Σ lead us to the TRS $(\Sigma', \mathcal{R}')$, s.t. for all Σ-ground terms t, $t \xrightarrow{!}_{\mathcal{R}'} t'$ iff $t \xrightarrow{!}_{\mathcal{R}} t'$. Hence, $(\Sigma', \mathcal{R}')$ gives an order-sorted specification with hidden operations and sorts for $(\Sigma, \mathcal{R})$.

Let $T(\Sigma, \mathcal{R})$ stand for the quotient term algebra of $(\Sigma, \mathcal{R})$, being initial in the category of its models. Let furthermore $A_{|\Sigma}$ be the algebra A' obtained from A by forgetting all operations and constants not in Σ, $\langle A \rangle_\Sigma$ give the subalgebra A' of A generated by Σ (cf. [Bergstra and Tucker, 1987]). Then, $\langle T(\Sigma', \mathcal{R}') \rangle_\Sigma = T(\Sigma, \mathcal{R})$, if we take Σ to be a one-sorted signature, but not $T(\Sigma', \mathcal{R}')_{|\Sigma} = T(\Sigma, \mathcal{R})$, due to auxiliary operations and constants.

We start with the definition of the conditional framework, followed by the transformation into an order-sorted specification and proofs of the properties of $\mathcal{R}'$ w.r.t. $\mathcal{R}$. Finally, we discuss complexity, possible extensions and applications of the transformation. All notations are compatible with the ones used in [Dershowitz and Jouannaud, 1990; Kirchner *et al.*, 1988; Dershowitz and Okada, 1990]. In particular, we call a TRS *orthogonal* [Dershowitz and Jouannaud, 1990], if it is left-linear all critical pairs are trivial, i.e. of the form (t, t). The reader is supposed to have basic knowledge in order-sorted rewriting.

2 The Conditional TRS

Let $\Sigma = (\mathcal{F}_i)_{i \in [0..k]}$ be an unsorted, finite signature, where $\mathcal{F}_i$ contains function symbols of arity i only, assuming $\mathcal{F}_i$ and $\mathcal{F}_j$ to be disjoint for two distinct $i, j \in [0..k]$, i.e. no overloading is allowed. Remark that overloaded signatures can easily be transformed into non-overloaded ones via bijective operator renaming.

Let furthermore $\mathcal{F}_{>0}$ be $\bigcup_{i\in[1..k]}\mathcal{F}_i$ and $\mathcal{F} = \bigcup_{i\in[0..k]}\mathcal{F}_i$. $\mathcal{T}(\Sigma,\mathcal{X})$ stands for the set of terms over Σ and a set of variables $\mathcal{X}$, $\mathcal{T}(\Sigma)$ for those over Σ without variables, called ground terms.

Let $\mathcal{R}$ be a finite, normal, ground-confluent and decreasing CTRS [Dershowitz and Okada, 1990] (for short *canonical* CTRS). For each $f \in \mathcal{F}_n, n \in [0..k]$, let $\mathcal{R}_f$ denote the set of rules defining f, i.e. all rules in $\mathcal{R}$ with f as top symbol on the left hand side of the rule. Let $\mathcal{R}_f$ be of the following form:

$$\mathcal{R}_f = \left\{ \begin{array}{l} l_f^1 : f(t_1^1, \ldots t_n^1) \rightarrow r_1 \text{ if } c_1, \\ \quad\vdots \\ l_f^m : f(t_1^m, \ldots t_n^m) \rightarrow r_m \text{ if } c_m \end{array} \right\}$$

Remark that the l_f^i, $i \in [1..m]$ are labels used to distinguish rules. A rewrite derivation respects *lexicographic rule priority* if whenever l_f^i is applied at occurrence ω of some term t in the derivation, then there is no $j < i$, s.t. l_f^j could be applied at t in ω instead.

3 The Corresponding Unconditional TRS

We now give a transformation of $(\Sigma, \mathcal{R})$ into an order-sorted unconditional specification $(\Sigma', \mathcal{R}')$. $\mathcal{R}'$ is a TRS provably confluent and terminating (see section 4).

3.1 Constructing Σ' from Σ

First we define Σ' to be $(\mathcal{S}', \mathcal{F}')$ with $\mathcal{F}'$ containing $\mathcal{F} \cup \mathcal{F}^{irr} \cup \mathcal{F}^{fz} \cup \mathcal{F}^{vars}$, where $\mathcal{F}^{vars} = \{d_x \mid \exists l \in \mathcal{R} : x \in Var(l)\}$, $\mathcal{F}^{irr}$ and $\mathcal{F}^{fz}$ are *mirror* function sets defined as $\{f^{irr} \mid f \in \mathcal{F}\}$ and $\{f^{fz} \mid f \in \mathcal{F}\}$, respectively, s.t. f, f^{irr} and f^{fz} have the same arity. Additionally, $\mathcal{F}'$ contains some operations defined later on, for which we assume that their names do not occur in $\mathcal{F} \cup \mathcal{F}^{irr} \cup \mathcal{F}^{fz} \cup \mathcal{F}^{vars}$. Let $\leq_{\mathcal{S}'}$ denote the sort ordering. $\mathcal{S}'$ contains the following ordered sorts:

$$IrrTerms \leq_{\mathcal{S}'} Terms \leq_{\mathcal{S}'} FrozenTerms,$$
$$Vars \leq_{\mathcal{S}'} FrozenTerms$$

Let $\mathcal{X}'$ be a set of $\mathcal{S}'$-sorted variables and $(x :: A)$ be the shorthand for 'x is of sort A'. In order to internalise the matching algorithm needed to simulate conditional rules, we represent any variable x contained in a rule of $\mathcal{R}$ by a constant d_x of sort $Vars$ in $\mathcal{F}'$. All functions f with $arity(f) = n$ in $\mathcal{F}$ lead to the function declarations $f : Terms^n \rightarrow Terms$, $f^{irr} : IrrTerms^n \rightarrow IrrTerms$ and $f^{fz} : FrozenTerms^n \rightarrow FrozenTerms$. $\mathcal{T}_{Terms}(\mathcal{X}')$, $\mathcal{T}_{IrrTerms}(\mathcal{X}')$ and $\mathcal{T}_{FrozenTerms}(\mathcal{X}')$ be a shorthand for $(\Sigma', \mathcal{X}')$-terms of the mentioned sorts.

3.2 Rule Transformation

If we want to apply the ith conditional axiom l_f^i of a function f to a term $f(x_1, \ldots, x_n)$, where $x_1, \ldots, x_n$ represent the current arguments of the functions, which are ground terms, we need to match $f(t_1^i, \ldots, t_n^i)$ against $f(x_1, \ldots, x_n)$. In order to mimic this with unconditional, left-linear rewrite rules, we need to compare possible variable images, since the pattern $f(t_1^i, \ldots, t_n^i)$ may be non-linear. Unfortunately, it is impossible to decide the difference of two variables x and y without negation, i.e. we need to replace them by constants d_x and d_y, respectively. Distinguishing variables is necessary to do the merging part of matching and to apply substitutions to variables, both using rewrite rules.

Furthermore, since we want a rule only to be applied if all arguments are irreducible, we need to attach at any node in the pattern the label irr. This motivates the definition of the following two term transformations, used later on to give the transformation of $\mathcal{R}_f$ for each $f \in \mathcal{F}$, $x \in \mathcal{X}'$:

$$\begin{array}{rl|rl} (.)^{irr} : & \mathcal{T}_{Terms}(\mathcal{X}') \to \mathcal{T}_{IrrTerms}(\mathcal{X}') & (.)^{cl} : & \mathcal{T}_{Terms}(\mathcal{X}') \to \mathcal{T}_{FrozenTerms}(\mathcal{X}') \\ f(t_1, \ldots, t_n)^{irr} = & f^{irr}(t_1^{irr}, \ldots, t_n^{irr}) & f(t_1, \ldots, t_n)^{cl} = & f^{irr}(t_1^{cl}, \ldots, t_n^{cl}) \\ x^{irr} = & x :: IrrTerms & x^{cl} = & d_x \end{array}$$

Let *solve* and *match* be operations representing the matching algorithm described in section 3.4 and $\leq$ be a matching equation constructor. Now we can represent the matching problem, which was presented in a simplified form using *matchpb* in the introduction, as the following term:

$$\sigma_i' = match(solve(f(t_1^i, \ldots, t_n^i)^{cl} \leq f^{irr}(x_1, \ldots, x_n), nil))$$

The already mentioned freezing of the right hand side r_i of l_f^i, i.e. attaching labels fz in term nodes and replacing variables by their corresponding constants, is performed by the following term transformation:

$$\begin{array}{rl} (.)^{fz} : & \mathcal{T}_{Terms}(\mathcal{X}') \to \mathcal{T}_{FrozenTerms}(\mathcal{X}') \\ f(t_1, \ldots, t_n)^{fz} = & f^{fz}(t_1^{fz}, \ldots, t_n^{fz}) \\ x^{fz} = & d_x \end{array}$$

The condition c_i of l_f^i is represented by the term $c_i' = (s_1^{fz} == s_1'^{cl}) \wedge (\ldots \wedge (s_n^{fz} == s_n'^{cl}) \ldots))$, assuming the empty conjunction, i.e. in case that l_f^i is unconditional, to be T. Remember that terms labelled by irr, like those produced by cl, are irreducible. Now, the set of rules $\mathcal{R}_f$ can be replaced by the following rule, assuming $x_i :: IrrTerms$ for $i \in [1..n]$:

$$\begin{array}{rl} l_f' : f(x_1, \ldots, x_n) \to & try\ \sigma_1'\ cond\ c_1'\ then\ r_1^{fz} \\ & else\ try \\ & \vdots \\ & else\ try\ \sigma_m'\ cond\ c_m'\ then\ r_m^{fz} \\ & else\ f^{irr}(x_1, \ldots, x_n), \end{array}$$

where *try-cond-then-else* results in the instantiated right hand side if there is a match and the instantiated condition evaluates to T, otherwise in the fourth argument. This definition prevents r_i from being instantiated and reduced, if the corresponding rule in $\mathcal{R}$ is not applicable. Due to the definition of decreasingness, $\sigma(r_i)$ may be greater than $\sigma(l_i)$ for any σ, s.t. $\sigma(c_i)$ reduces to F, i.e. reduction of $\sigma(r_i)$ may cause non-termination in this case. Let $\mathcal{R}'_{\mathcal{F}}$ denote the set $\{l'_f \mid f \in \mathcal{F}\}$. Remark that the size of $\mathcal{R}'_{\mathcal{F}}$ is linear w.r.t. the size of $\mathcal{R}$.

The difference between computing with the original set of rules and this set of unconditional rules originates in the point when the conditions are evaluated. The semantics of conditional rules suppose a rule to be applied only if its condition is satisfied. Here, in $\mathcal{R}'$, a function will only be evaluated if its arguments are irreducible, i.e. we force a bottom-up strategy. Therefore, the sort *IrrTerms* has to contain exactly all t^{irr} for each t in the original signature Σ. Consequently, we add for any operation symbol g with $arity(g) = n \geq 0$ in Σ which does not occur on the top of a left hand side in $\mathcal{R}$, the following rules, where $x_1, \ldots, x_n :: IrrTerms$:

$$g(x_1, \ldots, x_n) \rightarrow g^{irr}(x_1, \ldots, x_n)$$

3.3 Auxiliary Functions

In this section, we define formally the *if_then_else* construct and some boolean valued operations for deciding syntactic equality and inequality. They are given in order to prove the existence of orthogonal rules for these functions. Let the sort *Bool* be in $\mathcal{S}'$, $x :: Bool$ and consider the following operator declarations:

$$\begin{array}{rll}
F, T: & & \rightarrow Bool \\
\wedge: & Bool \times Bool & \rightarrow Bool \\
\vee: & Bool \times Bool & \rightarrow Bool \\
==: & FrozenTerms \times FrozenTerms & \rightarrow Bool \\
\neq\!=: & FrozenTerms \times FrozenTerms & \rightarrow Bool \\
if_then_else: & Bool \times Terms \times Terms & \rightarrow Terms
\end{array}$$

We start with boolean conjunction and disjunction:

$$\begin{array}{llll}
x \wedge T \rightarrow x, & x \wedge F \rightarrow F, & T \wedge x \rightarrow x, & F \wedge x \rightarrow F, \\
x \vee T \rightarrow T, & x \vee F \rightarrow x, & T \vee x \rightarrow T, & F \vee x \rightarrow x.
\end{array}$$

Next we define syntactic equality $==$ used for the translation of the conditions c_i. Let $x :: Bool$ and $x_i, y_i :: FrozenTerms$ for $i \in [1..max(m,n)]$. Assuming $c, d \in \mathcal{F}'_0 \cup \{T, F\}$, $f, g \in \mathcal{F}'_{>0}$, s.t. $c \neq d$, $f \neq g$, we can use the following axioms:

$$\begin{array}{ll}
== (c, c) & \rightarrow T, \\
== (c, d) & \rightarrow F, \\
== (f(x_1, \ldots, x_n), f(y_1, \ldots, y_n)) & \rightarrow (== (x_1, y_1) \wedge (\ldots \wedge == (x_n, y_n)) \ldots), \\
== (f(x_1, \ldots, x_n), g(y_1, \ldots, y_m)) & \rightarrow F.
\end{array}$$

Remark that these rules are schemas for all combinations of c, d and f, g, respectively. The definition of syntactic inequality $=\not=$ is dual. Note that the number of rules needed for $==$ and $=\not=$ is a quadratic function over the number of function symbols in $\mathcal{F}$. The *if_then_else* mixfix function is defined analogously as in the introduction, using T and F instead of *tt* and *ff*, respectively.

3.4 Matching by Term Rewriting

The complete unconditional term rewriting system $\mathcal{R}'$ for $\mathcal{R}$ is the set of all rules in $\mathcal{R}'_{\mathcal{F}}$ plus those described in this and the last subsection. Remark that the size of $\mathcal{R}'$ depends linearly on the size of $\mathcal{R}$ and quadratically on $\mathcal{F}$.

A *(variable disjoint) matching problem* $\mathcal{M}$ is of the form $\bigwedge_{j\in[1..m]} s_j \leq t_j$, s.t. $(\bigcup_{j\in[1..m]} Var(s_j)) \cap (\bigcup_{j\in[1..m]} Var(t_j)) = \emptyset$. $\mathcal{M}$ is then also called *variable disjoint*. A *matcher* for $\mathcal{M}$ is a substitution σ, s.t. $\bigwedge_{j\in[1..m]} \sigma(s_j) = t_j$. This is also written $Sol(\mathcal{M}) = \{\sigma\}$, or $Sol(\mathcal{M}) = \emptyset$ if σ does not exist. Note that matchers are unique in this framework. Our matching algorithm is derived from the set of transformation rules (in the style of [Jouannaud and Kirchner, 1991]) shown in Figure 1.

Delete	$\mathcal{M} \wedge t \leq t$
	$\Longrightarrow \mathcal{M}$
Decompose	$\mathcal{M} \wedge f(s_1, \ldots, s_n) \leq f(t_1, \ldots, t_n)$
	$\Longrightarrow \mathcal{M} \wedge s_1 \leq t_1 \wedge \ldots \wedge s_n \leq t_n$
Conflict	$\mathcal{M} \wedge f(s_1, \ldots, s_n) \leq g(t_1, \ldots, t_m)$
	$\Longrightarrow \mathbb{F}$
	if $f \neq g$
Merging	$\mathcal{M} \wedge x \leq t_1 \wedge x \leq t_2$
	$\Longrightarrow \mathcal{M} \wedge x \leq t_1$
	if $x \in \mathcal{X}$ and $t_1 = t_2$
MergingClash	$\mathcal{M} \wedge x \leq t_1 \wedge x \leq t_2$
	$\Longrightarrow \mathbb{F}$
	if $x \in \mathcal{X}$ and $t_1 \neq t_2$
SymbolClash	$\mathcal{M} \wedge f(t_1, \ldots, t_n) \leq x$
	$\Longrightarrow \mathbb{F}$
	if $x \in \mathcal{X}$

Fig. 1. MATCH : Matching using Transformation Rules

After normalisation with this set of rules, we obtain $\mathbb{F}$ if $Sol(\mathcal{M}) = \emptyset$. Otherwise, the normal form obtained is of the form $\bigwedge_{j\in[1..m]} x_j \leq u_j$, if $Sol(\mathcal{M}) = \{\sigma\}$ with $\sigma = \bigcup_{j\in[1..m]}\{x_j \mapsto u_j\}$, the empty conjunction being defined as $\mathbb{T}$.

Let *Presolved*, *MatchPb*, *Presuccess*, *Failure* $\in \mathcal{S}'$, s.t.:

$$Presuccess, Failure \leq_{\mathcal{S}'} Presolved \leq_{\mathcal{S}'} MatchPb.$$

We define the profile of the matching operations as follows:

$$
\begin{array}{rll}
match: & MatchPb & \to Presolved \\
merge: & Presolved & \to Presolved \\
solve: & MatchPb \times MatchPb & \to MatchPb \\
solve: & Presolved \times Presolved & \to Presolved \\
solve: & Presuccess \times Presuccess & \to Presuccess \\
\leq: & FrozenTerms \times IrrTerms & \to MatchPb \\
\leq: & Vars \times IrrTerms & \to Presuccess \\
nil: & & \to Presuccess \\
none: & & \to Failure
\end{array}
$$

We start with the axiomatization of *match*, *merge* and *solve*. Here we need to know the maximal number of variables in a rule figuring in $\mathcal{R}$, denoted by V_{max}, in order to have an upper bound for the number of variables in a matcher. We can assume V_{max} to be given, since the number of rules is finite and the rules have a finite number of variables. Assume $z :: MatchPb$. Informally, *match* can be given by $match(z) = none$ if $Sol(z) = \{\sigma\}$. Otherwise, if $Sol(z)$ is the singleton $\{\{x_1 \mapsto w_1, \ldots, x_k \mapsto w_k\}\}$, let $match(z)$ be $solve(d_{x_1} \leq w_1, solve(\ldots solve(d_{x_k} \leq w_k, nil) \ldots))$.

The matching algorithm itself is a two step process. First, an arbitrary matching problem constructed with *solve*, $\leq$ and *nil* is transformed into a first presolved form, s.t. all decompositions and clash detections are performed. These are performed by *solve* itself. Let us assume $x, x_1, \ldots x_n :: FrozenTerms$, $y, y_1, \ldots y_{max(m,n)} :: IrrTerms$, $z :: MatchPb$ and $c^{irr}, d^{irr} \in \mathcal{F}_0^{irr}$ s.t. $c^{irr} \neq d^{irr}$ and $f^{irr}, g^{irr} \in \mathcal{F}_{\geq 0}^{irr}$, s.t. $f^{irr} \neq g^{irr}$ or $arity(f^{irr}) = n \neq m = arity(g^{irr})$:

$$
\begin{array}{ll}
solve(c^{irr} \leq c^{irr}, z) & \to z, \\
solve(c^{irr} \leq d^{irr}, z) & \to none, \\
solve(x \leq y, none) & \to none, \\
solve(f^{irr}(x_1, \ldots, x_n) \leq g^{irr}(y_1, \ldots, y_m), z) & \to none, \\
solve(f^{irr}(x_1, \ldots, x_n) \leq f^{irr}(y_1, \ldots, y_n), z) & \to \\
\qquad solve(x_1 \leq y_1, solve(\ldots solve(x_n \leq y_n, z) \ldots)). &
\end{array}
$$

Remark that the right hand sides of the matching equations for *solve* can be assumed to be ground terms, since we are only interested in ground confluence of $\mathcal{R}'$. Therefore, *solve* always returns a matching problem of sort *Presolved*. The second step consists in the merge steps necessary to decide if there is an ambiguity in the image for a variable.

Consequently, *match* and *merge* become, assuming $v_1, v_2 :: Vars$, $w_1, w_2 :: IrrTerms$, $z :: Presolved$, $z_1 :: Presuccess$ and $z_2 :: Failure$:

$$
\begin{array}{llll}
match(z_1) & \to merge^{V_{max}}(z_1), & match(z_2) & \to z_2, \\
merge(solve(z, nil)) & \to solve(z, nil), & merge(none) & \to none,
\end{array}
$$

$$
\begin{array}{l}
merge(solve(v_1 \leq w_1, solve(v_2 \leq w_2, z))) \to \\
\qquad if(v_1 == v_2 \land w_1 == w_2) \; then \; merge(solve(v_1 \leq w_1, z)) \\
\qquad else \; if \; (v_1 =\!\!\not= v_2) \; then \; solve(v_2 \leq w_2, merge(solve(v_1 \leq w_1, z))) \\
\qquad else \; none.
\end{array}
$$

Here, $merge^{V_{max}}(z)$ stands for the V_{max} times application of *merge* on z. Therefore, we need to overload *if_then_else* to $Bool \times Presolved^2 \rightarrow Presolved$. Remark that as long as *merge* traverses the list, it is of sort *MatchPb*. Only after *merge* has disappeared, the result is of sort *Presuccess* or *Failure*. This definition of *merge* is also the reason why variables need to be replaced by constants: We need to decide the difference $v_1 =/= v_2$, which is not possible with 'real' variables.

This realisation of matching by term rewriting, especially the *merge* operation, may for implementation issues result in a considerable slow-down in comparison with $\mathcal{R}$, although its asymptotical complexity $O(n)$, where n is the size of the initial matching problem, seems to be the same. However, we claim that in both of our envisaged applications for the transformation – reuse of parallel, unconditional rewrite rule machines for CTRSs and soundness/correctness proofs for conditional narrowing – this does not play any role. In the first case, matching instructions are already present in the machine architecture and can therefore replace the rewrite rules. In the second case, the rewrite rules have a pure proof theoretic purpose and are never applied in reality.

3.5 How to Apply Substitutions

In *apply*, the first argument represents the substitution calculated by *match* and the second argument is a frozen term containing constants instead of variables. Remark that *apply* erases *fz*-marks. Assuming $f \in \mathcal{F}$, $x :: Presuccess$, $y_1, \ldots y_n :: FrozenTerms$, $z :: Terms$, $w :: IrrTerms$ and $v_1, v_2 :: Vars$, we define:

$$
\begin{array}{ll}
apply\ :\ Presuccess \times FrozenTerms & \rightarrow Terms \\
apply(nil, z) & \rightarrow z, \\
apply(x, f^{fz}(y_1, \ldots, y_n)) & \rightarrow f(apply(x, y_1), \ldots, apply(x, y_n)), \\
apply(x, f(y_1, \ldots, y_n)) & \rightarrow f(apply(x, y_1), \ldots, apply(x, y_n)), \\
apply(x, f^{irr}(y_1, \ldots, y_n)) & \rightarrow f^{irr}(apply(x, y_1), \ldots, apply(x, y_n)), \\
apply(solve(v_1 \leq w, x), v_2) & \rightarrow if\ (v_1 == v_2)\ then\ w\ else\ apply(x, v_2).
\end{array}
$$

Therefore, *apply*'s first axiom represents the application of the empty substitution, the second to fourth stand for the homomorphism property of substitutions and the last provides recursive search of a variable replacement in the term representation of a substitution.

3.6 Trying Alternatives

The axiomatisation of *try-cond-then-else* is done with the help of an auxiliary function called *trycond-cond-then-else*. The first operation actually only checks if there exists a matcher for the matching problem given as first argument and the latter tests the satisfaction of the condition, given in the second argument instantiated by the matcher, if there is one. If the instantiated condition evaluates to T, then the third argument, also instantiated by the matcher, is returned. In any other case, the fourth argument becomes the result of the operation. Hence, we get the following operator profiles:

$$\begin{aligned}&\mathit{try\_cond\_then\_else}\ :\\&\quad \mathit{MatchPb} \times \mathit{Bool} \times \mathit{FrozenTerms} \times \mathit{Terms} \to \mathit{Terms}\\&\mathit{trycond\_cond\_then\_else}\ :\\&\quad \mathit{Presuccess} \times \mathit{Bool} \times \mathit{FrozenTerms} \times \mathit{Terms} \to \mathit{Terms}\end{aligned}$$

Let x, x_1 :: *Presuccess*, x_2 :: *Failure*, y :: *Bool*, z_1 :: *FrozenTerms* and z_2 :: *Terms*. Then the axioms are as follows:

$$\begin{aligned}&\mathit{try}\ x_1\ \mathit{cond}\ y\ \mathit{then}\ z_1\ \mathit{else}\ z_2 \to\\&\quad \mathit{trycond}\ x_1\ \mathit{cond}\ \mathit{apply}(x_1, y)\ \mathit{then}\ z_1\ \mathit{else}\ z_2,\\&\mathit{try}\ x_2\ \mathit{cond}\ y\ \mathit{then}\ z_1\ \mathit{else}\ z_2 \to z_2,\\&\mathit{trycond}\ x\ \mathit{cond}\ T\ \mathit{then}\ z_1\ \mathit{else}\ z_2 \to \mathit{apply}(x, z_1),\\&\mathit{trycond}\ x\ \mathit{cond}\ F\ \mathit{then}\ z_1\ \mathit{else}\ z_2 \to z_2.\end{aligned}$$

4 Properties of $\mathcal{R}'$

In this section, we prove the correspondence of $\mathcal{R}'$ with $\mathcal{R}$ on $T(\Sigma)$, which can be summarised as follows:

$$\forall t \in T(\Sigma),\quad (t \xrightarrow{!}_{\mathcal{R}} t' \ \text{ iff } \ t \xrightarrow{!}_{\mathcal{R}'} t'^{irr})$$

We start with the correctness for our matching operations, followed by the proof of confluence of $\mathcal{R}'$, the existence of a derivation in $\mathcal{R}'$ for any derivation in $\mathcal{R}$ and finally termination of $\mathcal{R}'$ on $T(\Sigma)$. Let $\mathcal{R}'_{\mathcal{M}}$ stand for $\mathcal{R}' \setminus \mathcal{R}'_{\mathcal{F}}$.

4.1 Termination and Correctness of Matching

Lemma 1. *$\mathcal{R}'_{\mathcal{M}}$ is terminating.*

Proof. Using the precedence following below for an LPO-ordering, termination of these functions was proved with ELIOS-OBJ [Gnaedig, 1992]:

$$\begin{gathered}\mathit{try\_cond\_then\_else} > \mathit{trycond\_cond\_then\_else} >\\ \mathit{match} > \mathit{merge} > \mathit{solve} > \mathit{apply} > \mathit{if\_then\_else}, ==, =/= >\\ \wedge, \vee > T, F, \mathit{none}, \mathit{nil}\end{gathered}$$

Now, we show the soundness/completeness of our rules in $\mathcal{R}'_M$ w.r.t. **MATCH**:

Lemma 2. *Let for all $i \in [1..n], t_i \in \mathcal{T}(\Sigma)$ and $\mathcal{M} = \bigwedge_{i\in[1..n]} s_i \leq t_i$ be a variable disjoint matching problem in $\mathcal{T}(\Sigma, \mathcal{X})$. Let furthermore:*

$$\sigma_{\mathcal{M}} = \mathit{match}(\mathit{solve}(s_1^{cl} \leq t_1^{irr}, \mathit{solve}(\ldots, \mathit{solve}(s_n^{cl} \leq t_n^{irr}, \mathit{nil})\ldots)))$$

Then, **MATCH** *calculates a matcher $\rho = \{x_i \mapsto u_i\}_{i\in[1..m]}$ iff $\sigma_{\mathcal{M}} \xrightarrow{!}_{\mathcal{R}'_{\mathcal{M}}} \sigma$ and $\sigma = \mathit{solve}(d_{x_1} \leq u_1, \mathit{solve}(\ldots, \mathit{solve}(d_{x_m} \leq u_m, \mathit{nil})\ldots)))$. Furthermore, $\rho = \mathbb{F}$ iff $\sigma = \mathit{none}$.*

Proof. First, note that **MATCH** is locally confluent, sort decreasing and terminating, i.e. confluent, too. Therefore, we can calculate a solution of $\mathcal{M}$ using an arbitrary strategy with **MATCH**. Let **Decompose** and **Conflict** be applied first and only if none of these rules are applicable any more, we use **Delete** and finally **Merging** and **MergingClash**. Remark that **SymbolClash** is not needed since all t_i are ground. It is sufficient to prove that the normalisation of $\sigma_{\mathcal{M}}$ corresponds one-to-one with such a strategy and that the termination of the first implies the one of **MATCH**.

Clearly, **Decompose** is applicable to some $s \leq t$ iff *solve*'s third axiom is applicable to $s^{cl} \leq t^{irr}$. Furthermore, **Conflict** corresponds in the same strict way with *solve*'s second and fourth axiom, **Delete** with the first one. Together with the fact that *solve*'s last axiom gives the propagation of *none* to the top, we have $\sigma = none$ after *solve* if the set of **MATCH** rules **Decompose**, **Conflict** and **Delete** applied with the strategy corresponding with *solve*'s application gives $\mathbb{F}$. Furthermore, since these rules only depend on single equations, they cannot be applicable anymore on the corresponding problem if *solve* has terminated.

Now, and only now *match* becomes applicable, since its argument is of sort *Presolved*. Let $d_x \leq u$ be the first matching equation in $\sigma_{\mathcal{M}}$'s current form. Then, after the traversal of the first *merge*, all **Merging** and **MergingClash** steps possible with this $x \leq u$ in **MATCH** are done using *merge*'s last axiom. Since the maximal number of distinct variable constants d_x in $\sigma_{\mathcal{M}}$ is V_{max}, it is sufficient to let *merge* traverse the current term V_{max} times in order to perform all possible **Merging** and **MergingClash** steps, each time propagating the first equation to the end.

Since this does not produce any new matching equations, no more steps using **Decompose**, **Conflict** or **Delete** are applicable now, i.e. **MATCH** must have terminated on the corresponding matching problem in $\mathcal{T}(\Sigma, \mathcal{X})$, too. Consequently, the lemma holds.

4.2 Confluence of $\mathcal{R}'$

The confluence of $\mathcal{R}'$ is independent from the ground confluence of $\mathcal{R}$, since we forced bottom-up evaluation and lexicographic rule choice, which results in a unique normal form. Consequently, given a normalising rewriting sequence for some t in $\mathcal{R}$, all alternative rewrite sequences for t differ only in the evaluation ordering of incomparable subterms, which are independent and therefore do not change the resulting normal form.

Lemma 3. *All rules in $\mathcal{R}'$ are orthogonal and sort-decreasing, i.e. $\mathcal{R}'$ is confluent by construction.*

Proof. Orthogonal, sort-decreasing TRSs are *parallel closed* and every left-linear, parallel closed TRS is strongly confluent (cf. [Huet, 1980], Lemma 3.3).

4.3 Correspondence with $\mathcal{R}$

In this section, we will prove that if some ground term t has the normal form t' in $\mathcal{R}$, using bottom-up evaluation and lexicographic rule choice, then it has normal form $(t')^{irr}$ in $\mathcal{R}'$ and vice versa.

Let in the following l^i_f be defined as in section 2. The substitutions ρ^{irr} and ρ^{cl} denote $\{x_j \mapsto t_j^{irr}\}_{j\in J}$ resp. $\{x_j \mapsto t_j^{cl}\}_{j\in J}$, if $\rho = \{x_j \mapsto t_j\}_{j\in J}$. This allows us to state the correctness of $\mathcal{R}'$ as follows:

Theorem 4. *Let $t = f(t_1, \ldots, t_n) \in \mathcal{T}(\Sigma)$, s.t. all subterms are irreducible by $\mathcal{R}$. If $t \longrightarrow^{\rho}_{\mathcal{R}} t'$ at the top position with the first possible rule l^i_f in $\mathcal{R}$, then $f(t_1^{irr}, \ldots, t_n^{irr}) \stackrel{*}{\longrightarrow}_{\mathcal{R}'} t'' = \rho^{irr}(r_i)$. If no l^i_f is applicable, then $t \stackrel{*}{\longrightarrow}_{\mathcal{R}'} t^{irr}$.*

Proof. The proof is an induction on t w.r.t. the termination ordering $<$ for $\mathcal{R}$. First, remark that *apply* realizes the application of a substitution σ^{cl}, given by *match*, on some term s^{fz}. This corresponds with the application of a $\mathcal{T}(\Sigma, \mathcal{X})$-substitution σ to $s \in \mathcal{T}(\Sigma, \mathcal{X})$.

By the induction hypothesis together with Lemma 2 we get the correctness of the evaluation of c_i – remember that $\mathcal{R}$ is decreasing, i.e. $apply(\sigma'_i, c'_i)$ can be evaluated to T in $\mathcal{R}'$ and *try* σ'_i *cond* c'_i *then* r_i^{fz} *else* z yields $\rho^{irr}(r_i)$, since σ'_i evaluates to the term representation of ρ^{irr} (cf. Lemma 2).

Now, since l^i_f is the first applicable rule in $\mathcal{R}$, we know for all $k < i$, that σ'_i becomes *none* or $apply(\sigma'_k, c'_k)$ can by induction hypothesis be evaluated to F. Hence, *try_cond_then_else* discards all remaining alternatives in the instantiated right-hand side of l'_f and $f(t_1^{irr}, \ldots, t_n^{irr})$ is normalised to $\rho^{irr}(r_i)$.

Otherwise, if no l^i_f is applicable, then for all $k \in [1..m]$, either σ'_k evaluates to *none* or $apply(\sigma'_k, c'_k)$ can be evaluated to F. Hence, the axioms of *try_cond_then_else* give $f^{irr}(t_1^{irr}, \ldots, t_n^{irr}) = t^{irr}$.

Thus, we get the correctness of $\mathcal{R}'$ w.r.t. bottom-up evaluation with lexicographic rule priority in $\mathcal{R}$.

Corollary 5. *For every bottom-up normalisation $t_1 \longrightarrow_{\mathcal{R}} t_2 \ldots \longrightarrow_{\mathcal{R}} t_n$ in $\mathcal{R}$ with $t_i \in \mathcal{T}(\Sigma)$ for $i \in [1..n]$, respecting lexicographic rule priority, there exists an evaluation $t_1 \stackrel{*}{\longrightarrow}_{\mathcal{R}'} t_n^{irr}$ in $\mathcal{R}'$.*

This leads us immediately to a second corollary concerning the uniqueness of normal forms in $\mathcal{R}'$.

Corollary 6. *Let $t \in \mathcal{T}(\Sigma)$. If $t \stackrel{!}{\longrightarrow}_{\mathcal{R}} t''$ and $\exists t', t \stackrel{!}{\longrightarrow}_{\mathcal{R}'} t'$, then $t' = (t'')^{irr}$.*

Proof. Corollary 5 gives us $t \stackrel{*}{\longrightarrow}_{\mathcal{R}'} (t'')^{irr}$. Now, confluence of $\mathcal{R}'$ gives us $(t'')^{irr} \stackrel{*}{\longrightarrow}_{\mathcal{R}'} t'$. Hence, $t' = (t'')^{irr}$, since $(t'')^{irr}$ is irreducible by definition.

4.4 Termination of $\mathcal{R}'$

Before we give the proof of termination for all $\mathcal{R}'$-derivations starting with some $t \in T(\Sigma)$, let $\leadsto_{\mathcal{R}}$ be defined analogously to [Dershowitz and Okada, 1990] as follows: $s \leadsto_{\mathcal{R}} p$ if there is a rule $l : g \to d$ if $\bigwedge_{i\in I} s_i \xrightarrow{*}_{\mathcal{R}} s'_i$ in $\mathcal{R}$ and a substitution σ, s.t. $\sigma(g) = s$ and $p = \sigma(s_i)$ for some $i \in I$.

Theorem 7. *All $t \in T(\Sigma)$ evaluate in $\mathcal{R}'$ within a finite number of steps.*

Proof. Assume, for contradiction, there is an infinite derivation starting with t using $\mathcal{R}'$. The goal of this proof is to extract an infinite evaluation of t in $\mathcal{R}$. First of all, recall that all rules in $\mathcal{R}'_{\mathcal{F}}$ are inapplicable at top of some term u containing a strict subterm not of sort *IrrTerms*, since the variables of the left hand side are of sort *IrrTerms*. Hence, the first $\mathcal{R}'_{\mathcal{F}}$ rewriting step at or above ω after $t[s]_\omega \longrightarrow^{\omega}_{\mathcal{R}'_{\mathcal{F}}} t[s']_\omega$ can only take place after normalising all strict subterms of s' to some term of sort *IrrTerms*, i.e. without operators in $\mathcal{R}'_{\mathcal{M}}$. This allows us to assume w.l.o.g. the following form for the infinite derivation:

$$u_0 \longrightarrow^{\omega_0}_{\mathcal{R}'_{\mathcal{F}}} t_1 \xrightarrow{!}_{\mathcal{R}'_{\mathcal{M}}} u_1 \longrightarrow^{\omega_1}_{\mathcal{R}'_{\mathcal{F}}} t_2 \xrightarrow{!}_{\mathcal{R}'_{\mathcal{M}}} u_2 \longrightarrow^{\omega_2}_{\mathcal{R}'_{\mathcal{F}}} t_3 \xrightarrow{!}_{\mathcal{R}'_{\mathcal{M}}} \cdots,$$

where $u_0 = t$, and every strict subterm of $u_{i|_{\omega_i}}$, $i \geq 0$, is of sort *IrrTerms*. Remark that since $\mathcal{R}'_{\mathcal{M}}$ is terminating, there is no $k > 0$, s.t. only $\mathcal{R}'_{\mathcal{M}}$-steps occur after t_k.

Furthermore, let us assume that the derivation contains only $\mathcal{R}'_{\mathcal{F}}$-steps absolutely necessary to get the infinite derivation. Let furthermore the term transformation $.^{del}$, which transforms terms of sort *IrrTerms* into ones of sort *Terms* by erasing *irr*-marks and changing the sort of variables accordingly, be defined as follows:

$$\begin{array}{rlr} & (.)^{del} : Terms \to Terms & \\ f(t_1,\ldots,t_n)^{del} &= f(t_1^{del},\ldots,t_n^{del}) & (x :: IrrTerms)^{del} = x :: Terms \\ f^{irr}(t_1,\ldots,t_n)^{del} &= f(t_1^{del},\ldots,t_n^{del}) & (x :: Terms)^{del} = x :: Terms \end{array}$$

Now, we can construct an infinite descending $T(\Sigma)$-term sequence $(d_j)_{j\geq 0}$ in contradiction to the decreasingness of $\mathcal{R}$. We pass along the infinite derivation in order to extract it. Let d_0 be t.

Assume we have constructed $(d_j)_{0\leq j\leq k'}$ and the last u_i encountered was u_{k-1}. If $u_{k|_{\omega_k}}$ reduces to $u_{k|_{\omega_k}}^{irr}$ in some $u_{k''}$ for $k'' > k$, then we skip the part of the derivation until $u_{k''}$. This corresponds with the fact that no rule in $\mathcal{R}$ is applicable to $u_{k|_{\omega_k}}^{del}$. Remark that this can only happen a finite number of times in sequence, since otherwise the derivation were finite, ending with some $\mathcal{R}'$-irreducible t'^{irr}.

If $u_{k|_{\omega_k}}$ reduces to an instantiated right hand side $\sigma^{irr}(r)$, cf. Theorem 4, in some $u_{k''}$ for $k'' > k$, then we add u_k^{del} as $d_{k'+1}$. Furthermore, $d_{k'+2}$ becomes $u_{k''}^{del}$ and we skip all steps in the derivation until $u_{k''}$. Remark that this corresponds with a $\mathcal{R}$-rewrite step from $u_k{}^{del}$ to $(u_k[\sigma^{irr}(r)]_{\omega_k})^{del}$.

Otherwise, all remaining $\mathcal{R}'_{\mathcal{F}}$-steps must take place in an instantiated condition $\sigma^{irr}(c')$, where c' is introduced by the $\mathcal{R}'_{\mathcal{F}}$-step applied to u_k. All other cases are in contradiction to the absolute necessity of the $\mathcal{R}'_{\mathcal{F}}$-steps in order to get an infinite derivation. Hence, we can apply this sequence construction procedure recursively to the subderivation starting with $\sigma^{irr}(c')$ and append the result at the end of $(d_j)_{0 \leq j \leq k'}$.

Clearly, the extracted sequence is infinite. By construction, we get $d_j(\longrightarrow_{\mathcal{R}} \cup \rightsquigarrow_{\mathcal{R}})d_{j+1}$ for all $j \geq 0$. But this is in contradiction to the decreasingness of $\mathcal{R}$, as shown in [Dershowitz and Okada, 1990].

We conjecture the termination of $\mathcal{R}'$ for all Σ'-terms. However, this of no interest for our goal in this section:

Corollary 8. $\forall t \in T(\Sigma), \quad (t \xrightarrow{!}_{\mathcal{R}} t' \text{ iff } t \xrightarrow{!}_{\mathcal{R}'} t'^{irr})$

Proof. The right-to-left direction is a consequence of the ground-confluence of $\mathcal{R}$ and Corollary 5. From left to right, we can use Theorem 7 with Corollary 6.

5 Complexity

In the following we will briefly and informally discuss complexity issues concerning our transformation. The asymptotic complexity for the code size is quadratic w.r.t. the number of operator symbols in $\mathcal{F}$ and linear w.r.t. the size of the rule set in $\mathcal{R}$. The complexity of simplification in $\mathcal{R}'$ is in best/average/worst case equally linear w.r.t. the one of innermost reductions with lexicographic rule priority in $\mathcal{R}$, which should represent the worst case. Using the transformation naively may result in an even worse behaviour, because of the realisation of matching by rewriting, which replaces probably built-in functions. However, both kinds of term rewriting systems can easily be compiled into efficient functional or imperative languages. Hence, the difference between the runtime behaviour of the compiled $\mathcal{R}$ and the compiled $\mathcal{R}'$ should be smaller than in the naive use – especially when matching by rewriting is replaced by built-in functions.

6 Possible Extensions

We discuss some possible further work, which was not considered in this article, but which should be worth being mentioned.

Unsorted TRS. In order to obtain from $\mathcal{R}'$ an unsorted rewrite system $\mathcal{R}''$ the proposed rules have to be heavily expanded. Instead of the variable arguments in the left hand side of l'_f in section 3, we have to use (for any combination of operators g_i^{irr} in $\mathcal{F}^{irr}$) the following terms:

$$f(g_1^{irr}(y_{11}, \ldots, y_{1p_1}), \ldots, g_n^{irr}(y_{n1}, \ldots, y_{np_n})).$$

All theorems of this paper should extend without problems to this case, after observing that the unsorted l'_f is only applicable at the top of some term t if all strict subterms of t are completely marked by *irr*, provided the derivation starts with a term in $\mathcal{T}(\Sigma)$. Hence, bottom-up evaluation, which is essential for our theorems, is still forced in $\mathcal{R}''$.

All innermost derivations. In order to preserve all innermost derivations instead of only those, where the first rule figuring in $\mathcal{R}$ is applied, we might add another axiom for the *try-cond-then-else* operation, assuming $x_1 :: MatchPb$, $x_2 :: Presuccess$, $y_1 :: Bool$, $z_1, z_2 :: FrozenTerms$, $z_3 :: Terms$:

$$\begin{aligned}&try\ x_1\ cond\ y_1\ then\ z_1\\&else\ trycond\ x_2\ cond\ T\ then\ z_2\ else\ z_3 \rightarrow apply(x_2, z_2)\end{aligned}$$

Remark that this destroys the orthogonality of $\mathcal{R}'$. Maintaining all derivations, especially the outermost, is far more complicated and not a conservative extension of our technique.

Sorted CTRSs. Although operationally subsumed by conditions, sorts can help structuring CTRSs and offer partial functions on the model level. Therefore, an extension of the transformation for sorted CTRSs should be considered, too. In fact, our transformation may be extended to this case: Every sort S in $\Sigma = (\mathcal{F}, \mathcal{S})$ must simply be doubled with a subsort $S^{irr} \leq_{S'} S$. Additionally, $S^{irr} \leq_{S'} S'^{irr}$ holds whenever $S \leq_S S'$. Finally, we have to add for all sorts S in $\mathcal{S}$, the subsort declarations $S^{irr} \leq_{S'} IrrTerms$ and $S \leq_{S'} Terms$. Function declarations are mirrored in the same way, except that *try-cond-then-else* must have operator profiles reflecting the sort of the right hand sides of the conditional rules. The remaining parts of the transformation maybe kept unchanged.

Non-ground Terms. In order to handle non-ground terms, we need to use the matching included in the rewrite relation. This can be done along the lines of [Aida *et al.*, 1990] using extra-arguments. However, in general the resulting system is then clearly no more orthogonal, but the explicit strategy restriction in [Aida *et al.*, 1990] may be replaced by sorted rewriting following our approach.

Extra-variables. An extension of the used techniques to standard CTRSs with extra variables in the style of [Bertling and Ganzinger, 1989; Ganzinger, 1991b] seems to be possible by replacing $s_i^{fz} == s_i'^{cl}$ in c_i (cf. definition of l'_j, page 3.2) through $s_i'^{cl} \leq s_i^{fz}$ and concatenating stepwise the corresponding substitutions won by rewriting after instantiation of the current matching equation with the current substitution. But any further details are clearly beyond the scope of this paper. However, this extension may be particularly interesting for a development of narrowing using CTRSs with extra-variables, as recently done for orthogonal CTRSs by Hanus [Hanus, 1994].

7 Applications

We give two sample applications of the transformation, in order to underline our claim of usefulness of the transformation.

Parallel Implementation of CTRSs. This was the original problem motivating this work. There are projects of building parallel term rewriting machines. One major goal in the design is, of course, a maximum of simplicity for the architecture – a motto that has proven its usefulness by the creation of RISC machines. Hence, most of the projects on such machines deal with unconditional TRSs, but people would also like to be able to treat CTRSs. Our transformation provides a way to use TRS hardware for CTRSs, but once more, a naive use does not provide entire satisfaction, since either the TRS is order-sorted or it contains a big number of rules. This may be avoided by the use of two kinds of flags in term nodes: irr and fz. This is in the vein of decorated TRSs [Hintermeier *et al.*, 1994] and ultra fine grained parallelisation techniques for TRSs [Kirchner and Viry, 1992]. It may provide a very simple extension of existing architectures.

Narrowing. Any kind of innermost narrowing strategy that is complete for unconditional term rewriting systems (see [Hanus, 1994a] for an overview) can be reused to do complete narrowing with conditional term rewriting systems. The sorts used in our transformation automatically provide irreducible solutions. However, there are slight difficulties with the normalisation of non-ground terms, since the left-linear axiomatization of $==$ enumerates all ground instances when used with narrowing. A solution is to give up left-linearity and to replace the axiom $(==(c,c) \to T)$ by the non-linear $(==(x,x) \to T)$, together with a restriction to completely defined functions (cf. Fribourg's work on SLOG). However, we may extend the result to incompletely defined functions by reusing Comon's work on disunification. The immediate results of this transformation are relatively weak, e.g. in comparison to LSE-narrowing for CTRSs without extra-variables, as defined by Bockmayr and Werner (these Proceedings). After an extension of the transformation for CTRSs with extra-variables, it may be possible to achieve further results.

8 Conclusion

The achieved transformation proves in a constructive way the equivalence of the class of ground-confluent, decreasing CTRSs without extra-variables and (ground-)confluent, terminating TRSs. The transformed system is orthogonal, with one rule only for each function and preserves innermost derivations. The size of the transformed rule set is proportional to $m^2 * n$, where m is the cardinality of $\mathcal{F}$ and n is the number of rules in the CTRS. It may be used in proving CTRS properties or, practically, for the reuse of hardware designed for the unconditional case.

However, the transformation does not help us proving the ground-confluence of a CTRS, since not all derivations are preserved. In fact, applying the transformation to any non-confluent CTRS results in a confluent TRS. The key to this property of the transformation is the lexicographic rule priority w.r.t. the CTRS, which is realized by the TRS. Using the rule priority explicitly while giving the CTRS should be considered harmful. We think that ground confluence, which gives the non-ambiguity of operation results, should still be guaranteed by the use of appropriate techniques (cf. e.g. [Bergstra and Klop, 1986; Ganzinger, 1991a; Bachmair *et al.*, 1992]). Analogously, the transformation does not seem to help proving termination of the CTRS.

Acknowledgements: I would like to thank C. and H. Kirchner for their support, P. Viry, I. Alouini and V. Antimirov for discussion on this topic as well as D. Lugiez and M. Rusinowitch for their comments on the project.

References

[Aida *et al.*, 1990] H. Aida, G. Goguen, and J. Meseguer. Compiling concurrent rewriting onto the rewrite rule machine. In S. Kaplan and M. Okada, editors, *Proceedings 2nd International Workshop on Conditional and Typed Rewriting Systems, Montreal (Canada)*, volume 516 of *LNCS*, pages 320–332. Springer-Verlag, June 1990.

[Bachmair *et al.*, 1992] L. Bachmair, H. Ganzinger, C. Lynch, and W. Snyder. Basic paramodulation and superposition. In *Proceedings 11th International Conference on Automated Deduction, Saratoga Springs (N.Y., USA)*, pages 462–476, 1992.

[Bergstra and Klop, 1986] J. A. Bergstra and J. W. Klop. Conditional rewrite rules: Confluency and termination. *Journal of Computer and System Sciences*, 32(3):323–362, 1986.

[Bergstra and Tucker, 1980] J. A. Bergstra and J. V. Tucker. A characterisation of computable data types by means of a finite equational specification method. In J.W. de Bakker and J. van Leuwen, editors, *Proceedings 7th ICALP Conference, Noordwijkerhout*, volume 81 of *LNCS*, pages 76–90. Springer-Verlag, 1980.

[Bergstra and Tucker, 1987] J. A. Bergstra and J. V. Tucker. Algebraic specifications of computable and semicomputable data structures. *Theoretical Computer Science*, 50:137–181, 1987.

[Bertling and Ganzinger, 1989] H. Bertling and H. Ganzinger. Compile-time optimization of rewrite-time goal solving. In N. Dershowitz, editor, *Proceedings 3rd Conference on Rewriting Techniques and Applications, Chapel Hill (N.C., USA)*, volume 355 of *LNCS*, pages 45–58. Springer-Verlag, April 1989.

[Bosco *et al.*, 1989] P.G. Bosco, C. Cecchi, and C. Moiso. An extension of WAM for K-LEAF: a WAM-based compilation of conditional narrowing. In *Proceedings Sixth International Conference on Logic Programming, Lisboa (Portugal)*, pages 325–339. The MIT press, 1989.

[Burris, 1992] S. Burris. Discriminator varieties and symbolic computation. *Journal of Symbolic Computation*, 13(2):175–208, February 1992.

[Dauchet, 1989] M. Dauchet. Simulation of Turing machines by a left-linear rewrite rule. In N. Dershowitz, editor, *Proceedings 3rd Conference on Rewriting Techniques and Applications, Chapel Hill (N.C., USA)*, volume 355 of *LNCS*, pages 109–120. Springer-Verlag, April 1989.

[Dershowitz and Jouannaud, 1990] N. Dershowitz and J.-P. Jouannaud. *Handbook of Theoretical Computer Science*, volume B, chapter 6: Rewrite Systems, pages 244–320. Elsevier Science Publishers B. V. (North-Holland), 1990. Also as: Research report 478, LRI.

[Dershowitz and Okada, 1990] N. Dershowitz and M. Okada. A rationale for conditional equational programming. *Theoretical Computer Science*, 75:111–138, 1990.

[Dershowitz and Plaisted, 1988] N. Dershowitz and D. A. Plaisted. *Equational Programming*, pages 21–56. J. Richards, Oxford, 1988. Machine Intelligence 11: The Logic and Acquisition of knowledge.

[Ganzinger, 1991a] H. Ganzinger. A completion procedure for conditional equations. *Journal of Symbolic Computation*, 11:51–81, 1991.

[Ganzinger, 1991b] H. Ganzinger. Order-sorted completion: the many-sorted way. *Theoretical Computer Science*, 89(1):3–32, 1991.

[Giovannetti and Moisi, 1987] E. Giovannetti and C. Moisi. Notes on the elimination of conditions. In J.-P. Jouannaud and S. Kaplan, editors, *Proceedings 1st International Workshop on Conditional Term Rewriting Systems, Orsay (France)*, volume 308 of *LNCS*. Springer-Verlag, July 1987.

[Gnaedig, 1992] I. Gnaedig. ELIOS-OBJ: Theorem proving in a specification language. In B. Krieg-Brückner, editor, *Proceedings of the 4th European Symposium on Programming*, volume 582 of *LNCS*, pages 182–199. Springer-Verlag, February 1992.

[Goguen, 1987] J. A. Goguen. The rewrite rule machine project. In *Proceedings of the second international conference on supercomputing*, Santa Clara, California, May 1987.

[Hanus, 1994a] M. Hanus. The integration of functions into logic programming: From theory to practice. *Journal of Logic Programming*, 19&20:583–628, 1994.

[Hanus, 1994] M. Hanus. On extra variables in (equational) logic programming. Technical Report MPI-I-94-246, Max-Planck-Institut Saarbrücken, 1994.

[Hintermeier *et al.*, 1994] C. Hintermeier, C. Kirchner, and H. Kirchner. Dynamically-typed computations for order-sorted equational presentations. In S. Abiteboul and E. Shamir, editors, *Proc. 21st ICALP Conference*, volume 820 of *LNCS*, pages 450–461. Springer-Verlag, 1994.

[Huet and Lankford, 1978] G. Huet and D. S. Lankford. On the uniform halting problem for term rewriting systems. Technical Report 283, IRIA - Laboria, France, 1978.

[Huet, 1980] G. Huet. Confluent reductions: Abstract properties and applications to term rewriting systems. *Journal of the ACM*, 27(4):797–821, October 1980. Preliminary version in 18th Symposium on Foundations of Computer Science, IEEE, 1977.

[Josephson and Dershowitz, 1989] N. Alan Josephson and Nachum Dershowitz. An implementation of narrowing. *Journal of Logic Programming*, 6(1&2):57–77, March 1989.

[Jouannaud and Kirchner, 1991] J.-P. Jouannaud and Claude Kirchner. Solving equations in abstract algebras: a rule-based survey of unification. In Jean-Louis Lassez and G. Plotkin, editors, *Computational Logic. Essays in honor of Alan Robinson*, chapter 8, pages 257–321. The MIT press, Cambridge (MA, USA), 1991.

[Kaplan, 1987] S. Kaplan. A compiler for conditional term rewriting systems. In P. Lescanne, editor, *Proceedings 2nd Conference on Rewriting Techniques and Applications, Bordeaux (France)*, volume 256 of *LNCS*, pages 25–41, Bordeaux (France), May 1987. Springer-Verlag.

[Kirchner and Viry, 1992] Claude Kirchner and P. Viry. Implementing parallel rewriting. In B. Fronhöfer and G. Wrightson, editors, *Parallelization in Inference Systems*, volume 590 of *LNAI*, pages 123–138. Springer-Verlag, 1992.

[Kirchner *et al.*, 1988] Claude Kirchner, Hélène Kirchner, and J. Meseguer. Operational semantics of OBJ-3. In *Proceedings of 15th ICALP Conference*, volume 317 of *LNCS*, pages 287–301. Springer-Verlag, 1988.

[Sivakumar, 1989] G. Sivakumar. *Proofs and computations in conditional equational theories*. Phd thesis, University of Illinois, Urbana-Champaign (IL/USA), 1989.

Termination for Restricted Derivations and Conditional Rewrite Systems*

Charles Hoot

University of Illinois
hoot@cs.uiuc.edu

Abstract. In this paper, two topics will be examined. First, the general path ordering will be applied to three conditional rewrite systems to show termination. Second, the relationship between forward closures and innermost rewriting will be discussed. This leads one to consider completion limited to innermost derivations. The ability to easily extend innermost and outermost forward closures is examined.

1 Introduction

A rewrite system is terminating if all derivations are finite. A number of different methods of proving termination have been discovered. See Dershowitz[1] for a survey of termination methods. Many of these methods are simplification orderings (they satisfy the replacement property where if $s \succ t$ then $f(\ldots, s, \ldots) \succeq f(\ldots, t, \ldots)$). The advantage of simplification orderings is that they are usually easy to understand. Their major disadvantage is that they can not be used to show the termination of any rewrite system which has an embedding ($ffx \rightarrow fgfx$ for example.) In Dershowitz and Hoot[2] two methods were presented for showing termination where simplification orderings will not work. This paper is a follow-on to that one and has its structure.

The first method is the general path ordering, which allows one to combine various orderings lexicographically (with certain restrictions) to show termination. The general path ordering need not have the replacement property. It uses the weaker condition that the replacement property only needs to hold when $s \rightarrow t$ along with the strict subterm property (which requires that a term must be strictly larger than each of its proper subterms.) The first part of this paper uses the general path ordering to show the termination of several conditional rewrite systems.

The second method of showing termination in [2] focused on the use of forward closures. Forward closures are sets of derivations with particular properties. Syntactic categories of rewrite systems were presented for which forward closures suffice to show termination. The notion of restricting forward closures to innermost or outermost derivations was introduced and criteria for their applicability was given. In the second part of this paper, innermost forward

* This research was supported in part by the U. S. National Science Foundation under Grants CCR-90-07195, INT-90-169587, and CCR-90-24271.

closures are examined more closely. Motivation for using innermost forward closures is presented, which in turns leads to the question of how one computes innermost forward closures.

2 The General Path Ordering

The standard path ordering is a simplification ordering and, as such, does not allow one to show termination of self-embedding rewrite systems. The general path ordering, on the other hand, can be applied to such systems. It encompasses virtually all popular methods for showing termination including polynomial (and other) interpretations, the Knuth-Bendix ordering and its extensions, and the recursive path orderings and its variants. For additional details and examples of the use of the general path ordering, see [2] and [3].

The general path ordering combines mappings from terms to sets with well-founded orderings. These *component orderings* are composed lexicographically.

Definition 1. The component orderings used in this paper are:

a. $\langle\theta, \geq\rangle$ is a *precedence* when θ is a homomorphism which returns the outermost function symbol of a term and $\geq$ is a precedence ordering;
b. $\langle\theta, \geq\rangle$ is *strictly monotonic* when θ is a strictly monotonic homomorphism with the strict subterm property (with respect to $\geq$) and $\geq$ is a well-founded quasi-ordering;
c. $\langle\theta, \geq\rangle$ is *multiset extracting* when θ is an extraction function which depending on the outermost function symbol returns a multiset of immediate subterms $\mathcal{IS}(t) = \{t_1, t_2, \ldots\}$ of a term t, and $\geq$ is a multiset ordering $\succsim_{\mathcal{M}}$ induced by a well-founded ordering $\succsim$ on terms. (The notation P_S where S is a set of indices will be used to denote which terms are to be extracted. For example, $P_{\{2\}}(t)$ is the multiset containing the immediate subterm t_2.)

Definition 2 General Path Ordering. Let $\phi_0 = \langle\theta_0, \geq_0\rangle, \ldots, \phi_k = \langle\theta_k, \geq_k\rangle$ be component orderings, where for multiset extraction θ_x component orderings, $\geq_x$ is the general path ordering $\succsim$ itself. The induced *general path ordering* $\succsim$ is defined as follows:

$$s = f(s_1, \ldots, s_m) \succ g(t_1, \ldots, t_n) = t$$

if either of the two following cases hold:

(1) $s_i \succsim t$ for some s_i, $i = 1, \ldots, m$, or
(2) $s \succ t_1, \ldots, t_n$ and $\Theta(s) >_{lex} \Theta(t)$, where $\Theta(s) = \langle\theta_0(s), \ldots, \theta_k(s)\rangle$, and $>_{lex}$ is the lexicographic combination of the component orderings $>_x$,

while

$$s = f(s_1, \ldots, s_m) \approx g(t_1, \ldots, t_n) = t$$

in the general path ordering if

(3) $s \succ t_1, \ldots, t_n$, $t \succ s_1, \ldots, s_m$ and $\theta_0(s) \simeq_0 \theta_0(t), \ldots, \theta_k(s) \simeq_k \theta_k(t)$.

For more on the general path ordering see [3].

3 Application of the General Path Ordering to GCD

Consider the following recursive program for computing the greatest common divisor:

function $gcd(x, y)$
begin
  if $y = 0$ **then** x
  elseif $y > x$ **then** $gcd(y, x)$
  else $gcd(x - y, y)$
end.

This program can be translated into the following (infinite) conditional rewrite system:

$$
\begin{array}{rrll}
y\ gt\ x \downarrow t & gcd(x, y) \rightarrow gcd(y, x) & & \\
x\ ge\ s(y) \downarrow t & gcd(x, s(y)) \rightarrow gcd(x - s(y), s(y)) & & \\
& gcd(x, 0) \rightarrow x & & \\
& s(x)\ gt\ s(y) \rightarrow x\ gt\ y & & \\
& s^i(0)\ gt\ 0 \rightarrow t & i \geq 1 & \quad (1) \\
& s(x)\ ge\ s(y) \rightarrow x\ ge\ y & & \\
& s^i(0)\ ge\ 0 \rightarrow t & i \geq 0 & \\
& s(x) - s(y) \rightarrow x - y & & \\
& x - 0 \rightarrow x\ . & &
\end{array}
$$

Notice that without the conditions this rewrite system is non-terminating. In addition, the left-hand side of the second rule embeds in the right-hand side, so no simplification ordering can be used to show termination. Hence, one wants to try interpretations where $x - s(y)$ is less than x.

The reason for using an infinite set of rewrite rules schematically represented by $s^i(0)$ for *gt* and *ge* is so that a value-preserving homomorphism can be constructed easily. The more natural rule, $s(x)\ gt\ 0 \rightarrow t$, permits terms which do not have interpretations as natural numbers to be greater than zero. Alternatively, one could use membership or sorts to express the above restriction.

To show that the conditional rewrite system is terminating the general path ordering will be used with a value-preserving homomorphism, $\theta_{\mathcal{H}}$ which maps to a well-founded set. The range of the homomorphism consists of the natural numbers, *true*, and $\perp$. The well-founded ordering, $>_{\mathcal{H}}$, is the standard greater than ordering on the natural numbers combined with, $0 > \perp$, and $\perp >$ *true*. The homomorphism is:

$$
\begin{array}{rl}
gcd :: \lambda x, y. & \textbf{if } x = \bot \textbf{ or } y = \bot \textbf{ then } \bot \\
& \textbf{elseif } x = \mathit{true} \textbf{ or } y = \mathit{true} \textbf{ then } \mathit{true} \\
& \textbf{else } gcd(x, y). \\
- :: \lambda x, y. & \textbf{if } x = \bot \textbf{ or } y = \bot \textbf{ then } \bot \\
& \textbf{elseif } x = \mathit{true} \textbf{ or } y = \mathit{true} \textbf{ then } \mathit{true} \\
& \textbf{elseif } x \geq y \textbf{ then } x - y \\
& \textbf{else } \bot. \\
gt :: \lambda x, y. & \textbf{if } x = \bot \textbf{ or } y = \bot \textbf{ then } \bot \\
& \textbf{elseif } x = \mathit{true} \textbf{ or } y = \mathit{true} \textbf{ then } \bot \\
& \textbf{elseif } x > y \textbf{ then } \mathit{true} \\
& \textbf{else } \bot. \\
ge :: \lambda x, y. & \textbf{if } x = \bot \textbf{ or } y = \bot \textbf{ then } \bot \\
& \textbf{elseif } x = \mathit{true} \textbf{ or } y = \mathit{true} \textbf{ then } \bot \\
& \textbf{elseif } x \geq y \textbf{ then } \mathit{true} \\
& \textbf{else } \bot. \\
s :: \lambda x. & \textbf{if } x = \bot \textbf{ then } \bot \\
& \textbf{elseif } x = \mathit{true} \textbf{ then } \mathit{true} \\
& \textbf{else } x + 1. \\
0 :: 0. & \\
t :: \mathit{true}. &
\end{array}
\tag{2}
$$

The component orderings are combined in the following way:

$$
\begin{array}{l}
\phi_0 = \text{the precedence } gcd > gt > ge > - > s > 0 > t. \\
\phi_1 = \text{the extraction based on the outermost symbol } f \\
\quad \theta_1 = \begin{cases} P_{\{1\}} & f = -, gt, \text{ or } ge \\ \emptyset & \text{otherwise} \end{cases} \\
\quad \text{with } >_1 = \succ \text{ (applied recursively).} \\
\phi_2 = \text{the extraction based on the outermost symbol } f \\
\quad \theta_2 = \begin{cases} P_{\{2\}} & f = gcd \\ \emptyset & \text{otherwise} \end{cases} \\
\quad \text{with } >_2 = >_{\mathcal{H}}. \\
\phi_3 = \text{the extraction based on the outermost symbol } f \\
\quad \theta_3 = \begin{cases} P_{\{1\}} & f = gcd \\ \emptyset & \text{otherwise} \end{cases} \\
\quad \text{with } >_3 = >_{\mathcal{H}}
\end{array}
\tag{3}
$$

First, one must verify that the homomorphism $\theta_{\mathcal{H}}$ is value-preserving for each of the rewrite rules. Of particular interest are the three rules for gcd. The first rule is value-preserving independently of the condition. For the second rule, the joinability of the two terms in the condition requires that the interpretations be the same (provided that all the rules are value-preserving). In this case, the interpretation of $x\ ge\ s(y)$ is *true* only if $\theta_{\mathcal{H}}(x) \geq_{\mathcal{H}} \theta_{\mathcal{H}}(y) + 1$ with both $\theta_{\mathcal{H}}(x)$ and $\theta_{\mathcal{H}}(y)$ natural numbers. With this condition and knowledge of the gcd function, one can then show that the rewrite rule is value preserving if the condition is met. The third rule is the reason that $gcd(t,0)$ is mapped to *true* instead $\bot$ as one might have expected.

The proof of termination with the general path ordering using the component orderings specified above proceeds as for the unconditional case with the following exceptions. First, the conditions on the interpretation may be used in the proofs of termination for the rules. Second, the left-hand side of the rule must be larger than each of the terms in the condition. (The second term in each of the conditions is a ground term in normal form, so for this particular rewrite system one need only consider the first term.)

All of the non-conditional rules are handled by the precedence or case (1) of the general path ordering. For the first rule, the left and right-hand sides are equal under precedence. With the second component ordering, one compares $\theta_{\mathcal{H}}(y)$ with $\theta_{\mathcal{H}}(x)$. Normally, one would not be able to prove this, but, the joinability of the conditional part gives $\theta_{\mathcal{H}}(y) >_{\mathcal{H}} \theta_{\mathcal{H}}(x)$. For the second rule, the left and right-hand sides are equal under both the precedence and the second component ordering. With the third component ordering, one compares $\theta_{\mathcal{H}}(x)$ with $\theta_{\mathcal{H}}(x - s(y))$. By the condition, one knows that both $\theta_{\mathcal{H}}(x)$ and $\theta_{\mathcal{H}}(y)$ are natural numbers, so one needs to show that $\theta_{\mathcal{H}}(x) >_{\mathcal{H}} \theta_{\mathcal{H}}(x) - \theta_{\mathcal{H}}(y) - 1$. By the condition, it must be that $\theta_{\mathcal{H}}(x) \geq_{\mathcal{H}} \theta_{\mathcal{H}}(y) + 1$. Thus, one only needs to show $\theta_{\mathcal{H}}(y) + 1 >_{\mathcal{H}} 0$, but $\theta_{\mathcal{H}}(y)$ is a natural number, so this is true. (Note, one could have argued that if $\theta_{\mathcal{H}}(x) <_{\mathcal{H}} \theta_{\mathcal{H}}(y) + 1$ then $\theta_{\mathcal{H}}(x - s(y)) = \bot$, which is less than any natural number.)

An alternative formulation of the conditional rewrite system for gcd is obtained by noticing that the rules for *gt* and *ge* are nearly the same as the rules for subtraction.

$$\begin{array}{rl}
y - x \downarrow s^i(0) \quad gcd(x,y) \rightarrow gcd(y,x) & i \geq 1 \\
x - s(y) \downarrow s^i(0)\; gcd(x,s(y)) \rightarrow gcd(x - s(y), y) & i \geq 0 \\
gcd(x,0) \rightarrow x & \\
s(x) - s(y) \rightarrow x - y & \\
x - 0 \rightarrow x\,. &
\end{array} \tag{4}$$

One needs an infinite set of rules for similar reasons. Now the well-founded set for the value preserving homomorphism, $\theta_{\mathcal{H}_2}$, only needs the addition of $\bot$. The homomorphism is:

$$\begin{array}{rl}
gcd :: \lambda x, y. & \textbf{if } x = \bot \textbf{ or } y = \bot \textbf{ then } \bot \\
& \textbf{else } gcd(x,y). \\
- :: \lambda x, y. & \textbf{if } x = \bot \textbf{ or } y = \bot \textbf{ then } \bot \\
& \textbf{elseif } x \geq y \textbf{ then } x - y \\
& \textbf{else } \bot. \\
s :: \lambda x. & \textbf{if } x = \bot \textbf{ then } \bot \\
& \textbf{else } x + 1. \\
0 :: 0. &
\end{array} \tag{5}$$

The component orderings are as they were before except that the precedence does not need t, *gt*, and *ge*; and the ordering used for the components which extract subterms of gcd is $>_{\mathcal{H}_2}$. (Notice that the old ordering, $>_{\mathcal{H}}$, is an extension of $>_{\mathcal{H}_2}$.) The termination argument is similar.

The final version of gcd replaces the rule schematas with a condition which tests a term to see if it is an natural number.

$$
\begin{array}{rr}
nat(y - s(x)) \downarrow t & gcd(x, y) \to gcd(y, x) \\
nat(x - s(y)) \downarrow t & gcd(x, s(y)) \to gcd(x - s(y), y) \\
& gcd(x, 0) \to x \\
& s(x) - s(y) \to x - y \\
& x - 0 \to x \\
& nat(0) \to t \\
& nat(s(x)) \to nat(x)\ .
\end{array}
\tag{6}
$$

As in the original version, the well-founded set used with the value preserving homomorphism, $\theta_{\mathcal{H}_3}$, will include *true*. The homomorphism is:

$$
\begin{array}{rl}
gcd :: \lambda x, y. & \textbf{if } x = \bot \textbf{ or } y = \bot \textbf{ then } \bot \\
& \textbf{elseif } x = true \textbf{ or } y = true \textbf{ then } true \\
& \textbf{else } gcd(x, y). \\
- :: \lambda x, y. & \textbf{if } x = \bot \textbf{ or } y = \bot \textbf{ then } \bot \\
& \textbf{elseif } x = true \textbf{ or } y = true \textbf{ then } true \\
& \textbf{elseif } x \geq y \textbf{ then } x - y \\
& \textbf{else } \bot. \\
nat :: \lambda x. & \textbf{if } x = \bot \textbf{ or } x = true \textbf{ then } \bot \\
& \textbf{else } true. \\
s :: \lambda x. & \textbf{if } x = \bot \textbf{ then } \bot \\
& \textbf{elseif } x = true \textbf{ then } true \\
& \textbf{else } x + 1. \\
0 :: 0. & \\
t :: true. &
\end{array}
\tag{7}
$$

The component orderings are as they were before except that the precedence is $gcd > - > nat > s > 0 > t$; and the ordering used for the components which extract subterms of gcd is $>_{\mathcal{H}_3}$. The termination argument, though slightly more complicated, is similar.

4 Application of the General Path Ordering to the "91" Function

One well know example of a recursive function is the "91" function given by:

$$
\begin{aligned}
F_{91}(x) = \textbf{if } x > 100 \text{ then } x - 10 \\
\text{else } F_{91}(F_{91}(x + 11))\ .
\end{aligned}
\tag{8}
$$

This recursive function returns 91 if $x \leq 100$, otherwise it returns $x - 10$.

A conditional rewrite system corresponding to this recursive function is:

$$\begin{array}{rr} nat(x-101)\downarrow t & F(x)\rightarrow x-10 \\ nat(100-x)\downarrow t & F(x)\rightarrow F(F(s^{11}(x))) \\ & s(x)-s(y)\rightarrow x-y \\ & x-0)\rightarrow x \\ & nat(0)\rightarrow t \\ & nat(s(x))\rightarrow nat(x) \end{array} \tag{9}$$

A value preserving interpretation of this rewrite system can be constructed over the set of natural numbers augmented by $\perp$ and *true*. The mapping $\mathcal{H}$ from terms to $\mathcal{NAT}_{\perp,true}$ is given by the following definitions:

$$\begin{array}{rl} F :: \lambda x. & \textbf{if } x=\perp \textbf{ or } x=true \textbf{ then } \perp \\ & \textbf{elseif } x\leq 100 \textbf{ then } 91 \\ & \textbf{else } x-10. \\ - :: \lambda x,y. & \textbf{if } x=\perp \textbf{ or } y=\perp \textbf{ then } \perp \\ & \textbf{elseif } x=true \textbf{ or } y=true \textbf{ then } true \\ & \textbf{elseif } x\geq y \textbf{ then } x-y \\ & \textbf{else } \perp. \\ s :: \lambda x. & \textbf{if } x=\perp \textbf{ or } x=true \textbf{ then } \perp \\ & \textbf{else } x+1. \\ nat :: \lambda x. & \textbf{if } x=\perp \textbf{ or } x=true \textbf{ then } \perp \\ & \textbf{else } true. \\ 0 :: 0. & \\ t :: true. & \end{array} \tag{10}$$

In this case, a sufficient well-founded ordering over $\mathcal{NAT}_{\perp,true}$ is given by $>_{\mathcal{H}} = \; true < \perp < 0 < 1 < 2 < \ldots$ An instance of the general path ordering which proves that the conditional rewrite system (9) is terminating is given by:

$$\begin{array}{l} \phi_0 = \text{the precedence } F > - > s > 0 > nat > t \\ \phi_1 = \text{extract } P_{\{1\}} \text{ for } - \text{ and } nat \text{ with } \succ \text{ applied recursively} \\ \phi_2 = \text{extract } P_{\{1\}} \text{ for } F \\ \quad\text{and use the value preserving homomorphism } \mathcal{H} \text{ with } >_{\mathcal{H}}\ . \end{array} \tag{11}$$

5 Application of the General Path Ordering to Insertion Sort

As a final example, consider the following conditional rewrite system which sorts a list of integers into ascending order via an insertion sort. It is a modification of a similar rewrite system presented in [2] for sorting natural numbers.

$$
\begin{aligned}
p(s(x)) &\to x \\
s(p(x)) &\to x \\
s(x)\ eq\ s(y) &\to x\ eq\ y \\
p(x)\ eq\ p(y) &\to x\ eq\ y \\
p(x)\ eq\ y &\to x\ eq\ s(y) \\
x\ eq\ p(y) &\to s(x)\ eq\ y \\
0\ eq\ 0 &\to t \\
s(x)\ gt\ s(y) &\to x\ gt\ y \\
p(x)\ gt\ p(y) &\to x\ gt\ y \\
p(x)\ gt\ y &\to x\ gt\ s(y) \\
x\ gt\ p(y) &\to s(x)\ gt\ y \\
s^i(0)\ gt\ 0 &\to t \qquad\qquad i > 0 \\
sort(nil) &\to nil \\
sort(cons(x, y)) &\to insert(x, sort(y)) \\
insert(x, nil) &\to cons(x, nil) \\
x\ gt\ y \downarrow t\ insert(x, cons(y, z)) &\to cons(y, insert(x, z)) \\
x\ eq\ y \downarrow t\ insert(x, cons(y, z)) &\to cons(x, cons(y, z)) \\
y\ gt\ x \downarrow t\ insert(x, cons(y, z)) &\to cons(x, cons(y, z))\ .
\end{aligned}
\tag{12}
$$

With an appropriate value-preserving interpretation, it can be shown that the conditions on the final three rules are mutually exclusive. Hence, the entire rewrite system is locally confluent. The rules for *gt* include one infinite schema. In this case, completion is easier to do with the rule schema than the other approaches mentioned earlier. Notice also, that if one is willing to forego confluence, the rule $p(x)\ gt\ y \to x\ gt\ s(x)$ and the corresponding rule for *eq* may be discarded. In that case, termination can be shown by the standard recursive path ordering alone with right-to-left lexical status for both *gt* and *eq*.

With the additional two rules for *gt* and *eq*, the recursive path ordering with status is insufficient for showing termination. This is due to the combination of $p(x)\ gt\ y \to x\ gt\ s(x)$ with the rule $x\ gt\ p(y) \to s(x)\ gt\ y$. For one rule, there is a decrease in the first argument, and for the other the decrease is in the second. Unfortunately, with either multiset or lexicographic status, neither rule decreases. After a little consideration, it is apparent that both of these rules decrease the number of *p*'s. The following components allow one to show termination with the general path ordering:

$\phi_0 =$ the strictly monotonic homomorphism θ_0
with $>_0$ the usual greater-than for natural numbers
$\phi_1 =$ the precedence $sort > insert > gt > eq > cons > nil > p > s > t > 0$
$\phi_2 =$ the extraction based on the outermost symbol f

$$
\theta_2 = \begin{cases} P_{\{1\}} & f = sort, gt, eq \\ P_{\{2\}} & f = insert \\ \emptyset & otherwise\ , \end{cases}
$$

with $\succ$ applied recursively.

(13)

The monotonic homomorphism θ_0 counts p's in a term and is given by:

$$\begin{aligned} cons &:: \lambda x, y.\ x + y \\ insert &:: \lambda x, y.\ x + y \\ gt &:: \lambda x, y.\ x + y \\ ge &:: \lambda x, y.\ x + y \\ sort &:: \ \lambda x.\ \ x \\ s &:: \ \lambda x.\ \ x \\ p &:: \ \lambda x.\ \ x + 1 \\ 0 &:: \ \ 0 \\ t &:: \ \ 0 \\ nil &:: \ \ 0\,. \end{aligned} \tag{14}$$

This homomorphism is strict in its arguments and maps to the natural numbers. As specified above, the corresponding well-founded component ordering is the usual greater-than for natural numbers. The proof of termination proceeds with no difficulties.

6 Forward Closures

With forward closures, the idea is to restrict the application of rules to that part of a term created by previous rewrites.

Definition 3. The *forward closures* of a given rewrite system are a set of derivations inductively defined as follows:

- Every rule $l \rightarrow r$ is a forward closure.
- If $c \rightarrow \ldots \rightarrow d$ is a forward closure and $l \rightarrow r$ is a rule such that $d = u[s]$ for nonvariable s and $s\mu = l\mu$ for most general unifier μ, then $c\mu \rightarrow \ldots \rightarrow d\mu[l\mu] \rightarrow d\mu[r\mu]$ is also a forward closure.

One can define *innermost* (*outermost*) forward closures as those closures which are innermost (outermost) derivations. More generally, arbitrary redex choice strategies may be captured in an appropriate forward closure. If some forward closure has a rightmost term which initiates an infinite derivation, then the composition of that forward closure with the infinite derivation will be termed an *infinite forward closure*. Notice that every finite approximation of the infinite forward closure is a member of the set of forward closures defined above. Hence, the infinite forward closure is a limit object of the set.

In general, termination of forward closures (no infinite forward closures exist) does not ensure termination of a rewrite system. Previous results, however, give syntactic conditions which do guarantee termination.

Proposition 4 (Dershowitz and Hoot[2]). *A rewrite system which is locally-confluent and overlaying is terminating if, and only if, there are no infinite innermost forward closures.*

This allows one to consider just innermost forward closures. The following gives conditions for considering arbitrary restrictions (including outermost) on the strategy employed in computing forward closures.

Proposition 5 (Dershowitz and Hoot[3]). *A non-erasing non-overlapping system terminates if, and only if, there are no infinite arbitrary strategy forward closures.*

The statement of Proposition 4 can be refined in the following way:

Theorem 6. *A locally-confluent overlaying rewrite system is terminating if, and only if, there are no infinite leftmost (rightmost, deepest) innermost forward closures.*

Proof. A *term* t is *terminating* if all derivations from t are finite; t is *non-terminating* if some derivation from t is infinite; and t is on the *frontier* if t is non-terminating, but every proper subterm of t is terminating. If a term has no frontier subterms, then it must be terminating. Conversely, if a term has a frontier subterm, it is non-terminating.

The proof is similar to that for Proposition 4, which, in turn, is similar to the proof given by Geupel[4] for non-overlapping rewrite systems. Given an arbitrary infinite derivation, one constructs a new infinite derivation which is an instance of an infinite forward closure with the appropriate properties. First, select a frontier subterm from the derivation as the initial term t_0. Next, repeat the following two steps.

1. Select the leftmost innermost frontier subterm s_i of t_i. (A *frontier* term has at least one infinite derivation, but all of its subterms are terminating.)
2. Apply rules to s_i along an infinite derivation until a rule at the top of s_i is applied resulting in v_i. Let t_{i+1} be t_i with s_i replaced by v_i. This piece of the derivation is

$$t_i[s_i] \to^* t_i[v_i] = t_{i+1} \tag{15}$$

This gives the infinite derivation

$$t_0 = s_0 \to^* v_0 = t_1[s_1] \to^* t_1[v_1] \to \dots \tag{16}$$

This infinite derivation can be transformed into an instance of an infinite forward closure, by replacing all the terminating subterms with their normal forms. The normal forms are guaranteed to be unique due to the local-confluence property of the rewrite system. That this process of inner normalization results in a derivation is non-trivial and is guaranteed by the overlaying property. For details see the proof in [3]. The steps at the top of the frontier subterms are leftmost innermost derivation steps because the chosen frontier subterms were leftmost innermost. The only complication is that the application of a rule at the top of a frontier subterm may generate additional context below its position which is not in normal form. For example, consider the rewrite system

$$\begin{aligned} f(x,y) &\to g(f(h(x),y)) \\ h(x) &\to n(x)\,. \end{aligned} \tag{17}$$

It has the infinite derivation

$$\begin{aligned} f(x,y) &\to g(f(h(x),y)) \\ &\to g(g(f(h(h(x)),y))) \\ &\to g(g(g(f(h(h(h(x))),y)))) \\ &\to \dots \end{aligned} \tag{18}$$

Inner normalization gives the derivation

$$\begin{aligned} f(x,y) &\to^{+} g(f(n(x),y)) \\ &\to^{+} g(g(f(n(n(x)),y))) \\ &\to^{+} g(g(g(f(n(n(n(x))),y)))) \\ &\to \dots \end{aligned} \tag{19}$$

Notice that the derivation from $g(g(f(n(n(x)),y)))$ to $g(g(g(f(n(n(n(x))),y))))$ is

$$\begin{aligned} g(g(f(n(n(x)),y))) &\to g(g(g(f(h(n(n(x))),y)))) \\ &\to g(g(g(f(n(n(n(x))),y)))) \ . \end{aligned} \tag{20}$$

The extra step was required to normalize the context $(h(x))$ generated by the previous application at the top of the frontier subterm. Since terminating terms are confluent (where the additional context was generated), there must be a leftmost innermost derivation which accomplishes the inner normalization required. But after innermost normalization, steps to the right of the leftmost innermost frontier subterm are not required for any subsequent derivation steps. These normalization steps may be safely postponed indefinitely leading to a slightly different infinite derivation where all the derivation steps are leftmost and innermost. In addition, each of the rules was applied in context generated by previous rules and so the infinite derivation is also an instance of an infinite leftmost innermost forward closure.

The arguments for the rightmost and deepest are similar. □

This gives one the ability to further restrict the set of forward closures one examines when attempting to show termination of a locally-confluent overlay rewrite system. In addition, this conforms to the leftmost evaluation strategy chosen by many programming languages. Notice, however, that this argument will not work for arbitrary strategies of selecting from a set of innermost redexs. For example, if one were to choose the shallowist innermost redex, there can be normalization steps which are required for application of the step at the top of the shallowist frontier subterm. But, these normalization steps may be deeper than all of the redexs for some other frontier subterm and hence the above argument would fail. It seems likely that it could be shown by a different argument where one delays application of those normalization steps.

7 Using Innermost and Outermost Forward Closures

The proof for Proposition 4 presented in Dershowitz and Hoot[3] consists of two parts. First the following Theorem was proven:

Theorem 7. *A rewrite system has an infinite innermost derivation if, and only if, it has an infinite innermost forward closure.*

To complete the proof, one needs to show that innermost termination for locally-confluent overlaying rewrite systems implies termination. In general, if one has some category for which innermost termination suffices to show termination, then forward closures will show termination.

The other consequence of Theorem 7 is that the termination of (innermost) forward closures, ***is sufficient*** to guarantee innermost termination of a rewrite system. This can be exploited in a number of cases. For example, many programming languages are applicative, and hence innermost derivations may be all that one is interested in when proving termination. In addition, restricting completion to innermost derivations has the benefit of severely limiting the number of possible critical pairs to be considered. Only overlaps at the top position need to be considered [8]. For example, consider the following rewrite systems:

$$\begin{aligned} f(a) &\rightarrow f(a) \\ a &\rightarrow b\,, \end{aligned} \tag{21}$$

and

$$\begin{aligned} f(a) &\rightarrow f(c) \\ a &\rightarrow b\,. \end{aligned} \tag{22}$$

Both of these rewrite systems have no overlaps at the top and must be innermost locally-confluent. To show innermost termination, one can examine the innermost forward closures. For both rewrite systems, the *only* innermost forward closure is $a \rightarrow b$. Since this is terminating, the above two systems must be innermost terminating as well. Innermost confluence then follows from Newman's lemma.

Another advantage is seen for showing the termination of modular rewrite systems. Toyama, Klop and Barendregt[9] showed that termination is a modular property of left-linear, confluent, terminating rewrite systems. Unfortunately, confluence and termination alone are not sufficient to show termination of the combined system. But innermost termination is modular. (Innermost rewriting can be used to show that weak normalization is modular. See [7] for more information on this and other modularity topics.)

Using forward closures, it is easy to show this and even extend it to rewrite systems with shared constructor symbols. For different proof of this, see Gramlich[5].

Theorem 8. *If R_1 and R_2 are rewrite systems which only share constructors and are innermost terminating, then their union is also innermost terminating*

Proof. Since both R_1 and R_2 are innermost terminating their innermost forward closures are also innermost terminating. But since one may only extend a forward closure by rewriting in produced context, none of the rules in R_1 can extend a forward closure from R_2 and vice-versa. Hence, the set of innermost forward closures for the combined rewrite system is just the union of the innermost forward closures of the individual systems and must also be terminating. □

For example, consider the following rewrite system:

$$\begin{aligned} f(0,1,x) &\to f(x,x,x) \\ h &\to 0 \\ h &\to 1 \,. \end{aligned} \tag{23}$$

It can be decomposed into the sum of two rewrite systems, one consisting of the first rule, and the other of the remaining two rules. Each of those systems is terminating, and therefore, the combined system must be innermost terminating. If one actually computes the forward closures for the combined system, the only forward closures are the rules themselves!

Unfortunately, as the following example shows, forward closures may not terminate when the rewrite system is innermost terminating.

$$\begin{aligned} f(a,x) &\to g(x,x) \\ g(a,b(y)) &\to f(a,b(y)) \\ b(a) &\to a \,. \end{aligned} \tag{24}$$

The forward closures for the above include

$$\begin{aligned} g(a,b(y)) &\to f(a,b(y)) \to g(b(y),b(y)) \\ g(a,b(a)) &\to f(a,b(a)) \to f(a,a) \\ g(a,b(a)) &\to f(a,b(a)) \to g(b(a),b(a)) \,. \end{aligned} \tag{25}$$

The third of these leads to an infinite forward closure, but when restricted to innermost derivations, both the second and third are disallowed. None of the other forward closures lead to non-termination; and the above rewrite system is innermost terminating. (It is also an example in which shared rewriting terminates.)

For the above mentioned reasons, one may want to consider computation of innermost forward closures. The rest of the paper concerns the details of how one would compute the set of innermost forward closures.

Theorem 9. *If $s \to \ldots \to t$ is a forward closure of a rewrite system, but is not an innermost (outermost) forward closure then any extension $s\sigma \to \ldots \to t\sigma \to u$ is also not innermost (outermost).*

Proof. Consider the rewrite step which is not innermost (outermost). There must be a second redex below (above) which is innermost (outermost). Since extension of the forward closure can only substitute terms for variables uniformly, the viability of the second redex can not be affected. □

Thus, one only needs to consider innermost (outermost) forward closures as candidates for extension when generating innermost (outermost) forward closures. Unfortunately, as the following rewrite system shows, the extension of an innermost forward closure need not be innermost, even if the rule application is innermost.

$$
\begin{aligned}
qx &\rightarrow fgx \\
fx &\rightarrow hx \\
g(a) &\rightarrow b
\end{aligned}
\tag{26}
$$

As one can see, this rewrite system is overlaying, non-overlapping, left-linear, right-linear, non-erasing, and has unary function symbols. This virtually exhausts the standard syntactic categories for rewrite systems. It has the innermost forward closure $qx \rightarrow fgx \rightarrow hgx$. Its extension $qa \rightarrow fga \rightarrow hga \rightarrow hb$ is not innermost because of the redex ga in the second term.

On the other hand, for outermost forward closures the situation is not quite as bad.

Theorem 10. *If $s \rightarrow \ldots \rightarrow t$ is an outermost forward closure of an overlaying rewrite system with only unary function symbols, then any extension $s\sigma \rightarrow \ldots \rightarrow t\sigma \rightarrow u$ is also outermost provided that $t\sigma \rightarrow u$ is an outermost rewrite.*

Proof. Suppose there is some rewrite step that is outermost in the original forward closure, but is no longer outermost after the extension. It's context must be included in the new outermost redex as a proper subterm. But this contradicts the assumption that the rewrite system is overlaying. □

In particular, string rewriting systems are unary, and non-erasing. By Proposition 5 non-overlapping string rewrite systems are terminating if their outermost forward closures are terminating. By the previous theorem, such systems have the desirable characteristic that when one is computing outermost forward closures, only the last term in the forward closures need to be remembered. An example of this is Zantemas problem (circulated via email) with the single rule $0011 \rightarrow 111000$. A solution showing termination with outermost forward closures is given in Dershowitz and Hoot[3].

To compute innermost forward closures, one needs to check that when a forward closure is extended, each of the previous innermost rule applications remains innermost. This back checking can potentially lead to a lot of extra work. In general, one wants to avoid as much of the extra work as is possible.

When extending an innermost forward closure $t_0(\bar{x}) \ldots \rightarrow \ldots t_n(\bar{x})$ via the substitution σ, the following check must be made:

- If rule $l \rightarrow r$ is applied at position p of $t_i(\bar{x})$, then no rule can match $t_i(\bar{x})\sigma$ at a position below p.

The following is an algorithm for extending the innermost forward closure $t_0(\bar{x}) \rightarrow \ldots \rightarrow \ldots t_n(\bar{x})$, by a set of rules $l_j \rightarrow r_j$:

1. For each subterm v of $t_n(\bar{x})$ starting at the bottom, check to see if l_j unifies with v resulting in the substitution σ. (One must rename the variables in l_j first.)
   (a) If some subterm of $v\sigma$ matches a rule, discard σ.
   (b) If σ is a match, do not check above v for extensions.

2. Check each of the previous terms t_i to see if a redex has been added below the rule application for $t_i\sigma$. If so, discard σ.
3. Extend the forward closure to $t_0\sigma \rightarrow \ldots \rightarrow t_n[v]\sigma \rightarrow t_n[r]\sigma$.

Notice that the unifications done before previous extensions of the forward closure are at positions that one needs to check for a match. So there are two things one wants to be able to do. First, one wants to be able to determine if a unifier is a match. Second, one wants to use the work done in previous unifications to check for matches. Both requirements can be met easily with a slight modification of the standard rewrite unification algorithm [6], which will be denoted as unification*.

*Unification** will return a substitution which is a match if one exists, otherwise a unifying substitution is returned if one exists. Unification* works by blocking the application of unification rewrite rules which are inconsistent with generating a match as long as some other unification rewrite rule is applicable.

Definition 11. When unifying* a term $v(\hat{x})$ with the left-hand side of a rule $l(\hat{y})$, do not add a variable in $\hat{x}$ to the substitution if there is any other unification rewrite rule applicable.

A Unification* is a match for $l(\hat{y})$ with $v(\hat{x})$ if there is a substitution for each variable in $\hat{y}$ and nothing else.

Lemma 12. *If σ is a unifier of a term $v(\hat{x})$ with the left-hand side of a rule $l(\hat{y})$ and μ is a substitution applied to $v(\hat{x})$, then $l(\hat{y})$ matches $v(\hat{x})\mu$ if $(\sigma')_E\mu$ unifies* to a match for the remaining variables of $\hat{y}$ in σ', where σ' is the reduced unifier which just contains the part of σ with substitutions for variables in $\hat{x}$, and σ_E is the set of equations obtained by converting all of the substitutions in σ into equations.*

Proof. When unifying* $v(\hat{x})\mu$ with $l(\hat{y})$ The same unification rewrite rule applications can be used to produce the substitutions for the variables in $\hat{y}$ as before. But if one considers these substitutions as equations, they can never lead to a failure of unification* due to the substitution μ. There are only two ways that unification* can fail;

1. There is an occur check. But for each substitution $y_i \mapsto t$ there must be only one occurrence of y_i (otherwise it wouldn't have been part of the substitution). Since μ will not use y_i, there can be no occur check for $y_i = t$.
2. There is a mismatch of function symbols $f(\ldots) = g(\ldots)$. But since no substitution for y_i will be made in $y_i = t$, this can not happen either.

Therefore, the original unification* produces a substitution which is a match if the reduced substitution σ' with the substitution μ applied to it unifies* to a substitution for the variables of $\hat{y}$ in σ'. □

For example, consider the following rewrite system:

$$\begin{aligned} q(x) &\to r(f(h(x),a)) \\ h(a) &\to c \\ r(x) &\to b \\ f(x,y) &\to h(x)\ . \end{aligned} \tag{27}$$

The derivation $q(x) \to r(f(h(x),a))$ is an innermost forward closure. To extend this, one would first unify* rules with the terms $h(x)$ and a. $h(x)$ unifies* with the second rule giving the substitution $\sigma_1 = \{x \mapsto a\}$. This can then be used to extend the forward closure. Notice that σ_1 is not a match, so one would also attempt to unify* with $f(h(x),a)$. This unifies* with the last rule giving $\sigma_2 = \{x_2 \mapsto h(x), y_2 \mapsto a\}$, where x_2 and y_2 are the renamed versions of the variables in the rule. Since this is a match, no extensions can result from unifications above this term.

The substitution σ_2 gives the extension $q(x) \to r(f(h(x),a)) \to r(h(h(x)))$ To extend this forward closure, one would first unify* with $h(x)$. This gives the substitution $\sigma_3 = \{x \mapsto a\}$. To check that this is a valid extension, one computes the reduced substitution $(\sigma_1)_E = \{x = a\}$. Applying σ_3 to this results in $\{a = a\}$. This unifies* to the empty substitution. Therefore, by Lemma 12, one has a match and σ_3 does not result in a valid extension.

8 Conclusions

With appropriate interpretations the general path ordering can be used to show termination of conditional rewrite systems. Using a value preserving homomorphism allows one to obtain conditions in the interpretation from the conditional part of the rule. These conditions can then be used for showing termination of the rule itself.

Forward closures can be used to demonstrate that a rewrite system is innermost terminating. Completion when restricted to innermost derivations is less computationally demanding than in the general case. For the special case of non-overlapping string rewrite systems, outermost forward closures are easily computable and suffice to show termination. In general, however, computing innermost/outermost forward closures requires checking to guarantee that the extension of a forward closure (via a substitution) does not introduce a redex below/above a previously innermost/outermost rule application. This extra work can be limited by realizing that, with a slight modification, the unifications one did when attempting to extend the forward closures will give conditions for detecting the unwanted redexs.

References

1. Nachum Dershowitz. Termination of rewriting. *J. Symbolic Computation*, 3(1&2):69–115, February/April 1987. Corrigendum: *4*, 3 (December 1987), 409–410; reprinted in *Rewriting Techniques and Applications*, J.-P. Jouannaud, ed., pp. 69—115, Academic Press, 1987.

2. Nachum Dershowitz and Charles Hoot. Topics in termination. In C. Kirchner, editor, *Proceedings of the Fifth International Conference on Rewriting Techniques and Applications (Montreal, Canada)*, Lecture Notes in Computer Science, Berlin, June 1993. Springer-Verlag.
3. Nachum Dershowitz and Charles Hoot. Natural termination. *Theoretical Computer Science*, 142:179–207, May 1995.
4. Oliver Geupel. Overlap closures and termination of term rewriting systems. Report MIP-8922, Universität Passau, Passau, West Germany, July 1989.
5. Bernhard Gramlich. On termination, innermost termination and completeness of heirarchically structured rewrite systems. Technical Report SR-93-09, SEKI, Universitat Kaiserslautern, 1993.
6. A. Martelli, C. Moiso and G. F. Rossi. An algorithm for unification in equational theories. In *IEEE Symposium on Logic in Computer Science*, Salt Lake City, UT, September 1986.
7. Aart Middeldorp. Modular properties of term rewriting systems. *PhD thesis, Free University, Amsterdam*, 1990.
8. David A. Plaisted. Term rewriting systems. In D. M. Gabbay, C. J. Hogger, and J. A. Robinson, editors, *Handbook of Logic in Artificial Intelligence and Logic Programming*, volume 4, chapter 2. Oxford University Press, Oxford, 1993. To appear.
9. Yoshihito Toyama, Jan Willem Klop, and Hendrik Pieter Barendregt. Termination for the direct sum of left-linear term rewriting systems. In Nachum Dershowitz, editor, *Proceedings of the Third International Conference on Rewriting Techniques and Applications (Chapel Hill, NC)*, volume 355 of *Lecture Notes in Computer Science*, pages 477–491, Berlin, April 1989. Springer-Verlag.

Rewriting for Preorder Relations *

Paola Inverardi

Dipartimento di Matematica pura ed applicata, Universita' di L'Aquila
via Vetoio, 67010 Coppito (L'Aquila), Italy

Abstract. Preorders are often used as a semantic tool in various fields of computer science. Examples in this direction are the preorder semantics defined for process algebra formalisms, such as testing preorders and bisimulation preorders. Preorders turn out to be useful when modelling divergence or partial specification. In this paper we present some results on the possibility of associating to a preorder theory, presented via a set of equality axioms and ordering ones, an equivalent rewriting relation which, together with a proof strategy, allows the decidability of the preorder relation. Our approach has been developed in the framework of a project whose main goal is to develop a verification system for process algebra formalisms based on equational reasoning.

1 Introduction

Equational and preorder theories are often used as semantic tools in various fields of computer science. An example in this direction is given by process algebra formalisms [5, 10, 21]. Process algebras are generally recognized to be a convenient tool for describing concurrent, or more generally reactive, systems at different levels of abstraction [21, 5, 7]. In general, these formalisms are equipped with one or more notions of equivalence and/or preorder (behavioural relations) which are used to study the relationships between the different descriptions of the same system. Besides their operational definition based on the labelled transition systems interpretation of the language, these relations have also been characterized by means of sets of (in)equational laws. In the past few years there has been a growing interest in the field of the analysis and verification of properties for process algebra specifications. Often, this amounts to prove the behavioural equivalence of different specifications of the same system, i.e. if the two specifications can be considered to be equivalent with respect to a certain behaviour. The same applies for preorder semantics which are used to model divergence [23] and partial specification [1]. Many verification systems for process algebra formalisms support equational reasoning and make use of rewriting techniques. In fact, a term rewriting approach can be adopted both to execute the operational semantics of these languages, as advocated in a general framework in [12, 20], and to verify behavioural relations defined over process algebra specifications. Correct and complete rewriting relations for the verification of equationally presented behavioural relations, like observational congruence, branching bisimulation, and

* Work partially supported by "PF Sistemi Informatici e Calcolo Parallelo" of CNR

trace equivalence have been presented [8, 14, 15]. For preorder relations, such as testing preorders [7] and bisimulation preorders [23], the situation is different since one has to deal with theories that besides an equality part include also ordering axioms thus demanding a different rewriting treatment.

Therefore, we have studied [19, 13] the possibility of associating a rewriting relation to a preorder theory, presented via a set of equality and preorder axioms, plus a proof strategy, which allows the decidability of the preorder relation.

More recently, other approaches to deal with ordering relations by using rewriting techniques, have been presented [18, 4]. Although these approaches identify the same kind of difficulties, they are different from the one discussed in this paper because they propose to split the ordering relation into two *commuting* well-founded rewriting relations $\sqsubseteq$ and $\sqsupseteq$ so that deciding whether two terms t_1, t_2 are in the preorder amounts to proving that $t_1 \sqsupseteq t$ and $t_2 \sqsubseteq t$. In our setting, we want instead to take advantage as much as possible of the fact that some of the preorder axioms are equality axioms, for which the symmetric property holds. Future work can include combining these two approaches in order to overcome some of the problems which arise in our framework, like, for example, the difficulty of orienting some of the (derived) preorder axioms consistently with the syntactic well founded ordering used in the completion process. In the following we present results for a class of preorder relations which can be presented via a set of equality axioms plus a set of ordering ones. We assume that the reader is familiar with the basic concepts of term rewriting.

2 Basic definitions

We assume that the reader is familiar with the basic concepts of term rewriting systems. We summarize the most relevant definitions below, and refer to [9] for more details. Let $\mathcal{F} = \bigcup_n \mathcal{F}_n$ be a set of function symbols, where $\mathcal{F}_n$ is the set of symbols of *arity* n. Let $\mathcal{T}$ denote the set $\mathcal{T}(\mathcal{F}, \mathcal{X})$ of (finite, first order) terms with function symbols $\mathcal{F}$ and variables $\mathcal{X}$.

A binary relation $\succ$ is a *(strict) partial ordering* if it is irreflexive and transitive. A partial ordering $\succ$ on $\mathcal{T}$ is *well-founded* if there is no infinite descending sequence $t_1 \succ t_2 \succ \ldots$ of terms in $\mathcal{T}$. A relation $\succ$ on $\mathcal{T}$ is *monotonic* if $s \succ t$ implies $f(..s..) \succ f(..t..)$ for all terms in $\mathcal{T}$ (*replacement property*). A partial ordering $\succ$ on $\mathcal{T}$ is a *simplification ordering* if it is monotonic and $f(..t..) \succ t$ for all f in $\mathcal{F}$ and for all terms in $\mathcal{T}$ (*subterm property*).

An *equational theory* is any set $E = \{(s,t) \mid s,t \in \mathcal{T}\}$. Elements (s,t) are called *equations* and written $s = t$. Let $\sim_E$ be the smallest symmetric relation that contains E and is closed under monotonicity and substitution. Let $=_E$ be the reflexive-transitive closure of $\sim_E$.

A *term rewriting system* (TRS) or *rewrite system* R is any set $\{(l_i, r_i) \mid l_i, r_i \in \mathcal{T}, l_i \notin \mathcal{X}, Var(r_i) \subseteq Var(l_i)\}$. The pairs (l_i, r_i) are called *rewrite rules* and written $l_i \rightarrow r_i$. The *rewrite relation* $\rightarrow_R$ over $\mathcal{T}$ is defined as the smallest relation containing R that is closed under monotonicity and substitution. A term t *rewrites* to a term s, written $t \rightarrow_R s$, if there exists a rule $l \rightarrow r$ in R, a substitution σ and a subterm $t|_u$, called *redex*, at the position u such that $t|_u = l\sigma$ and $s = t[r\sigma]_u$. A term t is said to *overlap* a term t' if t unifies with a non-variable subterm of t' (after renaming the variables in t so as not to conflict with those in t').

If $l_i \to r_i$ and $l_j \to r_j$ are two rewrite rules (with distinct variables), u is the position of a non-variable subterm of l_i, and σ is a most general unifier of $l_i|_u$ and l_j, then the equation $(l_i\sigma)[r_j\sigma]_u = r_i\sigma$ is a *critical pair* formed from those rules.

Let $\xrightarrow{+}$, $\xrightarrow{*}$ and $\xleftrightarrow{*}$, denote the transitive, reflexive-transitive and reflexive-symmetric-transitive closure of $\to$, respectively. A TRS R is *terminating* if there is no infinite sequence $t_1 \to_R t_2 \to_R \ldots$ of rewrite steps in R. A TRS R is *confluent* if whenever $s \;{}_R\!\xleftarrow{*}\; t \xrightarrow{*}_R q$, there exists a term t' such that $s \xrightarrow{*}_R t' \;{}_R\!\xleftarrow{*}\; q$, and R is *locally confluent* if whenever $s \;{}_R\!\leftarrow t \to_R q$, there exists a term t' such that $s \xrightarrow{*}_R t' \;{}_R\!\xleftarrow{*}\; q$. A term t is *in R-normal form* if there is no term s such that $t \to_R s$. A term s is an *R-normal form* of t if $t \xrightarrow{*}_R s$ and s is in R-normal form, in this case we write $t \xrightarrow{!}_R s$. A TRS R is *canonical* if it is terminating and confluent.

3 Preorder relations

Below is a simple example of the kind of preorder theories we want to deal with:

Example 1.

$$\mathcal{A} \qquad X + H(Y) = X + Y; \qquad\qquad X + nil = X \qquad\qquad X \sqsupseteq \Omega$$

The intuitive meaning of the Ω operator is that of partial specification, and the preorder axiom $X \sqsupseteq \Omega$ states that any specification is greater (i.e. more specified) in the preorder than Ω. The above presentation A axiomatises a (operationally) defined preorder relation P. If A is correct and complete with respect to P, then A represents a proof system for the given relation P. This is actually what happens with testing and bisimulation preorders which are first operationally defined and then given a preorder axiomatisation. Therefore, if one wants to prove that two process specifications P_1, P_2 are in the preorder P, it is sufficient to prove either $\mathcal{A} \vdash P_1 \sqsupseteq P_2$ or $\mathcal{A} \vdash P_2 \sqsupseteq P_1$. This is not dissimilar from what happens in the equational case except that in the preorder case we are not concerned with proving an equality, i.e. with a symmetric relation. The problem we want to address is to provide a rewriting based proof strategy which can be used to decide if two processes are in a given axiomatically presented preorder relation. Notice that preorder relations are much weaker than equality relations. For example, the fact that $t \sqsupseteq t_1$ and $t \sqsupseteq t_2$ does not permit to deduce anything about the relation between t_1 and t_2. This is way it does not make sence to talk about normal forms. Therefore, in the preorder setting the notions of confluence, weak confluence, etc. assume different meanings than the usual one. In particular, as it will be clear in the following, similar notions are introduced in order to establish sufficient conditions on the oriented rules of a given preorder axiomatisation, so that the resulting search space for any term t is complete. Moreover, since this notion of complete search space for a term t implies that any term $t\prime$ such that $t \sqsupseteq t\prime$ has to be reachable via the rewriting relation this will lead us to revise the notion of critical pair. It is worth noting that, due to the above semantic differences, our goal cannot be reached in the well-known equational setting, even if, at a very first glance, the treatment of a preorder theory as the union of a set of equality axioms and a set of ordering rules can resemble the problem of completion of a set of rules modulo a set of equations [11, 22, 16].

A preorder relation $\sqsupseteq$ over $\mathcal{T}(\mathcal{F},\mathcal{X})$ is a reflexive and transitive binary relation. It is usually presented by means of a finite set of preorder axioms $A = \{A_1 \sqsupseteq A_2, A_3 \sqsupseteq A_4, \ldots, A_n \sqsupseteq A_{n+1}\}$ where a preorder axiom, $M \sqsupseteq N$, is an ordered pair of terms $M, N \in \mathcal{T}(\mathcal{F},\mathcal{X})$. A preorder relation $\sqsupseteq$ can always be presented as the union of a set of equality axioms $=$ and a set of ordering axioms $>$ since it holds that: $t_1 = t_2$ iff $t_1 \sqsupseteq t_2$ and $t_2 \sqsupseteq t_1$, $t_1 > t_2$ iff $t_1 \sqsupseteq t_2$ and $not(t_2 \sqsupseteq t_1)$. A preorder theory can then be represented as $T = \langle E, P\rangle$, where $E = \{E_1 = E_2, \ldots, E_n = E_{n+1}\}$ represents the set of equality axioms, and $P = \{P_1 > P_2, \ldots, P_k > P_{k+1}\}$ represents the set of ordering axioms. In the following, we will deal with theories which can be finitely characterized as $T = \langle E, P\rangle$. We denote the preorder relation, its equality part and its ordering part with $\sqsupseteq_{\langle E,P\rangle}, =_E, >_P$, respectively.

Let $T = \langle E, P\rangle$ be a preorder theory. The preorder relation $\sqsupseteq_{\langle E,P\rangle}$ is generated from T by applying the following deductive system:

1. $M \sqsupseteq_{\langle E,P\rangle} N \qquad \forall (M = N) \in E$
2. $N \sqsupseteq_{\langle E,P\rangle} M \qquad \forall (M = N) \in E$
3. $M \sqsupseteq_{\langle E,P\rangle} N \qquad \forall (M > N) \in P$
4. $M \sqsupseteq_{\langle E,P\rangle} M \qquad \forall M \in \mathcal{T}(\mathcal{F},\mathcal{X})$ — reflexivity
5. $M_1 \sqsupseteq_{\langle E,P\rangle} M_2 \wedge M_2 \sqsupseteq_{\langle E,P\rangle} M_3 \quad then \quad M_1 \sqsupseteq_{\langle E,P\rangle} M_3$ — transitivity
6. $M_1 \sqsupseteq_{\langle E,P\rangle} N_1 \wedge M_2 \sqsupseteq_{\langle E,P\rangle} N_2 \wedge \ldots \wedge M_n \sqsupseteq_{\langle E,P\rangle} N_n \quad then$ $f(M_1, \ldots, M_n) \sqsupseteq_{\langle E,P\rangle} f(N_1, \ldots, N_n)$ — congruence
7. $M \sqsupseteq_{\langle E,P\rangle} N \quad \forall \sigma : \mathcal{X} \rightarrow \mathcal{T}(\mathcal{F},\mathcal{X}) \quad \sigma(\mathcal{M}) \sqsupseteq_{\langle \mathcal{E},\mathcal{P}\rangle} \sigma(\mathcal{N})$ — substitutivity

In determining a rewriting relation for T we want to take advantage of the symmetric property of the axioms in E. That is, when trying to prove whether $t_1 \sqsupseteq_{\langle E,P\rangle} t_2$ we first try $t_1 =_E t_2$, and if this is not true we try $t_1 >_P t_2$. Note that, this amounts to require that the equational theory E completely characterizes all the pairs of terms that are in the preorder relation $\sqsupseteq_{\langle E,P\rangle}$ and for which the symmetric property holds. Thus, in the following, we assume to deal with preorder theories for which the following condition holds:
Let $=_E'$ be the equivalence relation generated by the above axioms except axiom 3. Then $=_E \equiv =_E'$.

Note that the above condition on E and P is not strong. In fact, it only requires the ordering relation $>_P$ to be consistent with the equality relation $=_E$. If this is not true, the ordering relation $>_P$ contains circular chains of derivations like, $[A] >_P [B_1] >_P \ldots >_P [B_k] >_P [A]$, where $[A]$ denotes the equivalence class of the term A with respect to $=_E$ and $A \neq_E B_1 \neq_E \ldots \neq_E B_k \neq_E A$. That is the ordering part of the relation contributes in establishing equalities. On the other hand, the above hypothesis does not prevent the ordering relation from intersecting with the equality relation, i.e. there may exist pairs of terms such that $A >_P B$ and $A =_E B$. The preorder theories we are interested in, satisfy the above condition. The intuition underlying their axiomatic presentation is that the equality axioms establish which terms can be considered as equivalent with respect to their potential behaviour i.e. the amount of information they contain. The ordering axioms, instead, relate, in a monotonic way, classes of terms which contain a decreasing/increasing amount of information. Summarizing $=_E$ is the relation associated to the axioms E for which the reflexive, symmetric and transitive properties hold; while $>_P$, obtained only by means of the axioms in P, can be seen as a monotonic ordering relation among the equivalence classes of terms induced by $=_E$. In the following section, we address the problem of associating to a preorder theory $\langle E, P\rangle$ an *equivalent* rewriting relation $\twoheadrightarrow_R$, in the sense that it allows the construction for any term of a complete search space. More precisely, this means that if $t \sqsupseteq_{\langle E,P\rangle} q$ then there exists a derivation $t \twoheadrightarrow_R t_2 \twoheadrightarrow_R \ldots \twoheadrightarrow_R q$.

4 Rewriting systems for preorder theories

Let us now see how to associate a suitable rewriting relation to a given preorder theory. Since in order to verify the preorder relation $\sqsupseteq_{\langle E,P\rangle}$ we make use of the two relations $=_E$ and $=_P$, it could be thought to adopt a modular approach [17] by separately considering the (completion of the) two relations. The following example shows that this is not possible.

Example 2. Given the set of rules

T	$X + H(Y) = X; X + nil = X;$	$H(X) \sqsupseteq \Omega$
R_E	$X + H(Y) \rightarrow_E X; X + nil \rightarrow_E X$	
R_P	$H(X) \rightarrow_P \Omega$	

Consider $t = (x+nil)+nil$ and $t_1 = (x+\Omega)+nil$ then $t \sqsupseteq t_1$: $(x+nil)+nil =_E x+nil =_E (x+H(y))+nil \sqsupseteq (x+\Omega)+nil$ and this derivation cannot be obtained by $R_E \cup R_P$.

Thus we have to consider the possible interactions between the two relations. For the two different relations we introduce two types of rules: $\rightarrow_E$ and $\rightarrow_P$. The rules $\rightarrow_E \in R_E$ allow terms which are in the relation $=_E$ to be rewritten while the rules $\rightarrow_P \in R_P$ allow terms which are in the relation $<_P$ to be rewritten. The whole rewriting system and the rewriting relation are indicated with $R = R_E \cup R_P$ and $\rightarrow_R$, respectively. The usual notation for rewriting relations extends straightforwardly to $\overset{*}{\leftrightarrow}_E, \overset{+}{\leftrightarrow}_E, \overset{*}{\leftrightarrow}_E, \overset{*}{\leftrightarrow}_P, \overset{+}{\leftrightarrow}_P, \overset{*}{\leftrightarrow}_P, \overset{*}{\leftrightarrow}_R, \overset{+}{\leftrightarrow}_R$. Following [2] we denote with $\overset{+}{\leftrightarrow}_{P/E}$ the $\overset{*}{\rightarrow}_R$ steps where at least one $\rightarrow_P$ step (besides $\overset{*}{\rightarrow}_E$) occurs. Our goal is to determine conditions on the rewriting relation $\rightarrow_R$ and on the rewriting system R which guarantee the decidability of the preorder theory

T. In the equational setting, confluence is one of the properties required for the related rewriting relation. Confluence allows for deterministic derivations and the definition of unique normal forms. In the preorder setting we have to weaken this condition, since the notion of unique normal form is useless. Analogously to the purely equational case, we would anyhow rely on a kind of confluence.

Given a system $R = R_E \cup R_P$ let us consider the following situations:

i) $N \overset{*}{\leftarrow}_E M \overset{*}{\rightarrow}_E P$ ii) $N \overset{*}{\leftarrow}_E M \overset{+}{\rightarrow}_{P/E} P$

iii) $N \overset{+}{\leftarrow}_{P/E} M \overset{+}{\rightarrow}_{P/E} P$

All the other cases reduce to these ones. Note that we do not consider the cases a) $N \overset{*}{\leftarrow}_E M \overset{+}{\rightarrow}_P P$ because is a special case of ii), $N \overset{+}{\leftarrow}_P M \overset{+}{\rightarrow}_P P$ and b) $N \overset{+}{\rightarrow}_P M \overset{+}{\rightarrow}_{P/E} P$ because are special cases of iii) and $N \overset{+}{\leftarrow}_P M \overset{*}{\rightarrow}_E P$, $N \overset{+}{\leftarrow}_{P/E} M \overset{*}{\rightarrow}_E P$ and $N \overset{+}{\leftarrow}_{P/E} M \overset{+}{\rightarrow}_P P$ because these are symmetric to a), ii) and b), respectively. Now it is easy to see that ordinary confluence can be required only for i), while for iii) it does not make any sense since we cannot deduce any relation between the two terms N and P. On the contrary, confluence for ii) means that from N it is still possible to reach, in the preorder theory, a term Q such that $P \overset{*}{\rightarrow}_E Q$ and $Q <_P N$. This observation motivates the following definition:

Definition 1. A rewriting system $R = R_E \cup R_P$ is *preorder confluent* (pc) if and only if $\forall M, N, P$:
a) If $N \overset{*}{\leftarrow}_E M \overset{*}{\rightarrow}_E P$ then there exists Q such that $N \overset{*}{\rightarrow}_E Q \overset{*}{\leftarrow}_E P$;
b) If $N \overset{*}{\leftarrow}_E M \overset{+}{\rightarrow}_P P$ then there exists Q such that $N \overset{+}{\rightarrow}_{P/E} Q \overset{*}{\leftarrow}_E P$.

The above definition says that for the equational sub-relation, part a) above, it is still possible to require the usual confluence condition while for the ordering sub-relation, part b), it is only possible to require that from the term N a term Q, equationally equivalent to P, has to be reached. Note that in the above definition we do not take into account case ii), since it holds the following:

Proposition 2. *Let $R = R_E \cup R_P$ preorder confluent then if $N \overset{*}{\leftarrow}_E M \overset{+}{\rightarrow}_{P/E} P$ there exists Q such that $N \overset{+}{\rightarrow}_{P/E} Q \overset{*}{\leftarrow}_E P$.*

Proof. The proof is by induction on the length of the derivation $\overset{*}{\leftarrow}_R$.

Definition 3. A rewriting system $R = R_E \cup R_P$ is *preorder locally confluent* if and only if:
i) R_E is locally confluent and
ii) $\forall M, N, P$ such that $N \leftarrow_E M \rightarrow_P P$ then there exists a term P' such that $N \overset{+}{\rightarrow}_{P/E} P' \overset{*}{\leftarrow}_E P$.

Proposition 4. *A terminating rewriting relation $R = R_E \cup R_P$ is preorder confluent if and only if it is preorder locally confluent.*

Proof. if part: The following properties are straighforwardly verified:
a) $\forall M, N, P$ such that $N \leftarrow_E M \rightarrow_E P$ then there exists a term Q such that $N \stackrel{*}{\rightarrow}_E Q \stackrel{*}{\leftarrow}_E P$,
b) $\forall M, N, P$ such that $N \leftarrow_E M \rightarrow_P P$ then there exists a term P' such that $N \stackrel{+}{\rightarrow}_{P/E} P' \stackrel{*}{\leftarrow}_E P$,
only if part: We have to show:
a) $\forall M, N, P$ such that $N \stackrel{*}{\leftarrow}_E M \stackrel{*}{\rightarrow}_E P$ there exists a term Q such that $N \stackrel{*}{\rightarrow}_E Q \stackrel{*}{\leftarrow}_E P$, same as in the equational case [11];
b) $\forall M, N, P$ such that $N \stackrel{*}{\leftarrow}_E M \stackrel{+}{\rightarrow}_P P$ there exists a term Q such that $N \stackrel{+}{\rightarrow}_{P/E} Q \stackrel{*}{\leftarrow}_E P$.
The proof is by noetharian induction.

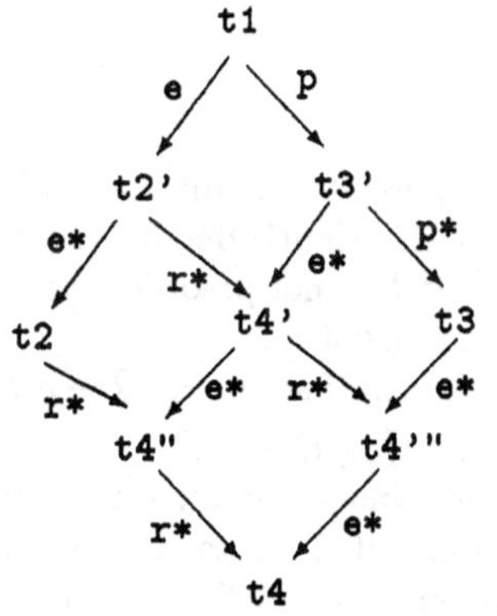

□

It is worth recalling that in order to prove by rewriting that two terms are in a preorder relation, it is no more possible to resort to the reduction to normal form. This means that, in principle, one should try all possible derivations from a term t in order to find all the terms which are in relation with t. When a rewriting relation is preorder locally confluent, it is possible to devise a proof technique which restricts the non-determinism of the search process.

Definition 5. Given $R = R_E \cup R_P$ and a term M, the normal form of M, denoted by $nf(M)$, is the term obtained by rewriting M into an R_E-normal form.

Proposition 6. *Let $R = R_E \cup R_P$ be preorder locally confluent and terminating. If $t_1 \stackrel{+}{\rightarrow}_{P/E} t_2$ then there exists a derivation: $t_1 \stackrel{*}{\rightarrow}_E nf(t1) \rightarrow_P x1 \stackrel{*}{\rightarrow}_E nf(x1) \rightarrow_P \ldots \stackrel{*}{\rightarrow}_E nf(t_2)$.*

Proof. The proof is by noetharian induction. If $t_1 \stackrel{+}{\rightarrow}_{P/E} t_2$ then we can have:
i) $t_1 \rightarrow_E x \stackrel{+}{\rightarrow}_{P/E} t_2$ by inductive hypothesis there exists $x \stackrel{*}{\rightarrow}_E nf(x) \rightarrow_P x_1 \stackrel{*}{\rightarrow}_E nf(x_1) \rightarrow_P \ldots \stackrel{*}{\rightarrow}_E nf(t_2)$ thus concluding $t_1 \rightarrow_E x \stackrel{*}{\rightarrow}_E nf(x) \rightarrow_P x_1 \stackrel{*}{\rightarrow}_E nf(x1) \rightarrow_P \ldots nf(t_2)$.

ii) $t_1 \rightarrow_P x \xrightarrow{+}_{P/E} t_2$ by inductive hypothesis there exists $x \xrightarrow{*}_E nf(x) \rightarrow_P x_1 \xrightarrow{*}_E nf(x_1) \rightarrow_P \dots \xrightarrow{*}_E nf(t_2)$, if t_1 in normal form then we conclude: $t_1 = nf(t_1) \rightarrow_P x \xrightarrow{*}_E nf(x) \rightarrow_P x_1 \xrightarrow{*}_E nf(x_1) \rightarrow_P \dots \xrightarrow{*}_E nf(t_2)$.
If t_1 is not in normal form then $t_1 \rightarrow_E y \xrightarrow{*}_E nf(y)$ but since $t_1 \rightarrow_P x$ for the preorderly local confluence there exists z such that $y \xrightarrow{+}_{P/E} z \xleftarrow{*}_E x$, y is a successor of t_1 therefore there exists $y \xrightarrow{*}_E nf(y) \rightarrow_P y_1 \xrightarrow{*}_E nf(y_1) \rightarrow_P \dots \xrightarrow{*}_E nf(z)$, with $nf(z) = nf(x)$. Thus there exists $t_1 \rightarrow_E y \xrightarrow{*}_E nf(y) \rightarrow_P y_1 \xrightarrow{*}_E nf(y_1) \rightarrow_P \dots \xrightarrow{*}_E nf(z) \rightarrow_P x1 \xrightarrow{*}_E nf(x_1) \rightarrow_P \dots \xrightarrow{*}_E nf(t_2)$ i.e. the thesis $t_1 \xrightarrow{*}_E nf(t_1) \rightarrow_P y_1 \dots \xrightarrow{*}_E nf(t_2)$. □

Proposition 6 allows the definition of an optimized search strategy to inspect the search space and decide whether $t_1 \sqsupseteq t_2$, for any terms t_1, t_2.

5 Completion for preorder relations

Let us analyze, given a preorder theory $T = \langle E, P \rangle$, how to obtain a preorder confluent equivalent rewriting relation by adapting the Knuth-Bendix critical pair approach to our framework. Let $R = R_E \cup R_P$. If R is not preorder locally confluent, either one or both conditions below are not satisfied:
a) $\forall M, N, P$ such that $N \leftarrow_E M \rightarrow_E P$ there exists Q such that $N \xrightarrow{*}_E Q \xleftarrow{*}_E P$,
b) $N \xleftarrow{*}_E M \xrightarrow{+}_P P$ then there exists Q such that $N \xrightarrow{+}_{P/E} Q \xrightarrow{*}_E P$.
Condition a) can be satisfied by completing, via the Knuth-Bendix algorithm, the set of rules R_E into a confluent rewriting system, if it exists. For b), we look for all the possible situations in which preorder local confluence can be compromised and try to add new rules to retrieve it. The interesting situations happen when a term M can be rewritten with both $\rightarrow_E$ and $\rightarrow_P$. Analogously to what happens in [18], when considering rewrite systems $R = R_E \cup R_P$ with R_E non-left linear the variable overlapping case becomes also interesting in our setting.

Example 3. $R_E = X + X \rightarrow_E X$ $\qquad$ $R_P = X \rightarrow_P \Omega$
If we compute critical pairs as usual (i.e. non considering variable subterms) no critical pairs are generated. Even so, R is not complete. In the theory there exists, for example, the derivation $P = P+P \sqsupseteq P+\Omega$ which cannot be obtained in the system.

Thus we have to review the notion of critical pair. First of all we distinguish critical pairs depending on whether a rule in $\rightarrow_E$ overlaps on a rule in $\rightarrow_P$ or vice versa.

Definition 7. If $\gamma_1 \rightarrow_E \delta_1$ and $\gamma_2 \rightarrow_P \delta_2$ are two rewrite rules (suitably renamed such that their sets of variables are disjoint) in R, u is a position in γ_1, and σ mgu of $(\gamma_1|_u, \gamma_2)$, then the equation $\langle \delta_1\sigma, \gamma_1\sigma[\delta_2\sigma]_u \rangle$ is a critical pair of type $\langle E, P \rangle$. If $\gamma_1|_u$ is a non-left- linear variable subterm, the resulting equation $\langle \delta_1\sigma, \gamma_1\sigma[\delta_2]_u \rangle$ is a critical pair of type $\langle Evar, P \rangle$.

Definition 8. If $\gamma_1 \rightarrow_E \delta_1$ and $\gamma_2 \rightarrow_P \delta_2$ are two rewrite rules (suitably renamed such that their sets of variables are disjoint) in R, u is the position of a non-variable subterm of γ_2, and σ mgu of $(\gamma_2|_u, \gamma_1)$ then the equation $\langle \gamma_2\sigma[\delta_1\sigma]_u, \delta_2\sigma\rangle$ is a critical pair of type $\langle P, E\rangle$.

The above definition does not cover all the critical cases which arise when considering variable overlapping. The Lemma we are going to introduce will help take into account the (infinitely many) cases when the overlap takes place below the variable position in γ_1.

Lemma 9. *Let $\langle \delta_1\sigma, \gamma_1\sigma[\delta_2]_u\rangle$ be a critical pair of type $\langle Evar, P\rangle$, obtained by overlapping a non-left- linear variable subterm of the rule $\gamma_1 \rightarrow_E \delta_1$ with the left-hand side of the rule $\gamma_2 \rightarrow_P \delta_2$. Then for any term $t[\delta 1\rho]$ with $\rho = \sigma\tau$ and for any position $u^i \geq u$ such that $\gamma_1\rho|_{u^i} = \gamma_j\alpha$, where $\gamma_j \rightarrow_P \delta_j$ it holds that $t[\delta_1\rho] >_P t[\gamma_1\rho[\delta_j\alpha]_{u^i}]$.*

Proof. In general there is an infinite number of terms which give rise to the above situation. Given a critical pair of type $\langle Evar, P\rangle$ the critical peaks will always be caused by applying the same rules but where the overlap with a preorder rule occurs at deeper and deeper positions than the original u. The left-hand side of a peak is originated in the following way: $t[\gamma_1\rho] \rightarrow_E t[\delta_1\rho]$ where $\gamma_1\rho$ contains at a certain depth $j \geq u$, a subterm $\gamma_1\rho|_j = \gamma_i\alpha$ where $\gamma_i \rightarrow_P \delta_i$ is a preorder rule. The right-hand side is: $t[\gamma_1\rho] \rightarrow_P t[\gamma_1\rho[\delta_i\alpha]_j]$. Therefore it holds that $t[\delta_1\rho] >_P t[\gamma_1\rho[\delta_i\alpha]_j]$. □

Now we are in the position of stating the main theorem to establish the preorderly local confluence for a rewriting system $R = R_E \cup R_P$.

Theorem 10. *A rewriting system $R = R_E \cup R_P$ is preorder locally confluent if and only if*
i) R_E is locally confluent;
ii) for any critical pair $\langle P, Q\rangle$ of type $\langle E, P\rangle$ and of type $\langle P, E\rangle$ there exists $P \overset{*}{\leftarrow}_R Q$ and
iii) for any critical pair of type $\langle Evar, P\rangle = \langle \delta_1\sigma, \gamma_1\sigma[\delta_2]_u\rangle$ then for any term $\delta_1\rho$ with $\rho = \sigma\tau$ and for any position $u^i \geq u$ such that $\gamma_1\rho|_{u^i} = \gamma_j\alpha$, where $\gamma_j \rightarrow_P \delta_j$, it holds $\delta_1\rho \overset{*}{\leftarrow}_R \gamma_1\rho[\delta_j\alpha]_{u^i}$

Proof. It follows from the peak situations described above and Lemma 9.

From the above theorem we cannot straightforwardly derive a completion algorithm for generic preorder theories since condition iii) imposes to deal with a rather general kind of rules. Instead, if we restrict to consider preorder theories with left-linear equational part, we have not to consider condition iii) and we can simply recover preorder local confluence by turning critical pairs into preorder rules. In the following we shortly discuss the possibility of extending the completion approach to preorder theories with non-left- linear equational theory. When dealing with preorder theories whose equational part is non-left-linear, in order to guarantee preorder local confluence, Theorem 10 requires to fulfill condition iii). This condition suggests the introduction of a more general form of rule, whose results depend on the actual match with the term to be rewritten. For any non-reducible-to-identity critical pair of type $\langle Evar, P\rangle = \langle \delta_1\sigma, \gamma_1\sigma[\delta_2]_u\rangle$ the idea is to introduce, beside the rule from the critical pair, a sort of rule schema: $\delta_1\sigma \Rightarrow_p \{\gamma_1\sigma\tau[\delta_2\alpha]^i_u\}$ where $\{\ldots\}$ indicates a set of right-hand side terms obtained in the following way:$t \Rightarrow_p \{t_1, t_2, \ldots, t_k\}$ if $t = \delta_1\sigma\tau$ and $t_i = \gamma_1\sigma\tau[\delta_2\alpha]^i_u$ for any position $u^i \geq u$ such that $\gamma_1\rho|u^i = \gamma_2\alpha$, with $1 \leq i \leq k$. Then, $t \Rightarrow_p \{t_1, t_2, \ldots, t_k\}$ stands for a set of k preorder rewriting steps: $t \rightarrow_P t_1, \ldots, t \rightarrow_P t_k$. Then it is possible to reformulate the Knuth-Bendix completion algorithm taking into account this new notion of rewriting.

6 Conclusion

In this paper we have presented some results to equip a preorder theory with a decision procedure based on the associated rewriting relation. In particular we have addressed the problem for preorder theories whose axiomatic presentation is given by means of a set of equality axioms and a set of ordering ones. The original motivation for this work was, in fact, to provide a rewriting decision procedure for testing preorders [7] in the context of a verification tool for process algebras based on equational reasoning [8]. In this respect our proposal is rather specific when compared with the more general approaches of [18, 4]. Besides this we tried to define a set of notions which could allow us to reuse, as much as possible, existing tools and consolidated techniques in the usual equational rewriting. This goal has been only partially achieved even when dealing with left-linear equational theories. One of the encountered problems is related to the compatibility between the termination ordering used for proving the termination of the rewriting relation and the ordering forced by the preorder relation. That is, when in the completion algorithm we derive a critical pair which could give rise to a new rule for R_P, we already have it oriented, thus we can only verify if the rule with its built-in orientation mantains the termination of the whole system. In many practical cases it happens that this is not true. Note that this amounts to require that the semantic ordering $>_P$ is contained in the syntactic ordering $\succ$ used for the completion of R (and therefore of R_E). This is, in general, a too strong condition. To this respect it would be interesting to apply to our ordering axioms the rewriting techniques developed in [18, 4] which appear to avoid the above mentioned problem since they allow for the definition of two complementary rewriting relations.

7 Acknowledgements

Thanks to Tommaso Leggio for common work on the topic of the paper, to Monica Nesi for fruitful discussions and to Harald Ganzinger for informing me about the papers [18, 4].

References

1. Aceto L., Hennessy M. 'Termination, Deadlock and Divergence', *Journal of ACM*, Jan. 1992.
2. Bachmair, L., Dershowitz, N. 'Commutation, Transformation, and Termination' in Proceedings of the *8th* CADE, LNCS 230, (1986), 52-60.
3. Bachmair, L., Dershowitz, N. 'Completion for Rewriting modulo a Congruence', in *Theoretical Computer Science*, North-Holland, (1989), 67, 173-201.
4. Bachmair, L., Ganzinger, H. 'Ordering Chaining for Total Orderings' in Proceedings of the *12th* CADE, LNCS 814, (1994).
5. Bergstra, J.A., Klop, J.W. 'Process Algebra for Synchronous Communication', in *Information and Control*, 60, No. 1/3, (1984), 109-137.
6. Camilleri, A., Inverardi, P., Nesi, M. 'Combining Interaction and Automation in Process Algebra Verification', LNCS 494, Springer-Verlag, (1991).
7. De Nicola, R., Hennessy, M. 'Testing Equivalences for Processes', in *Theoretical Computer Science*, North-Holland, 34, (1984), 83-133.
8. De Nicola, R., Inverardi, P., Nesi, M. 'Using the Axiomatic Presentation of Behavioural Equivalences for Manipulating CCS Specifications', in Proc. Workshop on Automatic Verification Methods for Finite State Systems, LNCS 407, (1990), 54-67.
9. Dershowitz N., Jouannaud J.-P., 'Rewrite Systems', in *Handbook of Theo. Computer Science, Vol. B: Formal Models and Semantics*, J. van Leeuwen (ed.), North-Holland, 1990, pp. 243–320.
10. Hoare, C.A.R. 'Communicating Sequential Processes', Prentice Hall Int., London, (1985).
11. Huet G., 'Confluent Reductions: Abstract Properties and Applications to Term Rewriting Systems', in *Journal of ACM,* Vol. 27, n. 4, Oct. 1980, pp.797-821.
12. Hussmann, H. 'Nondeterministic Algebraic Specifications', PhD Thesis, University of Passau, English Literal Translation as TUM - 19104, March 1991.
13. Inverardi P., Leggio, T., 'Rewriting Preorders', I.E.I. IR B4-3 January, 1993.
14. Inverardi P., Nesi M., 'A Rewriting Strategy to Verify Observational Congruence', *Information Processing Letters*, 1990, Vol. 35, pp. 191–199.
15. Inverardi, P., Nesi, M. 'Deciding Observational Congruence of Finite-State CCS Expressions by Rewriting', I.E.I. IR n B4-10, (revised 1993), to appear in *Theoretical Computer Science*, North-Holland, (1995).
16. Jouannaud, J.P., Kirchner, H. 'Completion of a Set of Rules modulo a Set of Equations' in , *SIAM J. Comput.*, Vol.15, No.4, (1986), 1155-1194.
17. Klop J.W., de Vrijer R.C. , 'Term Rewriting Systems', Cambridge University Press, forthcoming.
18. Levy, J., Augusti', J. 'Bi-rewriting, a Term Rewriting Technique for Monotonic Order Relations' in Proceedings of the *5th* RTA, LNCS 690, (1993).
19. Leggio, T. 'Riscrittura di preordini' Tesi di Laurea in Scienze dell'Informazione, Universita' di Pisa, February 1992.

20. Meseguer, J. 'Conditioned Rewriting Logic as a Unified Model of Concurrency', in *Theoretical Computer Science* Vol. 96, N. 1, 1992, North-Holland.
21. Milner, R. 'Communication and Concurrency', Prentice Hall, (1989).
22. Peterson, G.E., Stickel, M.E. 'Complete Sets of Reductions for Some Equational Theories', in *Journal of ACM*, Vol.28, No.2, (1981), 233-264.
23. D.J.Walker. 'Bisimulation and Divergence', in *Information and Computation* 85, (1990), 202-242.

Strong Sequentiality of Left-Linear Overlapping Rewrite Systems

Jean-Pierre Jouannaud* and Walid Sadfi

Laboratoire de Recherche en Informatique
CNRS / Université Paris-Sud
Bâtiment 490
91405 Orsay Cedex, France

jouannaud@lri.fr sadfi@ensi.rnrt.tn

Abstract. Confluent term rewriting systems can be seen as a model for functional computations, in which redexes corresponding to instances of left hand sides of rules are repeatedly replaced by their corresponding right hand side instance. Lazy sequential strategies reduce a given redex in a term if and only if this redex must be reduced by any other sequential strategy computing the normal form of the term. Lazy strategies always terminate when a normal form exist, and are indeed optimal. In a landmark paper, Huet and Levy showed that such a strategy always exists for left linear non-overlapping rewrite systems that are *strongly sequential*, a property that they proved decidable for such systems. This paper generalises the result to the case of left-linear, possibly overlapping rewrite systems.

1 Introduction

Confluent term rewriting systems can be seen as a model for functional computations, in which redexes corresponding to instances of left hand sides of rules are repeatedly replaced by their corresponding right hand side instance. As in other functional languages, the question arises whether normal forms can be computed effectively on a sequential machine, and when they can, whether this can be done by performing the minimum possible amount of computation.

Non-ambiguous linear term rewriting systems (also called orthogonal systems) enjoy the confluence property, hence are a good candidate for such a model. Besides, normal forms can always be computed when they exists by a parallel outermost strategy. Such a strategy, however, reduces redexes that are not needed, hence is not optimal. Let us call optimal, or lazy, a sequential strategy in which a redex in a term t is reduced if and only if this redex must be reduced by any other sequential strategy computing the normal form of t, and sequential, a rewrite system for which an optimal sequential strategy exists. Huet and Lévy have shown that not all orthogonal rewrite systems are sequential [3], and moreover that sequentiality is undecidable[2].

* Partially supported by the ESPRIT BRA COMPASS.

[2] Huet and Lévy 's paper was aimed at generalizing the most important syntactic properties of λ-calculus to left-linear non-overlapping term rewriting systems. Their original paper was published in 79 as an INRIA report. This paper was immediately

They have also introduced a decidable sufficient condition for sequentiality, called strong sequentiality, for which the optimal strategy amounts to reduce redexes at particular positions called indexes. Their decision procedure for strong sequentiality yields as an additional output an efficient implementation of index reduction in the form of a matching dag which generalizes to trees the famous construction by Knuth, Morris and Pratt of an automaton for efficiently matching strings.

Although Huet and Levy's framework was extended to several kinds of rewritings (priority rewriting [8], order-sorted rewriting [5], head-constructors systems [4], sufficient sequentiality [7]), orthogonality could not be *really* removed until Toyama introduced left-linear root-balanced systems [9]. The root-balanced property together with left-linearity ensure the Church-Rosser property, hence it replaces the non-ambiguity property. Besides, Toyama claimed that strong-sequentiality was decidable for left-linear ambiguous systems, and suggested that Klop and Middeldorp's reformulation of Huet and Lévy's proof would work without any modification. This is not true, however, as explained in this paper.

Our goal is therefore to prove Toyama's statement that strong sequentiality remains decidable for left-linear, possibly ambiguous rewrite systems by generalizing Klop and Middeldorp's proof.

2 Preliminaries

In this section we briefly present some notations and definitions borrowed from Dershowitz and Jouannaud [1, 2].

The set $\mathcal{T}(\mathcal{F},\mathcal{V})$ (abbreviated to $\mathcal{T}$) denotes the set of terms built from a denumerable set of function symbols, $\mathcal{F}$, and a denumerable set of variable symbols, $\mathcal{V}$, where $\mathcal{F}\cap\mathcal{V}=\emptyset$ and such that if $f\in\mathcal{F}$ has arity n and $t_1,...,t_n\in\mathcal{T}(\mathcal{F},\mathcal{V})$ then $t=f(t_1,...,t_n)\in\mathcal{T}(\mathcal{F},\mathcal{V})$.

The set of ***positions*** in a term t is denotes by $\mathcal{P}os(t)$ and is inductively defined as follows: $\mathcal{P}os(t)=\{\varepsilon\}$ if $t\in\mathcal{V}$, and $\mathcal{P}os(t)=\{\varepsilon\}\cup\{i.pos(t_i)\ :\ 1\leq i\leq n\}$ if $t=f(t_1,...,t_n)$. We write $|p|$ for the length of a position p, $t|_p$ for the subterm of t at position p , $t(p)$ for the root symbol of $t|_p$, and $t[s]_p$ the term obtained by replacing $t|_p$ by the term s in t. The ***depth*** $\rho(t)$ of a term t is the maximal length of a position in t. The ***depth*** $\rho(R)$ of a rewrite system R is the maximal depth of the left hand sides of rules in R. Positions are partially ordered by the ***prefixe ordering*** $\leq$: $p\leq q$ if $\exists p'$ such that $q=p.p'$, and we define accordingly $p/q=p'$. If $p\leq q$ and $p\neq q$ we write $p<q$ and if neither $p\leq q$ nor $q\leq p$ we write $p\parallel q$. Finally, we write $t[s_1|...|s_n]_{p_1|...|p_n}$ as an alternative for $t[s_1]_{p_1}[s_2]_{p_2}...[s_n]_{p_n}$ when $p_1\parallel...\parallel p_n$, and $t[s]_P$ when the s_i are all equal to s, P being the set of parallel positions where to perform the replacement.

A ***rewrite rule*** is a pair $l\longrightarrow r$ such that the left-hand side l is not a variable and all variables wich occur in the right-hand side r also occur in l. The rewrite

considered as important, and generated an active stream of research partly reported in this introduction. It was however notoriously difficult to read, which may explain why it was only published more than 10 years later in a journal format. At the same time, a comprehensive reformulation of the sequentiality question was published by Klop and Middeldorp [6].

rule $l \longrightarrow r$ is called *left linear* if l does not contain multiple occurences of the same variable symbol. A *term rewriting system* or *a rewrite system*, is a finite set of rewrite rules. It is called *left-linear* if all its rewrite rules are left-linear. A redex is an instance by a substitution σ of some left hand side of rule l. We use a postfix notation for substitution application, hence the redex is written as $l\sigma$. A term is in *normal-form* if it contains no redex.

A rewrite system R defines a *rewrite relation* $\longrightarrow_R$ on $\mathcal{T}$ as usual: $t \longrightarrow_R s$ if $\exists \sigma, \exists l_i \longrightarrow r_i \in R, \exists p \in \mathcal{P}os(t)$ such that $t|_p = l_i\sigma$ and $s = t[r_i\sigma]_p$. We call $t \xrightarrow[R]{p} s$ a *rewrite step*, and may omit the subscripts. The reflexive-transitive closure of $\longrightarrow_R$ is denoted by $\xrightarrow[R]{*}$. $\xrightarrow[R]{k}$ denotes a reduction of k ($k \geq 0$) steps. In the sequel we omit the subscript R. A term t has a normal form if $t \xrightarrow{*} s$ for some normal form s.

A rewrite system R is *confluent* (respectively *locally confluent*) if $\forall s, t, u \in \mathcal{T}$ with $s \xrightarrow{*} t$ and $s \xrightarrow{*} u$ (respectively $s \longrightarrow t$ and $s \longrightarrow u$) then $\exists v \in \mathcal{T}$ such that $t \xrightarrow{*} v$ and $u \xrightarrow{*} v$.

Two rules $l \longrightarrow r$ and $l' \longrightarrow r'$ *overlap* if there exists a non-variable position $p \in pos(l)$ and a substitution σ such that $l|_p\sigma = l'\sigma$. The pair $(l\sigma[r'\sigma]_p, r\sigma)$ is called a *critical pair*. After Toyama, we say that a critical pair (s,t) is *root-balanced joinable* if $s \xrightarrow[R]{k} s'$ and $t \xrightarrow[R]{k} t'$ for some term t' and $k \geq 0$.

In this paper, we will only consider left-linear rewriting systems. When all its critical pairs are root-balanced joinable, a left-linear rewriting system will be called *balanced*, and *orthogonal* when it has no overlapping rules.

3 Strong Sequentiality

In this section, we focus on those sequential strategies that compute a normal form for every term that posseses one, even in presence of non-terminating reductions issuing from the term. Such a strategy is called *normalizing*. This section is devoted to the introduction of the key concept of strong sequentiality. Most of the material is borrowed from [3, 6], but is applied to left-linear systems instead of orthogonal ones. Only those proofs that suffer changes are given here.

Let Ω be a new constant symbol representing an unknown part of a term, $\mathcal{T}^\Omega$ the set $\mathcal{T}(\mathcal{F} \cup \Omega, \mathcal{V})$ of Ω-terms, and $\mathcal{P}os_\Omega(t)$ the set of Ω-positions in a term t. The set of terms in $\mathcal{T}^\Omega \backslash \mathcal{T}$ which are in normal form for R is denoted by NF_R^Ω.

Definition 1. The *prefix ordering* $\preceq$ on $\mathcal{T}^\Omega$ is defined as follows:

(i) $\Omega \preceq t$ for all $t \in \mathcal{T}^\Omega$.

(ii) $f(t_1, ..., t_n) \preceq f(s_1, ..., s_n)$ if $t_i \preceq s_i$ for $i = 1, ..., n$

We write $t \prec s$ if $t \preceq s$ and $t \neq s$.

Definition 2. Two Ω-terms s and u are *compatible* if there exists $t \in \mathcal{T}^\Omega$ such that $s \preceq t$ and $u \preceq t$.

A *redex scheme* is a left-hand side of rule in which all variables are replaced by Ω.

An Ω-term t is *redex-compatible* if it is compatible with some redex scheme.
The R^{Ω}-rewriting relation is defined as follows:

$$t \longrightarrow^{p}_{R^{\Omega}} t[\Omega]_p \text{ iff } t|_p \neq \Omega \text{ and } t|_p \text{ is redex compatible.}$$

Lemma 3. *R^{Ω}-reductions are confluent and terminating.*

The proof of this lemma is straightforward. Let $t\downarrow_{R^{\Omega}}$ denote the normal form of t with respect to Ω-reduction. Note that $t\downarrow_{R^{\Omega}}$ is well-defined according to the previous lemma.

The following definition appears as a characteristic property of indexes in [6], and as a definition in [9].

Definition 4. Let $\bullet$ be a new constant symbol. The position p such that $t|_p = \Omega$ is an *index* of t if $(t[\bullet]_p\downarrow_{R^{\Omega}})|_p = \bullet$. We use $I(t)$ for the set of indexes of t.

We recall now a few standard results about indexes that do not assume the non-overlapping restriction.

Proposition 5. .

(a) *if $p.q \in I(t)$ then $p \in I(t[\Omega]_p)$ and $q \in I(t|_p)$.*
(b) *if $p \in I(t)$ then $\forall t' \succeq t \ : \ p \in I(t'[\Omega]_p)$.*
(c) *if $p \in I(t)$ then $\forall q \parallel p$ such that $t|_q\downarrow_{R^{\Omega}} = \Omega \ : \ p \in I(t[\Omega]_q)$*

The converse of the first statement does not hold in general, i.e. the following transitivity property for indices is not true: if $p \in I(t)$ and $q \in I(s)$ then $p.q \in I(t[s]_p)$. However Klop and Middeldorp have shown a partial transitivity result which does not depend upon the non-ambiguity assumption, hence remains true for overlapping left-linear rewrite systems [6]:

Proposition 6. *Let $t \in \mathcal{T}^{\Omega}$, $p_1, p_2, p_3 \in pos(t)$ such that $p_1 \leq p_2 < p_3$ and $|p_2/p_1| \geq \rho_R - 1$.*
If $p_2 \in I(t[\Omega]_{p_2})$, and $p_3 \in I(t|_{p_1})$ then $p_3 \in I(t)$.

Our interest in indexes lies in a fundamental result by Toyama which generalizes Huet and Lévy's result to the case of balanced rewrite systems. Before to state it, we need the notion of strong sequentiality, which roughly says that the computation of the normal form of a term can be done at particular positions called indexes.

Definition 7. A left-linear rewrite system is *strongly sequential* if there exists no index-free term in NF^{Ω}_{R}.

Definition 8. Given a stronly sequential rewrite system R, the associated *index reduction* is a reduction strategy which at each step, rewrites a redex of a term t at a position p such that p is an index of $t[\Omega]_p$.

Theorem 9. *[9] Index reduction is a normalizing strategy for balanced stronly sequential rewrite systems.*

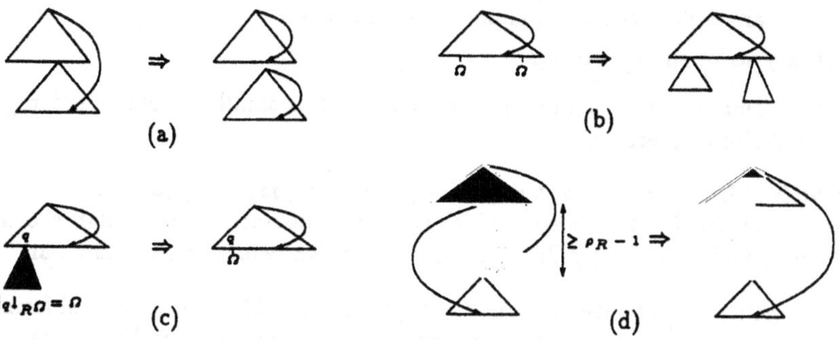

Fig. 1. Transitivity properties of indexes

We are left with the question that will be adressed in turn in the following section. Is strong sequentiality decidable for ambiguous left rewrite system ? Toyama stated that it was the case for the more general class of left-linear rewrite systems, by refering to Klop and Middeldorp's proof for orthogonal rewrite systems. This proof however, does not apply as such, although Toyama's intuition was right as we show now.

4 Decidability of Strong Sequentiality

The basic idea of Klop and Middeldorp hidden in Huet and Lévy's work is to construct a particular index-free term when the rewrite system is not sequential. Before to give the construction, we need to show that if there exists any free term at all, then there exists one which can be constructed by means of gluing special terms together called preredexes. We first state the key intermediate results that lead to this construction. Most of them apply to left-linear rewrite systems without any change in the proof. We will mention any change needed in a proof or a definition explicitely.

Definition 10. : A term $t \in \mathrm{NF}_R^\Omega$ is called *omegated* if $t{\downarrow_{R\Omega}} = \Omega$.

Proposition 11. *[6]: R is strongly sequential iff there exists no omegated index-free term.*

By induction on the derivation of particular omegated terms, called decomposable terms, to their normal form Ω, we are now going to slice them down into small pieces called atomic preredexes, a notion which refines the notion of preredex by Huet and Lévy. In order to perform this construction, we need some elaboration.

Definition 12. : A *preredex* is a term which is smaller or equal to a redex scheme in the ordering $\preceq$. A *proper* preredex is neither a redex scheme nor Ω.

A *maximal atomic preredex* is a maximal proper preredex s satisfying the following properties:

(i) for any redex scheme t compatible with s, $s \prec t$,

(ii) any strict subterm of s is not a preredex.

An *atomic preredex* is a proper preredex which is smaller or equal under $\preceq$ to a maximal atomic preredex.

Example 1. Let $R = \{f(hhx, hgy, gz) \rightarrow, f(hgx, hhy, gz) \rightarrow, hggx \rightarrow\}$. Then $f(h\Omega, h\Omega, \Omega)$ is a preredex which is not atomic since, in particular, it contains the preredex $h\Omega$. $f(\Omega, \Omega, g\Omega)$ is an atomic preredex, which is also maximal.

For orthogonal rewrite systems, every proper preredex is in R-normal form, but this is not true for the case of overlapping rules, as examplified by the rewrite system

$$\{f(g(b, h(a)), x) \longrightarrow ..., g(b, x) \longrightarrow ...\}$$

for which the term $f(g(b, h(\Omega)), \Omega)$ is a proper preredex but is not in R-normal form. This is one of the reasons for the introduction of atomic preredexes. Note also that preredexes may overlap, even for non-overlapping rewrite systems, but this is not the case of atomic ones. Our notion of atomic preredex gives rise to a notion of decomposition adapted from Klop and Middeldorp:

Definition 13. A set $\{(p_i, s_i)\}_{i \in J}$ of pairs of positions and atomic preredexes is called a decomposition headed by (p, s) iff

(i) $\forall i \in J, p \leq p_i$

(ii) $\forall i \in J$ *s.t.* $p_i \neq p$, $\exists j \in J$ *s.t.* $s_j|_{p_i/p_j} = \Omega$.

Pairs in a decomposition are naturally ordered by the prefix ordering $\leq$ on their first component.

The term $\tau(D)$ associated to the above decomposition D is defined inductively as follows:

(i) $\tau(D) = s$ if D is reduced to its head,

(ii) $\tau(D) = \tau(D \setminus \{(p, s), (p_i, s_i) \text{ s.t. } s|_{p_i/p} = \Omega\} \cup \{(p, s[s_i]_{p_i/p})\})$.

A term t is called decomposable iff there exists a decomposition D such that $t = \tau(D) \in \mathrm{NF}_R^{\Omega}$.

Example 2. Let $R = \{f(hhx, hgy, gz) \rightarrow, f(hgx, hhy, gz) \rightarrow, hggx \rightarrow\}$. Then $f(h\Omega, hh\Omega, g\Omega)$ is decomposable into four atomic peredexes: $\{(\Lambda, f(\Omega, \Omega, g\Omega)), (1, h\Omega), (1.1, h\Omega), (2, h\Omega)\}$.

Note that a decomposition can be seen as a tree (traversed depth-first) by ordering the nodes (p_i, t_i) lexicographically with respect to their first component. Note also that the position p_i/p in the above definition must correspond to an Ω-position in the unique maximal atomic preredex which is bigger than s in the ordering $\preceq$. We finaly want to stress that the term associated with a given decomposition may not belong to NF_R^{Ω} because of possible ovelapping rewrite rules. This is a difference with the standard case, which forces us to state explicitely that decomposable terms are in normal form.

We now show a strong relationship between decomposable and omegated terms. It is clear that decomposable terms are omegated terms, by a bottom-up reduction of the atomic preredexes in their decomposition to Ω. This property admits a sort of converse: any omegated term can be recursively transformed into a decomposable term by eliminating unnecessary subterms. This process which can be performed by induction on an R^{Ω}-reduction to Ω is the key to the following result:

Proposition 14. *A left-linear rewrite system is strongly sequential iff there is no index-free decomposable term.*

Proof. Since decomposable terms are particular omegated terms, we simply need to show that there exists a index-free decomposable term whenever the rewrite system is not stronly sequential. Consider an index-free omegated term of minimal size. We will show that it is indeed a decomposable term. To this end, let us consider an innermost R_{Ω} reduction

$$t \xrightarrow[l \to \Omega]{p} u \xrightarrow[R^{\Omega}]{*} \Omega$$

from t to Ω. Since the reduction is innermost, l is a preredex which does not contain another preredex. Besides, since $t \in \mathrm{NF}_R^{\Omega}$, l is not an instance of a left hand side of a rule in R. Hence, there exists a unique atomic preredex r compatible with l and maximal with respect to $\preceq$. Let us now consider the term $t[r]_p$ which is smaller or equal in size than t, and assume for a moment that it is index-free. Then, $r = l$ by minimality assumption on t. An induction on lengths of derivations easily yields the result that all rules have an atomic preredex on their left hand side. We can then readily associate a decomposition to the derivation, resulting in the needed property.

The proof that $t[r]_p$ is index-free results from the construction of r. Indeed, an Ω-position q in r may come either from an Ω-position in l or from a variable position in a redex scheme g compatible with l. In the former case, were q an index in $t[r]_p$, it would have been an index in $t[l]_p$ by property (b) of proposition 5. In the latter case, q cannot be an index, since $t[r[\bullet]_q]_p$ rewrites to $t[\Omega]_p$ with the rule $r[\bullet]_q \to \Omega$, because $r[\bullet]_q$ is compatible with g.

Klop and Middeldorp's decision procedure of strong sequentiality is based on the construction of a minimal index-free term, that is an index-free term with the minimal number of non-Ω-positions. Unfortunately, their construction does not work as such when there are overlapping rules, because substituting a term in R-normal form to an R-leaf in another term in R-normal form may result in a term which is not in Ω-normal form. Our construction is different in two respects. First, we try to construct a minimal decomposable term, by using atomic preredexes only. Second, the construction must work differently, so as to cope with the above mentionned problem.

The tentative construction of a minimal index-free decomposable term proceeds by stacking atomic preredexes on top of one another at non-index positions and eliminating terms reducible by R so as to obtain all possible decomposable terms in turn, then check whether they are or are not index-free. In order to stop the process, it will be necessary to look for the repetition of particular patterns in the term associated to a sequence of atomic preredexes.

Since a decomposition D is indeed a tree, we can consider branches in this tree. The *branch* of head (p, s) and tail (q, t) is the subset $\{(o, r) \ : \ p \leq o \leq q\}$ of D. (q, t) must be be a leaf of D, that is no pair $(o, r) \in D$ satisfies $o < q$. The size of a branch is the length of the word q/p.

Definition 15. A branch $\{(p_i, s_i)\}_{1 \leq i \leq n}$ is *special* iff $|p_n/p_1| \geq \rho(R) - 1$ and $|p_n/p_1| < \rho(R) - 1$

A decomposition D headed by (p_1, s_1) is a *pattern* iff it has the following properties:

(i) It has a non-trivial special branch $\{(p_i, s_i)\}_{1 \leq i \leq n}$.
(ii) Any branch headed by (p_1, s_1) is special or has a size at most $\rho(R) - 1$.

Because there are finitely many atomic preredexes, there are finitely many patterns. The next step consists in extracting patterns from large decompositions.

Definition 16. Given a decomposition D and a pair $(p_1, s_1) \in D$, we define a pattern of D headed by (p_1, s_1) as the longest pattern $B \subseteq D$.

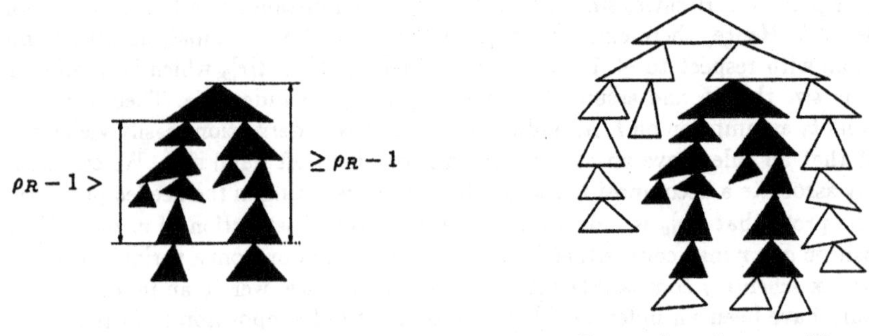

Pattern — Pattern in a decomposition

Fig. 2. Patterns

Note that our definition of a pattern is quite different from the definition of a special tower in [6]. Given now a decomposable index-free term which is big enough so as to contain at least twice the same pattern along a path from the root to a leaf, we will construct another smaller index-free decomposable term. This will result in a decision procedure for strong sequentiality as a corollary. This procedure will simply construct a tree of bigger and bigger terms not containing twice the same pattern on a path from the head, by attaching atomic preredexes at an index position of the term labelling a leaf in the tree, eliminating those whose associated term is not in R-normal form, and stop whenever an index free-term is reached (hence the rewrite system is not strongly sequential), or when all leaves contain twice the same pattern on a path from the head (hence the rewrite system is strongly sequential). Since there are finitely many such patterns, this procedure will enumerate finitely many decompositions, hence always terminate.

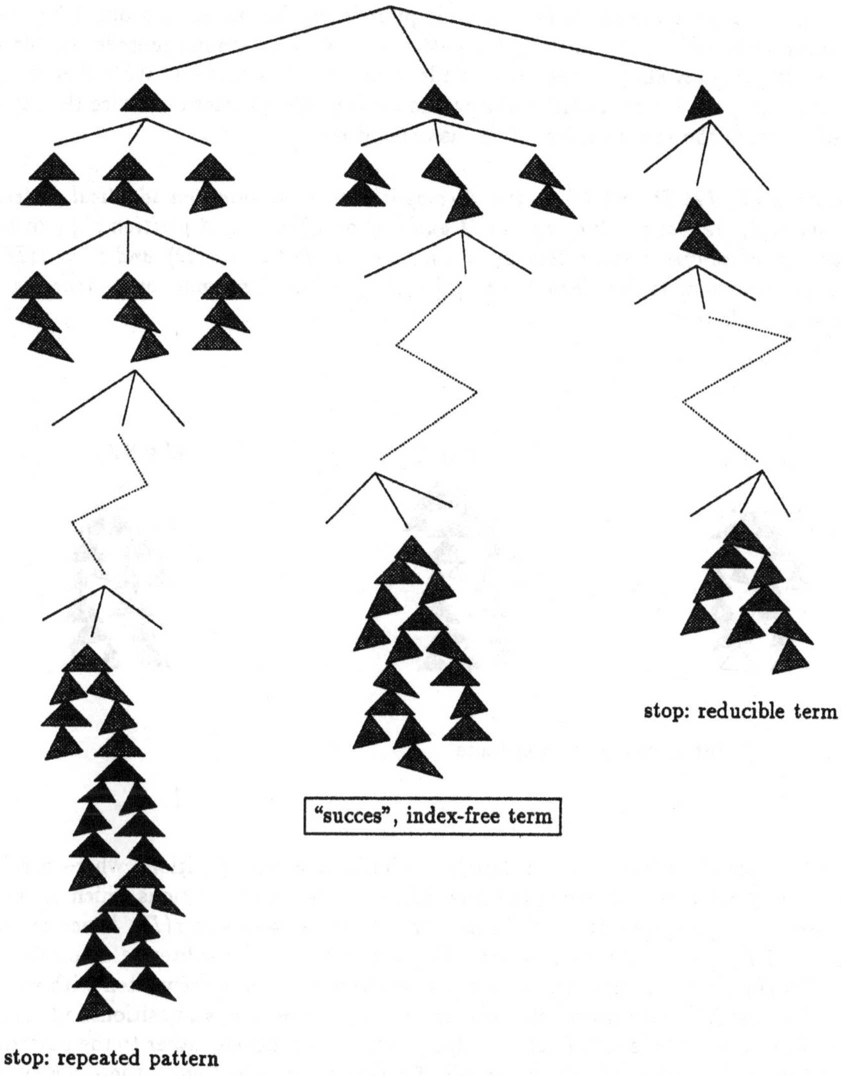

Fig. 3. Strong sequentiality decision algorithm

Since patterns are made from atomic preredexes, we can extend our definition of a decomposition, by allowing for patterns as well as atomic preredexes. More generally, we may allow for any term built from atomic preredexes. We will actually be interested in decompositions having one distinguished pattern or twice the same pattern from the root to a leaf of the associated term.

Lemma 17. *Let D and D' be two decompositions containing an identical pattern $\{(p.p_i, s_i)\}_{1\leq i\leq n}$ at position $p.p_1$ in D and $\{(p'.p_i, s_i)\}_{1\leq i\leq n}$ at position $p'.p_1$ in D' (we can of course assume that $p_1 = \epsilon$). Assume that $t = \tau(D)$ and $t' = \tau(D')$ are in R-normal form. Then $t'' = t[t'|_{p'.p_1}]_{p.p_1}$ is in R-normal form, hence is a decomposable term.*

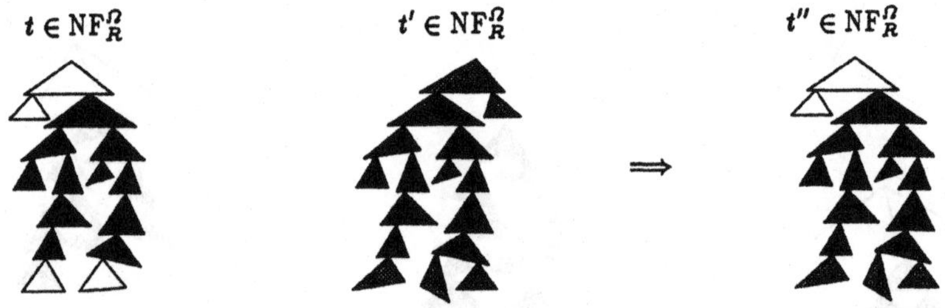

Fig. 4. Substitution of a pattern instance

Proof. Since t'' is built with atomic preredexes, the only positions where a rule may apply are the positions of the preredexes. Subterms at positions which are not smaller than $p.p_1$ are either subterms of $\tau(D')$, or subterms of $\tau(D)$, hence are in normal form by assumption. We are therefore left with the subterms at positions smaller than $p.p_1$. Assume the subterm at position $p.p_i$ can be rewritten with some rule $l \rightarrow r \in R$. By definition of a pattern in a decomposition, all positions in the set $p.p_i.P$, where P is the set of non-variable positions in l, belong either to the pattern, or to the subterms of t'' below the tail of a special branch. Hence, these positions were already in t. Since the two terms t and t'' do not differ at these positions, and because the rules are left-linear, t would be reducible, a contradiction.

We can see the importance of our definition of pattern whose role is to ensure the previous lemma. This is where our construction differs most significantly from Klop and Middeldorp's work.

Lemma 18. *Let $t = \tau(D)$ a decomposable term such that the decomposition D contains two identical patterns headed by (p_1, s_1) and $(p.p_1, s_1)$ with $p.p_1 > p_1$. Then $t = \tau(D)$ is not a minimal index-free decomposable term.*

Proof.

Fig. 5. Contraction of identical patterns

Suppose t is a minimal decomposable index-free term and let $t' = t[t|_{p.p_1}]_{p_1}$. By lemma 17, t' is in normal form. We therefore simply need to show that it is index-free. The proof is based on the way t' is constructed and properties of indexes including the partial transitivity property stated in proposition 6. Our construction differs from the construction by Klop and Middeldorp in the fact that we need to replace all subterms of the head of the first occurrence of the pattern by all subterms of the head of the second occurrence of the pattern instead of replacing only one subterm of the tail (taking advantage of the absence of critical pairs). These differences do not have any impact on the proof that t' is index-free if so is t. This ends the proof.

We can now conclude:

Theorem 19. *Strong sequentiality of left linear possibly overlapping rewrite system is a decidable property.*

The decision procedure, of course, is exponential. The possibility of having overlapping rules forced us to introduce the notion of atomic preredexes. Building on that, we were able to define decompositions, decomposable terms, and patterns in an abstract way, which leaves away many details of Klop and Middeldorp's constructions. As a result, our proofs are mostly simpler, and easier to understand. We think that this is not negligible an achievement for a problem which is notoriously considered as quite difficult both technically and conceptually.

References

1. Nachum Dershowitz and Jean-Pierre Jouannaud. Notations for rewriting. *EATCS Bulletin*, 43:162–172, 1990.

2. Nachum Dershowitz and Jean-Pierre Jouannaud. Rewrite systems. In J. van Leeuwen, editor, *Handbook of Theoretical Computer Science*, volume B, pages 243–309. North-Holland, 1990.
3. Gérard Huet and Jean-Jacques Lévy. Call by need computations in non-ambiguous linear term rewriting systems. In Jean-Louis Lassez and Gordon Plotkin, editors, *Computational Logic: Essays in Honor of Alan Robinson*. MIT Press, 1991.
4. J.R. Kennaway. Sequential evaluation strategies for parallel-or and related reduction systems. *Annals of Pure and Applied Logic*, 43:31–56, 1989.
5. Delia Kesner. Free sequentiality in orthogonal order-sorted rewriting systems with constructors. In *Proc. 11th Int. Conf. on Automated Deduction, Saratoga Springs, NY, LNAI 607*, 1992.
6. Jan Willem Klop and Aart Middeldorp. Sequentiality in orthogonal term rewriting systems. *Journal of Symbolic Computation*, 12:161–195, 1991.
7. Michio Oyamaguchi. Sufficient sequentiality: a decidable condition for call-by-need computations in term rewriting systems. Mie University, 1987.
8. Laurence Puel and Ascander Suarez. Compiling pattern matching by term decomposition. In *Proc. Int. Conf. on LISP and Functional Programming*, 1990.
9. Yoshihito Toyama. Strong sequentiality of left linear overlapping term rewriting systems. In *Proc. 7th IEEE Symp. on Logic in Computer Science*, Santa Cruz, CA, 1992.

A Conflict Between Call-by-Need Computation and Parallelism

Richard Kennaway

School of Information Systems, University of East Anglia, Norwich NR4 7TJ, U.K.

Abstract. In functional language implementation, there is a folklore belief that there is a conflict between implementing call-by-need semantics and parallel evaluation. In this note we illustrate this by proving that reduction algorithms of a certain general and commonly used form which give call-by-need semantics offer very little parallelism.
The analysis of lazy pattern-matching which leads to the above result also suggests an efficient sequential algorithm for the evaluation of a class functional programs satisfying certain constraints, an algorithm which respects the mathematical semantics of the program considered as a term rewrite system.

1 Introduction

Huet and Lévy [Huet and Lévy, 1979, Huet and Lévy, 1991] have considered the problem of call by need computation of normal forms in orthogonal term rewrite systems. Call by need here means that no redex is ever reduced unless it must be reduced in order to compute the normal form. In general, such a redex cannot be effectively determined. They consider a condition on reduction systems called *strong sequentiality*, having the properties that it is decidable whether any orthogonal system is strongly sequential, and in such a system, for every term having a normal form but not in normal form at least one needed redex can be effectively found. They give algorithms for performing these computations.

However, implementations of functional languages such as ML or Miranda (which for the purposes of this paper we regard as being based on term rewriting) usually use an algorithm of a rather different form. The long range goal of reducing a term to normal form is broken into subgoals of reducing various subterms to "root stable form" — that is, to a term which cannot subsequently be reduced to a redex. The goal of reduction to normal form only appears at the outermost level of the reduction algorithm. Such an algorithm can be sketched as:

> To reduce a term to normal form, first reduce it to root stable form, then apply this algorithm recursively to all the immediate subterms.

If the language is to have call-by-need semantics, this leads to the question studied in this paper: which systems allow call-by-need computation of root stable forms?

* The author was partially supported by SEMAGRAPH II, ESPRIT working group 6345, by a SERC Advanced Fellowship, and by SERC Grant no. GR/F 91582.

Call-by-need here means that when computing a root stable form for a term, no redex is reduced unless it is necessary to reduce it in order to find such a form. Provided the system is orthogonal, there will always exist at least one such redex, and perhaps several. If there are several, all of them can be reduced in parallel. However, it may not be possible to find such redexes without performing some sort of lookahead, which defeats the purpose of call-by-need.

We consider how to find such ***root-needed*** redexes given only the left hand sides of the rules. This limitation formalises one notion of prohibiting look-ahead. A redex which can be determined to be root-needed given only the left hand sides will be called ***strongly*** root-needed.

We show that in any orthogonal system, a term having a root stable form but not in root stable form has at most one strongly root-needed redex. Informally, this means that in an implementation of a call-by-need functional language based on reduction to root stable form, no parallelism can be extracted from a program if the algorithm for selecting redexes has no knowledge of the right-hand sides of the rules.

We indicate how in a strongly root-sequential system, strongly root-needed redexes may be computed. Strong root-sequentiality is a very strong condition, and the resulting algorithm is simpler than the Huet-Lévy algorithm for strongly sequential systems.

2 A typical reduction algorithm

We flesh out the outline reduction algorithm described in the introduction. The typical reduction algorithm for a call-by-need functional language can be outlined as the following two partial algorithms NF (reduce a term to normal form) and RF (reduce a term to root stable form):

NF
1. If t is root stable, and has the form $F(t_1,\ldots,t_n)$, then $NF(t) = F(NF(t_1),\ldots,NF(t_n))$.
2. Otherwise, $NF(t) = NF(RF(t))$.

RF
1. If t is a redex and reduces to t', then $RF(t) = RF(t')$.
2. If t is root stable, then $RF(t) = t$.
3. Otherwise, let $t = F(t_1,\ldots,t_n)$. $RF(t) = RF(F(t'_1,\ldots,t'_n))$, where $t'_i = RF(t_i)$ if a root stable form of t_i must be computed in order to determine which rule will eventually match t, and otherwise t'_i is t_i.

Observe that the algorithm NF uses the algorithm RF, but not vice versa. In fact, most of the work done by the algorithm NF consists in finding root stable forms of various terms.

Both of these algorithms are call-by-need — no redex is ever reduced unless it is necessary to reduce it in order to reach the required form. However, the properties of root-stability and root-neededness which they make use of are in general undecidable, hence our description of these as partial algorithms. An actual algorithm must use computable approximations to these properties.

3 Strongly root-needed redexes

3.1 Definitions

We assume familiarity with the basic concepts of term rewriting, and in particular, with the notions of orthogonality and residuals. For an introduction, see [Klop, 1991]. [Huet and Lévy, 1979] is the fundamental reference for call-by-need computation in orthogonal (there called regular) term rewrite systems; [Huet and Lévy, 1991] is a revised and more easily accessible version. We restrict attention throughout to orthogonal term rewrite systems.

In order to be precise about concepts of subterms and occurrences, we state some formal definitions. An *occurrence* or *position* (we use either word as convenient) is a finite sequence of integers. An infix dot is used to denote concatenation, and ϵ is the empty sequence. The set of occurrences or positions of a term t, denoted $\mathrm{Occ}(t)$, is defined inductively. If t is a variable, then $\mathrm{Occ}(t) = \epsilon$. If $t = F(t_1, \ldots, t_n)$, then $\mathrm{Occ}(t) = \epsilon \cup \bigcup_{i=1}^{n} \{i \cdot j | j \in \mathrm{Occ}(t_i)\}$. For any member of $\mathrm{Occ}(t)$, the subterm of t at that position is defined by $t/\epsilon = t$, and $F(t_1, \ldots, t_n)/i \cdot u = t_i/u$ (where i is the first element of the position $i \cdot u$).

If $t/u = F(t_1, \ldots, t_n)$, then F is said to be the function symbol of t at u; u is an occurrence of F in t. If u is the empty occurrence, F is the *root* function symbol of t.

We write $M[u := N]$ for the result of replacing the subterm of M at u by N.

We call a term *root-stable* if it cannot be reduced to a redex. A redex of a term is *root-needed* if every reduction of the term to root-stable form reduces some residual of that redex. As we remarked earlier, these properties are in general undecidable, and we shall define decidable approximations to them.

We shall assume that the systems we consider contain a nullary function symbol *error*, which does not occur in the left hand side of any rule. The reason for this requirement will become clear later. It is not a strong restriction, and could be avoided by using free variables as is done in [Huet and Lévy, 1979] and [Huet and Lévy, 1991] and described there as "a technical trick" (see Lemma 4.11 of [Huet and Lévy, 1991]). Klop and Middeldorp [Klop and Middeldorp, 1991] use the symbol • for the same purpose. The use of an explicit symbol such as *error* or • allows a clearer exposition. We also add to the system another nullary symbol Ω, not appearing in the left- or right-hand side of any rule, and intuitively representing "bottom" or "undefinedness". A *proper* term is one not containing Ω. Based on the intuition of "undefinedness", we define a partial ordering of terms.

- For every term M, $\Omega \leq M$.
- For any terms $M_1, \ldots, M_n, N_1, \ldots, N_n$, and any function symbol F,

$$F(M_1, \ldots, M_n) \leq F(N_1, \ldots, N_n) \Leftrightarrow M_1 \leq N_1 \wedge \ldots \wedge M_n \leq N_n$$

We write $M \sqcup N$ for the least upper bound of M and N with respect to this ordering, when it exists. $M \leq N$ is read "M is weaker than N" or "N is stronger than M".

A term which might be a redex but for occurrences of Ω is *redex-compatible*. Formally, a term M is *redex-compatible* if $M \leq N$ for some instance N of the left-hand side L of some rule.

ω-reduction is the reduction relation $\rightarrow_\omega$ on terms defined by:

- $M \rightarrow_\omega \Omega$ if M is a redex-compatible term other than Ω.

A redex w.r.t. $\rightarrow_\omega$ in a term M is an *ω-redex*. It may be denoted by (u, L), where u is the position of the ω-redex in M, and L is a left-hand side compatible with M/u. Note that in general, M may have several different ω-redexes at the same position, since a term can be compatible with several different rules, even in an orthogonal system. The *obstruction set* of an ω-redex (u, L) of M, denoted $O_{M,(u,L)}$, is the set of positions $u \cdot v$ of M such that $M/u \cdot v = \Omega$ and v is an occurrence of a function symbol in L.

$\omega(M)$ denotes the normal form of M w.r.t. $\rightarrow_\omega$. A term of the form $\omega(M)$ is an *ω-normal form*. (When we speak of a normal form without mentioning the reduction relation, we will always mean a normal form with respect to the original rules of the TRS.) These definitions of $\rightarrow_\omega$ and $\omega(M)$ coincide with those of Huet and Lévy, although they are expressed differently. A term containing no redexes by $\rightarrow_\omega$ except possibly at the root is an $\overline{\omega}$-normal form.

We can now define our decidable approximations to root-stability and root-neededness. M is *strongly root-stable* (written srs(M)) if $\omega(M) \neq \Omega$. That this is indeed an approximation to root-stability will be proved in the next section.

A predicate P on terms is *monotonic* if it is monotonic when considered as a function from terms ordered by $\leq$ to truth-values ordered by *false* < *true*. A monotonic predicate is *sequential at* a term M if:

$$(\neg P(M) \wedge \exists N > M.P(N)) \Rightarrow \exists u.(M/u = \Omega \wedge \forall N \geq M.P(N) \Rightarrow N/u \neq \Omega)$$

An occurrence u as above is an *index* of P in M. P is *sequential* if it is sequential at every term which is in normal form (by the original rules of the system, not ω-reduction). $\mathcal{I}_P(M)$ is the set of indexes of P in M, and in particular, $\mathcal{I}_{\text{srs}}(M)$ is the set of indexes of strong root-stability in M.

These definitions are analogous to those of Huet and Lévy. The predicate srs corresponds to their predicate *nf′*, and $\mathcal{I}_{\text{srs}}(M)$ to $\mathcal{I}(M)$, which we would call $\mathcal{I}_{nf'}(M)$. Informally, if we consider an occurrence of Ω in M to mean a lack of information about the subterm there, such an occurrence is in $\mathcal{I}_{\text{srs}}(M)$ if more information would be required about that subterm of M in order to find a strongly root-stable form of M.

Let $\Omega(M)$ be the result of substituting Ω for every outermost redex of M. A *strongly root-needed* redex of M is a redex at an occurrence in $\mathcal{I}_{\text{srs}}(\Omega(M))$.

$\mathcal{I}_{\text{srs}}(\Omega(M))$ may be empty, but in section 4 we will impose a restriction on rewrite systems that will ensure that it is nonempty for every term which is not srs but which is reducible to an srs term.

3.2 Equivalent characterisation

We now establish a simpler characterisation of indexes for srs, by showing that instead of considering all terms $N \geq M$, we need only look at those obtained from M by replacing one or more occurrences of Ω by *error*. We shall write $M[\Omega := \mathit{error}]$ for the result of replacing every occurrence of Ω in M by *error*, and $M[\overline{u} := \mathit{error}]$ for the result of making this substitution at every occurrence of Ω other than u.

Proposition 1. *If M is* srs, *then for every occurrence u of M, so is $M[u := error]$.*

Proof. Since *error* does not occur in any left-hand side, every ω-reduction sequence of $M[u := error]$ is also an ω-reduction sequence of M. Therefore if $M[u := error]$ ω-reduces to Ω so does M. □

Theorem 2. *For M in normal form, $u \in \mathcal{I}_{srs}(M)$ iff $M/u = \Omega$, $M[\overline{u} := error]$ is not* srs, *and $M[\Omega := error]$ is* srs.

Proof. If $u \in \mathcal{I}_{srs}(M)$ then the conclusion is immediate from the definition of indexes and Proposition 1.

Suppose the right hand side of the equivalence holds. Let M' be any term resulting from M by substituting *error* for some occurrences of Ω other than u. Then $M' \leq M[\overline{u} := error]$. Since the latter term is not srs, neither is M'. Therefore $u \in \mathcal{I}_{srs}(M)$. □

Theorem 3. *If M is* srs *then M is root-stable for every possible choice of right-hand sides of the rules of the TRS.*

Proof. If M is not root-stable for some choice of right-hand sides, then with that choice, there is a reduction of M to a redex N. In any reduction sequence $M \to^* N$, by performing ω-reductions instead we can obtain a sequence $M {\to_\omega}^* N'$ for some $N' \leq N$. Since N is a redex, $N' \to_\omega \Omega$, showing that M is not srs. □

The converse is not quite true, as demonstrated by rules of the form $F \to \ldots, H(A, B) \to \ldots$. The term $H(F, F)$ is not srs, but no choice of right-hand sides for these rules makes it reducible to a redex.

3.3 Uniqueness of indexes

Here we prove our first main result: in an orthogonal system, no term can have more than one strongly root-needed redex.

Lemma 4. *Let M be in normal form, $u \in \mathcal{I}_{srs}(M)$, and let $M \to_\omega M[v := \Omega]$. If $v < u$ then $v \in \mathcal{I}_{srs}(M[v := \Omega])$, otherwise $u \in \mathcal{I}_{srs}(M[v := \Omega])$.*

Proof. The case where $v < u$ holds for any monotonic predicate ([Klop and Middeldorp, 1991], Proposition 4.1(2)).

Now suppose $v \not< u$. We know that $M[v := \Omega]$ is a non-srs normal form, and that $M[v := \Omega][\Omega := error]$ is srs (since it is equal to $M[\Omega := error][v := error]$, and $M[\Omega := error]$ is srs). By Theorem 2 it only remains to show that $M[v := \Omega][\overline{u} := error] = M[v := error][\overline{u} := error]$ is not srs.

By hypothesis, $M[\overline{u} := error]$ is not srs. Because it is in normal form, and its only occurrence of Ω is at u, $M[\overline{u} := error]$ can only contain ω-redexes at positions which are prefixes of u, and the ω-reduction to Ω must proceed by reduction of ω-redexes at occurrences $u_1, \ldots, u_n$, such that $u > u_1 > \ldots > u_n = \epsilon$. Let $M_0 = M$, $u_0 = u$, and $M_i = M[u_i := error](1 \leq i \leq n)$. By the first case of this lemma, $u_i \in \mathcal{I}_{srs}(M_i)$

when $1 \leq i \leq n$. Choose the smallest i such that $v \geq i$. (There must be one, since $v > \epsilon = u_n$.) Then the first $i-1$ steps of the ω-reduction of $M[\overline{u} := error]$ can be applied also to $M[v := error][\overline{u} := error]$, and yield the term $M_{i-1}[v := error]$. M_{i-1} has an ω-redex at u_i. If $M_{i-1}[v := error]$ does also, then reducing it in the latter term gives M_i, because $v \geq u_i$. Hence to prove that $M[v := error][\overline{u} := error]$ is not srs, it is sufficient to show that $M_{i-1}[v := error]$ has an ω-redex at u_i.

Suppose that the occurrence v participates in the pattern-matching of the ω-redex at u in M_{i-1}. v is the position of an ω-redex of M_{i-1}, so by suitable substitution of terms for occurrences of Ωin M_{i-1} below v, it can be made into the position of a redex. Since $v \not< u_{i-1}$, u_{i-1} is not substituted for. Thus, independently of that substitution, a term can be substituted for u_{i-1} so as to make a redex at u_i. But the redexes thus constructed at u and v conflict at v, which by orthogonality is impossible.

Therefore v does not participate in the pattern-matching of the ω-redex at u in M_{i-1}. Hence $M_{i-1}[v := error]$ also has an ω-redex at u. This concludes the proof. □

Theorem 5. *For any normal form M, if $u \in \mathcal{I}_{srs}(M)$ then $M[u := error]$ is* srs.

Proof. By induction on the size of M. Suppose that M is a normal form and $u \in \mathcal{I}_{srs}(M)$.

If $M = \Omega$ then the result is trivial.

If M is not Ω, then since M is not srs, it must contain at least one ω-redex.

Suppose that there is an ω-redex in M at an address $v \not\leq u$. Let $M \rightarrow_\omega M'$ by ω-reduction at v. By Lemma 4, u is an index of M'. By induction, $M'[u := error]$ is srs. $M[u := error] \rightarrow_\omega M'[u := error]$, and hence $M[u := error]$ is srs.

Otherwise, every ω-redex of M is at a prefix of u. We shall prove, by induction on the size of M, that u is in the obstruction set of every such ω-redex. Suppose there exists an ω-redex (v, R) of M such that $v < u$ and $u \notin O_{M,(v,R)}$. Since M is a normal form, $O_{M,(v,R)}$ is nonempty, and therefore (v, R) cannot be an ω-redex of $M[\overline{u} := error]$. By hypothesis, $M[\overline{u} := error]$ is ω-reducible to Ω. Since $M[\overline{u} := error]$ contains no redexes, the obstruction set of each of its ω-redexes must be $\{u\}$.

Let (w, S) be an ω-redex of $M[\overline{u} := error]$. If $w \leq v$, then by suitable substitution of terms at u and at $O_{M[\overline{u}:=error],(v,R)}$, a term would be obtained having a redex at v, and another redex at w which would pattern-match at u, and hence also at v. Such redexes would conflict, which by orthogonality is impossible.

Therefore $v < w < u$. By Lemma 4, w is an index of srs for $M[w := \Omega]$. Since all the ω-redexes of $M[w := \Omega]$ are at prefixes of w, and $M[w := \Omega]$ is smaller than M, by induction w is in the obstruction set of each of these ω-redexes. In particular, it is in the obstruction set of (v, R). However, since (v, R) and (w, S) have disjoint obstruction sets in M, a suitable substitution of terms for the occurrences in their obstruction sets will make both into redexes. By orthogonality, they cannot conflict, and so w cannot play any part in the pattern-matching of (v, R). But then w cannot be in the obstruction set of (v, R) in $M[w := \Omega]$, contrary to what we have just proved. This implies that $M[\overline{u} := error]$ contains no ω-redexes, contradicting the fact that it is not srs. Therefore u must be in the obstruction set of every ω-redex of M, implying that $M[u := error]$ is srs. □

Corollary 6. *$\mathcal{I}_{\text{srs}}(M)$ contains at most one element.*

Corollary 7. *No term can have more than one strongly root-needed redex.*

3.4 Transitivity of indexes

Transitivity is a property of indexes which makes them easier to find. It implies that when one has found an index u for some prefix M' of a term M, one can continue the search by looking for an index of M/u. Not all predicates have transitive indexes, but srs does.

Theorem 8. (Transitivity of indexes.) *Let M, N, and $M[u := N]$ be terms in normal form. Let $u \in \mathcal{I}_{\text{srs}}(M)$ and $v \in \mathcal{I}_{\text{srs}}(N)$. Then $u \cdot v \in \mathcal{I}_{\text{srs}}(M[u := N])$.*

Proof. Let the hypotheses of the theorem be satisfied. Let σ be an error substitution of $M[u := N]$ which does not substitute for $u \cdot v$. Then since v is an index for N, $\sigma(M[u := N]) \rightarrow_{\omega}^{*} \sigma(M[u := \Omega]) = \sigma(M)$. Since u is an index for M and is not in the domain of σ, $\sigma(M)$ is not srs. Therefore if $M[u := N]$has an index, then $u \cdot v$ is such an index.

To show that $M[u := N]$ has an index, it remains to show that $M[u := N][\Omega = error]$ is srs. This term is equal to $M'[u := N']$, where $M' = M[\Omega = error]$ and $N' = N[\Omega = error]$. Since M and N have indexes, M' and N' are srs. Furthermore, since M and N are in normal form, M' and N' are in ω-normal form. If $M'[u := N']$ is not in ω-normal form, it can only be because it contains a redex which lies partly in M' and partly in N'. Choose such a redex, at an address w. (In fact, because of orthogonality, there can only be one such redex.) Let N'' be the prefix of N' containing those occurrences of N' which are pattern-matched by the redex.

Now consider an ω-reduction of N to Ω. Consider the first step which reduces an ω-redex rooted within N''. Before reducing it, the redex originally at w must still be present. If a substitution of terms for occurrences of Ω is made so as to make the ω-redex into a redex, we obtain a pair of conflicting redexes, which by orthogonality is not possible. Therefore the supposed redex at w cannot exist, and $M'[u := N']$ must be in normal form. □

Theorem 9. (Decomposition of indexes.) *If $M/u = \Omega$, and $u \cdot v \in \mathcal{I}_{\text{srs}}(M[u := N])$, then $u \in \mathcal{I}_{\text{srs}}(M)$. If in addition N has an index, then v is it.*

Proof. Given the hypotheses, Proposition 4.1(2) of [Klop and Middeldorp, 1991] implies that $u \in \mathcal{I}_{\text{srs}}(M)$.

Corollary 6 and Theorem 8 then imply that v is the only possible index of N. □

It is possible for N in the preceding theorem to have no index. Consider rules whose left hand sides are $F(G(A, x))$ and $G(B, C)$. Take $M = F(\Omega)$, $N = G(\Omega, \Omega)$. Then M has index 1, $M[1 := N]$ has an index $1 \cdot 1$, but N has no index for srs.

4 Existence of strongly root-needed redexes

We now consider the problem of guaranteeing the existence of indexes of srs. This will be done by imposing a condition analogous to Huet and Lévy's strong sequentiality.

Definition 10. An orthogonal system is strongly root-sequential if srs is a sequential predicate.

More concretely, if M is a term in normal form which is not strongly root stable, and there is a strongly root-stable $N \geq M$, then strong root-sequentiality requires that $\mathcal{I}_{\text{srs}}(M)$ be nonempty. Strong root-sequentiality requires a perhaps infinite number of terms to have indexes. We will show that it is sufficient to test only a very small subset of these.

Definition 11. A *simple* ω-redex is a closed ω-redex M such that (i) for every left-hand side N of which M is an ω-redex, $M \leq N$, (ii) no proper subterm of M is an ω-redex (i.e. M is an $\overline{\omega}$-normal form), and (iii) M is not a redex.

As an example, consider a system with left-hand sides $F(A, x)$ and $F(B, C)$. The simple ω-redexes of this set of left-hand sides are Ω, $F(\Omega, \Omega)$, and $F(B, \Omega)$. $F(\Omega, C)$ is not a simple ω-redex, since the first condition fails in relation to the left-hand side $F(A, x)$. Note that if there are finitely many rules in a system, there are only finitely many simple ω-redexes.

Theorem 12. *Strong root-stability is sequential iff it is sequential at every simple ω-redex.*

Proof. The "only if" part is trivial.

We prove the converse in two stages. First, we show that if srs is sequential at each ω-redex in $\overline{\omega}$-normal form, it is sequential. Then we show that if it is sequential at each simple ω-redex, it is sequential at each ω-redex in $\overline{\omega}$-normal form.

Assume that srs is sequential at each ω-redex in $\overline{\omega}$-normal form. We shall prove, by induction on the size of terms, using the $\overline{\omega}$-normal forms as the base case, that srs is sequential.

Let M be a term in normal form, but not in ω-normal form, such that $\neg\text{srs}(M)$, and there exists an srs term $\geq M$. Assume that at all smaller such terms, srs is sequential. We must prove that M has an index for srs. By hypothesis, there is a occurrence u of M such that M/u is an ω-redex. If the only such u is ϵ, then M is an ω-redex in $\overline{\omega}$-normal form, and by hypothesis has an index for srs. Otherwise, choose u to be nonempty. This implies that both $M[u := \Omega]$ and M/u are smaller than M. $M[u := \Omega]$ is not srs, is weaker than some srs term, and is smaller that M, so by induction, it has an index v for srs. If $v \neq u$, then v is an index for srs in $M[u := N]$ for any non-srs term N; in particular, v is an index for srs in M. If $v = u$, then there must be a term $N \geq M/u$ which is srs (otherwise there could be no srs term stronger than M). M/u is smaller than M, so by induction, M/u has an index w. By transitivity, $u \cdot w$ is an index for M.

Now we must prove that every ω-redex in $\overline{\omega}$-normal form has an index, given that every simple ω-redex does. Again, we proceed by induction on the size of terms.

Let M be a ω-redex in $\overline{\omega}$-normal form, and assume that all terms strictly weaker than M have indexes. If M is not simple, then there is a left-hand side N of which M is an ω-redex, and an occurrence u of M, such that $M/u \neq \Omega$ and N/u is a variable. $M[u := \Omega]$ is smaller than M, so by induction it has an index v. (Note that $M[u := \Omega]$ need not be in $\overline{\omega}$-normal form.) v cannot be u, since $M[u := \mathit{error}]$ is still an ω-redex of N, and so not srs. The term $M[u := \Omega][v := \mathit{error}] = M[v := \mathit{error}][u := \Omega]$ is srs, and therefore $M[v := \mathit{error}]$ is srs. Since v is the index of $M[u := \Omega]$, $M[u := \Omega][\overline{v} := \mathit{error}]$ is not srs. But $M[u := \Omega][\overline{v} := \mathit{error}] = M[u := \mathit{error}][\overline{v} := \mathit{error}]$, and therefore for any term M', $M[u := M'][\overline{v} := \mathit{error}]$ is not srs. In particular, $M[\overline{v} := \mathit{error}]$ is not srs. This demonstrates that v is the index of srs in M. □

Theorem 13. *A system is strongly root-sequential iff every simple ω-redex has exactly one index.*

Proof. Immediate from the previous results. □

Corollary 14. *In a strongly root-sequential system, every term which is not* srs *and is reducible to an* srs *term has exactly one strongly root-needed redex.*

We note here a possible relationship with lambda calculus. It is well-known that in the pure lambda calculus, a function of several arguments can only be strict in at most one of them [Berry, 1978]. The problem of determining which term rewrite systems can be translated into lambda calculus is open [Dershowitz *et al.*, 1991]. It is known that not all orthogonal systems can be translated, but that all strongly sequential orthogonal constructor systems can be translated [Berarducci and Böhm, 1992]. The above result suggests to us the conjecture that all strongly root-sequential systems can be translated. Root-sequentiality is stronger than strong sequentiality, but we are not restricted to constructor systems, so this conjecture is independent of the result of [Berarducci and Böhm, 1992].

The next theorem justifies the terminology "strongly root-needed". The lemma follows from standard properties of orthogonal TRSs. See e.g. [Huet and Lévy, 1991].

Lemma 15. *If M is not root-stable, every reduction of M to root-stable form reduces at least one redex at the root.*

Theorem 16. *If a redex of M is strongly root-needed then it is root-needed for every possible choice of right-hand sides of the rules of the TRS.*

Proof. Suppose that for some choice of right-hand sides, M has a non-root-needed redex at occurrence u. If u is not outermost, then u is not strongly root-needed. Suppose u is outermost. There is a reduction of M to root-stable form not reducing any residual of u. We can eliminate from that sequence any reduction taking place within any residual of the subterm at u, to get another reduction of M to root-stable form. This reduction can be applied to $M[u := \mathit{error}]$. Some initial segment of this reduction must reduce $M[u := \mathit{error}]$ to a redex, and then reduce that redex. From this segment we can derive an ω-reduction reducing $M[u := \mathit{error}]$ to Ω. Therefore $M[u := \mathit{error}]$ is not srs, u is not an index of $\Omega(M)$, and u is not a strongly root-needed redex of M. □

5 Deciding strong root-sequentiality

5.1 A test for strong root-sequentiality

Theorem 17. *An orthogonal term rewrite system is strongly root-sequential iff for every simple ω-redex M, there is a left-hand side $N \geq M$ such that for every occurrence u of Ω in M except exactly one, N/u is a variable. The one exception is the index of* srs *in M.*

Proof. Let the system be strongly root-sequential. Let M be a simple ω-redex. Then M has an index u. Let $M' = M[\overline{u} := error]$. Because u is an index of M, $\omega(M') = \Omega$. Since M contains no proper subterms which are ω-redexes, neither does M'. Therefore M' must itself be an ω-redex of some left-hand side N. Then for every occurrence v of *error* in M', N/v must be a variable. If N/u were also a variable, then M would be a redex of N, contrary to hypothesis. Therefore this N satisfies the conclusion of the theorem.

Conversely, suppose the condition holds for every simple ω-redex M. Then the occurrence instantiated by the given left-hand side is clearly an index for M'. Thus every simple ω-redex has an index, so by Theorem 13 the system is strongly root-sequential. □

Corollary 18. *Let R_1 and R_2 be two term rewrite systems. Suppose that:*

- *$R_1 \cup R_2$ is orthogonal,*
- *R_1 and R_2 are both strongly root-sequential,*
- *No left hand side of either system has the same root symbol as any left hand side of the other system.*

Then $R_1 \cup R_2$ is strongly root-sequential.

This result implies that strong root-sequentiality is modular (i.e. is preserved by the disjoint union of rewrite systems), but is much stronger than that, as it allows a large degree of overlap in the function symbols of the two systems.

5.2 Examples

The following set of rules is not strongly root-sequential.

$$\begin{aligned} And(True, True) &\rightarrow True \\ And(False, True) &\rightarrow False \\ And(True, False) &\rightarrow False \\ And(False, False) &\rightarrow False \end{aligned}$$

For example, the term $And(\Omega, \Omega)$ has no index, since both $And(error, \Omega)$ and $And(\Omega, error)$ are strongly root-stable. In contrast, the following set of rules is strongly root-sequential:

$$\begin{aligned} And(x, True) &\rightarrow x \\ And(x, False) &\rightarrow False \end{aligned}$$

The rules for Peano arithmetic are strongly root-sequential:

$$\begin{aligned} Add(0, x) &\rightarrow x \\ Add(Succ(x), y) &\rightarrow Succ(Add(x, y)) \\ Mult(0, x) &\rightarrow 0 \\ Mult(Succ(x), y) &\rightarrow Add(y, Mult(x, y)) \end{aligned}$$

The following rules are not strongly root-sequential:

$$\begin{aligned} F(A, B, x) &\rightarrow C \\ F(D, y, E) &\rightarrow G \end{aligned}$$

The ω-redex $F(\Omega, \Omega, \Omega)$ has no indexes, as demonstrated by the two srs terms $F(error, \Omega, \Omega)$ and $F(\Omega, error, error)$. However, the following rules are strongly root-sequential:

$$\begin{aligned} F(A, B, x) &\rightarrow C \\ F(D, y, E) &\rightarrow G \\ F(H, u, v) &\rightarrow J \end{aligned}$$

6 Computation in strongly root-sequential systems

6.1 An algorithm for finding strongly root-needed redexes

To compute in a term rewrite system, an algorithm is required for finding the next redex to reduce, and reducing it. After each reduction, the next redex should be found without having to rescan the term from the root, by reusing as much information as possible from the search for the previous redex. It should not be necessary to read any node of the term more than once. The code required to do this should be at the most linear in the size of the rewrite system.

We present an algorithm for computation in strongly root-sequential term rewrite systems having all of the above properties.

Let t be a term in a strongly root-sequential system which has a strongly root-stable form but is not strongly root-stable. t is represented as a syntax tree, or more precisely, a graph in which repeated subexpressions may be shared. Our algorithm for discovering a strongly root-needed redex of t will be represented as a finite-state machine which manipulates a stack and can read the contents of any node of t.

The states of the machine are *stop*, Ω, and (a set in $1-1$-correspondence with) the class of all ω-redexes which can be generated by the following rules. If q is a state, and is neither *stop* nor a redex, then q must have an index u for srs. For every function symbol f, if replacing the occurrence of Ω at u in q by $f(\Omega, \ldots, \Omega)$ yields an ω-redex q', then q' is also a state. Note that since indexes are unique, every state is generated in exactly one way. Since every ω-redex which is not a redex has an index, every left-hand side is a state (more precisely, the ω-redex obtained from any left-hand side by replacing variables by Ω). Every other state besides *stop* and Ω is weaker than some left-hand side.

Assume there are N_s states. As far as implementation is concerned, states can be represented simply as integers from 1 to N_s, since all the information this algorithm needs about the structure of ω-redexes and left hand sides is implicit in the lookup functions about to be described.

Each cell of the stack contains a reference to a node of t and a state. As the current state of the automaton will always be the state stored in the top cell of the stack, no separate record of the current state need be kept.

During the search for a redex, the following invariants hold while the stack is nonempty and the state is not *stop*.

1. The set of nodes on the stack is a connected set of nodes of t, with the root of t at the bottom of the stack.
2. There is a final segment of the stack, such that the set of nodes it contains is the pattern of a state,
3. For the longest such final segment, that state is the topmost state on the stack.
4. Only the topmost state can be a redex.

Initially, the stack is empty, and the current state is Ω.

Given a current state q and a stack S of length l:

If $q = stop$, stop. At this point, every node on the stack is strongly root-stable. For all the nodes which are descendants of nodes on the stack, but are not themselves on the stack, computations to reduce the subterms rooted at those nodes to normal form can be initiated in parallel. If there are no such nodes, t is in normal form and the computation is over.

If q is a left-hand side, stop. We have found a redex of q, at the node $S[l-k+1]$, where k is the number of non-variable nodes in q.

Otherwise, we push a new node and state onto the stack, calculated as follows:

```
If q = Ω then
        n' := the root of t;
else
        (i, j) := NextNodeRef(q);
        (n, q') = S[l − i];
        n' := the j'th descendant of n;
endif
f := the function symbol at n';
q'' := NewState(q, f);
push (n', q'') onto S.
```

NextNodeRef and NewState are lookup tables which can be computed in advance from the set of left-hand sides of the system. If $\text{NextNodeRef}(q) = (i, j)$, this means that the next node to be inspected is the jth descendant of the node stored in $S[l-i]$. The result of inspecting the function symbol f at that node is to determine a new state q'' via the lookup table NewState.

NextNodeRef and NewState are defined thus. Consider any simple ω-redex s. Let $u \cdot j$ be its index for srs. When s is the topmost state, the node n of t corresponding to the position u of s will be on the stack, at a certain distance i from the top of the stack (a distance which depends only on s). NextNodeRef(q) is defined to be the pair (i, j).

Consider the ω-term obtained by adding the node n' to q. If this does not contain any ω-redex, then NewState(q, f) = *stop*). If it does, then NewState(q, f) is the largest ω-redex it contains (which might or might not be the whole ω-term).

Note that the state q' plays no part in the computation of (n', q''). The state information in stack cells below the topmost is used when reducing redexes.

Having found a redex q at the node $S[l - k + 1]$, the next step is to reduce it. Each free variable of q corresponds to some descendant of some node stored on the stack. Because the nodes of a redex-pattern are always read in the same order, we already know where the relevant nodes on the stack are. That is, we can precompute a lookup table which will map every pair (q, x), where q is a left-hand side and x is one of its free variables, to a pair (i, j), such that the node of t matched by x is the j'th descendant of the node stored in $S[l - i]$.

Having found all the subterms of t matched by the free variables of q, we create all the nodes required by the corresponding right-hand side, remove the top k elements of the stack, and restore the state of the automaton to the state at the new top element of the stack, or to Ω if the stack is empty. (This needs modification to deal with collapsing rules, i.e. those whose right-hand side is a variable, but the usual and well-known technique of indirection nodes [Peyton Jones, 1987] may be used. The details would be too great a digression.)

We then search for the next redex as already described.

6.2 Complexity

Let the number of function symbols be N_f. The lookup tables NextNodeRef and NewState have respectively $N_s - N_l - 2$ and $(N_s - N_l - 1) \times N_f$ entries, where N_l is the number of rules.

The method of constructing the set of states shows that N_s is bounded by the total size of all the left-hand sides of the rule system. A more compact representation of NewState can be obtained by taking advantage of a sparseness property. Call a symbol f *expected* by state q if NewState$(q, f) \neq$ NewState(Ω, f). In real functional programs, for most (q, f), f is unexpected by q, and a representation of NewState which takes advantage of this can reduce its size to be proprtional to N_s. Examples can be contrived in which NewState is not sparse. However, for constructor systems, every state q has exactly one predecessor (q', f), i.e. a pair such that q = NewState(q', f). This implies that the total number of (q, f) where q expects f is no more than N_s.

Thus in practice we expect, and for constructor systems, we know, that the size of the lookup tables is proportional to the total size of the left-hand sides.

The code for reducing a redex simply constructs a copy of the appropriate right-hand side, which can be done in time and space proportional to the size of that right-hand side.

Therefore the total size of the code is linear in the total size of the program.

6.3 Computation in transitive systems

Every transitive system [Toyama *et al.*, 1993] can be translated into a strongly root-sequential system, by a transformation which adds sequentialising information at each place where the condition of Theorem 17 is violated [Byun, 1994]. The translation of a system R to an srs system R' makes each term t of R a term of R', and if t has a normal form in R, then it has the same normal form in R'. Conversely, if t is a term of R having a normal form t' in R', and if t' is a term of R, then t is reducible to t' in R. This relationship between R and R' justifies the use of R' as an implementation of R.

It is less clear whether strongly sequential systems in general can be translated to srs systems.

7 Conclusion

Corollary 7 shows that no parallelism is available when reducing a term to strongly root-stable form, if the evaluation algorithm is determined only by the left-hand sides of the system. Theorem 17 shows that under the same constraint on the evaluation algorithm, for fully lazy evaluation to be possible requires that the program satisfy some quite strong constraints.

The only parallelism available in the general algorithm described in the introduction is at the top level: when the whole term has been reduced to root-stable form, evaluation of its immediate subterms can then proceed in parallel. This is the only source of parallelism; however, it usually yields very little. The final result of a functional program is usually either a list of characters or a list of system commands. The strongly root-stable form of the initial term will thus be $Cons(t_1, t_2)$, where t_1 reduces to a character or a single system command, and t_2 reduces to the whole of the rest of the output. These two tasks will rarely be of comparable size; in fact, t_1 may already have been reduced to normal form in the process of reducing the whole term to $Cons(t_1, t_2)$. When the result is a stream of system commands, the implementation may also enforce evaluation of t_1 before t_2 in order to make the events which these commands cause outside the functional environment happen in the right order.

Divide-and-conquer algorithms are a more fruitful source of parallelism. The general form of such an algorithm combines three functions: one to decompose a big problem into two or more smaller subproblems, one to solve small problems, and one to combine the results of several subproblems into the result of a larger problem. The last of these functions will normally be strict in all its arguments — this is where divide-and-conquer parallelism comes from. However, no such function can appear in a strongly root-sequential system.

The general conclusion is not that parallelism is not available in lazy functional programming, but that it must be obtained from information other than the rule-patterns. Sources of such information include knowledge about the possible form of the final result (for example, that it is a list of basic values), type information, or abstract interpretation.

8 Acknowledgments

We acknowledge Aart Middeldorp's useful comments.

References

[Berarducci and Böhm, 1992] A. Berarducci and C. Böhm. A self-interpreter of lambda calculus having a normal form. Technical Report Rapporto tecnico 16, Dipartimento di Matematica Pura ed Applicata, Università di L'Aquila, 1992.

[Berry, 1978] G. Berry. Séquentialité de l'évaluation formelle des λ-expressions. In *Proc. 3-e Colloque International sur la Programmation*, Paris, 1978.

[Byun, 1994] S. Byun. *The Simulation of Term Rewriting Systems by the Lambda Calculus.* PhD thesis, University of East Anglia, Norwich, U.K., 1994.

[Dershowitz *et al.*, 1991] N. Dershowitz, J.-P. Jouannaud, and J. W. Klop. Open problems in rewriting. In *Proc. 4th International Conference on Rewriting Techniques and Applications*, LNCS 488, pages 445–456. Springer-Verlag, 1991.

[Huet and Lévy, 1979] G. Huet and J.-J. Lévy. Call by need computations in non-ambiguous linear term rewriting systems. Technical Report Rapport de Recherche 359, INRIA, 1979.

[Huet and Lévy, 1991] G. Huet and J.-J. Lévy. Computations in orthogonal rewriting systems: I and ii. In J.-L. Lassez and G. D. Plotkin, editors, *Computational Logic: Essays in Honor of Alan Robinson*, pages 394–443. MIT Press, 1991.

[Klop and Middeldorp, 1991] J.W. Klop and A. Middeldorp. Sequentiality in orthogonal term rewriting systems. *J. Symbolic Computation*, 12:161–195, 1991.

[Klop, 1991] J. W. Klop. Term rewriting systems. In S. Abramsky, D. Gabbay, and T. Maibaum, editors, *Handbook of Logic in Computer Science, vol.2*, pages 2–116. Oxford University Press, 1991.

[Peyton Jones, 1987] S. L. Peyton Jones. *The implementation of functional programming languages.* Prentice-Hall, 1987.

[Toyama *et al.*, 1993] Y. Toyama, S. Smetsers, M.C.J.D. van Eekelen, and M.J. Plasmeijer. *The functional strategy and transitive term rewriting systems*, chapter 5, pages 61–75. Wiley, 1993. eds. M.R. Sleep and M.J. Plasmeijer and M.C.J.D. van Eekelen.

The Complexity of Testing Ground Reducibility for Linear Word Rewriting Systems with Variables

Gregory Kucherov and Michaël Rusinowitch

INRIA-Lorraine and CRIN
615, rue du Jardin Botanique, BP 101
54506 Vandœuvre-lès-Nancy, France
email: {kucherov,rusi}@loria.fr

Abstract. In [9] we proved that for a word rewriting system with variables $\mathcal{R}$ and a word with variables w, it is undecidable if w is ground reducible by $\mathcal{R}$, that is if all the instances of w obtained by substituting its variables by non-empty words are reducible by $\mathcal{R}$. On the other hand, if $\mathcal{R}$ is linear, the question is decidable for arbitrary (linear or non-linear) w. In this paper we futher study the complexity of the above problem and prove that it is *co-NP*-complete if both $\mathcal{R}$ and w are restricted to be linear. The proof is based on the construction of a deterministic finite automaton for the language of words reducible by $\mathcal{R}$. The construction generalizes the well-known Aho-Corasick automaton for string matching against a set of keywords.

1 Introduction

This paper continues with the study undertaken in [9] where the following question has been proved to be generally undecidable: Given a set of words with variables $\mathcal{R}$ and a word with variables (subject pattern) w, do all the instances of w obtained by substituting its variables by non-empty words contain an instance of a word from $\mathcal{R}$ as a factor? From the point of view of rewriting systems [2], the question amounts to testing a well-known *ground reducibility* property [12] which is at the core of the *inductive completion* methods for proving theorems in the initial model of equational specifications. Here $\mathcal{R}$ is identified with the set of left-hand sides of a rewriting system, and the rewriting systems under consideration, called *word rewriting systems with variables* (WRSV), are rewriting systems over a signature of a single associative binary symbol (concatenation) and a finite set of constants (letters). The undecidability result above strengthens a result from [6] of undecidability of ground reducibility problem for term rewriting systems over an associative function and a number of non-associative ones. On the other hand, the result can be regarded as the undecidability result for a particular fragment of positive $\forall\exists$-theory of free semigroups, and in this sense generalizes the result of [11].

If the rewriting system $\mathcal{R}$ is restricted to be left-linear, the ground reducibility problem is decidable regardless of the linearity of w. In this paper we study the complexity of the problem and prove that it is *co-NP*-complete when w is restricted to be linear too. This case can be viewed as the inclusion problem for regular languages represented by regular expressions of a particular class, the problem known to be *PSPACE*-complete for general regular expressions [3]. Proving that the problem is in *co-NP* requires an analysis of the structure of a deterministic finite automaton recognizing the words reducible by $\mathcal{R}$. The key idea is to prove that although the total number of states in such an automaton is exponential on the size of $\mathcal{R}$, the length of a loop-free path going from the initial to a final state can be bounded polynomially. This allows us to construct a non-deterministic polynomial-time algorithm for checking the existence of an irreducible instance of a given subject word w. The automaton construction generalizes the well-known Aho-Corasick automaton for string matching against a set of keywords. Using similar techniques we prove that the related problem of testing finiteness of the set of irreducible words is also *co-NP*-complete.

2 Notations

Given a finite alphabet A and an alphabet of variables $\mathcal{X}$, we consider words over $A \cup \mathcal{X}$ called ***words with variables*** (or ***patterns***) and words over A called simply ***words*** (or ***strings***). A pattern is called ***linear*** if every variable occurs at most once in it. A substitution σ is a mapping from $\mathcal{X}$ to A^+.[1] It can be extended to a homomorphism from $(\mathcal{X} \cup A)^+$ to A^+ (regarded as monoids) such that $\sigma(a) = a$ for every $a \in A$.

Word concatenation will be denoted either by $\cdot$ or just by jaxtaposition. ε denotes the empty string and $|w|$ the length of a word w. A position of a symbol (a letter or a variable) in a pattern w is a nonnegative integer in $\{0, ..., |w|-1\}$. If p_1, p_2 are positions in w and $p_1 < p_2$, then by $w[p_1 \leftarrow p_2]$ we denote the pattern obtained by deleting all symbols at the positions $\{p_1, \ldots, p_2 - 1\}$.

Given a set $\mathcal{R}$ of words with variables, $\mathit{Inst}(\mathcal{R})$ denotes the set of instances of patterns from $\mathcal{R}$, that is the set $\{\sigma(w) | w \in \mathcal{R}\}$ for all possible substitutions σ. We define $\mathit{Red}(\mathcal{R}) = \{u_1 \cdot v \cdot u_2 | v \in \mathit{Inst}(\mathcal{R}),\ u_1, u_2 \in A^*\}$, and $NF(\mathcal{R}) = A^+ \setminus \mathit{Red}(\mathcal{R})$. Our notation comes from the term rewriting system vocabulary. Think of $\mathit{Red}(\mathcal{R})$ as the set of strings ***reducible*** by a WRSV (see introduction) with $\mathcal{R}$ being the set of left-hand sides of the rules. For this reason sometimes we call elements of $\mathcal{R}$ ***rules*** without adding "the left-hand side of". Similarly, $NF(\mathcal{R})$ stands for the set of all ***normal forms*** (= irreducible words) for $\mathcal{R}$. A word with variables w is called ***ground reducible*** by $\mathcal{R}$ iff $\mathit{Inst}(\{w\}) \subseteq \mathit{Red}(\mathcal{R})$, or equivalently, iff $\mathit{Red}(\{w\}) \subseteq \mathit{Red}(\mathcal{R})$.

3 Preliminary results

In [9] we proved the following result.

[1] Having in mind term rewriting system applications, we do not allow a variable to be substituted by the empty string. This assumption is technical and does not affect the soundness of the results.

Theorem 1 *For a WRSV $\mathcal{R}$ and a word with variables w, it is undecidable if w is ground reducible by $\mathcal{R}$.*

We even showed that the problem remains undecidable for a fixed and very simple word $w = axa$ where a is a letter and x a variable.

If $\mathcal{R}$ consists of linear patterns, every $v \in \mathcal{R}$ can be written as $v = u_0x_1u_1x_2\ldots x_nu_n$ where $u_0, u_n \in A^*$, $u_i \in A^+$, $1 \le i \le n-1$, $x_i \in \mathcal{X}$, $1 \le i \le n$ and all x_i are different. Obviously in this case $Red(\{v\}) = A^*u_0A^+u_1A^+\ldots A^+u_nA^*$ is a regular language and so is $Red(\mathcal{R}) = \bigcup_{v\in\mathcal{R}} Red(\{v\})$. If w is a linear pattern in addition, testing ground reducibility of w by $\mathcal{R}$ amounts to testing the inclusion of regular languages which is of course decidable.

We show now that the problem remains decidable for linear $\mathcal{R}$ and arbitrary (non-linear) w. We need a notation to identify positions in $\sigma(w)$. Assume that $w = u_0x_1u_1x_2\ldots x_nu_n$ (where variables x_i may be equal). If p_i is the position of x_i then by p_i^σ we denote the position in $\sigma(w)$ which corresponds to the beginning of the substring corresponding to x_i. Formally, we define recurrently $p_1^\sigma = p_1$, and $p_{i+1}^\sigma = p_i^\sigma + |\sigma(x_i)| + |u_i|$ for $2 \le i \le n$.

Lemma 1 *For a linear WRSV $\mathcal{R}$ and an arbitrary word with variables w, it is decidable if w is ground reducible w.r.t. $\mathcal{R}$.*

Proof: The idea of the proof is somewhat similar to that for ordinary term rewriting systems [7, 10]. We show that a constant $C(\mathcal{R}, w)$ can be computed such that if w is not ground reducible w.r.t. $\mathcal{R}$, then there exists a $\mathcal{R}$-irreducible instance $\sigma(w)$, and $|\sigma(x)| \le C(\mathcal{R}, w)$ for each variable x in w. We prove that if for some variable x in w, $|\sigma(x)|$ exceeds the bound, then we can always modify $\sigma(x)$ by reducing its length and preserving the irreducibility of $\sigma(w)$.

Let $\mathcal{A}$ be a deterministic automaton recognizing $NF(\mathcal{R})$. To each position p in a word $u \in NF(\mathcal{R})$ the automaton associates a state denoted $\mathcal{A}(u,p)$. We will use the usual pumping lemma trick: if p_1, p_2 are positions in u, $p_1 < p_2$ and $\mathcal{A}(u,p_1) = \mathcal{A}(u,p_2)$, then $u[p_1 \leftarrow p_2]$ is also recognized by $\mathcal{A}$ and therefore is not $\mathcal{R}$-reducible.

We show now that $C(\mathcal{R}, w)$ can be set to $|\mathcal{A}|^n$, where $|\mathcal{A}|$ is the number of states of $\mathcal{A}$ and n is the maximal number of occurrences of a variable in w. Assume that $\sigma(w)$ is not $\mathcal{R}$-reducible and suppose $|\sigma(x)| > C(\mathcal{R}, w)$ where x is a variable in w. Assume that x occurs at positions $p_1, \ldots, p_m$ in w ($m \le n$). The idea is to find two distinct positions p', p'' in $\sigma(x)$ such that $\sigma'(w)$ still belongs to $NF(\mathcal{R})$ where σ' is the substitution defined by $\sigma'(x) = \sigma(x)[p' \leftarrow p'']$, and $\sigma'(y) = \sigma(y)$ if $y \ne x$. To do this, we choose p', p'' that satisfy the following property: for every j, $1 \le j \le m$, $\mathcal{A}(\sigma(w), p_j^\sigma + p') = \mathcal{A}(\sigma(w), p_j^\sigma + p'')$. Note that with every position p in $\sigma(x)$ we can associate a m-tuple of states $< \mathcal{A}(\sigma(w), p_1^\sigma + p), \ldots, \mathcal{A}(\sigma(w), p_m^\sigma + p) >$. It is clear that there are at most $|\mathcal{A}|^m$ different tuples of this form. Since p has at least $|\mathcal{A}|^n + 1$ possible values where $n \ge m$, by the pigeon hole principle we conclude that positions p', p'' with the desired property must exist. □

It follows from the proof above that if w is linear, $C(\mathcal{R}, w)$ is just the number of states of $\mathcal{A}$. However, it is easy to see that in this case the proof remains valid if $C(\mathcal{R}, w)$ is taken to be a constant which bounds *the number of states along any loop-free path in $\mathcal{A}$ going from the initial to a final state.* This refinement, important for the rest of the paper, is summarized as follows.

Corollary 1 *Let $\mathcal{R}$ be a linear WRSV and w a linear pattern. Assume that $\mathcal{A}$ is a deterministic automaton recognizing $NF(\mathcal{R})$ and K is the maximal length of a loop-free path going from the initial to a final state in $\mathcal{A}$. If w is not ground reducible by $\mathcal{R}$, then there is an irreducible instance $\sigma(w)$ such that $|\sigma(x)| \leq K$ for every variable x of w.*

4 Complexity Results

4.1 Complexity of Testing Ground Reducibility

Now we give a complexity analysis of the ground reducibility problem for a linear WRSV $\mathcal{R}$ and a linear subject pattern w. We show, namely, that this problem is *co-NP*-complete. As usual, the proof consists of two parts. We first prove that the problem is *co-NP*-hard by reducing to it the MONOTONE-ONE-IN-THREE-SAT problem.

Lemma 2 *Testing ground reducibility of a linear subject pattern by a linear WRSV is co-NP-hard.*

Proof: Let X be a finite set of variable symbols and $\mathcal{C} = \mathcal{C}_1 \wedge \mathcal{C}_2 \wedge \ldots \wedge \mathcal{C}_m$ be a conjunction of clauses each consisting of 3 variables of X (positive literals). It is known [3] that the following problem, labeled MONOTONE-ONE-IN-THREE-SAT, is *NP*-complete: Given $\mathcal{C}$, does there exist a truth assignment $\sigma : X \rightarrow \{\mathtt{t}, \mathtt{f}\}$ such that every clause $\mathcal{C}_i$ contains exactly one variable mapped to $\mathtt{t}$ under σ?

Assume that $\{x_1, \ldots, x_n\}$ is the set of variables in $\mathcal{C}$. We encode $\mathcal{C}$ into the following string over the alphabet $A = \{1, \ldots, n\} \cup \{\mathtt{t}, \mathtt{f}\} \cup \{\#\}$ and a variable set $\mathcal{Y} = \{y_1, y_2, \ldots\}$:

$$w = \#\mathbf{C}_1\#\mathbf{C}_2\#\ldots\#\mathbf{C}_n\#$$

where each $\mathbf{C}_i$ is obtained from $\mathcal{C}_i$ by replacing an occurrence of a variable x_i in $\mathcal{C}_i$ by two symbols $y_j\, i$ where y_j is a fresh variable occurring nowhere else in w. (Two different occurrences of x_i in $\mathcal{C}$ are replaced by $y_j\, i$ and $y_k\, i$ where $y_j \neq y_k$.) By construction, w is a linear pattern.

We construct now a linear WRSV $\mathcal{R}$ over A and $\mathcal{X} = \{x, \ldots\}$ which applies to all instances of w but those which correspond to the solutions of MONOTONE-ONE-IN-THREE-SAT for $\mathbf{C}$.

Firstly, we put into $\mathcal{R}$ the following two patterns:

$$\#\# \quad (1)$$

$$\#x_1\#\ldots\#x_{n+1}\# \quad (2)$$

These patterns reduce any instance of w that contains more than $n+1$ occurrences of # and thus guarantee that every variable in an irreducible instance must be substituted by a string from $(\{\mathtt{t}, \mathtt{f}\} \cup \{1, \ldots, n\})^{+}$.

Then, for every i, j, $1 \leq i, j \leq n$, we add the pattern

$$ij \tag{3}$$

Also, we add the four patterns

$$\mathtt{tt},\ \mathtt{tf},\ \mathtt{ft},\ \mathtt{ff} \tag{4}$$

To "prevent" variables in w to take values other than $\mathtt{t}, \mathtt{f}$ we add for every i_1, i_2, i_3, i_4, $1 \leq i_1, i_2, i_3, i_4 \leq n$, the eight patterns schematized by the expression

$$i_1\{\mathtt{t}, \mathtt{f}\}i_2\{\mathtt{t}, \mathtt{f}\}i_3\{\mathtt{t}, \mathtt{f}\}i_4 \tag{5}$$

Patterns (3)-(5) "force" every variable y_j in w to take one of the values $\{\mathtt{t}, \mathtt{f}\}$ under σ.

For every i, $1 \leq i \leq n$, we add the following two patterns. They make reducible any instance of w in which two variables corresponding to the same variable x_i in C are substituted by different values $\{\mathtt{t}, \mathtt{f}\}$.

$$\mathtt{t}ix\mathtt{f}i \tag{6}$$

$$\mathtt{f}ix\mathtt{t}i \tag{7}$$

$$\mathtt{t}i\mathtt{f}i \tag{8}$$

$$\mathtt{f}i\mathtt{t}i \tag{9}$$

Finally, for every i_1, i_2, i_3 $1 \leq i_1, i_2, i_3 \leq n$, we add the patterns

$$\mathtt{f}i_1\mathtt{f}i_2\mathtt{f}i_3 \tag{10}$$

$$\mathtt{t}i_1\mathtt{t}i_2\mathtt{f}i_3 \tag{11}$$

$$\mathtt{t}i_1\mathtt{f}i_2\mathtt{t}i_3 \tag{12}$$

$$\mathtt{f}i_1\mathtt{t}i_2\mathtt{t}i_3 \tag{13}$$

$$\mathtt{t}i_1\mathtt{t}i_2\mathtt{t}i_3 \tag{14}$$

Clearly, the instances of w that remain irreducible by the patterns above correspond exactly to the solutions of MONOTONE-ONE-IN-THREE-SAT for C. We have constructed $\mathcal{O}(n^4)$ patterns of constant length and two patterns (including w) of length $\mathcal{O}(n)$ over an alphabet of $\mathcal{O}(n)$ symbols. It is obvious that the whole construction can be done in polynomial time. □

We are now to prove that the existence of an irreducible instance of a linear pattern w w.r.t. a linear WRSV $\mathcal{R}$ can be tested in polynomial time on the non-deterministic Turing machine. In order to simplify the presentation of this part we will allow variables to be substituted by an empty string. We leave the reader to make sure that this assumption does not affect the complexity.

Throughout the rest of this section we assume that $\mathcal{R} = \{p_1, \ldots, p_n\}$ is a set of patterns where each p_j is of the form $u_1^j x_1^j u_2^j \ldots x_{n_j-1}^j u_{n_j}^j$ where $u_i^j \in A^+$, $1 \le i \le n_j$ and $x_i^j \in \mathcal{X}$, $1 \le i \le n_j - 1$ are pairwise distinct. (Note that a linear variable at the beginning and/or at the end of a pattern can be omitted.) A non-deterministic algorithm that tests whether there exists an irreducible instance of w w.r.t. $\mathcal{R}$ consists of the following two steps.

1. guess a substitution σ assigning to every variable a string no longer than K, where K is the constant from corollary 1,
2. test if $\sigma(w)$ is irreducible by $\{p_1, \ldots, p_n\}$

Let $C = \max\{|u_i^j| \,|1 \le j \le n, 1 \le i \le n_j\}$ and $N = \max\{n_j \,|1 \le j \le n\}$. Step 2 can be done deterministically in time $\mathcal{O}(n(|\sigma(w)| + CN))$ by using for each p_j the Knuth-Morris-Pratt string matching algorithm successively for $u_1^j, \ldots, u_{n_j}^j$. In order to prove that the whole algorithm is in polynomial time it is sufficient to show that K can be bounded polynomially on the size of $\mathcal{R}$. In this case the size of a guess at step 1 and the run time of step 2 would be polynomial on $(|\mathcal{R}| + |w|)$.

Recall that K is the length of the longest loop-free path going from the initial to a final state in some deterministic automaton recognizing $NF(\mathcal{R})$. Note that we can equivalently reason about a deterministic automaton for *Red*($\mathcal{R}$) since it can be obtained from the one for $NF(\mathcal{R})$ by changing the set of final states to its complement. From semantical considerations it is clear that in a deterministic automaton for *Red*($\mathcal{R}$):

- every transition from a final state leads to a final state,
- from every reachable state there is a path to a final state.

Therefore, K can be taken to be the length of the longest loop-free path in some deterministic automaton recognizing *Red*($\mathcal{R}$).

In the rest of this section we construct a deterministic automaton for *Red*($\mathcal{R}$) for which we show that although the total number of states is exponential on $|\mathcal{R}|$, the number of states along a loop-free path can be bounded polynomially.

Example 1 For $k > 0$, consider the system

$$\mathcal{R} = \{\#a\#x\#a\#, \#aa\#x\#aa\#, \ldots, \#a^k\#x\#a^k\#\}$$

over the two-letter alphabet $A = \{a, \#\}$. It can be shown that the minimal deterministic automaton for *Red*($\mathcal{R}$) has the number of states exponential on k. Informally, if the automaton reaches some non-final state after reading a word w, then this state should "memorize" the set $\{i \,|\, 1 \le i \le k, \#a^i\# \text{ is a subword of } w\}$. States corresponding to different sets cannot be factorized since for any two of them, there is a word which leads to a final state from one but not from another. The number of different such sets is 2^k.

On the other hand, the longest loop-free paths are of polynomial length. It can be proved that the words spelled out by the longest loop-free paths in the automaton are of the form $w = \#a^{i_1}\#\#a^{i_2}\# \ldots \#a^{i_k}\#\#a^k\#$, where $(i_1, i_2, \ldots, i_k)$

is an arbitrary permutation of $(1, \ldots, k)$. The proof consists of two parts. First we show that the automaton does not go twice through the same state during its run on w. To show this, we prove that for any two distinct prefixes w_1, w_2 of w, there is a word $v \in A^*$ such that $w_1 v \in Red(\mathcal{R})$ and $w_2 v \notin Red(\mathcal{R})$, or vice versa. This implies that no two prefixes of w take the automaton to the same state. At the second step, we show that any word longer than w has two distinct prefixes w_1, w_2 such that for every $v \in A^*$, $w_1 v \in Red(\mathcal{R})$ iff $w_2 v \in Red(\mathcal{R})$. This means that any such word makes the automaton visit twice the same state. Both steps of the proof can done by exhaustive case analysis. We omit further details.

In conclusion, the length of the longest loop-free paths is $\frac{k^2+5k+4}{2}$. □

To explain the structure of the automaton let us start with a very particular case when every pattern $p_i \in \mathcal{R}$ is just a string $v_i \in A^+$. We come up then with a well-known problem of matching against a set of keywords [1]. The well-known Aho-Corasick algorithm is a generalization of the Knuth-Morris-Pratt algorithm to the multiple-keyword case [1]. Similar to the Knuth-Morris-Pratt algorithm, the Aho-Corasick algorithm preprocesses (in time $\mathcal{O}(|\mathcal{R}|)$) the set $\mathcal{R}$ into an automaton which allows one to perform pattern matching by scanning the input string in linear time without backtracking. Let us recall very briefly the idea of the construction. Think of the algorithm as scanning the input string and moving a pointer in each v_i. Clearly, a state in the automaton is associated with a combination of pointer positions. However, it is not necessary to consider (a potentially exponential number of) all possible combinations, as the following argument shows. The position of each pointer is uniquely determined by the suffix w of length $\max\{|v_i| \mid 1 \leq i \leq n\}$ of the scanned part of the input string. The pointer position in v_i is then defined as the longest prefix of v_i which is a suffix of w. Moreover, it is sufficient to know the longest suffix of w which is at the same time a prefix of some v_i. But this shows that a state can be identified with a prefix of some v_i which shows that the number of states is bounded by $|\mathcal{R}|$. We essentially use this idea in the construction below.

We describe a deterministic automaton $\mathcal{A}$ that recognizes the strings matched by at least one of the patterns. Let us first introduce some notations. Given a word w, $pref(w)$ (respectively $suff(w)$) denotes the set of prefixes (respectively suffixes) of w. ε denotes the empty string. Given two words v, w, $S(v, w)$ stands for the longest word from $suff(v) \cap pref(w)$. Finally, if q is a state and v a word, $q \cdot v$ denotes the state reached by $\mathcal{A}$ from the state q after processing the word v.

The set of states Q of $\mathcal{A}$ is a set of triples

$$< (i_1, \ldots, i_n), \pi, (\mu_1, \ldots, \mu_n) >$$

where $1 \leq i_j \leq n_j + 1$, $\pi \in \bigcup_{j=1}^{n} pref(u^j_{i_j})$ and $\mu_j \in pref(u^j_{i_j}) \cup \{*\}$.
The initial state of the automaton is :

$$q_0 =< (1, \ldots, 1), \varepsilon, (\varepsilon, \ldots, \varepsilon) >$$

Assume that $a \in A$, $q =< (i_1, \ldots, i_n), \pi, (\mu_1, \ldots, \mu_n) >$, and $i_j \leq n_j$ for all j, $1 \leq j \leq n$. We define $q \cdot a =< (i'_1, \ldots, i'_n), \pi', (\mu'_1, \ldots, \mu'_n) >$, where $i'_1, \ldots, i'_n, \pi', \mu'_1, \ldots, \mu'_n$ are computed as follows. For every j, $1 \leq j \leq n$, first compute

$$\alpha_j = \begin{cases} S(\pi a, u^j_{i_j}) & \text{if } \mu_j = * \\ S(\mu_j a, u^j_{i_j}) & \text{if } \mu_j \neq * \end{cases} \tag{15}$$

There are two cases:

1. (***local transition***) if there is no j, $1 \leq j \leq n$ such that $\alpha_j = u^j_{i_j}$, then
   (a) $i'_j = i_j$ for all j, $1 \leq j \leq n$,
   (b) π' is the longest string of $\{\alpha_1, \ldots, \alpha_n\}$,
   (c) for every j, $1 \leq j \leq n$,

$$\mu'_j = \begin{cases} * & \text{if } \alpha_j = S(\pi', u^j_{i_j}), \\ \alpha_j & \text{otherwise} \end{cases} \tag{16}$$

2. (***global transition***) if there exists j, $1 \leq j \leq n$ such that $\alpha_j = u^j_{i_j}$, then
   (a) for every j, $1 \leq j \leq n$, $i'_j = \begin{cases} i_j + 1 & \text{if } \alpha_j = u^j_{i_j}, \\ i_j & \text{otherwise} \end{cases}$
   (b) π' is the longest string of $\{\alpha_j \mid 1 \leq j \leq n,\ \alpha_j \neq u^j_{i_j}\}$,
   (c) for every j, $1 \leq j \leq n$,

$$\mu'_j = \begin{cases} \varepsilon & \text{if } \alpha_j = u^j_{i_j}, \\ * & \text{if } \alpha_j \neq u^j_{i_j} \text{ and } \alpha_j = S(\pi', u^j_{i_j}), \\ \alpha_j & \text{otherwise} \end{cases} \tag{17}$$

If a (global) transition results in a state $q =< (i_1, \ldots, i_n), \pi, (\mu_1, \ldots, \mu_n) >$ such that $i_j = n_j + 1$ for some j, $1 \leq j \leq n$, then q is a final state. Every transition from a final state leads to the same state. Clearly, the constructed automaton is deterministic and complete.

Let us explain informally the construction above. As for the Aho-Corasick algorithm, think of the automaton as simulating the process of moving a pointer in every pattern $p_j = u^j_1 x^j_1 u^j_2 \ldots x^j_{n_j - 1} u^j_{n_j}$. At each moment the pointer is located in some $u^j_{i_j}$, and the pointed prefix of $u^j_{i_j}$, say ν_j, is a suffix of the scanned part of the input string. After reading a letter a from the input, the pointer moves one position right if a is the letter which follows ν_j in $u^j_{i_j}$, and moves left or stays at the same position otherwise. In the latter case the pointed prefix becomes equal to the longest prefix of $u^j_{i_j}$ which is a suffix of $\nu_j a$. In both cases, the new pointed prefix is $S(\nu_j a, u^j_{i_j})$. Once the pointer gets to the end of $u^j_{i_j}$, the following word $u^j_{i_j+1}$ is entered, that is the pointer is placed at the beginning of it.

The first component $(i_1, \ldots, i_n)$ of the state of the automaton indicates, for every p_j, the word $u^j_{i_j}$ that the pointer is currently located in. We call i_j *the j-th coordinate* of the state. The second component π is maintained to be the longest among all pointed prefixes. Unlike the Aho-Corasick algorithm, π does not generally determine the pointer position in each of $u^1_{i_1}, \ldots, u^n_{i_n}$ because they have not been generally entered at the same moment. Formally, saying that π determines the pointer position in $u^j_{i_j}$ means that the pointed prefix is equal to $S(\pi, u^j_{i_j})$. In order to keep track of the pointer positions, a third component $(\mu_1, \ldots, \mu_n)$ is added to the state. If π determines the pointer position in $u^j_{i_j}$, then the corresponding μ_j is set to $*$, otherwise μ_j is assigned the pointed prefix of $u^j_{i_j}$. The states of the automaton are defined recursively. To compute a "new" state $q \cdot a$ from a "current" state q, auxiliary words $\alpha_1, \ldots, \alpha_n$ are first computed which correspond exactly to the new pointer positions in $u^1_{i_1}, \ldots, u^n_{i_n}$.

The next two lemmas show the correctness of the construction, i.e. that the automaton recognizes precisely the words reducible by $\mathcal{R}$.

Lemma 3 *Assume that $v \in A^*$ and $q = q_0 \cdot v$. Assume that no proper prefix of v is accepted by the automaton (i.e. either q is not final or it is a final state reached by the automaton for the first time during its run on v).*

(1) If $q = < (i_1, \ldots, i_n), \pi, (\mu_1, \ldots, \mu_n) >$, then for every j, $1 \leq j \leq n$, there exists a decomposition

$$v = \beta^j_1 u^j_1 \beta^j_2 \ldots u^j_{i_j - 1} \beta^j_{i_j}, \quad \beta^j_1, \ldots, \beta^j_{i_j} \in A^* \tag{18}$$

where $\beta^j_1, \beta^j_2, \ldots, \beta^j_{i_j}$ satisfy the following properties

(i) for every k, $1 \leq k \leq i_j - 1$, u^j_k does not occur in $\beta^j_k u^j_k$ as a factor except at the suffix position,

(ii) if $i_j \neq n_j + 1$, then $u^j_{i_j}$ does not occur in $\beta^j_{i_j}$ as a factor,

(iii) if $i_j \neq n_j + 1$, then $S(\beta^j_{i_j}, u^j_{i_j}) = \begin{cases} S(\pi, u^j_{i_j}) & \text{if } \mu_j = * \\ \mu_j & \text{if } \mu_j \neq * \end{cases}$

(2) Conversely, for every j, $1 \leq j \leq n$, let i_j, $1 \leq i_j \leq n_j$, and $\beta^j_1, \ldots, \beta^j_{i_j}$ be such that v admits decomposition (18) that satisfies conditions (i),(ii). Then $q = < (i_1, \ldots, i_n), \pi, (\mu_1, \ldots, \mu_n) >$, where $\pi, \mu_1, \ldots, \mu_n$ verify the following conditions

(iv) π is the longest string of $\{S(\beta^j_{i_j}, u^j_{i_j}) \mid 1 \leq j \leq n\}$,

(v) $\mu_j = \begin{cases} \varepsilon & \text{if } \beta^j_{i_j} = \varepsilon \\ * & \text{if } \beta^j_{i_j} \neq \varepsilon \text{ and } S(\pi, u^j_{i_j}) = S(\beta^j_{i_j}, u^j_{i_j}) \\ S(\beta^j_{i_j}, u^j_{i_j}) & \text{otherwise} \end{cases}$

Proof: First we note that part (2) of the lemma is stated correctly since a decomposition of v satisfying (i),(ii) is unique and therefore i_j's and $\beta^j_{i_j}$'s are well-defined.

We use induction on the length of v.

(1) For $v = \varepsilon$ the lemma trivially holds. Assume that the lemma holds for a word v and $q_0 \cdot v = < (i_1, \ldots, i_n), \pi, (\mu_1, \ldots, \mu_n) >$. Let $a \in A$ and

$q_0 \cdot va =< (i'_1, \ldots, i'_n), \pi', (\mu'_1, \ldots, \mu'_n) >$. We have to show that va can be decomposed according to the lemma where $i_1, \ldots, i_n, \pi, \mu_1, \ldots, \mu_n$ are replaced by $i'_1, \ldots, i'_n, \pi', \mu'_1, \ldots, \mu'_n$ respectively.

Consider the a-transition from $q_0 \cdot v$ to $q_0 \cdot va$ and suppose it is a local transition, i.e. $i'_j = i_j$ for all j. By assumption, $q_0 \cdot v$ is not final and thus $q_0 \cdot va$ is not final either (i.e. $i'_j \neq n_j$ for all j). Take some j, $1 \le j \le n$. By induction hypothesis $v = \beta^j_1 u^j_1 \beta^j_2 \ldots u^j_{i_j - 1} \beta^j_{i_j}$, and conditions (i)-(iii) are verified. Let us show that the decomposition $va = \beta^j_1 u^j_1 \beta^j_2 \ldots u^j_{i_j - 1} \delta^j_{i_j}$, where $\delta^j_{i_j} = \beta^j_{i_j} a$ satisfies the lemma. Condition (i) is trivially verified. By induction hypothesis (condition (iii))

$$S(\beta^j_{i_j}, u^j_{i_j}) = \begin{cases} S(\pi, u^j_{i_j}) & \text{if } \mu_j = * \\ \mu_j & \text{if } \mu_j \neq * \end{cases}$$

This implies that

$$S(\beta^j_{i_j} a, u^j_{i_j}) = \begin{cases} S(\pi a, u^j_{i_j}) & \text{if } \mu_j = * \\ S(\mu_j a, u^j_{i_j}) & \text{if } \mu_j \neq * \end{cases}$$

The expression on the right is exactly α_j defined by (15). Since the transition is local, then α_j is a proper prefix of $u^j_{i_j}$, and therefore $u^j_{i_j}$ is not a suffix of $\beta^j_{i_j} a$. Thus, condition (ii) is also verified. By reading (16) from right to left, we have

$$\alpha_j = \begin{cases} S(\pi', u^j_{i_j}) & \text{if } \mu'_j = * \\ \mu'_j & \text{if } \mu'_j \neq * \end{cases}$$

Thus,

$$S(\delta^j_{i_j}, u^j_{i_j}) = \begin{cases} S(\pi', u^j_{i_j}) & \text{if } \mu'_j = * \\ \mu'_j & \text{if } \mu'_j \neq * \end{cases}$$

which proves condition (iii).

Assume now that the transition under consideration is global. For those j's that satisfy $i_j = i'_j$, the same decomposition and proof as in the case of local transition apply. Consider j such that $\alpha_j = u^j_{i_j}$ and $i'_j = i_j + 1$. Then the decomposition $va = \beta^j_1 u^j_1 \beta^j_2 \ldots u^j_{i_j - 1} \delta^j_{i_j} u^j_{i_j} \beta^j_{i_j + 1}$ satisfies the lemma, where $\delta^j_{i_j} u^j_{i_j} = \beta^j_{i_j} a$ and $\beta^j_{i_j + 1} = \varepsilon$. Note that this decomposition is correct since $S(\beta^j_{i_j} a, u^j_{i_j}) = u^j_{i_j}$, that is $u^j_{i_j}$ is indeed a suffix of $\beta^j_{i_j} a$. Condition (i) of the lemma follows from the induction hypothesis (condition (ii)) that $\beta^j_{i_j}$ does not contain $u^j_{i_j}$ as a factor. Condition (ii) is trivial as $\beta^j_{i_j + 1} = \varepsilon$. Condition (iii) is also trivial as $\mu'_j = \varepsilon$ by (17).

(2) This part can be proved using similar arguments. □

Let $v \in A^*$, $|v| > 0$. For some j, $1 \le j \le n$, consider the decomposition of v according to lemma 3. The remarks below follow from the proof above.

Remark 1 *If q is a current state and the last transition was local, then $S(\beta^j_{i_j}, u^j_{i_j}) = \alpha_j$ where α_j's are computed according to (15) and correspond to the last transition.*

Remark 2 $\beta^j_{i_j} = \varepsilon$ *iff the last transition was global and modified the j-th coordinate of the state from $i_j - 1$ to i_j. Otherwise $|\beta^j_{i_j}|$ is equal to the number of transitions made after that modification.*

Lemma 4 *The language accepted by the automaton described above is $Red(\mathcal{R}) = \bigcup_{j=1}^{n} A^* u^j_1 A^* \ldots A^* u^j_{n_j} A^*$*

Proof: Let $w \in A^*$ be accepted by the automaton. Take the shortest prefix v of w accepted by the automaton. Assume that $q_0 \cdot v = q$, $q = <(i_1, \ldots, i_n), \pi, (\mu_1, \ldots, \mu_n)>$, and $i_j = n_j + 1$ for some j, $1 \le j \le n$. From part (1) of lemma 3 it follows that v can be decomposed as $v = \beta^j_1 u^j_1 \beta^j_2 \ldots \beta^j_{n_j} u^j_{n_j}$ (by remark 2, $\beta^j_{n_j+1} = \varepsilon$). Therefore v is reducible by $p_j = u^j_1 x^j_1 u^j_2 \ldots x^j_{n_j-1} u^j_{n_j}$ and so is w.

If $w \in Red(\mathcal{R})$, take the shortest reducible prefix v of w, and let $p_j = u^j_1 x^j_1 u^j_2 \ldots x^j_{n_j-1} u^j_{n_j}$ be a pattern which applies to v. Find a decomposition $v = \beta^j_1 u^j_1 \beta^j_2 \ldots \beta^j_{n_j} u^j_{n_j}$ such that for every k, $1 \le k \le n_j$, u^j_k does not occur in $\beta^j_k u^j_k$ as a factor except at the suffix position. This decomposition can be obtained by taking iteratively for each k, $1 \le k \le n_j$, the leftmost occurrence of u^j_k which follows the occurrence of u^j_{k-1}. By part 2 of lemma 3, v takes the automaton to a final state, and therefore w is also accepted. □

The following lemma shows that after a bounded number of steps every μ_j gets equal to $*$ unless the j-th coordinate of the state is changed.

Lemma 5 *Let $v \in A^*$ and $q_0 \cdot v = <(i_1, \ldots, i_n), \pi, (\mu_1, \ldots, \mu_n)>$. Assume that $v = \beta^j_1 u^j_1 \beta^j_2 \ldots u^j_{i_j-1} \beta^j_{i_j}$ is the decomposition of v according to lemma 3 for some j, $1 \le j \le n$. Then $|\beta^j_{i_j}| \ge |u^j_{i_j}|$ implies $\mu_j = *$.*

Proof: Since $|\beta^j_{i_j}| > 0$, the last transition did not change the j-th coordinate of the state (remark 2). Together with (16), (17) this implies that proving $\mu_j = *$ amounts to proving $\alpha_j = S(\pi, u^j_{i_j})$ where α_j corresponds to the last transition on the path induced by v. On the other hand, $\alpha_j = S(\beta^j_{i_j}, u^j_{i_j})$ according to remark 1. Hence, we have to prove that $S(\pi, u^j_{i_j}) = S(\beta^j_{i_j}, u^j_{i_j})$.

Recall that both π and $\beta^j_{i_j}$ is a suffix of v. If π is longer than $\beta^j_{i_j}$, then every suffix of $\beta^j_{i_j}$ is also a suffix of π. On the other hand, since $|\beta^j_{i_j}| \ge |u^j_{i_j}|$, every prefix of $u^j_{i_j}$ which is a suffix of π is also a suffix of $\beta^j_{i_j}$. Therefore, $S(\pi, u^j_{i_j}) = S(\beta^j_{i_j}, u^j_{i_j})$. If $\beta^j_{i_j}$ is longer than π, then every suffix of π is also a suffix of $\beta^j_{i_j}$. On the other hand, π is longer than or equal to $S(\beta^j_{i_j}, u^j_{i_j})$ by definition of π. This implies again $S(\pi, u^j_{i_j}) = S(\beta^j_{i_j}, u^j_{i_j})$. □

Now we are in position to establish a bound for the loop-free paths in the automaton.

Lemma 6 *Assume that* $M = \sum_{j=1}^{n} n_j$ *and* $C = \max\{|u^j_{i_j}| \,|\, 1 \leq j \leq n, 1 \leq i_j \leq n_j\}$. *Then the maximal length of a loop-free transition sequence of* $\mathcal{A}$ *is bounded by* $(n+1)MC$.

Proof: Consider an arbitrary loop-free path in the automaton. It is clear that any chain of transitions modifies the first tuple of the state at most $\sum_{j=1}^{n}(n_j - 1) + 1$ times before reaching an accepting state.

Let us fix the tuple of coordinates to $(i_1, \ldots, i_n)$. By lemma 5 and remark 2, after at most C transitions every μ_j gets equal to $*$ and keeps this value unless i_j is modified. As soon as both the first and the third component is fixed, every state is uniquely associated with the value of π. Since π is a prefix of some word of $u^1_{i_1}, \ldots, u^n_{i_n}$, there are at most nC such states.

To sum up, the length of a loop-free path in the automaton is bounded by $M(C + nC) = (n+1)MC$. □

Thus, the length of a loop-free path in a deterministic automaton which recognizes $Red(\mathcal{R})$ ($NF(\mathcal{R})$) can be bounded polynomially (quadratically) on $|\mathcal{R}|$. In conclusion, we obtain

Lemma 7 *Testing ground reducibility of a linear subject pattern by a linear WRSV is in co-NP.*

Finally, lemmas 2 and 7 prove the main result.

Theorem 2 *Testing ground reducibility of a linear subject pattern by a linear WRSV is co-NP-complete.*

4.2 Complexity of Testing Finiteness of $NF(\mathcal{R})$

We use the technique of the previous section to show that if a WRSV $\mathcal{R}$ is restricted to be linear, the problem of finiteness of the set $NF(\mathcal{R})$ of irreducible words is also *co-NP*-complete.

Assume we are given a linear WRSV $\mathcal{R}$. From lemma 6 the length of all loop-free paths in the automaton $\mathcal{A}$ constructed in the previous section is bounded by a polynomial $p(|\mathcal{R}|)$. This implies that $p(|\mathcal{R}|) - 1$ bounds the length of irreducible words in the case when their number is finite. Conversely, if every word of length $p(|\mathcal{R}|)$ is reducible, then this is trivially the case for all longer words. Thus, to test nondeterministically if $NF(\mathcal{R})$ is infinite, guess a word of the length $p(|\mathcal{R}|)$ and check if it is irreducible. This proves that testing finiteness of $NF(\mathcal{R})$ is in *co-NP*.

Now we prove that the problem is complete for *co-NP*.

Lemma 8 *Testing finiteness of* $NF(\mathcal{R})$ *for a linear WRSV* $\mathcal{R}$ *is co-NP-complete.*

Proof: By the remark above it remains to show that the problem is *co-NP*-hard. We encode a formula $\mathcal{C} = \mathcal{C}_1 \wedge \mathcal{C}_2 \wedge \ldots \wedge \mathcal{C}_m$ into a WRSV $\mathcal{R}$ over the alphabet $A = \{1, \ldots, n\} \cup \{\mathtt{t}, \mathtt{f}\} \cup \{\#\}$ and a variable set $\mathcal{X} = \{x, \ldots\}$. For technical reasons we assume that all $\mathcal{C}_i$ are different.

We first modify the WRSV constructed in the proof of lemma 2. We replace pattern 2 by the patterns

$$\mathtt{t}\#, \mathtt{f}\# \tag{19}$$

$$\#i, \quad \text{for all } 1 \leq i \leq n \tag{20}$$

We add further the patterns

$$\#\{\mathtt{t}, \mathtt{f}\}i\#, \quad \#\{\mathtt{t}, \mathtt{f}\}i\{\mathtt{t}, \mathtt{f}\}j\#, \quad \text{for all } 1 \leq i, j \leq n \tag{21}$$

It should be clear that words that remain irreducible are factors of words from the regular language $((\{\mathtt{t}, \mathtt{f}\}\{1, \ldots, n\})^3\#)^*$.

Assume that $\mathbf{C}_i$ encodes $\mathcal{C}_i$ in the same way as in the proof of lemma 2. Define $C_i \subseteq (\{\mathtt{t}, \mathtt{f}\}\{1, \ldots, n\})^3$ to be the set of all instances of $\mathcal{C}_i$ which can be obtained by applying some truth assignement. We add to $\mathcal{R}$ all the words from

$$(\{\mathtt{t}, \mathtt{f}\}\{1, \ldots, n\})^3 \setminus \bigcup_{i=1}^{n} C_i \tag{22}$$

Now, every 6-letter factor of an irreducible word occurring between two #'s belongs to some C_i. Finally, let $\bar{C}_i = (\{\mathtt{t}, \mathtt{f}\}\{1, \ldots, n\})^3 \setminus C_i$. We add the patterns

$$v\#v', \quad \text{for all } 1 \leq i \leq n-1, \ v \in C_i, \ v' \in \bar{C}_{i+1} \tag{23}$$

$$v\#v', \quad \text{for all } v \in C_n, \ v' \in \bar{C}_1 \tag{24}$$

If the set of irreducible words w.r.t. the constructed WRSV is infinite, then every sufficiently long irreducible word contains a factor $\#v_1\#v_2\#\ldots\#v_n\#$ that encodes a solution of the MONOTONE-ONE-IN-THREE-SAT problem for $\mathcal{C}$. Conversely, if $\#v_1\#v_2\#\ldots\#v_n\#$ encodes a solution of $\mathcal{C}$, then all the words $(v_1\#v_2\#\ldots\#v_n\#)^*$ are irreducible. We conclude that $NF(\mathcal{R})$ is infinite if and only if $\mathcal{C}$ has a solution, and therefore testing finiteness of $NF(\mathcal{R})$ for a given $\mathcal{R}$ is *co-NP*-hard. □

5 Remarks and Related Works

Note that theorem 2 remains valid even if the subject pattern w is assumed to be fixed. Necessary modifications of the proof of lemma 2 are suggested by the proof of lemma 8.

The ground reducibility problem we have considered in section 4.1 is the inclusion problem for regular languages represented by regular expressions of a particular class. The inclusion problem for general regular languages represented by regular expressions is *PSPACE*-complete [3]. Various complexity results for formal language theory can be found in [5, 4]. For example, it is proven that the inclusion of regular languages $L_1 \subseteq L_2$ remains $PSPACE$-complete even if L_1 is fixed. On the other hand, we are unaware about results on complexity of language inclusion (equivalence) for subclasses of regular languages similar to the one considered in this paper.

Recently we proposed an efficient algorithm for testing the reducibility of a word with respect to a linear WRSV [8]. This problem is equivalent to a string matching problem for a specific set of patterns (strings with *variable length don't-care symbols*), and has various practical applications.

Acknowledgements: We are grateful to Paliath Narendran for giving an initial impulse to this study and to Valentin Antimirov for enlightening discussions about Example 1.

References

1. A. V. Aho. Algorithms for finding patterns in strings. In J. van Leeuwen, editor, *Handbook of Theoretical Computer Science*. Elsevier Science Publishers B. V. (North-Holland), 1990.
2. N. Dershowitz and J.-P. Jouannaud. Rewrite systems. In J. van Leeuwen, editor, *Handbook of Theoretical Computer Science*. Elsevier Science Publishers B. V. (North-Holland), 1990.
3. M. Garey and D. Johnson. *Computers and Intractability. A guide to the theory of NP-completeness.* W. Freeman and Compagny, New York, 1979.
4. Harry B. Hunt III and Daniel J. Rosenkrantz. Computational parallels between the regular and context-free languages. *Theoretical Computer Science*, 7(1):99–114, February 1978.
5. Harry B. Hunt III, Daniel J. Rosenkrantz, and Thomas G. Szymanski. On the equivalence, containment, and covering problems for the regular and context-free languages. *Journal of Computer and System Sciences*, 12:222–268, 1976.
6. D. Kapur, P. Narendran, D. Rosenkrantz, and H. Zhang. Sufficient-completeness, ground-reducibility and their complexity. *Acta Informatica*, 28:311–350, 1991.
7. D. Kapur, P. Narendran, and H. Zhang. On sufficient completeness and related properties of term rewriting systems. *Acta Informatica*, 24:395–415, 1987.
8. G. Kucherov and M. Rusinowitch. Matching a set of strings with variable length don't cares. In E. Ukkonen, editor, *Proceedings of the 6th Symposium on Combinatorial Pattern Matching*, Helsinki, July 1995. to appear in Lect. Notes Comput. Sci. Series.
9. G. Kucherov and M. Rusinowitch. Undecidability of ground reducibility for word rewriting systems with variables. *Information Processing Letters*, 53:209–215, 1995.
10. G. Kucherov and M. Tajine. Decidability of regularity and related properties of ground normal form languages. *Information and Computation*, 117, 1995. to appear.
11. S.S. Marchenko. Undecidability of the positive ∀∃-theory of a free semigroup. *Sibirskii Matematicheskii Zhurnal*, 23(1):196–198, 1982. in Russian.
12. D. Plaisted. Semantic confluence and completion method. *Information and Control*, 65:182–215, 1985.

Coherence for Cartesian Closed Categories: A Sequential Approach

Akira Mori and Yoshihiro Matsumoto

Kyoto University, Department of Information Science, Kyoto 606-01, Japan

Abstract. A coherence theorem states that the arrows between two particular objects in free categories are unique. In this paper, we give a direct proof for cartesian closed categories (CCC's) without passing to typed lambda calculus. We first derive categorical combinators for CCC's together with their equations directly from the adjoint functors defining CCC's. Then categorical interpretation of the intuitionistic sequent calculus is regarded as the construction of free CCC's. Each arrow is generated along the proof derivation in form of categorical combinators. The system enjoys the cut-elimination theorem and we can use various proof-theoretic techniques such as Kleene's permutability theorem. The coherence is proved by showing that the reconstruction of derivations for the given class of arrows is deterministic and unique up to equivalence.

1 Introduction

A coherence theorem states that the particular class of diagrams in free categories commute, or in other words that the arrows between two particular objects are equivalent. In order to prove such a theorem, one must cope with two different problems in nature:

- enumeration problem: how to enumerate all the arrows $A \to B$,
- decision problem: how to decide when two arrows are equivalent.

The first problem requires an exhaustive search procedure while the second one requires a decision procedure for equivalence. Considering the relationship "natural deduction : typed lambda calculi = sequent calculus : free categories", we notice that the issue spreads over two different formalisms: the sequent calculus formalism and the natural deduction formalism. One should pick the sequent calculus formalism for the first problem since arrow enumeration is nothing but proof search in intuitionistic logic. The cut-elimination theorem and the subformula property readily yield an exhaustive search procedure. However, for the second problem, one should pick the natural deduction formalism since the normalization theorem immediately gives a decision procedure. Therefore, the usual proofs of coherence theorem proceed through a detour involving the translation from categorical arrows to deductive terms and back [Lambek, 1969, Babaev and Soloviev, 1979, Mints, 1992b, Jay, 1990].

In this paper, we take the first problem seriously and give a direct proof of a coherence problem for cartesian closed categories (CCC's).

Our coherence problem is in a weak form compared to [Mints, 1992b, Babaev and Soloviev, 1979] since the particular isomorphisms are excluded, however, the proof does not involve intertranslation between typed lambda terms and categorical arrows. It is based on the sequential generation of free CCC's via categorical combinators [Curien, 1986], and on proof-theoretic techniques such as Kleene's permutability theorem [Kleene, 1952]. Although many decidable reduction systems underlying CCC's have been proposed, the relation to the enumeration problem is unexplored. We maintain uniqueness of derivations during the search process and do not count on any reduction system.

For the preliminaries, we first show how an equational system of categorical combinators for CCC's which is similar to Curien's system [Curien, 1986] is derived systematically from the basic concepts of category theory. Categorical combinators appear as *units* and *counits* of adjoint functors defining CCC's, and their equations directly come from the definitions of categories, functors and natural transformations, and from the *triangular identities* of adjoint functors. Following the ideas of Lawvere's *hyperdoctrines* [Lawvere, 1969, Lawvere, 1970] and Lambek's *multicategories* [Lambek, 1989], we obtain a sequential system in which all arrows of free CCC's are constructed along the proof derivations. The system is shown to be cut-free in the sense of Gentzen respecting the equations of combinators. This allows us to use various proof-theoretic techniques such as Kleene's permutability theorem [Kleene, 1952] that guarantees certain permutation of the inference rules used in the proof derivation.

Then we relate the coherence problem to the uniqueness of derivations. The coherence problem for CCC's is concerned with the so-called *balanced condition* to avoid infinite search. Using Kleene's permutability theorem, we show that the contraction rule is unnecessary under the balanced condition and that derivations can be reconstructed in a deterministic manner. We manage cut/contraction-free proof search maintaining uniqueness of derivations, which completes the proof of the coherence theorem.

We do not claim to have solved an open problem, but we believe that the proof-theoretic methods borrowed from the sequent calculus have potential for the categorical study of operational semantics such as the CAM (Categorical Abstract Machines) [Cousineau *et al.*, 1985]. We hope that the equational aspects of categories stressed in this paper suggests a new application of rewriting method in this direction.

The reader is referred to [Barr and Wells, 1990, Mac Lane, 1971, McLarty, 1992] and [Girard *et al.*, 1989, Gallier, 1993] for the details of category theory and proof-theoretic matters of intuitionistic sequent calculi, respectively.

2 Equational Presentation of Categories

In this section, we show how basic concepts of category theory such as functors, natural transformation and adjoint functors lead to an equational presentation of categories. We try to make this section self-contained.

First of all, the word "equational" is critical since the theory of categories is not equational in the usual sense. Arrows are indexed by pairs of objects and composition of arrows are only defined for the pair of arrows where the one's codomain meets the other's domain. Strictly speaking, it is an equational theory with dependent types, or a conditionally equational theory. In this paper, we stick to the diagrammatic reasoning and assume that an equation is regarded as the commutative diagram equipped with complete information about arrows and objects involved.

We will use ";" for arrow composition, whose order is *diagrammatic* and opposite to that of the usual "o".

Categoricity A category consists of a collection of objects and a collection of arrows. Each arrow is associated with two objects as its domain and codomain. An arrow f with domain A and codomain B is written as $f : A \to B$. A pair of arrows is said to be *composable* when one's codomain is identical to the other's domain. For $f : A \to B$ and $g : B \to C$, we have a composite $f ; g : A \to C$. And for each object A, there is an identity arrow $\mathbf{id}_A : A \to A$. Composition of arrows is required to satisfy the associativity and identity laws, which are expressed by the following equations.

$$f ; (g ; h) = (f ; g) ; h, \tag{1}$$

$$\mathbf{id} ; f = f, \tag{2}$$

$$f ; \mathbf{id} = f. \tag{3}$$

Functoriality A functor $F : \mathbf{C} \to \mathbf{D}$ is a mapping between categories which sends an object A of $\mathbf{C}$ to an object $F(A)$ of $\mathbf{D}$ and an arrow $f : A \to B$ of $\mathbf{C}$ to an arrow $F(f) : F(A) \to F(B)$ of $\mathbf{D}$. A functor preserves identities and composition and this is expressed by the following equations.

$$F(\mathbf{id}_A) = \mathbf{id}_{F(A)}, \tag{4}$$

$$F(f ; g) = F(f) ; F(g). \tag{5}$$

Naturality Given a pair of functors $F, G : \mathbf{C} \to \mathbf{D}$, a natural transformation τ from F to G is a family of arrows $\tau_A : F(A) \to G(A)$ in $\mathbf{D}$ assigned to each object A of $\mathbf{C}$, such that for every arrow $f : A \to B$ of $\mathbf{C}$ the following diagram commutes.

$$\begin{array}{ccc} F(A) & \xrightarrow{\tau_A} & G(A) \\ {\scriptstyle F(f)}\downarrow & & \downarrow{\scriptstyle G(f)} \\ F(B) & \xrightarrow{\tau_B} & G(B) \end{array}$$

Thus we have the following equation.

$$\tau_A ; G(f) = F(f) ; \tau_B. \tag{6}$$

Adjunction Adjoint functors are defined in the situation called *adjunctions*. An adjunction consists of 1) a pair of categories $\mathbf{C}$ and $\mathbf{D}$, 2) a pair of functors $F : \mathbf{C} \to \mathbf{D}$ and $G : \mathbf{D} \to \mathbf{C}$, and 3) a pair of natural transformations $\eta : I_{\mathbf{C}} \to GF$ and $\epsilon : FG \to I_{\mathbf{D}}$, such that the following triangular identities of natural transformations hold. ($I_{\mathbf{C}}$ is an identity functor on the category $\mathbf{C}$ and $\mathbf{id}_F$ is an identity natural transformation on the functor F.)

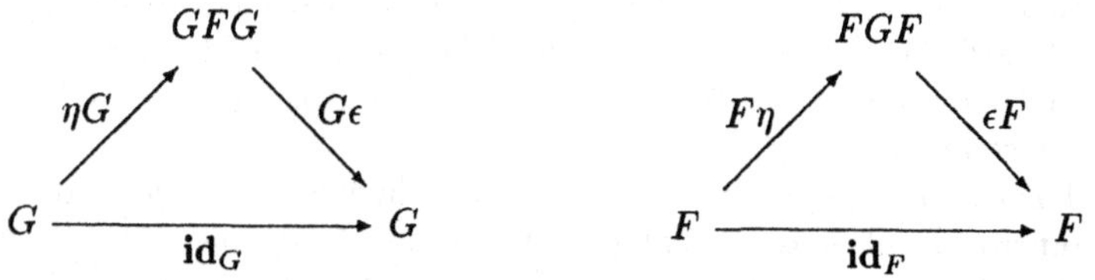

The functor F is called left adjoint to G and the functor G is called right adjoint to F. The natural transformation η is called the unit and the natural transformation ϵ is called the counit. The natural transformations ηG and $G\epsilon$ consist of families of $\mathbf{C}$–arrows $\langle \eta_{G(B)} : G(B) \to GF(G(B)) \rangle$ and $\langle G(\epsilon_B) : G(FG(B)) \to G(B) \rangle$ respectively for each $\mathbf{D}$–object B, both of which are guaranteed to be natural transformations. The same goes for $\epsilon F : FG(F(A)) \to F(A)$ and $F\eta : F(A) \to F(GF(A))$.

After all, we have the following equations for an adjoint pair of functors F and G above. One must read subscripts carefully.

$$\eta_{G(B)} \,;\, G(\epsilon_B) = \mathbf{id}_{G(B)}, \tag{7}$$

$$F(\eta_A) \,;\, \epsilon_{F(A)} = \mathbf{id}_{F(A)}. \tag{8}$$

These are all we need to describe categorical combinators of CCC's and the construction of free CCC's. Once the instances of functors, natural transformations and adjoint functors are provided, we can obtain their equational descriptions from the equations above.

3 Categorical Combinators for CCC's

In this section, we derive categorical combinators for CCC's and their equations from basic ingredients of category theory as in the previous section. A CCC is a category having the specified terminal object 1, binary products $\times$ and exponentiations $\supset$. It is well known that these can be defined in terms of adjoint functors.

Definition 1 (adjoint functors defining CCC's). A CCC $\mathbf{C}$ is defined by the following functors.

- The product functor $(-) \times (-) : \mathbf{C} \times \mathbf{C} \to \mathbf{C}$, which is right adjoint to the diagonal functor $\Delta : \mathbf{C} \to \mathbf{C} \times \mathbf{C}$. The symbol $\times$ in the expression $\mathbf{C} \times \mathbf{C}$ stands for the product of categories. The product category $\mathbf{C} \times \mathbf{D}$

is a category whose objects are pairs (C, D) of $\mathbf{C}$–object C and $\mathbf{D}$–object D, and in which an arrow from (C, D) to (C', D') is a pair (f, g) of $\mathbf{C}$–arrow $f : C \to C'$ and $\mathbf{D}$–arrow $g : D \to D'$. The identity and composition are defined in a trivial manner. The functor Δ takes an object C to the pair (C, C) and an arrow $f : C \to C'$ to the pair (f, f). The unit of this adjunction is $\eta_A : A \to A \times A$ and the counit is $\epsilon_{(A,B)} : (A \times B, A \times B) \to (A, B)$.

- The exponential functor $C \supset (-) : \mathbf{C} \to \mathbf{C}$, which is right adjoint to the right product functor $(-) \times C : \mathbf{C} \to \mathbf{C}$ for each $\mathbf{C}$–object C. The functor $(-) \times C$ takes an object A to the object $A \times C$ and an arrow $f : A \to B$ to the arrow $f \times \mathbf{id}_C : A \times C \to B \times C$. The unit is $\eta_A : A \to (C \supset A \times C)$ and the counit is $\epsilon_A : (C \supset A) \times C \to A$.
- The terminal object functor $! : \mathbf{1} \to \mathbf{C}$, which is right adjoint to the constant functor $S : \mathbf{C} \to \mathbf{1}$. The category $\mathbf{1}$ consists of only one object T and only one identity arrow $\mathbf{id}_T$. The unit is $\eta_A : A \to !(T)$ and the counit is $\epsilon_T : T \to T$ which is nothing but $\mathbf{id}_T$. Since $\mathbf{1}$ has only one object, we can identify the functor ! with a $\mathbf{C}$–object $!(T)$, which is to be a terminal object 1 of $\mathbf{C}$. □

As is clearly seen, the categorical combinators for CCC's appear as units and counits of these adjunctions and from Equation (1)–(8), we immediately obtain a set of equations for combinators. All we need is 1) category axioms, 2) functoriality of adjoint functors, 3) naturality of units and counits, and 4) triangular identities of adjunctions.

We assume by default that the operators bind with the following order of strength: $(-) \times (-) \geq (-) \supset (-) \geq (-) ; (-)$.

Definition 2 (categorical combinators for CCC's). The categorical combinators for CCC's and their equations are as follows.

$\mathbf{id}_A : A \to A$ (identity),
$\mathbf{dupl}_A : A \to A \times A$ (duplicator),
$\mathbf{fst}_{A,B} : A \times B \to A$ (first projection),
$\mathbf{snd}_{A,B} : A \times B \to B$ (second projection),
$\mathbf{mark}_{A,C} : A \to C \supset A \times C$, (place-marker)
$\mathbf{app}_{A,C} : (C \supset A) \times C \to A$, (application)
$\mathbf{kill}_A : A \to 1$ (killer).

<u>Categoricity</u>

$f ; (g ; h) = (f ; g) ; h$,
$\mathbf{id} ; f = f$,
$f ; \mathbf{id} = f$.

<u>Functoriality</u>

$f \times g ; h \times k = (f ; h) \times (g ; k)$,
$\mathbf{id} \times \mathbf{id} = \mathbf{id}$,
$\mathbf{id} \supset (f ; g) = \mathbf{id} \supset f ; \mathbf{id} \supset g$,
$\mathbf{id} \supset \mathbf{id} = \mathbf{id}$.

<u>Naturality</u>

$f ; \mathbf{dupl} = \mathbf{dupl} ; f \times f$,
$f \times g ; \mathbf{fst} = \mathbf{fst} ; f$,
$f \times g ; \mathbf{snd} = \mathbf{snd} ; g$,
$f ; \mathbf{mark} = \mathbf{mark} ; \mathbf{id} \supset (f \times \mathbf{id})$,
$(\mathbf{id} \supset f) \times \mathbf{id} ; \mathbf{app} = \mathbf{app} ; f$,
$f ; \mathbf{kill} = \mathbf{kill}$.

<u>Adjunctions</u>

$\mathbf{dupl} ; \mathbf{fst} \times \mathbf{snd} = \mathbf{id}$,
$\mathbf{dupl} ; \mathbf{fst} = \mathbf{id}$,
$\mathbf{dupl} ; \mathbf{snd} = \mathbf{id}$,
$\mathbf{mark} \times \mathbf{id} ; \mathbf{app} = \mathbf{id}$,
$\mathbf{mark} ; \mathbf{id} \supset \mathbf{app} = \mathbf{id}$,
$\mathbf{kill} = \mathbf{id}$.

The subscripts are omitted. For instance, the equation $\mathbf{mark}\,;\mathbf{id} \supset \mathbf{app} = \mathbf{id}$ should read $\mathbf{mark}_{C \supset A, C}\,;\mathbf{id}_C \supset \mathbf{app}_{A,C} = \mathbf{id}_{C \supset A}$. □

Although these combinators slightly differ from the ones by Curien [Curien, 1986] in that they do not have constructors for combinators, it is not difficult to see both systems are equivalent. Just take $\mathbf{dupl}\,; f \times g$ for $\langle f, g \rangle$ and $\mathbf{mark}\,; \mathbf{id} \supset f$ for $\Lambda(f)$, and conversely $\langle \mathbf{id}_A, \mathbf{id}_A \rangle$ for $\mathbf{dupl}_A$ and $\Lambda(\mathbf{id}_{A \times C})$ for $\mathbf{mark}_{A,C}$.

4 Free CCC

A free CCC is a category freely generated from a set of generating objects so as to satisfy the conditions imposed on CCC's. Let **Set** denote the category of small sets and functions, and **CCC** the category of small CCC's and functors preserving cartesian closed structures. A free CCC generated from a set A is abstractly defined to be a category $F(A)$ where $F : \mathbf{Set} \to \mathbf{CCC}$ is left adjoint to the forgetful functor $U : \mathbf{CCC} \to \mathbf{Set}$ that maps a CCC to its set of objects.

4.1 Functorial Generation

We give a construction of the free CCC $F(A)$ through the right adjoint functors in Definition 1. The elements of A are called *atoms* and the set of objects of $F(A)$ is denoted by $|F(A)|$. Note that the left adjoint functors Δ, $(-) \times A$, $!$ have no effect on the free generation.

Definition 3 (free CCC). $|F(A)|$ is defined inductively as follows.

1. $A \subset |F(A)|$
2. $1 \in |F(A)|$
3. $X \times Y \in |F(A)|$ and $X \supset Y \in |F(A)|$ if $X, Y \in |F(A)|$.

The arrow expressions of $F(A)$ are defined inductively as follows.

1. Each combinator whose domain and codomain belong to $|F(A)|$ is an arrow expression.
2. $f\,; g : X \to Z$ is an arrow expression if $f : X \to Y$ and $g : Y \to Z$ are both arrow expressions.
3. $f \times g : X \times S \to Y \times T$ is an arrow expression if $f : X \to Y$ and $g : S \to T$ are both arrow expressions.
4. $\mathbf{id}_X \supset f : X \supset Y \to X \supset Z$ is an arrow expression if $X \in |F(A)|$ and $f : Y \to Z$ is an arrow expression.

Then the arrow expressions of $F(A)$ are turned into the arrows of $F(A)$ (called canonical arrows) by picking the smallest equivalence relation $\equiv$ defined inductively by the following inference rules with the equations in Definition 2 (regarding $=$ as $\equiv$) and $f \equiv f$ as axioms.

$$\frac{f \equiv g}{g \equiv f} \quad \frac{f \equiv g \quad f \equiv h}{g \equiv h} \quad \frac{f \equiv g \quad h \equiv k}{f \mathbin{;} h \equiv g \mathbin{;} k} \quad \frac{f \equiv g}{\mathbf{id} \supset f \equiv \mathbf{id} \supset g} \quad \frac{f \equiv g \quad h \equiv k}{f \times h \equiv g \times k}$$

□

We will not distinguish notationally arrow expressions and canonical arrows. One should note again that the above is not an ordinary equational system since composition is only defined for the composable pair of arrows.

We did not include $f \supset g : Y \supset S \to X \supset T$ as an arrow expression for $f : X \to Y$ and $g : S \to T$. $(-) \supset (-)$ is not defined as a binary operation but just parameterized in the first argument. If we define, however,

$$f \supset g \triangleq \mathbf{mark} \mathbin{;} \mathbf{id} \supset (\mathbf{id} \times f \mathbin{;} \mathbf{app} \mathbin{;} g),$$

$(-) \supset (-)$ becomes a bi-functor $\mathbf{C}^{\mathrm{op}} \times \mathbf{C} \to \mathbf{C}$ which is *contravariant* in the first argument.

4.2 Sequential Generation of Free CCC's

We are going to present sequential construction of free CCC's based on the categorical interpretation of the intuitionistic sequent calculus $\mathcal{LJ}^{\wedge,\Rightarrow,I}$ (see Appendix A), whose logical operations are conjunction $\wedge$, implication $\Rightarrow$ and the absolute truth I.

The construction is twofold. First, following the idea of Lawvere's *hyperdoctrines* [Lawvere, 1969, Lawvere, 1970] which interpret logical operations by adjoint functors, we relate adjunctions and categorical combinators to the logical inference rules of the sequent calculus. Implication, conjunction and the absolute truth are interpreted by the exponential, product, and terminal object functors respectively. Then, following the idea of Lambek's *multicategories* [Lambek, 1989] which identify sequent calculi with free categories, we obtain a system in which all canonical arrows are constructed along the derivations.

Definition 4 (categorical interpretation $\mathcal{LJ}^{CCC}$). The symmetric monoidal presence of the product $\times$ is made explicit by writing $\otimes$, which is used to interpret the multiset of formulas in the lefthand side of a sequent (called the *context*). The distinction between $\times$ and $\otimes$ is purely operational. In terms of the typed lambda calculus, $x : A,\ y : B \vdash \langle x, y\rangle : A \times B$ (surjective pairing with $x : A$ and $y : B$) and $z : A \times B \vdash z : A \times B$ are denotationally the same but operationally different since the latter does not accept the substitution for its A-component. The distinction has been already made in the construction of a term and should be present in the derivation. As a result, the combinators **dupl**, **fst** and **snd** are divided. We use $\mathbf{Dupl}_A : A \to A \otimes A$ and $\mathbf{Fst}_{A,B} : A \otimes B \to A$, $\mathbf{Snd}_{A,B} : A \otimes B \to B$ for $\otimes$, which are used to interpret structural rules.

The domain of an arrow is factorized by $\otimes$ and each factor will be called the *component*. In the following, A, B and C denote components and E in the structural rule and the left logical rule may not be present. The symmetric monoidal isomorphisms of $\otimes$ can be used freely in the derivation. They are natural isomorphisms and no confusion is caused as we have explicit structural rules.

identity axioms: (A:atom)

$$A \xrightarrow{\mathbf{id}} A$$

cut:

$$\frac{D \xrightarrow{f} A \quad A \otimes E \xrightarrow{g} B}{D \otimes E \xrightarrow{f \otimes \mathbf{id}\,;\, g} B}\,(\mathsf{cut})$$

structural rules:

$$\frac{E \xrightarrow{f} B}{A \otimes E \xrightarrow{\mathbf{Snd}\,;\, f} B}\,(\mathsf{weakening}) \qquad \frac{(A \otimes A) \otimes E \xrightarrow{f} B}{A \otimes E \xrightarrow{\mathbf{Dupl} \otimes \mathbf{id}\,;\, f} B}\,(\mathsf{contraction})$$

logical rules:

$$\frac{A \otimes E \xrightarrow{f} C}{(A \times B) \otimes E \xrightarrow{\mathbf{fst} \otimes \mathbf{id}\,;\, f} C}\,(\times\mathsf{left1})$$

$$\frac{B \otimes E \xrightarrow{f} C}{(A \times B) \otimes E \xrightarrow{\mathbf{snd} \otimes \mathbf{id}\,;\, f} C}\,(\times\mathsf{left2})$$

$$\frac{E \xrightarrow{f} A \quad E \xrightarrow{g} B}{E \xrightarrow{\mathbf{dupl}\,;\, f \times g} A \times B}\,(\times\mathsf{right})$$

$$\frac{D \xrightarrow{f} A \quad B \otimes E \xrightarrow{g} C}{((A \supset B) \otimes D) \otimes E \xrightarrow{(\mathbf{id} \otimes f;\mathbf{app}) \otimes \mathbf{id};g} C}\,(\supset\mathsf{left}) \qquad \frac{E \otimes A \xrightarrow{f} B}{E \xrightarrow{\mathbf{mark}\,;\, \mathbf{id} \supset f} A \supset B}\,(\supset\mathsf{right})$$

$$E \xrightarrow{\mathbf{kill}} 1 \quad (\mathsf{1right})$$

□

$\mathcal{LJ}^{CCC}$ is essentially a categorical-combinatoric representation of the proof derivations in $\mathcal{LJ}^{\wedge,\Rightarrow,I}$, which parallels the relationship between natural deduction and typed lambda calculi.

The interpretation of logical rules can be schematized using abstract notations of functors and natural transformations as follows:

$$\frac{A \otimes E \xrightarrow{g} B}{FG(A) \otimes E \xrightarrow{\epsilon_A \otimes \mathbf{id}} A \otimes E \xrightarrow{g} B}\,G\ \mathsf{left} \qquad \frac{F(A) \xrightarrow{f} B}{A \xrightarrow{\eta_A} GF(A) \xrightarrow{G(f)} G(B)}\,G\ \mathsf{right}$$

The logical operation is being interpreted by a functor $G : \mathbf{D} \to \mathbf{C}$ which is right adjoint to $F : \mathbf{C} \to \mathbf{D}$. η and ϵ denote the unit and counit respectively. The left schema generates a $\mathbf{D}$–arrow from a $\mathbf{D}$–arrow, while the right schema generates a $\mathbf{C}$–arrows from a $\mathbf{D}$–arrow. There is a slight deviation with $\supset$left due to the fact that the object C of the exponential $C \supset (-)$ is a parameter.

Now we regard $\mathcal{LJ}^{CCC}$ as sequential generation of free CCC's, whose generating objects are the ones introduced by identity axioms and whose arrows are the equivalence classes of derived arrow expressions.

Proposition 5. *$\mathcal{LJ}^{CCC}$ generates all the canonical arrows defined in Definition 3, within the choice of $\times$ and $\otimes$, along derivations.*

The proof is a simple induction. For example, the arrows $f \times g$, and $f \supset g$ can be derived using left and right logical rules consecutively.

$$f \times g \equiv \mathbf{dupl} \,;\, (\mathbf{fst} \,;\, f) \times (\mathbf{snd} \,;\, g), \mathbf{id} \supset f \equiv \mathbf{mark} \,;\, \mathbf{id} \supset (\mathbf{app} \,;\, f).$$

In terms of the schemas before, the $G(g)$ is derived in the following way.

$$\dfrac{\dfrac{A \xrightarrow{\ g\ } B}{FG(A) \xrightarrow{\epsilon_A \,;\, g} B}\ G\text{ left}}{G(A) \xrightarrow{\eta_{G(A)}} GFG(A) \xrightarrow{G(\epsilon_A \,;\, g)} G(B)}\ G\text{ right} \quad ,$$

$$\begin{aligned} \eta_{G(A)} \,;\, G(\epsilon_A \,;\, g) &\equiv \eta_{G(A)} \,;\, G(\epsilon_A) \,;\, G(g) \\ &\equiv G(g). \end{aligned}$$

Now we are allowed to concentrate on $\mathcal{LJ}^{CCC}$ exploiting various results of proof theory. Most importantly, the cut-elimination theorem holds for $\mathcal{LJ}^{CCC}$ just as for $\mathcal{LJ}^{\wedge,\Rightarrow,I}$, but respecting the equivalence relation between categorical combinators in Definition 3.

Theorem 6 (cut-elimination theorem). *There is an algorithm which, given any derivation $A \xrightarrow{f} B$ in $\mathcal{LJ}^{CCC}$ produces a cut-free derivation $A \xrightarrow{g} B$ in $\mathcal{LJ}^{CCC}$ such that $g \equiv f$.*

The proof is done by the well-known double induction taking care of equivalence of the derivations. We are based on [Dummett, 1977] for the proof for the intuitionistic systems.

Suppose that we have an application of the cut rule, say

$$\dfrac{\begin{matrix}\vdots \\ D \xrightarrow{\ f\ } A\end{matrix} \qquad \begin{matrix}\vdots \\ A \otimes E \xrightarrow{\ g\ } B\end{matrix}}{D \otimes E \xrightarrow{f \otimes \mathbf{id} \,;\, g} B}\ \text{cut}$$

We call an application of the cut rule *trivial* if either of the premises is an identity axioms, or the second premise is 1right. A non-trivial application of the cut rule is called *principal* if the component A is introduced into the premises of the cut rule both by way of a logical rule.

The cut-elimination algorithm consists of three processes: 1) eliminating trivial cuts, 2) moving non-principal cuts upward until they become principal, and 3) transforming principal cuts to the ones which operates on the smaller components.

trivial cut The cut with an identity axiom is replaced by the other premise, and the cut into 1right is replaced by another 1right in view of the naturality of **kill**.

non-principal cut If the rule deriving the left premise f is not a right logical rule introducing A, it must be one of an identity axiom, a structural rule or a left logical rule. If is is an identity axiom, the cut is trivial. In other cases, the effect of the rule is to have f as the composite $h\,;f'$ where f' is the premise of the rule for f. The rule can be delayed so that the cut with the premise f' comes first. The equivalence of the derivations is ensured by the naturality of $\otimes$ and the associativity of composition. For example,

$$\frac{\dfrac{D \xrightarrow{f} A' \qquad B'\otimes D' \xrightarrow{g} A\supset B}{((A'\supset B')\otimes D)\otimes D' \xrightarrow{(\mathbf{id}\otimes f\,;\,\mathbf{app})\otimes \mathbf{id}\,;\,g} A\supset B}\,(\supset\mathsf{left}) \qquad (A\supset B)\otimes E \xrightarrow{h} C}{(((A'\supset B')\otimes D)\otimes D')\otimes E \xrightarrow{((\mathbf{id}\otimes f\,;\,\mathbf{app})\otimes \mathbf{id}\,;\,g)\otimes \mathbf{id}\,;\,h} C}\,(\mathsf{cut})$$

$$\Downarrow$$

$$\frac{D \xrightarrow{f} A' \qquad \dfrac{B'\otimes D' \xrightarrow{g} A\supset B \qquad (A\supset B)\otimes E \xrightarrow{h} C}{(B'\otimes D')\otimes E \xrightarrow{g\otimes \mathbf{id}\,;\,h} C}\,(\mathsf{cut})}{(((A'\supset B')\otimes D)\otimes D')\otimes E \xrightarrow{((\mathbf{id}\otimes f\,;\,\mathbf{app})\otimes \mathbf{id})\otimes \mathbf{id}\,;\,g\otimes \mathbf{id}\,;\,h} A\supset B}\,(\supset\mathsf{left})$$

The same delay is possible with the right premise g, however, one must take into consideration the naturality of **dupl**, **fst**, **snd**, **Dupl**, **Fst** and **Snd**, i.e., one must duplicate or discharge the derivation itself. For example,

$$\frac{D \xrightarrow{f} A \qquad \dfrac{(A\otimes A)\otimes E \xrightarrow{g} B}{A\otimes E \xrightarrow{\mathbf{Dupl}\otimes \mathbf{id}\,;\,g} B}\,(\mathsf{contraction})}{D\otimes E \xrightarrow{f\otimes \mathbf{id}\,;\,\mathbf{Dupl}\otimes \mathbf{id}\,;\,g} B}\,(\mathsf{cut})$$

$$\Downarrow$$

$$
\cfrac{D \xrightarrow{f} A \qquad \cfrac{D \xrightarrow{f} A \qquad (A \otimes A) \otimes E \xrightarrow{g} B}{(D \otimes A) \otimes E \xrightarrow{(f \otimes \mathbf{id}) \otimes \mathbf{id}\, ;\, g} B}\,(\mathsf{cut})}{\cfrac{(D \otimes D) \otimes E \xrightarrow{(f \otimes \mathbf{id}\, ;\, \mathbf{id} \otimes f) \otimes \mathbf{id}\, ;\, g} B}{D \otimes E \xrightarrow{(\mathbf{Dupl}\, ;\, f \otimes f) \otimes \mathbf{id}\, ;\, g} B}\,(\mathsf{contraction})}\,(\mathsf{cut})
$$

Thus any non-principal application of the cut rule can be moved upward so that only (possibly several) principal applications of the cut rule appear in the derivation. Notice that these transformations do not change the size of the components on which the cut rules operate.

principal cut A principal cut is transformed into possibly several applications of the cut rule operating on the smaller size of components. Using the previous schemas, it is illustrated as follows:

$$
\cfrac{\cfrac{F(A) \xrightarrow{f} B}{A \xrightarrow{\eta_A} GF(A) \xrightarrow{G(f)} G(B)}\,G\ \mathsf{right} \qquad \cfrac{B \otimes E \xrightarrow{g} C}{FG(B) \otimes E \xrightarrow{\epsilon_B \otimes \mathbf{id}} B \otimes E \xrightarrow{g} C}\,G\ \mathsf{left}}{F(A) \otimes E \xrightarrow{F(\eta_A\, ;\, G(f)) \otimes \mathbf{id}\, ;\, (\epsilon_B \otimes \mathbf{id}\, ;\, g)} C}\,\mathsf{cut}
$$

$$\Downarrow$$

$$
\cfrac{F(A) \xrightarrow{f} B \qquad B \otimes E \xrightarrow{g} C}{F(A) \otimes E \xrightarrow{f \otimes \mathbf{id}\, ;\, g} C}\,\mathsf{cut}
$$

$$
\begin{aligned}
& F(\eta_A\, ;\, G(f)) \otimes \mathbf{id}\, ;\, (\epsilon_B \otimes \mathbf{id}\, ;\, g) \\
& \quad \equiv F(\eta_A) \otimes \mathbf{id}\, ;\, (FG(f)\, ;\, \epsilon_B) \otimes \mathbf{id}\, ;\, g && \text{(functoriality of } F) \\
& \quad \equiv (F(\eta_A)\, ;\, \epsilon_{F(A)}) \otimes \mathbf{id}\, ;\, f \otimes \mathbf{id}\, ;\, g && \text{(naturality of } \epsilon) \\
& \quad \equiv f \otimes \mathbf{id}\, ;\, g && \text{(adjunction of } F \text{ and } G).
\end{aligned}
$$

Overall, the alternate repetition of these processes leads to the cut-free derivation. To see the termination, one must develop inductive arguments. The reader is referred to [Dummett, 1977] for the details.

One may have noticed that we have already forced a choice between $\times$ and $\otimes$; $\otimes$ in the $\supset$left rule and the $\times$ in the $\times$right rule are mandatory. Now that the cut rule is eliminated, we are able to use structural rules whenever possible, that is, to use $\otimes$ outside the scope of $\supset$, so that the choice becomes unique. For example, a canonical arrow $f : A \times (A \supset B \times C) \times (B \times 1) \to B \times B$ should be $f : A \otimes (A \supset B \times C) \otimes (B \otimes 1) \to B \times B$.

5 A Coherence Theorem for CCC's

In this section, we give a direct proof of a coherence theorem for CCC's in $\mathcal{LJ}^{CCC}$. The coherence problem we consider is in a weak form since 1 and $\times$ are excluded in view of isomorphisms. This is due to the technical difficulties and we are yet to see a simple proof for the stronger form of the theorem. Let us put a few definitions before stating the theorem to be proved.

Definition 7 (signatures). An occurrence of an atom in the expression $P \to Q$ is given a positive (resp. negative) signature $+$ (resp. $-$) if it appears even (resp. odd) number of times in the lefthand side of $\supset$ or $\to$.

For example, the signatures in the expression $((B^+ \supset A^-) \otimes B^-) \otimes (A^+ \times 1 \supset C^-) \to B^- \supset C^+$ are indicated by the superscripts. Note that 1 is not regarded as an atom and hence is not given a signature.

Definition 8 (balanced arrows). An arrow with the expression $P \to Q$ is said to be balanced if no atom occurs twice or more with the same signature in $P \to Q$.

Theorem 9 (a coherence theorem for CCC's). *If the expression $P \to Q$ is balanced and does not contain* 1 *and* $\times$, *the arrows $P \to Q$ in the free CCC is unique.*

The terminal object 1 and the product $\times$ inside the scope of $\supset$ is excluded in view of the isomorphisms $(A \times 1) \simeq (1 \times A) \simeq (1 \supset A) \simeq A$, $(A \supset 1) \simeq 1$ and $A \supset B \times C \simeq (A \supset B) \otimes (A \supset C)$, $A \times B \supset C \simeq A \supset (B \supset C)$.

The theorem including $\times$ inside $\supset$ was proved first by Babaev and Soloviev [Babaev and Soloviev, 1979], and later by Mints [Mints, 1992b] with some simplifications. Their proofs involve rather complicated arguments of typed lambda terms and rely on the non-trivial result from the proof theory of natural deduction. Here, we stick to category theory as close as possible and show a direct proof of a coherence theorem at the expense of the abovementioned simplification.

As we stated in Section 1, the difficulty of coherence problems lies in the fact that we must perform two different tasks cooperatively: enumeration of arrows and decision of equivalence. We believe that the first problem is more important and is properly handled in the sequent calculus formalisms by way of proof search. In the natural deduction formalism, arrow enumeration amounts to generating terms inhabiting a given type and heavily depends on the syntactic definitions of types and terms. It is difficult to handle syntactic constraints such as the balanced condition since the main focus is not on types but on terms in the typed lambda calculus.

In $\mathcal{LJ}^{CCC}$, arrows between given domain and codomain can be enumerated with the reverse use of inference rules until reaching identity axioms. For example, the arrows between $((B \supset A) \otimes B) \otimes (A \supset A \times C)$ and $B \supset C$ are enumerated as follows:

$$\dfrac{\dfrac{\dfrac{B \otimes B \longrightarrow B \qquad A \longrightarrow A}{(B \supset A) \otimes (B \otimes B) \longrightarrow A}\supset\text{left} \qquad \dfrac{C \longrightarrow C}{A \times C \longrightarrow C}\times\text{left2}}{(((B \supset A) \otimes B) \otimes (A \supset A \times C)) \otimes B \longrightarrow C}\supset\text{let}}{((B \supset A) \otimes B) \otimes (A \supset A \times C) \longrightarrow B \supset C}\supset\text{right}$$

The double line indicates the use of symmetric monoidal isomorphisms. Such a search is exhaustive thanks to the subformula property, however, what we want to show is the uniqueness of the enumeration of the specified class of arrows for the coherence problem. The most problematic fact here is that cut-free derivations are not unique at all since the order of the inference rules can be rearranged and there are a number of ways to reconstruct the derivations. To cope with this problem, we adapt Kleene's permutability theorem The following result is adapted from [Kleene, 1952] for $\mathcal{LJ}^{CCC}$ and guarantee certain permutation of the inference rules in the derivation.

Theorem 10 (Kleene's permutability theorem for $\mathcal{LJ}^{CCC}$). *A cut-free derivation $P \xrightarrow{f} Q$ in $\mathcal{LJ}^{CCC}$ can be rearranged within equivalence as follows:*

- *Weakening rules are moved downward so that they only occur 1) preceding the conclusion $P \longrightarrow Q$ with only other weakening rules intervening, or 2) preceding* ⊃right *in the following way:*

$$\dfrac{\dfrac{E \longrightarrow B}{A \otimes E \longrightarrow B}\ \mathsf{weakening}}{E \longrightarrow A \supset B}\ \supset\mathsf{right}$$

If a component introduced by a weakening rule is contracted later by a contraction rule, then these two rules cancel out. The equivalence of derivations is ensured first by the naturality of **Fst** and **Snd**, e.g., $f \otimes g\,;\mathbf{Snd} = \mathbf{Snd}\,; g$ and second by the triangular identities of $\otimes$, e.g., $\mathbf{Dupl}\,;\mathbf{Snd} = \mathbf{id}$.
- Under the condition above, the order of logical rules can be permuted freely 1) under the restriction of the scope of the connectives that the rules introduce, and 2) with the exception of the lower ⊃right pushing out the component inherited from the first premise of an upper ⊃left. This permutation is an instance of the functoriality of $\otimes$, e.g., $f \otimes \mathbf{id}\,;\mathbf{id} \otimes g = \mathbf{id} \otimes g\,; f \otimes \mathbf{id}$, and the naturality of **mark**, i.e., $f\,;\mathbf{mark} = \mathbf{mark}\,;\mathbf{id} \supset (f \times \mathbf{id})$ which allows the composition to go inside the scope of $\supset$.
- Contraction rules are moved downward so that they only occur 1) preceding the conclusion $P \longrightarrow Q$ with only other contraction rules, followed by weakening rules, intervening, or or 2) preceding ⊃right in the following way;

$$\dfrac{\dfrac{(A \otimes A) \otimes E \longrightarrow B}{A \otimes E \longrightarrow B}\ \mathsf{contraction}}{E \longrightarrow A \supset B}\ \supset\mathsf{right}$$

The equivalence of the derivation is ensured by the naturality of **Dupl**, i.e., $f\,;\mathbf{Dupl} = \mathbf{Dupl}\,; f \otimes f$.

Now by a categorical observation, we can say more about the order of inference rules in the derivation. A right logical rule G right is exactly the natural isomorphism of adjunctions $\mathsf{Hom}_{\mathcal{D}}(F(A), B) \simeq \mathsf{Hom}_{\mathcal{C}}(A, G(B))$:

$$\frac{F(A) \xrightarrow{g} B}{A \xrightarrow{f} G(B)}\ G\ \text{right}\quad ,$$

and invertible in the sense that whenever $A \xrightarrow{f} G(B)$ is derived there is a derivation $F(A) \xrightarrow{g} B$ from which $A \xrightarrow{f'} G(B)$ is derived immediately by the G right and $f' \equiv f$. Therefor without loss of information, we can concentrate on the arrow $P \longrightarrow A$ (A is atomic). We call such an arrow *atomic*.

Let us turn to the balanced condition. This condition is posed to put bounds on the search space. For example, the enumeration of the arrows $(N \supset N) \otimes N \rightarrow N$, which is not balanced, never terminates and the coherence for these arrows does not make sense [Szabo, 1975].

$$\cfrac{\cfrac{(N \supset N) \otimes N \xrightarrow{f_{n-1}} N \qquad N \xrightarrow{\mathbf{id}} N}{(N \supset N) \otimes ((N \supset N) \otimes N) \xrightarrow{\mathbf{id} \otimes f_{n-1}\,;\,\mathbf{app}} N}\ \supset\text{left}}{(N \supset N) \otimes N \xrightarrow[= f_n]{\mathbf{Dupl} \otimes \mathbf{id}\,;\,\mathbf{Assr}\,;\,\mathbf{id} \otimes f_{n-1}\,;\,\mathbf{app}} N}\ \text{contraction}\quad ,$$

$$f_0 = \mathbf{Snd},$$
$$f_n = \mathbf{Dupl} \otimes \mathbf{id}\,;\,\mathbf{Assr}\,;\,\mathbf{id} \otimes f_{n-1}\,;\,\mathbf{app}.$$

$\mathbf{Assr}_{A,B,C} : (A \otimes B) \otimes C \rightarrow A \otimes (B \otimes C)$ is a symmetric monoidal isomorphism. In terms of the typed lambda calculus, it is a generation of Church's numerals $\lambda z.\lambda s.s(s(\cdots(sz)\cdots)) : N$ ($s : N \supset N$ and $z : N$).

To clarify the effect of the balanced condition against the contraction rule, we introduce the following notion. It can be seen as a proof-theoretic version of Kelly-Mac Lane graphs for symmetric monoidal categories [Kelly and Mac Lane, 1972].

Definition 11 (connections). To cut-free derivations, one can naturally associate the pairs of positive and negative occurrences of atoms that are inherited from identity axioms. These pairs are called *connections* and displayed graphically as follows.

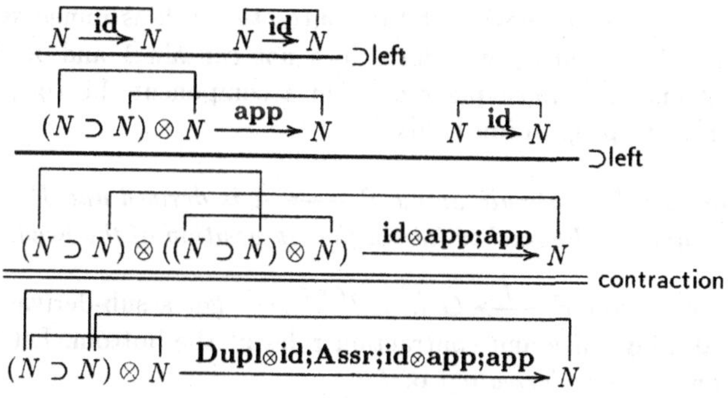

The crucial point is that the balanced condition implies the unique identification of the possible connections. We will prove later that the balanced condition excludes the contraction rule under some restrictions.

By a simple induction, we have the following result for an atomic arrow.

Proposition 12 (anchor, head). *If an atomic arrow $P \longrightarrow A$ is derived, there exists a unique negative occurrence of A (called the anchor) in P which is linked by a connection to the occurrence of A at the codomain. The component containing the anchor is called the head.*

This proposition suggests a uniform way of arrow enumeration. First apply right logical rules to get atomic arrows, followed by structural rules, and then the left logical rule for the head component. It corresponds to the notions of *uniform proofs* [Miller *et al.*, 1991] and the *backchaining rule* [Miller, 1991] introduced by Miller and others for a generic account of the resolution method in the sequent calculus. Below is an example of such uniform arrow enumeration. The lower derivation is a permutation of the upper and $f \equiv g$.

$$
\dfrac{B \longrightarrow B \qquad \dfrac{\dfrac{\dfrac{A \longrightarrow A \quad A \longrightarrow A}{A \longrightarrow A \times A}\,\times\text{right} \qquad \dfrac{\dfrac{\dfrac{C \longrightarrow C}{C \otimes B \longrightarrow C}\,\text{weakening}}{C \longrightarrow B \supset C}\,\supset\text{right}}{A \otimes C \longrightarrow B \supset C}\,\text{weakening}}{(A \otimes A) \otimes (A \times A \supset C) \longrightarrow B \supset C}\,\supset\text{left (double line)}}{A \otimes (A \times A \supset C) \longrightarrow B \supset C}\,\text{contraction}}{((B \supset A) \otimes B) \otimes (A \times A \supset C) \xrightarrow{f} B \supset C}\,\supset\text{left}
$$

$$\Downarrow$$

$$
\dfrac{\dfrac{\dfrac{\dfrac{\dfrac{B \longrightarrow B \quad A \longrightarrow A}{(B \supset A) \otimes B \longrightarrow A}\,\supset\text{left} \qquad \dfrac{B \longrightarrow B \quad A \longrightarrow A}{(B \supset A) \otimes B \longrightarrow A}\,\supset\text{left}}{(B \supset A) \otimes B \longrightarrow A \times A}\,\times\text{right} \qquad C \longrightarrow C}{((B \supset A) \otimes B) \otimes (A \times A \supset C) \longrightarrow C}\,\supset\text{left (double line)}}{(((B \supset A) \otimes B) \otimes (A \times A \supset C)) \otimes B \longrightarrow C}\,\text{weakening}}{((B \supset A) \otimes B) \otimes (A \times A \supset C) \xrightarrow{g} B \supset C}\,\supset\text{right}
$$

Now we are going to show that the contraction rule is unnecessary for the derivation of a balanced arrow which does not contain 1 and $\times$. Let us call the rightmost and innermost place of $\supset$ in a component the *tip* position. By induction, the next proposition holds.

Proposition 13. *If an atomic arrow $P \xrightarrow{f} A$ is derived and $P \longrightarrow A$ does not contain* 1 *and* $\times$, *the anchor lies in the tip position of the head.*

For any derivation $P \xrightarrow{f} Q$ in $\mathcal{LJ}^{CCC}$, we get a sub-derivation of f by omitting weakening rules and contraction rules at the bottom. Let us call this sub-derivation the *core derivation* of f.

Proposition 14. *In the core derivation $P' \xrightarrow{f'} A$ (A is atomic), the (negative) occurrence of an atom at the tip position in each component C of P' is linked by a unique connection.*

The proof is by induction and follows from the absence of the structural rule concerning the components in P'.

Theorem 15 (contraction elimination for balanced arrows). *A balanced arrow in a free CCC can be derived without using the contraction rule if it contains no* 1 *and* $\times$.

Proof. If such a balanced arrow is derived by a $\supset$right, the premises of the rule are balanced, too, Thus by the invertibility of the right logical rule, we can concentrate on the atomic arrow.

Take any derivation of a balanced atomic arrow $P \xrightarrow{f} A$ that contains no 1 and $\times$. We first claim that the head cannot be a contraction in the derivation f. If the head is a contraction, we have two components in the core derivation $P' \xrightarrow{f'} A$ that are contracted to the head of $P \xrightarrow{f} A$. Then from Proposition 14 and 12, the atom A has a positive occurrence in P' and hence in P. As A has another positive occurrence at the codomain of $P \xrightarrow{f} A$, the balanced condition is violated. A contradiction.

Now suppose that a component C other than the head is a contraction in the derivation f. Then in the core derivation $P' \xrightarrow{f'} A$, we have two components C_1 and C_2 that are contracted to C. We reconstruct $\supset$left rules that constitute the head in $P' \xrightarrow{f'} A$ and get a disjoint partition of the components of P'.

$$\dfrac{E_{n-1} \longrightarrow A_{n-1} \qquad \dfrac{E_n \longrightarrow A_n \qquad A \longrightarrow A}{A_n \supset A \otimes E_n \longrightarrow A}\supset\text{left}}{A_{n-1} \supset (A_n \supset A) \otimes E_{n-1} \otimes E_n \longrightarrow A}\supset\text{left}$$

$$\vdots$$

$$\dfrac{E_1 \longrightarrow A_1 \qquad A_2 \supset (\cdots \supset (A_n \supset A)\cdots) \otimes E_2 \otimes \cdots \otimes E_n \longrightarrow A}{A_1 \supset (A_2 \supset (\cdots \supset (A_n \supset A)\cdots)) \otimes E_1 \otimes E_2 \otimes \cdots \otimes E_n \longrightarrow A}\supset\text{left}$$

$$\vdots \text{ contractions and weakenings}$$

$$A_1 \supset (A_2 \supset (\cdots \supset (A_n \supset A)\cdots)) \otimes E_{i_1} \otimes E_{i_2} \otimes \cdots \otimes E_{i_m} \longrightarrow A$$

We then claim that the components C_1 and C_2 appear in the same E_k. Assume that C_1 appears in E_i and C_2 in E_j ($i \neq j$). From Proposition 14, each occurrence of the atom B at the tip position of C_1 and C_2 are linked by a connection to a positive occurrence of B in P'. Let us denote those positive occurrences of B by B_1^+ and B_2^+ respectively. We have four cases.

1. If B_1^+ appears in A_i and B_2^+ in A_j, then we have two positive occurrences of B in the head of the core derivation $P' \xrightarrow{f'} A$ and hence in the derivation $P \xrightarrow{f} A$. A contradiction.

2. If B_1^+ appears in A_i and B_2^+ in E_j, then we have one positive occurrence of B in the head and another elsewhere in the core derivation $P' \xrightarrow{f'} A$. Since the head cannot be a contraction, there are two positive occurrence of B in the derivation $P \xrightarrow{f} A$. A contradiction.
3. If B_1^+ appears in E_i and B_2^+ in A_j, the same as above. A contradiction.
4. If B_1^+ appears in E_i and B_2^+ in E_j, we have two distinct components D_1 and D_2 in the core derivation $P' \xrightarrow{f'} A$ that contain B_1^+ and B_2^+ respectively. To satisfy the balanced condition, D_1 and D_2 must be contracted. By repeating the same argument, we get a sequence of components $C_1, D_1, \dots$ in E_i and $C_2, D_2, \dots$ in E_j. In any case, the number of the components are finite and there must be a loop in the sequence. In such a loop, each component is linked with the connection coming from the tip position of the preceding component. If we try to reconstruct ⊃left rules that constitute one of those components in the loop, we have no way to break the loop because the contraction rule cannot be used for those components from the core derivation. A contradiction.

Since each case leads to a contradiction, we can assume that both C_1 and C_2 appear in the same sub-derivation $E_k \xrightarrow{h} A_k$. Lifting up all necessary contraction rules if present, we get a derivation $E' \xrightarrow{h'} A'$ from $E_k \xrightarrow{h} A_k$ where $E' \longrightarrow A'$ is a balanced atomic arrow that contains no 1 and $\times$. Since the size of the arrow is smaller than $P \xrightarrow{f} A$, we can apply inductive arguments. When the head is atomic, the derivation is by a sequence of the weakening rule and there is no room for the contraction rule (end up being canceled). □

The reason we drop 1 and $\times$ is that the theorem cannot be proved in their presence. The connections involving 1 are not defined, and two components of the expression $A \supset B \times C$ can be contracted without violating the balanced condition.

Now we almost proved the theorem. The arrow enumeration can be done with the unique choice of inference rules, in which two premises of the ⊃left rule are determined consulting the information of connections. The coherence follows immediately.

Proof of the main theorem. Since every inference rule except the contraction rule preserves the balance in the reverse direction, we can perform arrow enumeration for balanced arrows in a deterministic manner. We first apply ⊃right as many times as required to get atomic arrows, then reconstruct the ⊃left rules that constitute the head until the head becomes atomic. When applying ⊃left, we examine the possible connections determined by the balanced condition to split passive components into two premises of the rule. By repetition of this procedure in each branch of the ⊃left rule, we reach to the atomic arrows having atomic heads. The balanced condition are maintained so far and we get a unique sequence of weakening rules at each branch. Although we delay the use of the weakening rule until the head of the atomic arrow becomes atomic, every derivation with the prior use of the weakening rule must be found with the posterior use of the rule with no divergence. The overall process is deterministic and unique within equivalence. □

Finally, we show an example of arrow enumeration for the balanced arrow.

$$\cfrac{\cfrac{\cfrac{\cfrac{D \longrightarrow D}{(E \supset E) \otimes D \longrightarrow D}\,\text{weakening} \qquad \cfrac{C \longrightarrow C \quad A \longrightarrow A}{(C \supset A) \otimes C \longrightarrow A}\,\supset\text{left}}{(((D \supset (C \supset A)) \otimes C) \otimes (E \supset E)) \otimes D \longrightarrow A}\,\supset\text{left}}{((D \supset (C \supset A)) \otimes C) \otimes (E \supset E) \longrightarrow (D \supset A)}\,\supset\text{right}}{(D \supset (C \supset A)) \otimes C \longrightarrow (E \supset E) \supset (D \supset A)}\,\supset\text{right}$$

6 Concluding Remarks

We have presented a direct proof of a simplified form of the coherence theorem for cartesian closed categories without passing to the typed lambda calculus. The proof relies on the sequential generation of free CCC's which coincides with categorical interpretation of the intuitionistic sequent calculus. We have also stressed the equational nature of category theory. In particular, the process of cut-elimination and permutation of derivations can be seen as type-sensitive rewriting of combinators.

The connection between the coherence problem and the cut-elimination theorem has been felt for years. We have mixed the ideas of Lawvere and Lambek to regard free CCC's as a kind of sequent calculi. Various proof-theoretic techniques such as Kleene's permutability theorem are used to prove the theorem within the vocabularies of category theory.

The coherence theorem for CCC's we have taken avoids the terminal object 1 and the product $\times$ inside the exponentiation $\supset$. This is due to the difficulty in proving the absence of the contraction rule as in Theorem 15. In a sense, what we have proved is the uniqueness of proofs in implicational intuitionistic logic whose logical operation is only $\supset$. We are looking for a direct proof of a stronger form of the theorem (including 1 and $\times$ inside $\supset$) in which the uniqueness of arrow enumeration is shown independently of elimination of the contraction rule. Other lines of study include a weakly normalizing system for categorical combinators based on the cut-elimination and the permutability.

The categorical diagrams and proof figures in this paper are formatted using Paul Taylor's TEXmacro packages `diagrams.tex` and `prooftree.tex`.

References

[Babaev and Soloviev, 1979] A.A. Babaev and S.V. Soloviev. A coherence theorem for canonical maps in cartesian closed categories. *Zapisiki Nauchnykh Seminarov LOMI*, 88, 1979. Russian with English summary. English translation appears in *J. of Soviet Math.*, 20, 1982.

[Barr and Wells, 1990] Michael Barr and Charles Wells. *Category Theory for Computing Science*. International Series in Computer Science. Prentice Hall, 1990.

[Cousineau *et al.*, 1985] Guy Cousineau, P.-L. Curien, and Michael Mauny. The Categorical Abstract Machines. *Lecture Notes in Computer Science*, 201:50–64, 1985.

[Curien, 1986] P.-L. Curien. *Categorical combinators, Sequential algorithms and Functional programming*. Research Notes in Theoretical Computer Science. Pitman, 1986. The revised edition is published from Birkhäuser, in the series of *Progress in Theoretical Computer Science*, 1993.

[Dummett, 1977] Michael Dummett. *Elements of Intuitionism*, volume 2 of *Oxford Logic Guides*. Oxford University Press, 1977.

[Gallier, 1993] J. Gallier. Constructive logics. part I: A tutorial on proof systems and typed λ-calculi. *Theor. Comput. Sci.*, 110:249–339, 1993.

[Girard *et al.*, 1989] J.-Y. Girard, Y. Lafont, and P. Taylor. *Proofs and Types*, volume 7 of *Cambridge Tracts in Theoretical Computer Science*. Cambridge University Press, 1989.

[Jay, 1990] Collin Barry Jay. The structure of free closed categories. *J. Pure and Applied Algebra*, 66:271–285, 1990.

[Kelly and Mac Lane, 1972] G.M. Kelly and Saunders Mac Lane. Coherence in closed categories. *J. Pure and Applied Algebra*, 1(1):97–140, 1972.

[Kleene, 1952] S.C. Kleene. Permutability of inferences in Gentzen's calculi LK and LJ. *Memoirs of the American Mathematical Society*, 10:1–26, 1952.

[Lambek and Scott, 1986] J. Lambek and P.J. Scott. *Introduction to Higher Order Categorical Logic*. Cambridge University Press, 1986.

[Lambek, 1969] J. Lambek. Deductive systems and categories II: Standard constructions and closed categories. *Lecture Notes in Mathematics*, 86:76–122, 1969.

[Lambek, 1989] J. Lambek. Multicategories revisited. In J.W Gray and A. Scedrov, editors, *Categories in Computer Science and Logic*, pages 217–239. American Mathematical Society, 1989. Contemporary Mathematics Vol.92.

[Lawvere, 1969] F.W. Lawvere. Adjointness in foundations. *Dialectica*, 23:281–296, 1969.

[Lawvere, 1970] F.W. Lawvere. Equality in hyperdoctorines and the comprehension schema as an adjoint functor. In *Proc. NY Symposium on Applications of Categorical Logic*, pages 1–14. American Mathematical Society, 1970.

[Mac Lane, 1971] Saunders Mac Lane. *Categories for the Working Mathematician*, volume 5 of *Graduate Texts in Mathematics*. Springer-Verlag, 1971.

[McLarty, 1992] Colin McLarty. *Elementary Categories, Elementary Toposes*, volume 21 of *Oxford Logic Guides*. Oxford University Press, 1992.

[Miller *et al.*, 1991] Dale Miller, Gopalan Nadathur, Frank Pfenning, and Andre Scedrov. Uniform proofs as a foundation for logic programming. *Ann. Pure and Applied Logic*, 51:125–157, 1991.

[Miller, 1991] Dale Miller. A logic programming language with lambda-abstraction, function variables, and simple unification. *J. of Logic and Computation*, 1(4):497 – 536, 1991.

[Mints, 1992a] Gregorii E. Mints. Proof theory and category theory. In *Selected Papers in Proof Theory*, chapter 10, pages 183–212. Bibliopolis/North-Holland, 1992.

[Mints, 1992b] Gregorii E. Mints. A simple proof of the coherence theorem for cartesian closed categories. In *Selected Papers in Proof Theory*, chapter 11, pages 213–220. Bibliopolis/North-Holland, 1992.

[Szabo, 1975] M.E. Szabo. A counter-example to coherence in cartesian closed categories. *Canad.Math.Bull.*, 18(1):111–114, 1975.

Appendix

A The Sequent Calculus $\mathcal{LJ}^{\wedge,\Rightarrow,I}$

The lefthand side of a sequent is treated as a multiset and therefore we do not have the exchange rule as a structural rule. We say a sequent $\Gamma \vdash A$ is *derived* when it is reached from identity axioms using inference rules. We call the tree figure of such a inference a *derivation* of $\Gamma \vdash A$.

identity axioms: (A:atom)

$$A \vdash A$$

cut rule:

$$\frac{\Gamma \vdash A \qquad A,\, \Delta \vdash B}{\Gamma,\, \Delta \vdash B}\,(\mathsf{cut})$$

structural rules:

$$\frac{\Gamma \vdash B}{A,\, \Gamma \vdash B}\,(\mathsf{weakening}) \qquad \frac{A,\, A,\, \Gamma \vdash B}{A,\, \Gamma \vdash B}\,(\mathsf{contraction})$$

logical rules:

$$\frac{A,\, \Gamma \vdash C}{A \wedge B,\, \Gamma \vdash C}\,(\wedge\mathsf{left1}) \qquad \frac{\Gamma \vdash A \qquad \Gamma \vdash B}{\Gamma \vdash A \wedge B}\,(\wedge\mathsf{right})$$

$$\frac{B,\, \Gamma \vdash C}{A \wedge B,\, \Gamma \vdash C}\,(\wedge\mathsf{left2})$$

$$\frac{\Gamma \vdash A \qquad B,\, \Delta \vdash C}{A \Rightarrow B,\, \Gamma,\, \Delta \vdash C}\,(\Rightarrow\mathsf{left}) \qquad \frac{A,\, \Gamma \vdash B}{\Gamma \vdash A \Rightarrow B}\,(\Rightarrow\mathsf{right})$$

$$\Gamma \vdash I \; (\mathsf{I\,right})$$

Theorem 16 (cut-elimination theorem). *There is an algorithm which, given any derivation in $\mathcal{LJ}^{\wedge,\Rightarrow,I}$ produces a cut-free derivation in $\mathcal{LJ}^{\wedge,\Rightarrow,I}$.*

Theorem 17 (subformula property). *In $\mathcal{LJ}^{\wedge,\Rightarrow,I}$, A cut-free derivation of $\Gamma \vdash A$ only uses subformulas of the formulas in Γ or of A.*

Modular Properties of Constructor-Sharing Conditional Term Rewriting Systems

Enno Ohlebusch

Universität Bielefeld, 33501 Bielefeld, Germany,
e-mail: enno@techfak.uni-bielefeld.de

Abstract. First, using a recent modularity result [Ohl94b] for unconditional term rewriting systems (TRSs), it is shown that semi-completeness is a modular property of constructor-sharing join conditional term rewriting systems (CTRSs). Second, we do not only extend results of Middeldorp [Mid93] on the modularity of termination for disjoint CTRSs to constructor-sharing systems but also simplify the proofs considerably. Moreover, we refute a conjecture of Middeldorp [Mid93] which is related to the aforementioned results.

1 Introduction

A property is modular for disjoint (constructor-sharing, respectively) term rewriting systems if it is preserved under the combination of disjoint (constructor-sharing, respectively) TRSs. In his pioneering paper, Toyama [Toy87b] proved that confluence is modular for disjoint TRSs. Middeldorp [Mid93] extended this result to CTRSs. In contrast to these encouraging results, Kurihara & Ohuchi [KO92] refuted the modularity of confluence for constructor-sharing TRSs – systems which may share constructors. Constructors are function symbols that do not occur at the root position in left-hand sides of rewrite rules, the others are called defined symbols. Recently, we have shown [Ohl94b] that semi-completeness (confluence plus normalization) is modular for constructor-sharing TRSs. In the first part of this paper, we extend this result to CTRSs.

Unlike confluence, termination lacks a modular behavior for disjoint TRSs – see [Toy87a]. The first sufficient conditions ensuring the preservation of termination under disjoint union were obtained by investigating the distribution of collapsing and duplicating rules among the TRSs – see [Rus87, Mid89]. The results were extended, mutatis mutandis, to disjoint CTRSs [Mid93] and constructor-sharing TRSs [Ohl94c]. In the second part of this paper, we provide a relatively simple proof for analogous results for constructor-sharing CTRSs. Furthermore, a simple counterexample disproves a conjecture of Middeldorp [Mid93] which is related to the above-mentioned results.

More sufficient conditions for the modularity of termination of TRSs can for instance be found in [TKB89, KO92, Gra94a, Gra94b, Ohl94c, Mar95]. The reader is referred to [Mid90, Mid93, Mid94, Ohl93, Gra93] for modularity results of CTRSs. In recent investigations, one tries to weaken the constructor-sharing requirement. On the one hand, [MT93, Mid94, Ohl94a, KO94] consider composable systems – systems which have to contain all rewrite rules that define a defined symbol whenever that symbol is shared. On the other hand [KR93, KR94, Der95, FJ94] investigate hierarchical systems, where defined symbols of one TRS may occur as constructors in the other but not vice versa. We refer to the final section of this paper for more details. Moreover, we point out that [Ohl94a] contains a comprehensive survey of all modularity results obtained so far.

The paper is organized as follows. Section 2 briefly recalls the basic notions of (conditional) term rewriting. Section 3 contains required notions of combined systems with shared constructors. In Section 4, we prove the aforementioned modularity results. Finally, the last section contains concluding remarks.

2 Preliminaries

This section contains a concise introduction to term rewriting. The reader is referred to the surveys of Dershowitz & Jouannaud [DJ90] and Klop [Klo92] for more detail.

A *signature* is a countable set $\mathcal{F}$ of *function symbols* or *operators*, where every $f \in \mathcal{F}$ is associated with a natural number denoting its arity. Nullary operators are called *constants*. The set $\mathcal{T}(\mathcal{F},\mathcal{V})$ of *terms* built from a signature $\mathcal{F}$ and a countable set of *variables* $\mathcal{V}$ with $\mathcal{F} \cap \mathcal{V} = \emptyset$ is the smallest set such that $\mathcal{V} \subseteq \mathcal{T}(\mathcal{F},\mathcal{V})$ and if $f \in \mathcal{F}$ has arity n and $t_1,\ldots,t_n \in \mathcal{T}(\mathcal{F},\mathcal{V})$, then $f(t_1,\ldots,t_n) \in \mathcal{T}(\mathcal{F},\mathcal{V})$. We write f instead of $f(\,)$ whenever f is a constant. The set of variables appearing in a term $t \in \mathcal{T}(\mathcal{F},\mathcal{V})$ is denoted by $\mathcal{V}ar(t)$. For $t \in \mathcal{T}(\mathcal{F},\mathcal{V})$, we define $root(t)$ by $root(t) = t$ if $t \in \mathcal{V}$, and $root(t) = f$ if $t = f(t_1,\ldots,t_n)$. $|t|$ denotes the *size* of t, i.e., $|t| = 1$ if $t \in \mathcal{V}$, and $|t| = 1 + |t_1| + \ldots + |t_n|$ if $t = f(t_1,\ldots,t_n)$.

A *substitution* σ is a mapping from $\mathcal{V}$ to $\mathcal{T}(\mathcal{F},\mathcal{V})$ such that $\{x \in \mathcal{V} \mid \sigma(x) \neq x\}$ is finite. This set is called the *domain* of σ and will be denoted by $\mathcal{D}om(\sigma)$. Occasionally, we present a substitution σ as $\{x \mapsto \sigma(x) \mid x \in \mathcal{D}om(\sigma)\}$. The substitution with empty domain will be denoted by ϵ. Substitutions extend uniquely to morphisms from $\mathcal{T}(\mathcal{F},\mathcal{V})$ to $\mathcal{T}(\mathcal{F},\mathcal{V})$, that is, $\sigma(f(t_1,\ldots,t_n)) = f(\sigma(t_1),\ldots,\sigma(t_n))$ for every n-ary function symbol f and terms $t_1,\ldots,t_n$. We call $\sigma(t)$ an *instance* of t. We also write $t\sigma$ instead of $\sigma(t)$.

Let $\Box$ be a special constant. A *context* $C[,\ldots,]$ is a term in $\mathcal{T}(\mathcal{F} \cup \{\Box\},\mathcal{V})$ which contains at least one occurrence of $\Box$. If $C[,\ldots,]$ is a context with n occurrences of $\Box$ and $t_1,\ldots,t_n$ are terms, then $C[t_1,\ldots,t_n]$ is the result of replacing from left to right the occurrences of $\Box$ with $t_1,\ldots,t_n$. A context containing precisely one occurrence of $\Box$ is denoted by $C[\,]$. A term t is a *subterm* of a term s if there exists a context $C[\,]$ such that $s = C[t]$. A subterm t of s is *proper*, denoted by $s \rhd t$, if $s \neq t$. By abuse of notation we write $\mathcal{T}(\mathcal{F},\mathcal{V})$ for $\mathcal{T}(\mathcal{F} \cup \{\Box\},\mathcal{V})$, interpreting $\Box$ as a special constant which is always available but used only for the aforementioned purpose.

Let $\rightarrow$ be a binary relation on terms, i.e., $\rightarrow \subseteq \mathcal{T}(\mathcal{F},\mathcal{V}) \times \mathcal{T}(\mathcal{F},\mathcal{V})$. The reflexive transitive closure of $\rightarrow$ is denoted by $\rightarrow^*$. If $s \rightarrow^* t$, we say that s

reduces to t and we call t a *reduct* of s. We write $s \leftarrow t$ if $t \rightarrow s$; likewise for $s \ {}^*\!\!\leftarrow t$. The transitive closure of $\rightarrow$ is denoted by $\rightarrow^+$, and $\leftrightarrow$ denotes the symmetric closure of $\rightarrow$ (i.e., $\leftrightarrow = \rightarrow \cup \leftarrow$). The reflexive transitive closure of $\leftrightarrow$ is called *conversion* and denoted by $\leftrightarrow^*$. If $s \leftrightarrow^* t$, then s and t are *convertible*. Two terms t_1, t_2 are *joinable*, denoted by $t_1 \downarrow t_2$, if there exists a term t_3 such that $t_1 \rightarrow^* t_3 \ {}^*\!\!\leftarrow t_2$. Such a term t_3 is called a *common reduct* of t_1 and t_2. The relation $\downarrow$ is called *joinability*. A term s is a *normal form* w.r.t. $\rightarrow$ if there is no term t such that $s \rightarrow t$. A term s has a normal form if $s \rightarrow^* t$ for some normal form t. The set of all normal forms of $\rightarrow$ is denoted by $NF(\rightarrow)$. The relation $\rightarrow$ is *normalizing* if every term has a normal form; it is *terminating*, if there is no infinite reduction sequence $t_1 \rightarrow t_2 \rightarrow t_3 \rightarrow \ldots$. In the literature, the terminology *weakly normalizing* and *strongly normalizing* is often used instead of normalizing and terminating, respectively. The relation $\rightarrow$ is *confluent* if for all terms s, t_1, t_2 with $t_1 \ {}^*\!\!\leftarrow s \rightarrow^* t_2$ we have $t_1 \downarrow t_2$. It is well-known that $\rightarrow$ is confluent if and only if every pair of convertible terms is joinable. The relation $\rightarrow$ is *locally confluent* if for all terms s, t_1, t_2 with $t_1 \leftarrow s \rightarrow t_2$ we have $t_1 \downarrow t_2$. If $\rightarrow$ is confluent and terminating, it is called *complete* or *convergent*. The famous Newman's Lemma states that termination and local confluence imply confluence. If $\rightarrow$ is confluent and normalizing, then it is called *semi-complete*. Sometimes this property is called *unique normalization* because it is equivalent to the property that every term has a unique normal form.

A *term rewriting system* (TRS) is a pair $(\mathcal{F}, \mathcal{R})$ consisting of a signature $\mathcal{F}$ and a set $\mathcal{R} \subset \mathcal{T}(\mathcal{F}, \mathcal{V}) \times \mathcal{T}(\mathcal{F}, \mathcal{V})$ of *rewrite rules* or *reduction rules*. Every rewrite rule (l, r) must satisfy the following two constraints: (i) the left-hand side l is not a variable, and (ii) variables occurring in the right-hand side r also occur in l. Rewrite rules (l, r) will be denoted by $l \rightarrow r$. An instance of a left-hand side of a rewrite rule is a *redex* (reducible expression). The rewrite rules of a TRS $(\mathcal{F}, \mathcal{R})$ define a *rewrite relation* $\rightarrow_{\mathcal{R}}$ on $\mathcal{T}(\mathcal{F}, \mathcal{V})$ as follows: $s \rightarrow_{\mathcal{R}} t$ if there exists a rewrite rule $l \rightarrow r$ in $\mathcal{R}$, a substitution σ and a context $C[\,]$ such that $s = C[l\sigma]$ and $t = C[r\sigma]$. We say that s rewrites to t by *contracting* redex $l\sigma$. We call $s \rightarrow_{\mathcal{R}} t$ a *rewrite step* or *reduction step*. A TRS $(\mathcal{F}, \mathcal{R})$ has one of the above properties (e.g. termination) if its rewrite relation has the respective property. Let $(\mathcal{F}, \mathcal{R})$ be an arbitrary TRS. A function symbol $f \in \mathcal{F}$ is called a *defined symbol* if there is a rewrite rule $l \rightarrow r \in \mathcal{R}$ such that $f = root(l)$. Function symbols from $\mathcal{F}$ which are not defined symbols are called *constructors*. The set of normal forms of $(\mathcal{F}, \mathcal{R})$ will also be denoted by $NF(\mathcal{F}, \mathcal{R})$. We often simply write $\mathcal{R}$ instead of $(\mathcal{F}, \mathcal{R})$ if there is no ambiguity about the underlying signature $\mathcal{F}$. A rewrite rule $l \rightarrow r$ of a TRS $\mathcal{R}$ is *collapsing* if r is a variable, and *duplicating* if r contains more occurrences of some variable than l. A TRS $\mathcal{R}$ is *non-duplicating* (non-collapsing, respectively) if it does not contain duplicating (collapsing, respectively) rewrite rules.

In a *join conditional term rewriting system* (CTRS for short) $(\mathcal{F}, \mathcal{R})$, the rewrite rules of $\mathcal{R}$ have the form $l \rightarrow r \Leftarrow s_1 \downarrow t_1, \ldots, s_n \downarrow t_n$ with $l, r, s_1, \ldots, s_n, t_1, \ldots, t_n \in \mathcal{T}(\mathcal{F}, \mathcal{V})$. $s_1 \downarrow t_1, \ldots, s_n \downarrow t_n$ are the *conditions* of the rewrite rule. If a rewrite rule has no conditions, we write $l \rightarrow r$. We impose the same restrictions on conditional rewrite rules as on unconditional rewrite rules. That is, we allow *extra variables* in the conditions but not on right-hand sides of rewrite rules. The rewrite relation associated with $(\mathcal{F},\mathcal{R})$ is defined by: $s \to_{\mathcal{R}} t$ if there exists a rewrite rule $l \to r \Leftarrow s_1 \downarrow t_1, \ldots, s_n \downarrow t_n$ in $\mathcal{R}$, a substitution $\sigma : \mathcal{V} \to \mathcal{T}(\mathcal{F},\mathcal{V})$, and a context $C[\,]$ such that $s = C[l\sigma], t = C[r\sigma]$, and $s_j\sigma \downarrow_{\mathcal{R}} t_j\sigma$ for all $j \in \{1,\ldots,n\}$. For every CTRS $\mathcal{R}$, we inductively define TRSs $\mathcal{R}_i$, $i \in \mathbb{N}$, by:

$$\begin{aligned} \mathcal{R}_0 &= \{l \to r \mid l \to r \in \mathcal{R}\} \\ \mathcal{R}_{i+1} &= \{l\sigma \to r\sigma \mid l \to r \Leftarrow s_1 \downarrow t_1, \ldots, s_n \downarrow t_n \in \mathcal{R} \text{ and} \\ &\qquad s_j\sigma \downarrow_{\mathcal{R}_i} t_j\sigma \text{ for all } j \in \{1,\ldots,n\}\}. \end{aligned}$$

Note that $\mathcal{R}_i \subseteq \mathcal{R}_{i+1}$ for all $i \in \mathbb{N}$. Furthermore, $s \to_{\mathcal{R}} t$ if and only if $s \to_{\mathcal{R}_i} t$ for some $i \in \mathbb{N}$. The *depth* of a rewrite step $s \to_{\mathcal{R}} t$ is defined to be the minimal i with $s \to_{\mathcal{R}_i} t$. Depths of reduction sequences $s \to_{\mathcal{R}}^* t$, conversions $s \leftrightarrow_{\mathcal{R}}^* t$, and valleys $s \downarrow_{\mathcal{R}} t$ are defined analogously. All notions defined previously for TRSs extend to CTRSs.

A *partial ordering* $(A, >)$ is a pair consisting of a set A and a binary irreflexive and transitive relation $>$ on A. A partial ordering is called *well-founded* if there are no infinite sequences $a_1 > a_2 > a_3 > \ldots$ of elements from A. A *multiset* is a collection in which elements are allowed to occur more than once. If A is a set, then the set of all finite multisets over A is denoted by $\mathcal{M}(A)$. The *multiset extension* of a partial ordering $(A, >)$ is the partial ordering $(\mathcal{M}(A), >^{mul})$ defined as follows: $M_1 >^{mul} M_2$ if $M_2 = (M_1 \setminus X) \cup Y$ for some multisets $X, Y \in \mathcal{M}(A)$ that satisfy (i) $\emptyset \neq X \subseteq M_1$ and (ii) for all $y \in Y$ there exists an $x \in X$ such that $x > y$. Dershowitz and Manna [DM79] proved that the multiset extension of a well-founded partial ordering is a well-founded partial ordering.

3 Basic Notions of Constructor-Sharing CTRSs

Definition 3.1 Two CTRSs $(\mathcal{F}_1, \mathcal{R}_1)$ and $(\mathcal{F}_2, \mathcal{R}_2)$ are called *constructor-sharing* if they share at most constructors, more precisely if

$$\mathcal{F}_1 \cap \mathcal{F}_2 \cap \{root(l) \mid l \to r \Leftarrow s_1 \downarrow t_1, \ldots, s_n \downarrow t_n \in \mathcal{R}_1 \cup \mathcal{R}_2\} = \emptyset.$$

In this case, their union $(\mathcal{F}, \mathcal{R}) = (\mathcal{F}_1 \cup \mathcal{F}_2, \mathcal{R}_1 \cup \mathcal{R}_2)$ is called the *combined CTRS of* $(\mathcal{F}_1, \mathcal{R}_1)$ *and* $(\mathcal{F}_2, \mathcal{R}_2)$ *with shared constructors* $\mathcal{C}$, where $\mathcal{C} = \mathcal{F}_1 \cap \mathcal{F}_2$. Furthermore, we define $\mathcal{D}_1 = \mathcal{F}_1 \setminus \mathcal{C}$, $\mathcal{D}_2 = \mathcal{F}_2 \setminus \mathcal{C}$, and $\mathcal{D} = \mathcal{D}_1 \cup \mathcal{D}_2$. A property $\mathcal{P}$ of CTRSs is called *modular for constructor-sharing CTRSs* if for all CTRSs $\mathcal{R}_1$ and $\mathcal{R}_2$ which share at most constructors, their union $\mathcal{R}_1 \cup \mathcal{R}_2$ has the property $\mathcal{P}$ if and only if both $\mathcal{R}_1$ and $\mathcal{R}_2$ have the property $\mathcal{P}$.

If $(\mathcal{F}_1, \mathcal{R}_1)$ and $(\mathcal{F}_2, \mathcal{R}_2)$ are constructor-sharing CTRSs, then $(\mathcal{F}, \mathcal{R}_1)$ and $(\mathcal{F}, \mathcal{R}_2)$ are also CTRSs, where $\mathcal{F} = \mathcal{F}_1 \cup \mathcal{F}_2$. In order to avoid misunderstandings, we write $\Rightarrow_{\mathcal{R}_i}$ for the rewrite relation associated with $(\mathcal{F}_i, \mathcal{R}_i)$ and $\to_{\mathcal{R}_i}$ for the rewrite relation associated with $(\mathcal{F}, \mathcal{R}_i)$, where $i \in \{1, 2\}$. If $s, t \in \mathcal{T}(\mathcal{F}_i, \mathcal{V})$ and $s \Rightarrow_{\mathcal{R}_i} t$, then we clearly have $s \to_{\mathcal{R}_i} t$. A priori, it is not clear at all whether the converse is also true. For, if $s \to_{\mathcal{R}_i} t$, then there exists a rewrite rule

$l \to r \Leftarrow s_1 \downarrow t_1, \ldots, s_n \downarrow t_n$ in $\mathcal{R}_i$, a substitution $\sigma : \mathcal{V} \to \mathcal{T}(\mathcal{F}, \mathcal{V})$, and a context $C[\,]$ such that $s = C[l\sigma], t = C[r\sigma]$, and $s_j\sigma \downarrow_{\mathcal{R}_i} t_j\sigma$ for all $j \in \{1, \ldots, n\}$. And $\sigma : \mathcal{V} \to \mathcal{T}(\mathcal{F}, \mathcal{V})$ may substitute mixed terms for extra-variables occurring in the conditions. From now on we implicitly assume that $(\mathcal{F}_1, \mathcal{R}_1)$ and $(\mathcal{F}_2, \mathcal{R}_2)$ are constructor-sharing join CTRSs and that $(\mathcal{F}, \mathcal{R})$ denotes their combined system $(\mathcal{F}_1 \cup \mathcal{F}_2, \mathcal{R}_1 \cup \mathcal{R}_2)$. In the following, $\to = \to_{\mathcal{R}} = \to_{\mathcal{R}_1 \cup \mathcal{R}_2}$.

Definition 3.2 A reduction rule $l \to r \Leftarrow s_1 \downarrow t_1, \ldots, s_n \downarrow t_n \in \mathcal{R}$ is called *constructor-lifting* if $root(r)$ is a shared constructor, i.e., $root(r) \in \mathcal{C}$. A rewrite rule $l \to r \Leftarrow s_1 \downarrow t_1, \ldots, s_n \downarrow t_n \in \mathcal{R}$ is said to be *layer-preserving* if it is neither collapsing nor constructor-lifting.

Definition 3.3 In order to enhance readability, function symbols from $\mathcal{D}_1$ are called black, those from $\mathcal{D}_2$ white, and shared constructors as well as variables are called transparent. If a term s does not contain white (black) function symbols, we speak of a *black (white) term*. s is said to be *transparent* if it only contains shared constructors and variables. Consequently, a transparent term may be regarded as black or white, this is convenient for later purposes. s is called *top black (top white, top transparent)* if $root(s)$ is black (white, transparent). To emphasize that $\mathcal{F}_i = \mathcal{D}_i \cup \mathcal{C}$, we write $\mathcal{T}(\mathcal{D}_i, \mathcal{C}, \mathcal{V})$ instead of $\mathcal{T}(\mathcal{F}_i, \mathcal{V})$ at the appropriate places.

In the sequel, we often state definitions and considerations only for one color (the same applies mutatis mutandis for the other color).

Definition 3.4 If s is a top black term such that $s = C^b[s_1, \ldots, s_n]$ for some black context $C^b[, \ldots,] \neq \Box$ and $root(s_j) \in \mathcal{D}_2$ for $j \in \{1, \ldots, n\}$, then we denote this by $s = C^b[\![s_1, \ldots, s_n]\!]$. In this case we define the multiset $S^b_{\mathcal{P}}(s)$ of all *black principal* subterms of s to be $S^b_{\mathcal{P}}(s) = [s]$ and the set of all *white principal* subterms of s to be $S^w_{\mathcal{P}}(s) = [s_1, \ldots, s_n]$. The *topmost black homogeneous part* of s, denoted by $top^b(s)$, is obtained from s by replacing all white principal subterms with $\Box$. The *topmost white homogeneous part* of s is $top^w(s) = \Box$.
If s is a top transparent term such that

$$s = \begin{cases} C^t[s_1, \ldots, s_l] \text{ where } C^t[, \ldots,] \in \mathcal{T}(\mathcal{C}, \mathcal{V}) \text{ , } root(s_j) \in \mathcal{D}_1 \cup \mathcal{D}_2 \\ C^b[t_1, \ldots, t_m] \text{ where } C^b[, \ldots,] \in \mathcal{T}(\mathcal{D}_1, \mathcal{C}, \mathcal{V}) \text{ , } root(t_j) \in \mathcal{D}_2 \\ C^w[u_1, \ldots, u_n] \text{ where } C^w[, \ldots,] \in \mathcal{T}(\mathcal{D}_2, \mathcal{C}, \mathcal{V}) \text{ , } root(u_j) \in \mathcal{D}_1, \end{cases}$$

then this will be denoted by

$$s = \begin{cases} C^t[\![s_1, \ldots, s_l]\!] \text{ (note that } C^t[, \ldots,] \neq \Box) \\ C^b[\![t_1, \ldots, t_m]\!] \text{ (note that } C^b[, \ldots,] \neq \Box) \\ C^w[\![u_1, \ldots, u_n]\!] \text{ (note that } C^w[, \ldots,] \neq \Box). \end{cases}$$

In this situation, we define the multiset $S^b_{\mathcal{P}}(s)$ ($S^w_{\mathcal{P}}(s)$, respectively) of *black (white, respectively) principal* subterms of s to be $S^b_{\mathcal{P}}(s) = [u_1, \ldots, u_n]$ ($S^w_{\mathcal{P}}(s) = [t_1, \ldots, t_m]$, respectively). The *topmost black (white) homogeneous part* of s, denoted by $top^b(s)$ ($top^w(s)$), is obtained from s by replacing all white (black) principal subterms with $\Box$.

Example 3.5 Let $\mathcal{D}_1 = \{F, A\}$, $\mathcal{D}_2 = \{g, b\}$, and $\mathcal{C} = \{c\}$. The term $s = c(F(b), g(A))$ has representations

$$s = \begin{cases} C^t[\![F(b), g(A)]\!] \text{ with } C^t[,\ldots,] = c(\Box, \Box) \\ C^b[\![b, g(A)]\!] \text{ with } C^b[,\ldots,] = c(F(\Box), \Box) \\ C^w[\![F(b), A]\!] \text{ with } C^w[,\ldots,] = c(\Box, g(\Box)) \end{cases}$$

Definition 3.6 For a top black term s, the *rank* of s is defined by

$$rank(s) = \begin{cases} 1 & \text{, if } s \in \mathcal{T}(\mathcal{D}_1, \mathcal{C}, \mathcal{V}) \\ 1 + max\{rank(s_j) \mid 1 \le j \le n\}, & \text{if } s = C^b[\![s_1, \ldots, s_n]\!] \end{cases}$$

If s is a top transparent term, then the rank of s is defined by

$$rank(s) = \begin{cases} 0 & \text{, if } s \in \mathcal{T}(\mathcal{C}, \mathcal{V}) \\ max\{rank(t_j) \mid 1 \le j \le m\}, & \text{if } s = C^t[\![t_1, \ldots, t_m]\!] \end{cases}$$

Definition 3.7 Let s be a top black term such that $s = C^b[\![s_1, \ldots, s_n]\!]$ and $s \rightarrow_{\mathcal{R}} t$. We write $s \rightarrow^i_{\mathcal{R}} t$ if the rewrite rule is applied in one of the s_j and we write $s \rightarrow^o_{\mathcal{R}} t$ otherwise. Now let s be a top transparent term. If $s = C^t[\![s_1, \ldots, s_n]\!]$ and $s \rightarrow_{\mathcal{R}} t$, then $t = C^t[s_1, \ldots, s_{j-1}, t_j, s_{j+1}, \ldots, s_n]$ for some $j \in \{1, \ldots, n\}$. In this case we write $s \rightarrow^i_{\mathcal{R}} t$ if $s_j \rightarrow^i_{\mathcal{R}} t_j$ and $s \rightarrow^o_{\mathcal{R}} t$ if $s_j \rightarrow^o_{\mathcal{R}} t_j$. The reduction step $s \rightarrow^i_{\mathcal{R}} t$ is called *inner* reduction step and $s \rightarrow^o_{\mathcal{R}} t$ is called *outer* reduction step.

Definition 3.8 Let s be a top black term. A rewrite step $s \rightarrow t$ is *destructive at level 1* if the root symbols of s and t have different colors (i.e., $root(t) \in \mathcal{D}_2 \cup \mathcal{C} \cup \mathcal{V}$). A rewrite step $s \rightarrow t$ is *destructive at level* $m+1$ (for some $m \geq 1$) if $s = C^b[\![s_1, \ldots, s_j, \ldots, s_n]\!] \rightarrow^i C^b[s_1, \ldots, t_j, \ldots, s_n] = t$ with $s_j \rightarrow t_j$ destructive at level m. For a top transparent term s a rewrite step $s \rightarrow t$ is *destructive at level* m if it is of the form $s = C^t[\![s_1, \ldots, s_j, \ldots, s_n]\!] \rightarrow C^t[s_1, \ldots, t_j, \ldots, s_n] = t$ with $s_j \rightarrow t_j$ destructive at level m. Note that if a rewrite step is destructive, then the applied rewrite rule is collapsing or constructor-lifting.

Definition 3.9 As in [Mid90], we introduce some special notations in order to enable a compact treatment of "degenerate" cases of $t = C^b[\![t_1, \ldots, t_n]\!]$. To this end, the notion of context is extended. We write $C^b\langle,\ldots,\rangle$ for a black term containing zero or more occurrences of $\Box$ and $C^b\{,\ldots,\}$ for a black term different from $\Box$ itself, containing zero or more occurrences of $\Box$. If $t_1, \ldots, t_n$ are the (possibly zero) white principal subterms of some term t (from left to right), then we write $t = C^b\{\!\{t_1, \ldots, t_n\}\!\}$ provided that $t = C^b\{t_1, \ldots, t_n\}$. We write $t = C^b\langle\!\langle t_1, \ldots, t_n\rangle\!\rangle$ if $t = C^b\langle t_1, \ldots, t_n\rangle$ and either $C^b\langle,\ldots,\rangle \neq \Box$ and $t_1, \ldots, t_n$ are the white principal subterms of t or $C^b\langle,\ldots,\rangle = \Box$ and $t \in \{t_1, \ldots, t_n\}$.

Lemma 3.10 If $s \rightarrow^* t$, then $rank(s) \geq rank(t)$.
Proof: Routine. $\Box$

Definition 3.11 Let σ and τ be substitutions. We write $\sigma \propto \tau$ if $x\sigma = y\sigma$ implies $x\tau = y\tau$ for all $x, y \in \mathcal{V}$. The notation $\sigma \to^* \tau$ is used if $x\sigma \to^* x\tau$ for all $x \in \mathcal{V}$. Note that $\sigma \to^* \tau$ implies $t\sigma \to^* t\tau$ for all $t \in \mathcal{T}(\mathcal{F}, \mathcal{V})$. Moreover, σ is said to be in *normal form* or $\to$ *normalized* if $x\sigma \in NF(\to)$ for every $x \in \mathcal{V}$. A substitution σ is called *black* if $x\sigma$ is black for all $x \in \mathcal{D}om(\sigma)$ and it is said to be *top black* if $x\sigma$ is top black for all $x \in \mathcal{D}om(\sigma)$.

Proposition 3.12 Every substitution σ can be decomposed into $\sigma_2 \circ \sigma_1$ such that σ_1 is black, σ_2 is top white, and $\sigma_2 \propto \epsilon$.
Proof: Essentially the same as for disjoint conditional term rewriting systems (see [Mid93] and cf. also [Ohl94a]). □

4 Modular Properties

4.1 Semi-Completeness

Recently, we have shown that semi-completeness is a modular property of unconditional TRSs with shared constructors [Ohl94b]. Our first goal is to extend this result to constructor-sharing CTRSs. We use the structure of the proof of the modularity of confluence for disjoint CTRSs [Mid90, Mid93]. The proof idea is to construct two rewrite relations $\to_1$ and $\to_2$ on $\mathcal{T}(\mathcal{F}, \mathcal{V})$ such that their union is semi-complete, and reduction in the combined system $\mathcal{R}$ corresponds to joinability with respect to $\to_{1,2} = \to_1 \cup \to_2$. From these two properties and the equality $NF(\mathcal{F}, \mathcal{R}) = NF(\to_{1,2})$, the modularity of semi-completeness for CTRSs with shared constructors follows.

Definition 4.1 The rewrite relation $\to_1$ is defined by: $s \to_1 t$ if there exists a rewrite rule $l \to r \Leftarrow s_1 \downarrow t_1, \ldots, s_n \downarrow t_n$ in $\mathcal{R}_1$, a substitution $\sigma : \mathcal{V} \to \mathcal{T}(\mathcal{F}, \mathcal{V})$, and a context $C[\,]$ such that $s = C[l\sigma], t = C[r\sigma]$, and $s_j\sigma \downarrow_1^o t_j\sigma$ for $j \in \{1, \ldots, n\}$. Here the superscript o in $s_j\sigma \downarrow_1^o t_j\sigma$ means that $s_j\sigma$ and $t_j\sigma$ are joinable using only *outer* $\to_1$ reduction steps. The relation $\to_2$ is defined analogously. The union of $\to_1$ and $\to_2$ is denoted by $\to_{1,2}$.

Example 4.2 Consider the CTRSs $\mathcal{R}_1 = \{F(x, c) \to G(x) \Leftarrow x \downarrow c\}$ and $\mathcal{R}_2 = \{a \to c\}$. We have $F(a, c) \to_{\mathcal{R}} G(a)$ but neither $F(a, c) \to_1 G(a)$ nor $F(a, c) \to_2 G(a)$. However, the terms are joinable with respect to $\to_{1,2}$: $F(a, c) \to_2 F(c, c) \to_1 G(c) \leftarrow_2 G(a)$.

Lemma 4.3 If $s \to_{1,2} t$, then $s \to_{\mathcal{R}} t$.
Proof: Trivial. □

Lemma 4.4 Let s be a black term and let σ be a top white substitution such that $s\sigma \to_1^o t$. Then there is a black term u such that $t = u\sigma$.
Proof: Straightforward. □

Lemma 4.5 Let s, t be black terms and let σ be a top white substitution with $s\sigma \to_1^\circ t\sigma$. If τ is a substitution with $\sigma \propto \tau$, then $s\tau \to_1^\circ t\tau$.
Proof: The lemma is proved by induction on the depth of $s\sigma \to_1^\circ t\sigma$. The case of zero depth is straightforward. Let the depth of $s\sigma \to_1^\circ t\sigma$ equal $d+1$, $d \geq 0$. There is a context $C[\,]$, a substitution $\rho : \mathcal{V} \to \mathcal{T}(\mathcal{F}, \mathcal{V})$, and a rewrite rule $l \to r \Leftarrow s_1 \downarrow t_1, \ldots, s_n \downarrow t_n$ in $\mathcal{R}_1$ such that $s\sigma = C[l\rho]$, $t\sigma = C[r\rho]$ and $s_j\rho \downarrow_1^\circ t_j\rho$ is of depth $\leq d$ for every $j \in \{1, \ldots, n\}$. According to Proposition 3.12, ρ has a decomposition $\rho = \rho_2 \circ \rho_1$ such that ρ_1 is black, ρ_2 is top white, and $\rho_2 \propto \epsilon$. We define a substitution ρ' by $\rho'(x) = \tau(y)$ for every $x \in \mathcal{D}om(\rho_2)$ and $y \in \mathcal{D}om(\sigma)$ satisfying $\rho_2(x) = \sigma(y)$. ρ' is well-defined because $\sigma \propto \tau$. It follows from $\rho_2 \propto \epsilon$ and $\epsilon \propto \rho'$ that $\rho_2 \propto \rho'$. By Lemma 4.4, for any $j \in \{1, \ldots, n\}$, we may write

$$\rho_2(\rho_1(s_j)) = \rho_2(u_1) \to_1^\circ \cdots \to_1^\circ \rho_2(u_k) = \rho_2(v_l) \; {}_1^\circ\!\leftarrow \cdots {}_1^\circ\!\leftarrow \rho_2(v_1) = \rho_2(\rho_1(t_j))$$

for some black terms $u_1, \ldots, u_k, v_1, \ldots, v_l$. Now repeated application of the induction hypothesis yields

$$\rho'(\rho_1(s_j)) = \rho'(u_1) \to_1^\circ \cdots \to_1^\circ \rho'(u_k) = \rho'(v_l) \; {}_1^\circ\!\leftarrow \cdots {}_1^\circ\!\leftarrow \rho'(v_1) = \rho'(\rho_1(t_j))$$

Thus $\rho'(\rho_1(l)) \to_1^\circ \rho'(\rho_1(r))$. Let $\hat{C}[\,]$ be the context obtained from $C[\,]$ by replacing every white principal subterm which must be of the form $\sigma(x)$ for some variable $x \in \mathcal{D}om(\sigma)$ by the corresponding $\tau(x)$. (This is a slight abuse of notation because $\hat{C}[\,]$ contains in general more than one occurrence of $\Box$.) It is fairly simple to verify that $s\tau = \hat{C}[\rho'(\rho_1(l))]$ and $t\tau = \hat{C}[\rho'(\rho_1(r))]$. Hence $s\tau \to_1^\circ t\tau$. $\Box$

Lemma 4.6 The restriction of $\to_1$ to $\mathcal{T}(\mathcal{F}_1, \mathcal{V}) \times \mathcal{T}(\mathcal{F}_1, \mathcal{V})$ and $\Rightarrow_{\mathcal{R}_1}$ coincide.
Proof: "$\supseteq$" Trivial.
"$\subseteq$" Let $s, t \in \mathcal{T}(\mathcal{F}_1, \mathcal{V})$ with $s \to_1 t$. In order to show that $s \Rightarrow_{\mathcal{R}_1} t$, we proceed by induction on the depth of $s \to_1^\circ t$. The case of zero depth is straightforward. So suppose that the depth of $s \to_1^\circ t$ equals $d+1$, $d \geq 0$. Then there exists a rewrite rule $l \to r \Leftarrow s_1 \downarrow t_1, \ldots, s_n \downarrow t_n$ in $\mathcal{R}_1$, a substitution $\sigma : \mathcal{V} \to \mathcal{T}(\mathcal{F}, \mathcal{V})$, and a context $C[\,]$ such that $s = C[l\sigma], t = C[r\sigma]$, and $s_j\sigma \downarrow_1^\circ t_j\sigma$ with depth $\leq d$ for $j \in \{1, \ldots, n\}$. According to Proposition 3.12, σ can be decomposed into $\sigma_2 \circ \sigma_1$ such that σ_1 is black, σ_2 is top white, and $\sigma_2 \propto \epsilon$. Induction on the number of rewrite steps in $s_j\sigma \downarrow_1^\circ t_j\sigma$ in combination with Lemma 4.5 yields $\epsilon(\sigma_1(s_j)) \downarrow_1^\circ \epsilon(\sigma_1(t_j))$ for $j \in \{1, \ldots, n\}$. Since every term in the conversion $\sigma_1(s_j) \downarrow_1^\circ \sigma_1(t_j)$ is black, we obtain $\sigma_1(s_j) \Downarrow_{\mathcal{R}_1} \sigma_1(t_j)$ by repeated application of the induction hypothesis. Consequently, we have $\sigma_1(l) \Rightarrow_{\mathcal{R}_1} \sigma_1(r)$. Now $s \Rightarrow_{\mathcal{R}_1} t$ follows from $s = C[l\sigma] = C[l\sigma_1]$ and $t = C[r\sigma] = C[r\sigma_1]$ because s and t are black. $\Box$

Proposition 4.7 If $(\mathcal{F}_1, \mathcal{R}_1)$ and $(\mathcal{F}_2, \mathcal{R}_2)$ are semi-complete constructor-sharing conditional term rewriting systems, then the relation $\to_{1,2}$ is semi-complete.
Proof: We define two unconditional TRSs $(\mathcal{F}_1, \mathcal{S}_1)$ and $(\mathcal{F}_2, \mathcal{S}_2)$ by

$$\mathcal{S}_i = \{u \to v \mid u, v \in \mathcal{T}(\mathcal{F}_i, \mathcal{V}), root(u) \notin \mathcal{F}_1 \cap \mathcal{F}_2 \text{ and } u \to_i v\}.$$

First of all note that $(\mathcal{F}_1, \mathcal{S}_1)$ and $(\mathcal{F}_2, \mathcal{S}_2)$ are constructor-sharing TRSs. By Lemma 4.6, the restriction of $\rightarrow_i$ to $\mathcal{T}(\mathcal{F}_i, \mathcal{V}) \times \mathcal{T}(\mathcal{F}_i, \mathcal{V})$ and $\Rightarrow_{\mathcal{R}_i}$ coincide. It is easy to show that $\rightarrow_{\mathcal{S}_i}$ and the restriction of $\rightarrow_i$ to $\mathcal{T}(\mathcal{F}_i, \mathcal{V}) \times \mathcal{T}(\mathcal{F}_i, \mathcal{V})$ are also the same. Hence $\rightarrow_{\mathcal{S}_i}$ and $\Rightarrow_{\mathcal{R}_i}$ coincide on $\mathcal{T}(\mathcal{F}_i, \mathcal{V}) \times \mathcal{T}(\mathcal{F}_i, \mathcal{V})$. In particular, the TRS $(\mathcal{F}_i, \mathcal{S}_i)$ is semi-complete because $(\mathcal{F}_i, \mathcal{R}_i)$ is semi-complete. It follows from the modularity of semi-completeness for constructor-sharing unconditional TRSs [Ohl94b] that $(\mathcal{F}_1 \cup \mathcal{F}_2, \mathcal{S}_1 \cup \mathcal{S}_2)$ is also semi-complete.

We next show that the relations $\rightarrow_{\mathcal{S}_i}$ and $\rightarrow_i$ are also the same on $\mathcal{T}(\mathcal{F}, \mathcal{V}) \times \mathcal{T}(\mathcal{F}, \mathcal{V})$.

"$\subseteq$" Straightforward.

"$\supseteq$" Without loss of generality let $i = 1$. If $s \rightarrow_1 t$, then there exists a rewrite rule $l \rightarrow r \Leftarrow s_1 \downarrow t_1, \ldots, s_n \downarrow t_n$ in $\mathcal{R}_1$, a substitution $\sigma : \mathcal{V} \rightarrow \mathcal{T}(\mathcal{F}, \mathcal{V})$, and a context $C[\,]$ such that $s = C[l\sigma], t = C[r\sigma]$, and $s_j\sigma \downarrow_1^o t_j\sigma$ for $j \in \{1, \ldots, n\}$. Note that particularly $l\sigma \rightarrow_1 r\sigma$. According to Proposition 3.12, σ has a decomposition $\sigma = \sigma_2 \circ \sigma_1$ such that σ_1 is black, σ_2 is top white, and $\sigma_2 \propto \epsilon$. Now we apply Lemma 4.5: $\sigma_1(l)$ and $\sigma_1(r)$ are black terms and σ_2 is a top white substitution with $\sigma_2(\sigma_1(l)) \rightarrow_1 \sigma_2(\sigma_1(r))$ and ϵ is a substitution with $\sigma_2 \propto \epsilon$. Consequently, we obtain $\sigma_1(l) = \epsilon(\sigma_1(l)) \rightarrow_1 \epsilon(\sigma_1(r)) = \sigma_1(r)$. Since $\sigma_1(l)$ and $\sigma_1(r)$ are black terms and $root(\sigma_1(l)) = root(l) \notin \mathcal{F}_1 \cap \mathcal{F}_2$, it follows that $\sigma_1(l) \rightarrow \sigma_1(r)$ is a rewrite rule of $\mathcal{S}_1$. Thus $s = C[\sigma_2(\sigma_1(l))] \rightarrow_{\mathcal{S}_1} C[\sigma_2(\sigma_1(r))] = t$.

With the above results, it further follows from

$$\rightarrow_{\mathcal{S}_1 \cup \mathcal{S}_2} \;=\; \rightarrow_{\mathcal{S}_1} \cup \rightarrow_{\mathcal{S}_2} \;=\; \rightarrow_1 \cup \rightarrow_2 \;=\; \rightarrow_{1,2}$$

that $\rightarrow_{1,2}$ is semi-complete on $\mathcal{T}(\mathcal{F}, \mathcal{V}) \times \mathcal{T}(\mathcal{F}, \mathcal{V})$. □

Definition 4.8 If $\rightarrow_{1,2}$ is semi-complete, then every term t has a unique normal form w.r.t. $\rightarrow_{1,2}$. In the sequel, this normal form will be denoted by $t^{\rightarrow}$. Furthermore, for any substitution σ, $\sigma^{\rightarrow}$ denotes $\{x \mapsto \sigma(x)^{\rightarrow} \mid x \in \mathcal{D}om(\sigma)\}$.

Lemma 4.9 Let $\rightarrow_{1,2}$ be semi-complete. If s and t are black terms and σ is a top white $\rightarrow_{1,2}$ normalized substitution such that $s\sigma \downarrow_{1,2} t\sigma$, then $s\sigma \downarrow_1^o t\sigma$.

Proof: We show that $s\sigma \rightarrow_{1,2} u$ implies $s\sigma \rightarrow_1^o u$. Since $u = v\sigma$ for some black term v by Lemma 4.4, the lemma then follows by a straightforward induction on the length of the valley. In order to prove the claim, we use induction on the depth of $s\sigma \rightarrow_{1,2} u$. The case of zero depth is trivial. So suppose that the depth of $s\sigma \rightarrow_{1,2} u$ equals $d + 1$, $d \geq 0$. Since σ is a top white $\rightarrow_{1,2}$ normalized substitution, there exists a rewrite rule $l \rightarrow r \Leftarrow s_1 \downarrow t_1, \ldots, s_n \downarrow t_n$ in $\mathcal{R}_1$, a substitution $\rho : \mathcal{V} \rightarrow \mathcal{T}(\mathcal{F}, \mathcal{V})$, and a context $C[\,]$ such that $s\sigma = C[l\rho]$, $u = C[r\rho]$, and $s_j\rho \downarrow_{1,2} t_j\rho$ with depth $\leq d$ for $j \in \{1, \ldots, n\}$. By Proposition 3.12, ρ can be decomposed into $\rho_2 \circ \rho_1$ such that ρ_1 is black, ρ_2 is top white, and $\rho_2 \propto \epsilon$. Note that for every $x \in \mathcal{D}om(\rho_2) \cap \mathcal{V}ar(l\rho_1)$, we have $\rho_2(x) \in NF(\rightarrow_{1,2})$. Nevertheless, we do not have $\rho_2(x) \in NF(\rightarrow_{1,2})$ in general because of possible extra variables. Since $\rightarrow_{1,2}$ is semi-complete, $\rho_2 \rightarrow_{1,2}^* \rho_2^{\rightarrow}$. Thus $\rho_2^{\rightarrow}(\rho_1(s_j)) \;{}_{1,2}^{*}\!\!\leftarrow s_j\rho \downarrow_{1,2} t_j\rho \rightarrow_{1,2}^* \rho_2^{\rightarrow}(\rho_1(t_j))$. The confluence of $\rightarrow_{1,2}$ guarantees $\rho_2^{\rightarrow}(\rho_1(s_j)) \downarrow_{1,2} \rho_2^{\rightarrow}(\rho_1(t_j))$ for every $j \in \{1, \ldots, n\}$. By Proposition 3.12, $\rho_2^{\rightarrow}$ can be decomposed into $\rho_4 \circ \rho_3$ such that ρ_3 is black, ρ_4 is top white, and $\rho_4 \propto \epsilon$. Evidently, $\rho_3(\rho_1(s_j))$, $\rho_3(\rho_1(t_j))$ are black terms and ρ_4 is a top white $\rightarrow_{1,2}$ normalized substitution. Hence the induction hypothesis yields $\rho_4(\rho_3(\rho_1(s_j))) \downarrow_1^o \rho_4(\rho_3(\rho_1(t_j)))$. In other words, $\rho_2^{\rightarrow}(\rho_1(s_j)) \downarrow_1^o \rho_2^{\rightarrow}(\rho_1(t_j))$, and we obtain as a consequence that $\rho_2^{\rightarrow}(\rho(l)) \rightarrow_1^o \rho_2^{\rightarrow}(\rho_1(r))$ and $C[\rho_2^{\rightarrow}(\rho_1(l))] \rightarrow_1^o C[\rho_2^{\rightarrow}(\rho_1(r))]$. Clearly, $s\sigma = C[\rho_2^{\rightarrow}(\rho_1(l))]$ and $u = C[\rho_2^{\rightarrow}(\rho_1(r))]$ because $\rho_2(x) \in NF(\rightarrow_{1,2})$ for every $x \in \mathcal{D}om(\rho_2) \cap \mathcal{V}ar(l\rho_1)$. This proves the claim. □

Lemma 4.10 Let $\rightarrow_{1,2}$ be semi-complete and let $s_1, \ldots, s_n, t_1, \ldots, t_n$ be black terms. If σ is a substitution with $s_j\sigma \downarrow_{1,2} t_j\sigma$ for every $j \in \{1, \ldots, n\}$, then $\sigma^{\rightarrow}(s_j) \downarrow_1^o \sigma^{\rightarrow}(t_j)$ for every $j \in \{1, \ldots, n\}$.
Proof: We have $\sigma^{\rightarrow}(s_j) \; {}_{1,2}^{*}\!\leftarrow s_j\sigma \downarrow_{1,2} t_j\sigma \rightarrow_{1,2}^{*} \sigma^{\rightarrow}(t_j)$. The confluence of $\rightarrow_{1,2}$ implies $\sigma^{\rightarrow}(s_j) \downarrow_{1,2} \sigma^{\rightarrow}(t_j)$. Proposition 3.12 yields a decomposition of $\sigma^{\rightarrow}$ into $\sigma_2 \circ \sigma_1$ such that σ_1 is black and σ_2 is top white. Evidently, $\sigma_1(s_j)$, $\sigma_1(t_j)$ are black terms and σ_2 is a top white $\rightarrow_{1,2}$ normalized substitution. According to Lemma 4.9, we eventually derive $\sigma^{\rightarrow}(s_j) = \sigma_2(\sigma_1(s_j)) \downarrow_1^o \sigma_2(\sigma_1(t_j)) = \sigma^{\rightarrow}(t_j)$.
□

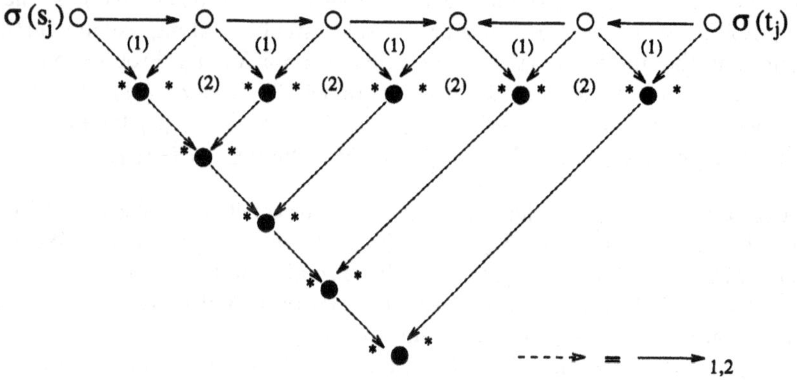

Fig. 1.

Proposition 4.11 Let $(\mathcal{F}_1, \mathcal{R}_1)$ and $(\mathcal{F}_2, \mathcal{R}_2)$ be semi-complete constructor-sharing CTRSs. If $s \rightarrow_{\mathcal{R}} t$, then $s \downarrow_{1,2} t$.

Proof: We proceed by induction on the depth of $s \rightarrow t$. The case of zero depth is trivial. So suppose that the depth of $s \rightarrow t$ equals $d+1$, $d \geq 0$. Then there is a context $C[\,]$, a substitution $\sigma : \mathcal{V} \rightarrow \mathcal{T}(\mathcal{F}, \mathcal{V})$, and a rewrite rule $l \rightarrow r \Leftarrow s_1 \downarrow t_1, \ldots, s_n \downarrow t_n$ in $\mathcal{R}$ such that $s = C[l\sigma]$, $t = C[r\sigma]$, and $s_j\sigma \downarrow t_j\sigma$ is of depth less than or equal to d for every $j \in \{1, \ldots, n\}$. Fig. 1 depicts how the induction hypothesis and confluence of $\rightarrow_{1,2}$ yield $s_j\sigma \downarrow_{1,2} t_j\sigma$ for every $j \in \{1, \ldots, n\}$ (where (1) signals an application of the induction hypothesis and (2) stands for an application of Proposition 4.7). W.l.o.g. we may assume that the applied rewrite rule stems from $\mathcal{R}_1$. By Lemma 4.10, we have $\sigma^{\rightarrow}(s_j) \downarrow_1^o \sigma^{\rightarrow}(t_j)$ for $j \in \{1, \ldots, n\}$ and thus $\sigma^{\rightarrow}(l) \rightarrow_1 \sigma^{\rightarrow}(r)$. Finally, we obtain $s \downarrow_{1,2} t$ from

$$s = C[l\sigma] \rightarrow_{1,2}^* C[\sigma^{\rightarrow}(l)] \rightarrow_1 C[\sigma^{\rightarrow}(r)] \;{}_{1,2}^*\!\leftarrow C[r\sigma] = t.$$

□

Proposition 4.12 Let $(\mathcal{F}_1, \mathcal{R}_1)$ and $(\mathcal{F}_2, \mathcal{R}_2)$ be semi-complete constructor-sharing CTRSs. Then the relations $\leftrightarrow_{\mathcal{R}}^*$ and $\downarrow_{1,2}$ coincide.
Proof: This is a consequence of Lemma 4.3 and Propositions 4.11 and 4.7. □

Lemma 4.13 If $(\mathcal{F}_1, \mathcal{R}_1)$ and $(\mathcal{F}_2, \mathcal{R}_2)$ are semi-complete constructor-sharing CTRSs, then $NF(\mathcal{F}, \mathcal{R}) = NF(\rightarrow_{1,2})$.
Proof: "$\subseteq$" Trivial.
"$\supseteq$" If $NF(\rightarrow_{1,2}) \not\subseteq NF(\mathcal{F}, \mathcal{R})$, then there is a term s with $s \in NF(\rightarrow_{1,2})$ and $s \notin NF(\mathcal{F}, \mathcal{R})$. W.l.o.g. we may assume that s is of minimal size (i.e., $|s|$ is minimal). Hence s is a redex and every proper subterm of s is irreducible by $\rightarrow_{\mathcal{R}}$. Therefore, there exists a rewrite rule $l \rightarrow r \Leftarrow s_1 \downarrow t_1, \ldots, s_n \downarrow t_n$ in $\mathcal{R}$ and a substitution $\sigma : \mathcal{V} \rightarrow \mathcal{T}(\mathcal{F}, \mathcal{V})$ such that $s = l\sigma, t = r\sigma$, and $s_j\sigma \downarrow_{\mathcal{R}} t_j\sigma$ for all $j \in \{1, \ldots, n\}$. Note that for every variable $x \in \mathcal{D}om(\sigma) \cap \mathcal{V}ar(l)$, we have $\sigma(x) \in NF(\mathcal{F}, \mathcal{R})$ because $\sigma(x)$ is a proper subterm of s. W.l.o.g. we may further assume that the applied rewrite rule originates from $\mathcal{R}_1$. By Proposition 4.12, $s_j\sigma \downarrow_{1,2} t_j\sigma$ which, in conjunction with Lemma 4.10, yields $\sigma^{\rightarrow}(s_j) \downarrow_1^o \sigma^{\rightarrow}(t_j)$. It follows $s = \sigma(l) = \sigma^{\rightarrow}(l) \rightarrow_1^o \sigma^{\rightarrow}(r)$ because $\sigma(x) = \sigma^{\rightarrow}(x)$ for every $x \in \mathcal{V}ar(l)$. This means that $s \notin NF(\rightarrow_1)$, contradicting $s \in NF(\rightarrow_{1,2})$. □

Theorem 4.14 Semi-completeness is modular for constructor-sharing CTRSs.
Proof: Let $(\mathcal{F}_1, \mathcal{R}_1)$ and $(\mathcal{F}_2, \mathcal{R}_2)$ be CTRSs with shared constructors. We have to show that their combined system $(\mathcal{F}, \mathcal{R})$ is semi-complete if and only if both $(\mathcal{F}_1, \mathcal{R}_1)$ and $(\mathcal{F}_2, \mathcal{R}_2)$ are semi-complete. In order to show the if case, we consider a conversion $t_1 \;{}_{\mathcal{R}}^*\!\leftarrow s \rightarrow_{\mathcal{R}}^* t_2$. According to Proposition 4.12 we have $t_1 \downarrow_{1,2} t_2$. Since $\rightarrow_{1,2}$ is semi-complete, $t_1 \rightarrow_{1,2}^* t_3$ and $t_2 \rightarrow_{1,2}^* t_3$, where t_3 is the unique normal form of s, t_1, and t_2. Now Lemma 4.3 implies $t_1 \rightarrow_{\mathcal{R}}^* t_3 \;{}_{\mathcal{R}}^*\!\leftarrow t_2$. Thus $(\mathcal{F}, \mathcal{R})$ is confluent. It remains to show normalization of $\rightarrow_{\mathcal{R}}$. Let $s \in \mathcal{T}(\mathcal{F}, \mathcal{V})$. Since $\rightarrow_{1,2}$ is normalizing, $s \rightarrow_{1,2}^* t$ for some $t \in NF(\rightarrow_{1,2})$. By Lemma 4.3, $s \rightarrow_{\mathcal{R}}^* t$. It follows from Lemma 4.13 that $t \in NF(\mathcal{F}, \mathcal{R})$. Hence $(\mathcal{F}, \mathcal{R})$ is also normalizing. This all proves that $(\mathcal{F}, \mathcal{R})$ is semi-complete. The only-if case follows straightforwardly from Lemma 4.15. □

Lemma 4.15 Let $(\mathcal{F}, \mathcal{R})$ be the combined system of two constructor-sharing CTRSs $(\mathcal{F}_1, \mathcal{R}_1)$ and $(\mathcal{F}_2, \mathcal{R}_2)$ such that $(\mathcal{F}, \mathcal{R})$ is semi-complete. If s is a black term and $s \rightarrow_{\mathcal{R}} t$, then $s \Rightarrow_{\mathcal{R}_1} t$.
Proof: We show the following stronger claim, where the rewrite relation associated with $(\mathcal{F}_1 \cup \{\square\}, \mathcal{R}_1)$ is also denoted by $\Rightarrow_{\mathcal{R}_1}$.
Claim: If s is a black term and σ is a top white $\rightarrow_{\mathcal{R}}$ normalized substitution such that $s\sigma \rightarrow_{\mathcal{R}} t\sigma$, then $s\sigma^{\square} \Rightarrow_{\mathcal{R}_1} t\sigma^{\square}$, where $\sigma^{\square}$ denotes the substitution $\{x \mapsto \square \mid x \in \mathcal{D}om(\sigma)\}$.

Since $\mathcal{R}$ is semi-complete, every term t has a unique normal form $t\downarrow$ w.r.t. $\mathcal{R}$. Furthermore, for any substitution σ, let $\sigma\downarrow$ denote the substitution $\{x \mapsto \sigma(x)\downarrow \mid x \in \mathcal{D}om(\sigma)\}$. The claim is proved by induction on the depth of $s\sigma \to t\sigma$. The case of zero depth is straightforward. Let the depth of $s\sigma \to t\sigma$ equal $d+1$, $d \geq 0$. There is a context $C[\,]$, a substitution $\rho : \mathcal{V} \to \mathcal{T}(\mathcal{F}, \mathcal{V})$, and a rewrite rule $l \to r \Leftarrow s_1 \downarrow t_1, \ldots, s_n \downarrow t_n$ in $\mathcal{R}_1$ such that $s\sigma = C[l\rho]$, $t\sigma = C[r\rho]$ and $s_j\rho \downarrow t_j\rho$ is of depth $\leq d$ for every $j \in \{1, \ldots, n\}$. By Proposition 3.12, ρ can be decomposed into $\rho_2 \circ \rho_1$ such that ρ_1 is black and ρ_2 is top white. Note that for every $x \in \mathcal{D}om(\rho_2) \cap \mathcal{V}ar(l\rho_1)$, we have $\rho_2(x) \in NF(\to)$. Nevertheless, we do not have $\rho_2(x) \in NF(\to)$ in general because of possible extra variables. Since $\to$ is semi-complete, $\rho_2 \to^* \rho_2\downarrow$. Thus $\rho_2\downarrow(\rho_1(s_j))\ {}^*\!\leftarrow s_j\rho \downarrow t_j\rho \to^* \rho_2\downarrow(\rho_1(t_j))$. The confluence of $\to$ guarantees $\rho_2\downarrow(\rho_1(s_j)) \downarrow \rho_2\downarrow(\rho_1(t_j))$ for every $j \in \{1, \ldots, n\}$. By Proposition 3.12, $\rho_2\downarrow$ can be decomposed into $\rho_4 \circ \rho_3$ such that ρ_3 is black and ρ_4 is top white. Evidently, $\rho_3(\rho_1(s_j))$ and $\rho_3(\rho_1(t_j))$ are black terms and ρ_4 is a top white $\to$ normalized substitution. Repeated application of the induction hypothesis yields $\rho_4^{\Box}(\rho_3(\rho_1(s_j))) \Downarrow_{\mathcal{R}_1} \rho_4^{\Box}(\rho_3(\rho_1(t_j)))$. We obtain as a consequence that $\rho_4^{\Box}(\rho_3(\rho_1(l))) \Rightarrow_{\mathcal{R}_1} \rho_4^{\Box}(\rho_3(\rho_1(r)))$. Clearly, $s\sigma = C[\rho_2\downarrow(\rho_1(l))]$ and $t\sigma = C[\rho_2\downarrow(\rho_1(r))]$ because $\rho_2(x) \in NF(\to)$ for every $x \in \mathcal{D}om(\rho_2) \cap \mathcal{V}ar(l\rho_1)$. Let $\hat{C}[\,]$ be the context obtained from $C[\,]$ by replacing every white principal subterm which must be of the form $\sigma(x)$ for some variable $x \in \mathcal{D}om(\sigma)$ with $\Box$. It is fairly simple to verify that $s\sigma^{\Box} = \hat{C}[\rho_4^{\Box}(\rho_3(\rho_1(l)))]$ and $t\sigma^{\Box} = \hat{C}[\rho_4^{\Box}(\rho_3(\rho_1(r)))]$. Thus $s\sigma^{\Box} \Rightarrow_{\mathcal{R}_1} t\sigma^{\Box}$. This proves the claim. $\Box$

It has been shown by Middeldorp [Mid90, Mid93] that confluence is also modular for semi-equational disjoint CTRSs. Analogously, it should be possible to prove that Theorem 4.14 also holds true for semi-equational CTRSs.

4.2 Termination

In contrast to the unconditional case (see [Rus87]), termination is not modular for non-duplicating disjoint CTRSs. This is witnessed by the following example which stems from [Mid93].

Example 4.16 Consider the CTRSs

$$\mathcal{R}_1 = \{\, F(x) \to F(x) \Leftarrow x \downarrow A,\ x \downarrow B \,\}$$

$$\mathcal{R}_2 = \begin{cases} g(x, y) \to x \\ g(x, y) \to y \end{cases}$$

Both systems are terminating and non-duplicating but their combined system is non-terminating.

Note that the CTRS $\mathcal{R}_2$ in Example 4.16 is not confluent. Middeldorp [Mid90, Mid93] has shown that this is essential. He proved the following theorem.

Theorem 4.17 If $\mathcal{R}_1$ and $\mathcal{R}_2$ are terminating disjoint CTRSs, then their combined system $\mathcal{R} = \mathcal{R}_1 \cup \mathcal{R}_2$ is terminating provided that one of the following conditions is satisfied:

1. Neither $\mathcal{R}_1$ nor $\mathcal{R}_2$ contain collapsing rules.
2. Both systems are confluent and non-duplicating.
3. Both systems are confluent and one of the systems contains neither collapsing nor duplicating rules.

Furthermore, he conjectured that the disjoint union of two terminating join CTRSs is terminating if one of them contains neither collapsing nor duplicating rules and the other is confluent. The next example disproves this conjecture. The function symbols have been chosen in resemblance to other known counterexamples.

Example 4.18 Let

$$\mathcal{R}_1 = \left\{ \begin{array}{ccccc} 0 & & & & 1 \\ \downarrow & \searrow & & \swarrow & \downarrow \\ A & & 2 & & B \end{array} \right. \qquad F(x) \rightarrow F(x) \Leftarrow x \downarrow A,\ x \downarrow B$$

and

$$\mathcal{R}_2 = \left\{ \begin{array}{l} g\,(x, y, y) \rightarrow x \\ g\,(y, y, x) \rightarrow x. \end{array} \right.$$

Clearly, $\mathcal{R}_1$ is non-collapsing, non-duplicating, and terminating (there is no term $t \in \mathcal{T}(\mathcal{F}_1, \mathcal{V})$ which rewrites to both A and B). Note that $\mathcal{R}_1$ is not confluent. Moreover, the CTRS $\mathcal{R}_2$ is evidently terminating and confluent. However, their disjoint union $\mathcal{R} = \mathcal{R}_1 \cup \mathcal{R}_2$ is not terminating. Since

$$B \;{}_{\mathcal{R}}\!\leftarrow 1 \;{}_{\mathcal{R}}\!\leftarrow g(0,0,1) \rightarrow_{\mathcal{R}} g(0,2,1) \rightarrow_{\mathcal{R}} g(0,2,2) \rightarrow_{\mathcal{R}} 0 \rightarrow_{\mathcal{R}} A,$$

there is the cyclic reduction "sequence" $F(g(0,0,1)) \rightarrow_{\mathcal{R}} F(g(0,0,1))$.

It will next be shown that the above theorem also holds, mutatis mutandis, in the presence of shared constructors. We point out that our proof (though based on the ideas of [Mid93]) is considerably simpler than that of [Mid93].

As in the previous subsection, let $(\mathcal{F}_1, \mathcal{R}_1)$ and $(\mathcal{F}_2, \mathcal{R}_2)$ be constructor-sharing join CTRSs. It is not difficult to verify that the CTRS $(\mathcal{F}_i \cup \{\Box\}, \mathcal{R}_i)$ is terminating if and only if $(\mathcal{F}_i, \mathcal{R}_i)$ is terminating. Again, we also denote the rewrite relation associated with $(\mathcal{F}_i \cup \{\Box\}, \mathcal{R}_i)$ by $\Rightarrow_{\mathcal{R}_i}$ (by abuse of notation).

Proposition 4.19 Let $(\mathcal{F}_2, \mathcal{R}_2)$ be layer-preserving.

1. If $s \rightarrow^o t$ by some rule from $\mathcal{R}_1$, then $top^b(s) \Rightarrow_{\mathcal{R}_1} top^b(t)$.
2. If $s \rightarrow^o t$ by some rule from $\mathcal{R}_2$ or $s \rightarrow^i t$, then $top^b(s) = top^b(t)$.

Proof: We proceed by induction on the depth of $s \rightarrow t$. The case of zero depth is straightforward. So suppose that the depth of $s \rightarrow t$ equals $d+1$, $d \geq 0$. The induction hypothesis covers the statement that $u \rightarrow v$ implies $top^b(u) \Rightarrow^*_{\mathcal{R}_1} top^b(v)$ whenever $u \rightarrow v$ is of depth less than or equal to d.
(1) If $s \rightarrow^o t$ by some rule from $\mathcal{R}_1$, then $s = C^b\{\!\{u_1, \ldots, u_p\}\!\}$ and $t = \hat{C}^b\langle\!\langle u_{i_1}, \ldots, u_{i_q}\rangle\!\rangle$, where $i_1, \ldots, i_q \in \{1, \ldots, p\}$. Moreover, there is a context $C[\,]$,

a substitution σ and a rewrite rule $l \to r \Leftarrow s_1 \downarrow t_1, \ldots, s_n \downarrow t_n \in \mathcal{R}_1$ such that $s = C[l\sigma]$, $t = C[r\sigma]$ and $s_j\sigma \downarrow t_j\sigma$ is of depth less than or equal to d for every $j \in \{1, \ldots, n\}$. We first show that $top^b(s_j\sigma) \Downarrow_{\mathcal{R}_1} top^b(t_j\sigma)$ for every $j \in \{1, \ldots, n\}$. Fix j. Let w be the common reduct of $s_j\sigma$ and $t_j\sigma$. Clearly, it suffices to show that $top^b(s_j\sigma) \Rightarrow^*_{\mathcal{R}_1} top^b(w)$ and $top^b(t_j\sigma) \Rightarrow^*_{\mathcal{R}_1} top^b(w)$. W.l.o.g. we consider only the former claim. The claim is proved by induction on the length of $s_j\sigma \to^* w$. The case of zero length is trivial, so let $s_j\sigma \to v \to^l w$ with $l \geq 0$. The induction hypothesis (on l) yields $top^b(v) \Rightarrow^*_{\mathcal{R}_1} top^b(w)$. Furthermore, the induction hypothesis (on d) yields $top^b(s_j\sigma) \Rightarrow^*_{\mathcal{R}_1} top^b(v)$. This proves the claim. Thus $top^b(w)$ is a common reduct of $top^b(s_j\sigma)$ and $top^b(t_j\sigma)$ w.r.t. $\Rightarrow_{\mathcal{R}_1}$. According to Proposition 3.12, $\sigma = \sigma_2 \circ \sigma_1$, where σ_1 is a black substitution and σ_2 is top white. Recall that $\sigma_2^{\Box}$ denotes the substitution $\{x \mapsto \Box \mid x \in \mathcal{D}om(\sigma_2)\}$. It is clear that $top^b(s_j\sigma) = \sigma_2^{\Box}(\sigma_1(s_j))$ and $top^b(t_j\sigma) = \sigma_2^{\Box}(\sigma_1(t_j))$. Hence $\sigma_2^{\Box}(\sigma_1(s_j)) \Downarrow_{\mathcal{R}_1} \sigma_2^{\Box}(\sigma_1(t_j))$ and thus $\sigma_2^{\Box}(\sigma_1(l)) \Rightarrow_{\mathcal{R}_1} \sigma_2^{\Box}(\sigma_1(r))$. Let $\hat{C}[\,]$ be the context obtained from $C[\,]$ by replacing all white principal subterms with $\Box$. Now (1) follows from $top^b(s) = \hat{C}[\sigma_2^{\Box}(\sigma_1(l))]$ and $top^b(t) = \hat{C}[\sigma_2^{\Box}(\sigma_1(r))]$.
(2) Let $s \to^o t$ by some rule from $\mathcal{R}_2$ or $s \to^i t$. Since $\mathcal{R}_2$ is layer-preserving, we may write $s = C^b\langle\!\langle u_1, \ldots, u_j, \ldots, u_p \rangle\!\rangle$ and $t = C^b\langle\!\langle u_1, \ldots, u'_j, \ldots, u_p \rangle\!\rangle$, where $u_j \to u'_j$. Hence $top^b(s) = top^b(t)$. □

In the preceding proposition, the assumption that $(\mathcal{F}_2, \mathcal{R}_2)$ has to be layer-preserving cannot be dropped, as is witnessed by the next example (cf. [Mid90, Mid93]).

Example 4.20 Let $\mathcal{R}_1 = \{F(x) \to G(x) \Leftarrow x \downarrow A\}$ and $\mathcal{R}_2 = \{h(x) \to x\}$. Then $F(h(A)) \to^o G(h(A))$ by the only rule of $\mathcal{R}_1$ but $top^b(F(h(a)))) = F(\Box)$ is a normal form w.r.t. $\Rightarrow_{\mathcal{R}_1}$.

Our next goal is to show an analogous statement to Proposition 4.19 (1) without the layer-preservingness requirement on $(\mathcal{F}_2, \mathcal{R}_2)$ but under the additional assumption that $\to_{1,2}$ is semi-complete.

Definition 4.21 Let the rewrite relation $\to_{1,2}$ be semi-complete. For $t = C^b\langle\!\langle t_1, \ldots, t_m \rangle\!\rangle$, we define $top^b_{\to}(t)$ by:

$$top^b_{\to}(t) = top^b(C^b\langle t_1^{\to}, \ldots, t_m^{\to} \rangle).$$

In other words, first the white principal subterms in t are replaced with their unique $\to_{1,2}$ normal form, and then the topmost black homogeneous part of the term obtained is taken.

Lemma 4.22 Let $\to_{1,2}$ be semi-complete. If s and t are black terms and σ is a top white substitution such that $s\sigma \to^o t\sigma$ by some rewrite rule from $\mathcal{R}_1$, then $\sigma^{\to}(s) \to^o_1 \sigma^{\to}(t)$.
Proof: There is a context $C[\,]$, a substitution $\rho : \mathcal{V} \to \mathcal{T}(\mathcal{F}, \mathcal{V})$ and a rewrite rule $l \to r \Leftarrow s_1 \downarrow t_1, \ldots, s_n \downarrow t_n \in \mathcal{R}_1$ such that $s\sigma = C[l\rho]$, $t\sigma = C[r\rho]$ and $s_j\rho \downarrow t_j\rho$ for $j \in \{1, \ldots, n\}$. Fix j. From Proposition 4.12 we know that

$s_j\rho \downarrow_{1,2} t_j\rho$. According to Proposition 3.12, ρ can be decomposed into $\rho_2 \circ \rho_1$ such that ρ_1 is black and ρ_2 is top white. Since $\to_{1,2}$ is semi-complete, it follows as in the proof of Lemma 4.9 that $\rho_2^{\to}(\rho_1(s_j)) \downarrow_{1,2} \rho_2^{\to}(\rho_1(t_j))$. Applying Lemma 4.10 to the black terms $\rho_1(s_1), \ldots, \rho_1(s_n), \rho_1(t_1), \ldots, \rho_1(t_n)$ and the substitution $\rho_2^{\to}$ yields $\rho_2^{\to}(\rho_1(s_j)) \downarrow_1^o \rho_2^{\to}(\rho_1(t_j))$. Therefore, $\rho_2^{\to}(\rho_1(l)) \to_1^o \rho_2^{\to}(\rho_1(r))$. Let $\hat{C}[\,]$ be the context obtained from $C[\,]$ by replacing all white principal subterms with their respective $\to_{1,2}$ normal form. It is clear that $\sigma^{\to}(s) = \hat{C}[\rho_2^{\to}(\rho_1(l))]$ and $\sigma^{\to}(t) = \hat{C}[\rho_2^{\to}(\rho_1(r))]$. Thus $\sigma^{\to}(s) \to_1^o \sigma^{\to}(t)$. □

Proposition 4.23 Let $\to_{1,2}$ be semi-complete. If $s \to^o t$ by some rule from $\mathcal{R}_1$, then $top_{\to}^b(s) \Rightarrow_{\mathcal{R}_1} top_{\to}^b(t)$.

Proof: We may write $s = C^b\{\!\{s_1, \ldots, s_n\}\!\}$ and $t = \hat{C}^b\langle\!\langle s_{i_1}, \ldots, s_{i_m}\rangle\!\rangle$ for some black contexts $C^b\{, \ldots, \}$, $\hat{C}^b\langle, \ldots, \rangle$, and $i_1, \ldots, i_m \in \{1, \ldots, n\}$. Let $x_1, \ldots, x_n$ be distinct fresh variables and define $\sigma = \{x_j \mapsto s_j \mid 1 \le j \le n\}$, $s' = C^b\{x_1, \ldots, x_n\}$, and $t' = \hat{C}^b\langle x_{i_1}, \ldots, x_{i_m}\rangle$. Since σ is top white, we obtain $\sigma^{\to}(s') \to_1^o \sigma^{\to}(t')$ by Lemma 4.22. According to Proposition 3.12, $\sigma^{\to}$ has a decomposition $\sigma^{\to} = \sigma_2 \circ \sigma_1$, where σ_1 is black and σ_2 is top white. It follows from Lemma 4.5 that $\sigma_2^{\square}(\sigma_1(s')) \to_1^o \sigma_2^{\square}(\sigma_1(t'))$ because $\sigma_2 \propto \sigma_2^{\square}$. To verify that $\sigma_2^{\square}(\sigma_1(s')) \Rightarrow_{\mathcal{R}_1} \sigma_2^{\square}(\sigma_1(t'))$ is relatively simple. Now $top_{\to}^b(s) \Rightarrow_{\mathcal{R}_1} top_{\to}^b(t)$ is a consequence of

$$top_{\to}^b(s) = top^b(C^b\{s_1^{\to}, \ldots, s_n^{\to}\}) = top^b(\sigma^{\to}(s')) = top^b(\sigma_2(\sigma_1(s'))) = \sigma_2^{\square}(\sigma_1(s'))$$

and $top_{\to}^b(t) = \sigma_2^{\square}(\sigma_1(t'))$. □

The above preparatory considerations and the following two lemmata (the proofs of which are omitted) pave the way for the main results of this subsection.

Lemma 4.24 If $\to$ is well-founded on a set $T \subseteq \mathcal{T}(\mathcal{F}, \mathcal{V})$, then $> = (\to \cup \rhd)^+$ is a well-founded partial ordering on T. □

Lemma 4.25 Let $s \to^o t$ by an application of a non-duplicating rewrite rule from $\mathcal{R}_1$. Then $S_{\mathcal{P}}^{w}(t) \subseteq S_{\mathcal{P}}^{w}(s)$. □

Theorem 4.26 Let $\mathcal{R}_1$ and $\mathcal{R}_2$ be terminating constructor-sharing CTRSs such that their combined system $\mathcal{R} = \mathcal{R}_1 \cup \mathcal{R}_2$ is not terminating. Then the following statements hold (where $d, \bar{d} \in \{1, 2\}$ with $d \neq \bar{d}$):

1. There exists an infinite $\mathcal{R}$ rewrite derivation $D : s_1 \to s_2 \to s_3 \to \ldots$ of minimal rank such that D contains infinitely many $s_j \to^o s_{j+1}$ reduction steps where s_j reduces to s_{j+1} by some rule from $\mathcal{R}_d$.
2. $\mathcal{R}_{\bar{d}}$ is not layer-preserving.
3. If both systems are confluent, then D contains infinitely many duplicating $s_j \to^o s_{j+1}$ reduction steps such that s_j reduces to s_{j+1} by some rule from $\mathcal{R}_d$.

Proof: First of all, the rank of a derivation $D : s_1 \to s_2 \to s_3 \to \ldots$ is defined to be $rank(D) = rank(s_1)$. Let D be an infinite $\mathcal{R}$ rewrite derivation of minimal rank, say $rank(D) = k$. Then $rank(s_j) = rank(D)$ for all indices j. Moreover, $\to_{\mathcal{R}}$ is terminating on $\mathcal{T}^{<k} = \{t \in \mathcal{T}(\mathcal{F}, \mathcal{V}) \mid rank(t) < k\}$. If s_1 is top transparent, say $s_1 = C^t[\![t_1, \ldots, t_n]\!]$, then there must be an infinite rewrite derivation starting from some t_l, $l \in \{1, \ldots, n\}$ with $rank(t_l) = k$. Therefore, we may assume without loss of generality that s_1 is top black or top white, say top black. Fix $j \in \mathbb{N}$. It is not difficult to verify that the term s_j cannot be top white and that $rank(u) < k$ for each white principal subterm $u \in S^w_{\mathcal{P}}(s_j)$.
(1) Suppose that there are only finitely many $\to^o$ reduction steps using a rule from $\mathcal{R}_1$ in D. Then we find an index $j \in \mathbb{N}$ such that the rewrite derivation

$$D' : s_j \to s_{j+1} \to s_{j+2} \to \ldots$$

contains no such reduction steps at all. Thus, if $s_j = C^b[\![t_1, \ldots, t_n]\!]$, then there must be an infinite rewrite derivation starting from some $t_l \in S^w_{\mathcal{P}}(s_j)$. But this contradicts the minimality assumption on $rank(D)$ since $rank(t_l) < rank(s_j)$.
(2) Suppose that $\mathcal{R}_2$ is layer-preserving, i.e., it contains neither collapsing nor constructor-lifting rules. By Proposition 4.19, we have:

- If $s_j \to^o s_{j+1}$ by some rewrite rule from $\mathcal{R}_1$, then $top^b(s_j) \Rightarrow_{\mathcal{R}_1} top^b(s_{j+1})$.
- If $s_j \to^o s_{j+1}$ by some reduction rule from $\mathcal{R}_2$ or $s_j \to^i s_{j+1}$, then the equality $top^b(s_j) = top^b(s_{j+1})$ holds.

From (1) we know that infinitely many reduction steps of the former kind occur in D. This yields a contradiction to the termination of $\Rightarrow_{\mathcal{R}_1}$.
(3) Let $> = (\to_{\mathcal{R}} \cup \rhd)^+$. According to Lemma 4.24, $(\mathcal{T}^{<k}, >)$ is a well-founded ordering. Let $(\mathcal{M}(\mathcal{T}^{<k}), >^{mul})$ denote its well-founded multiset extension. Note that $S^w_{\mathcal{P}}(s_j) \in \mathcal{M}(\mathcal{T}^{<k})$. As in the proof of (1), we may suppose that there is no outer reduction step using a duplicating rule from $\mathcal{R}_1$ in D. We distinguish between three cases:

- If $s_j \to^o s_{j+1}$ by some rule from $\mathcal{R}_1$, then it follows from Lemma 4.25 that $S^w_{\mathcal{P}}(s_{j+1}) \subseteq S^w_{\mathcal{P}}(s_j)$ because the reduction step is non-duplicating. Clearly, this implies $S^w_{\mathcal{P}}(s_j) \geq^{mul} S^w_{\mathcal{P}}(s_{j+1})$.
- If $s_j \to^i s_{j+1}$ by some reduction rule from $\mathcal{R}_1$, then there is a white principal subterm u of s_j such that $u = C^w[\![u_1, \ldots, u_l, \ldots, u_n]\!] \to C^w[u_1, \ldots, v_l, \ldots, u_n] = v$. Evidently, $v \in S^w_{\mathcal{P}}(s_{j+1})$. It follows from $S^w_{\mathcal{P}}(s_{j+1}) = (S^w_{\mathcal{P}}(s_j) \setminus [u]) \cup [v]$ that $S^w_{\mathcal{P}}(s_j) >^{mul} S^w_{\mathcal{P}}(s_{j+1})$.
- If $s_j \to s_{j+1}$ by some rule from $\mathcal{R}_2$, then there is a white principal subterm $u \in S^w_{\mathcal{P}}(s_j)$ such that $u \to v$ for some v, i.e., $s_j = C^b[\![, \ldots, u, \ldots,]\!] \to C^b[, \ldots, v, \ldots,] = s_{j+1}$. Thus we have $S^w_{\mathcal{P}}(s_{j+1}) = (S^w_{\mathcal{P}}(s_j) \setminus [u]) \cup S^w_{\mathcal{P}}(v)$. It follows from $u \to v$ in conjunction with $v = w$ or $v \rhd w$ for any principal subterm $w \in S^w_{\mathcal{P}}(v)$ that $u > w$ for any $w \in S^w_{\mathcal{P}}(v)$. As a consequence, we obtain $S^w_{\mathcal{P}}(s_j) >^{mul} S^w_{\mathcal{P}}(s_{j+1})$.

We conclude from the well-foundedness of $(\mathcal{M}(\mathcal{T}^{<k}), >^{mul})$ that only a finite number of inner reduction steps as well as reduction steps using a rule from $\mathcal{R}_2$

occur in D. W.l.o.g. we may suppose that there are no reduction steps of that kind in D. Consequently, for all $j \in \mathbb{N}$, we have $s_j \rightarrow^o s_{j+1}$ by some rewrite rule from $\mathcal{R}_1$. Now $\rightarrow_{1,2}$ is semi-complete because $(\mathcal{F}_1, \mathcal{R}_1)$ and $(\mathcal{F}_2, \mathcal{R}_2)$ are complete. Proposition 4.23 yields $top^b_{\downarrow}(s_j) \Rightarrow_{\mathcal{R}_1} top^b_{\downarrow}(s_{j+1})$ for every $j \in \mathbb{N}$. This is a contradiction to the termination of $\Rightarrow_{\mathcal{R}_1}$. □

Corollary 4.27 If $\mathcal{R}_1$ and $\mathcal{R}_2$ are terminating constructor-sharing CTRSs, then their combined system $\mathcal{R}$ is terminating provided that one of the following conditions is satisfied:

1. Neither $\mathcal{R}_1$ nor $\mathcal{R}_2$ contain either collapsing or constructor-lifting rules.
2. Both systems are confluent and non-duplicating.
3. Both systems are confluent and one of the systems contains neither collapsing, constructor-lifting, nor duplicating rules.

Proof: This is an immediate consequence of Theorem 4.26. □

Corollary 4.28

1. Termination is modular for layer-preserving constructor-sharing CTRSs.
2. Completeness is modular for layer-preserving constructor-sharing CTRSs.
3. Completeness is modular for non-duplicating constructor-sharing CTRSs.

Proof: (1) is an immediate consequence of Corollary 4.27. (2) and (3) follow from Theorem 4.14 in conjunction with Corollary 4.27. □

Theorem 4.17 is also true for semi-equational CTRSs, see [Mid90]. Again, it is plausible that Corollaries 4.27 and 4.28 also hold for semi-equational CTRSs.

5 Conclusions

We have shown that semi-completeness is a modular property of constructor-sharing CTRSs and that completeness is modular for non-duplicating constructor-sharing CTRSs. It is definitely worthwhile to try to extend these results to more general kinds of combinations of CTRSs. In particular, the question arises whether the above results also hold for composable CTRSs. Composable systems may share constructors as well as defined function symbols – if they share a defined symbol, then they must contain every rewrite rule which defines that defined symbols. A typical example is the combination of

$$\mathcal{R}_1 = \begin{cases} 0 + x & \rightarrow x \\ S(x) + y & \rightarrow S(x+y) \\ 0 * x & \rightarrow 0 \\ S(x) * y & \rightarrow (x * y) + y \end{cases}$$

with

$$\mathcal{R}_2 = \begin{cases} 0 + x & \rightarrow x \\ S(x) + y & \rightarrow S(x+y) \\ fib(0) & \rightarrow 0 \\ fib(S(0)) & \rightarrow S(0) \\ fib(S(S(x))) & \rightarrow fib(S(x)) + fib(x) \end{cases}$$

$\mathcal{R}_1$ and $\mathcal{R}_2$ share the constructors 0 and S as well as the defined symbol $+$. $\mathcal{R}_1$ and $\mathcal{R}_2$ are composable since the two defining rewrite rules for $+$ occur in both of them. Recently, we have shown that the above results also hold for composable unconditional TRSs – see [Ohl94a]. This gives reason to expect that the aforementioned question has an answer in the affirmative. Note, however, that the proofs presented in this paper do not carry over to composable systems. Moreover, we point out that there are two closely related results obtained by Middeldorp [Mid94]. He proved that semi-completeness and completeness are modular for composable conditional constructor systems without extra variables (it is yet unknown if the same is true when extra variables are allowed). Hereby, a CTRS $(\mathcal{F}, \mathcal{R})$ is called a conditional constructor system if all function symbols occurring at non-root positions in left-hand sides of conditional rewrite rules are constructors.

Another extension of combinations with shared constructors are hierarchical combinations (see for example [Ohl94a] for a precise definition). The standard example of a hierarchical combination is the following, where the base system

$$\mathcal{R}_1 = \begin{cases} 0 + x & \rightarrow x \\ S(x) + y \rightarrow S(x + y) \end{cases}$$

is extended with

$$\mathcal{R}_2 = \begin{cases} 0 * x & \rightarrow 0 \\ S(x) * y \rightarrow (x * y) + y \end{cases}$$

Here the defined symbol $+$ occurs as a constructor in the right-hand side of the second rule of $\mathcal{R}_2$ and $*$ does not appear in $\mathcal{R}_1$. [KR93, KR94, Der95, FJ94] state sufficient conditions which allow to conclude termination of the combined system from the termination of the constituent systems. It goes without saying that it should also be investigated which of the known modularity results for hierarchical combinations of TRSs can be carried over to CTRSs.

Decreasing (finite) CTRSs have been investigated by many researchers because all basic properties (like reducibility for instance) are decidable and a critical pair lemma holds for those systems (cf. [DOS88]). But in contrast to the slightly less general simplifying property (see [Ohl93, Ohl94a]), decreasingness is not modular (not even for disjoint CTRSs) – Toyama's counterexample [Toy87a] to the modularity of termination for disjoint TRSs applies because every terminating TRS can be regarded as a decreasing CTRS. Using the modular reduction relation (see [KO91]), it is possible, however, to compute (unique) normal forms w.r.t. the combined system of n finite, decreasing, confluent, and pairwise constructor-sharing CTRSs. Let us make this more precise. Consider n pairwise constructor-sharing CTRSs $(\mathcal{F}_1, \mathcal{R}_1), \dots, (\mathcal{F}_n, \mathcal{R}_n)$, and let $(\mathcal{F}, \mathcal{R}) = (\bigcup_{j=1}^n \mathcal{F}_j, \bigcup_{j=1}^n \mathcal{R}_j)$. Roughly speaking, the modular reduction relation requires that in reducing a term the same CTRS $(\mathcal{F}, \mathcal{R}_j)$, $j \in \{1, \dots, n\}$, is used for as long as possible. It can be shown that the modular reduction relation is complete whenever each of the n CTRSs is semi-complete. Suppose that the n systems are decreasing and confluent (hence complete). Since every CTRS $(\mathcal{F}_j, \mathcal{R}_j)$, $j \in \{1, \dots, n\}$, is semi-complete, the modular reduction relation is complete and moreover their combined system $(\mathcal{F}, \mathcal{R})$ is semi-complete according to Theorem 4.14. Thus every term $t \in \mathcal{T}(\mathcal{F}, \mathcal{V})$ has a unique normal form $t\downarrow$ with respect to $(\mathcal{F}, \mathcal{R})$. Now if in addition every $\mathcal{R}_j$ is finite, then, due to the decreasingness of the constituent systems, $t\downarrow$ is computable by computing the unique normal form of t w.r.t. the modular reduction relation. Rigorous proofs of these facts can be found in [Ohl94a].

Acknowledgements: The author thanks Aart Middeldorp for discussions about the subtleties of CTRSs.

References

[Der95] N. Dershowitz. Hierarchical Termination. 1995. This volume.

[DJ90] N. Dershowitz and J.-P. Jouannaud. Rewrite Systems. In L. van Leeuwen, editor, *Handbook of Theoretical Computer Science*, volume B, chapter 6. North-Holland, 1990.

[DM79] N. Dershowitz and Z. Manna. Proving Termination with Multiset Orderings. *Communications of the ACM* **22***(8)*, pages 465–476, 1979.

[DOS88] N. Dershowitz, M. Okada, and G. Sivakumar. Canonical Conditional Rewrite Systems. In *Proceedings of the 9th Conference on Automated Deduction*, pages 538–549. Lecture Notes in Computer Science **310**, Springer Verlag, 1988.

[FJ94] M. Fernández and J.-P. Jouannaud. Modular Termination of Term Rewriting Systems Revisited. In *ADT Workshop*, 1994.

[Gra93] B. Gramlich. Sufficient Conditions for Modular Termination of Conditional Term Rewriting Systems. In *Proceedings of the 3rd International Workshop on Conditional Term Rewriting Systems 1992*, pages 128–142. Lecture Notes in Computer Science **656**, Springer Verlag, 1993.

[Gra94a] B. Gramlich. Abstract Relations between Restricted Termination and Confluence Properties of Rewrite Systems. *Fundamenta Informaticae*, 1994. To appear.

[Gra94b] B. Gramlich. Generalized Sufficient Conditions for Modular Termination of Rewriting. *Applicable Algebra in Engineering, Communication and Computing* **5**, pages 131–158, 1994.

[Klo92] J.W. Klop. Term Rewriting Systems. In S. Abramsky, D. Gabbay, and T. Maibaum, editors, *Handbook of Logic in Computer Science*, volume 2, pages 1–116. Oxford University Press, 1992.

[KO91] M. Kurihara and A. Ohuchi. Modular Term Rewriting Systems with Shared Constructors. *Journal of Information Processing* **14***(3)*, *IPS of Japan*, pages 357–358, 1991.

[KO92] M. Kurihara and A. Ohuchi. Modularity of Simple Termination of Term Rewriting Systems with Shared Constructors. *Theoretical Computer Science* **103**, pages 273–282, 1992.

[KO94] M. Kurihara and A. Ohuchi. Termination of Combination of Composable Term Rewriting Systems. In *Proceedings of the 7th Australian Joint Conference on Artificial Intelligence*, 1994. To appear.

[KR93] M.R.K. Krishna Rao. Completeness of Hierarchical Combinations of Term Rewriting Systems. In *Proceedings of the 13th Conference on the Foundations of Software Technology and Theoretical Computer Science*, pages 125–139. Lecture Notes in Computer Science **761**, Springer Verlag, 1993.

[KR94] M.R.K. Krishna Rao. Simple Termination of Hierarchical Combinations of Term Rewriting Systems. In *Proceedings of the International Symposium on Theoretical Aspects of Computer Software*, pages 203–223. Lecture Notes in Computer Science **789**, Springer Verlag, 1994.

[Mar95] M. Marchiori. Modularity of Completeness Revisited. In *Proceedings of the 6th International Conference on Rewriting Techniques and Applications*. Lecture Notes in Computer Science , Springer Verlag, 1995. To appear.

[Mid89] A. Middeldorp. A Sufficient Condition for the Termination of the Direct Sum of Term Rewriting Systems. In *Proceedings of the 4th IEEE Symposium on Logic in Computer Science*, pages 396–401, 1989.

[Mid90] A. Middeldorp. *Modular Properties of Term Rewriting Systems*. PhD thesis, Vrije Universiteit te Amsterdam, 1990.

[Mid93] A. Middeldorp. Modular Properties of Conditional Term Rewriting Systems. *Information and Computation* **104***(1)*, pages 110–158, 1993.

[Mid94] A. Middeldorp. Completeness of Combinations of Conditional Constructor Systems. *Journal of Symbolic Computation* **17**, pages 3–21, 1994.

[MT93] A. Middeldorp and Y. Toyama. Completeness of Combinations of Constructor Systems. *Journal of Symbolic Computation* **15***(3)*, pages 331–348, 1993.

[Ohl93] E. Ohlebusch. Combinations of Simplifying Conditional Term Rewriting Systems. In *Proceedings of the 3rd International Workshop on Conditional Term Rewriting Systems 1992*, pages 113–127. Lecture Notes in Computer Science **656**, Springer Verlag, 1993.

[Ohl94a] E. Ohlebusch. *Modular Properties of Composable Term Rewriting Systems*. PhD thesis, Universität Bielefeld, 1994.

[Ohl94b] E. Ohlebusch. On the Modularity of Confluence of Constructor-Sharing Term Rewriting Systems. In *Proceedings of the 19th Colloquium on Trees in Algebra and Programming*, pages 261–275. Lecture Notes in Computer Science **787**, Springer Verlag, 1994.

[Ohl94c] E. Ohlebusch. On the Modularity of Termination of Term Rewriting Systems. *Theoretical Computer Science* **136**, pages 333–360, 1994.

[Rus87] M. Rusinowitch. On Termination of the Direct Sum of Term Rewriting Systems. *Information Processing Letters* **26**, pages 65–70, 1987.

[TKB89] Y. Toyama, J.W. Klop, and H.P. Barendregt. Termination for the Direct Sum of Left-Linear Term Rewriting Systems. In *Proceedings of the 3rd International Conference on Rewriting Techniques and Applications*, pages 477–491. Lecture Notes in Computer Science **355**, Springer Verlag, 1989.

[Toy87a] Y. Toyama. Counterexamples to Termination for the Direct Sum of Term Rewriting Systems. *Information Processing Letters* **25**, pages 141–143, 1987.

[Toy87b] Y. Toyama. On the Church-Rosser Property for the Direct Sum of Term Rewriting Systems. *Journal of the ACM* **34***(1)*, pages 128–143, 1987.

Church-Rosser Property and Unique Normal Form Property of Non-Duplicating Term Rewriting Systems

Yoshihito Toyama[1] and Michio Oyamaguchi[2]

[1] School of Information Science, JAIST,
Tatsunokuchi, Ishikawa 923-12, Japan
(email: toyama@jaist.ac.jp)

[2] Faculty of Engineering, Mie University,
Kamihama-cho, Tsu-shi 514, Japan
(email: mo@info.mie-u.ac.jp)

Abstract. We propose a new type of conditional term rewriting systems: *left-right separated* conditional term rewriting systems, in which the left-hand side and the right-hand side of a rewrite rule have separate variables. By developing a concept of weight decreasing joinability we first present a sufficient condition for the Church-Rosser property of left-right separated conditional term rewriting systems which may have overlapping rewrite rules. We next apply this result to show sufficient conditions for the unique normal form property and the Church-Rosser property of unconditional term rewriting systems which are non-duplicating, non-left-linear, and overlapping.

1 Introduction

The original idea of the conditional linearization of non-left-linear term rewriting systems was introduced by De Vrijer [4], Klop and De Vrijer [6] for giving a simpler proof of Chew's theorem [2, 9]. They developed an interesting method for proving the unique normal form property for some non-Church-Rosser, non-left-linear term rewriting system R. The method is based on the fact that the unique normal form property of the original non-left-linear term rewriting system R follows the Church-Rosser property of an associated left-linear conditional term rewriting system R^L which is obtained from R by *linearizing* a non-left-linear rule, for example $Dxx \to x$, into a left-linear conditional rule $Dxy \to x \Leftarrow x = y$. Klop and Bergstra [1] proved that non-overlapping left-linear semi-equational conditional term rewriting systems are Church-Rosser. Hence, combining these two results, Klop and De Vrijer [4, 5, 6] showed that the term rewriting system R has the unique normal form property if R^L is non-overlapping. However, as their conditional linearization technique is based on the Church-Rosser property for the traditional conditional term rewriting system R^L, its application is restricted in non-overlapping R^L (though this limitation may be slightly relaxed with R^L containing only trivial critical pairs).

In this paper, we introduce a new conditional linearization based on a *left-right separated* conditional term rewriting system R_L. The point of our linearization is that a non-left-linear rule $Dxx \to x$ is translated into a left-linear conditional rule $Dxy \to z \Leftarrow x = z, y = z$ in which the left-hand side and the right-hand side have separate variables. By considering this new system R_L instead of a traditional conditional system R^L we can easily relax the non-overlapping limitation of conditional systems originated from Klop and Bergstra [1] if the original system R is non-duplicating. Here, R is non-duplicating if for any rewrite rule $l \to r$, no variable has more occurrences in r than it has in l.

By developing a new concept of weight decreasing joinability we first present a sufficient condition for the Church-Rosser property of a left-right separated conditional term rewriting system R_L which may have overlapping rewrite rules. We next apply this result to our conditional linearization, and show a sufficient condition for the unique normal form property of the original system R which is non-duplicating, non-left-linear, and overlapping.

Moreover, our result can be naturally applied to proving the Church-Rosser property of some non-duplicating non-left-linear overlapping term rewriting systems such as right-ground systems. More recently, Oyamaguchi and Ohta [7] proved that non-E-overlapping right-ground term rewriting systems are Church-Rosser by using the joinability of E-graphs, and Oyamaguchi extended this result into some overlapping systems [8]. The results by conditional linearization in this paper strengthen some part of Oyamaguchi's results by E-graphs [7, 8], and vice verse.

2 Reduction Systems

Assuming that the reader is familiar with the basic concepts and notations concerning reduction systems in [3, 5], we briefly explain notations and definitions.

A reduction system (or an abstract reduction system) is a structure $A = \langle D, \to \rangle$ consisting of some set D and some binary relation $\to$ on D (i.e., $\to \subseteq D \times D$), called a reduction relation. A reduction (starting with x_0) in A is a finite or infinite sequence $x_0 \to x_1 \to x_2 \to \cdots$. The identity of elements x, y of D is denoted by $x \equiv y$. $\xrightarrow{\equiv}$ is the reflexive closure of $\to$, $\leftrightarrow$ is the symmetric closure of $\to$, $\xrightarrow{*}$ is the transitive reflexive closure of $\to$, and $\xleftrightarrow{*}$ is the equivalence relation generated by $\to$ (i.e., the transitive reflexive symmetric closure of $\to$). We write $x \leftarrow y$ if $y \to x$; likewise $x \xleftarrow{*} y$.

If $x \in D$ is minimal with respect to $\to$, i.e., $\neg \exists y \in D[x \to y]$, then we say that x is a normal form; let NF be the set of normal forms. If $x \xrightarrow{*} y$ and $y \in NF$ then we say x has a normal form y and y is a normal form of x.

Definition 1. $A = \langle D, \to \rangle$ is Church-Rosser (or confluent) iff
$\forall x, y, z \in D[x \xrightarrow{*} y \wedge x \xrightarrow{*} z \Rightarrow \exists w \in D, y \xrightarrow{*} w \wedge z \xrightarrow{*} w]$.

Definition 2. $A = \langle D, \to \rangle$ has unique normal forms iff
$\forall x, y \in NF[x \xleftrightarrow{*} y \Rightarrow x \equiv y]$.

The following fact observed by Klop and De Vrijer [6] plays an essential role in our linearization too.

Proposition 3. *[Klop and De Vrijer] Let* $A_0 = \langle D, \underset{0}{\rightarrow} \rangle$ *and* $A_1 = \langle D, \underset{1}{\rightarrow} \rangle$ *be two reduction systems with the sets of normal forms* NF_0 *and* NF_1 *respectively. Then* A_0 *has unique normal forms if each of the following conditions holds:*

(i) $\underset{1}{\rightarrow}$ *extends* $\underset{0}{\rightarrow}$,
(ii) A_1 *is Church-Rosser,*
(iii) NF_1 *contains* NF_0.

Proof. Easy. □

3 Weight Decreasing Joinability

This section introduces the new concept of weight decreasing joinability. In the later sections this concept is used for analyzing the Church-Rosser property of conditional term rewriting systems with extra variables occurring in conditional parts of rewrite rules.

Let N^+ be the set of positive integers. $A = \langle D, \rightarrow \rangle$ is a weighted reduction system if $\rightarrow = \bigcup_{w \in N^+} \rightarrow_w$, that is, positive integers (weights w) are assigned to each reduction step to represent costs.

Definition 4. A proof of $x \overset{*}{\leftrightarrow} y$ is a sequence $\mathcal{P}$: $x_0 \leftrightarrow_{w_1} x_1 \leftrightarrow_{w_2} x_2 \cdots \leftrightarrow_{w_n} x_n$ $(n \geq 0)$ such that $x \equiv x_0$ and $y \equiv x_n$. The weight $w(\mathcal{P})$ of the proof $\mathcal{P}$ is $\sum_{i=1}^{n} w_i$. If $\mathcal{P}$ is a 0 step sequence (i.e., $n = 0$), then $w(\mathcal{P}) = 0$.

We usually abbreviate a proof $\mathcal{P}$ of $x \overset{*}{\leftrightarrow} y$ by $\mathcal{P}$: $x \overset{*}{\leftrightarrow} y$. The form of a proof may be indicated by writing, for example, $\mathcal{P}$: $x \overset{*}{\rightarrow} \cdot \overset{*}{\leftarrow} y$, $\mathcal{P}'$: $x \leftarrow \cdot \overset{*}{\rightarrow} \cdot \leftarrow y$, etc. We use the symbols $\mathcal{P}, \mathcal{Q}, \cdots$ for proofs.

Definition 5. A weighted reduction system $A = \langle D, \rightarrow \rangle$ is weight decreasing joinable iff for all $x, y \in D$ and any proof $\mathcal{P}$: $x \overset{*}{\leftrightarrow} y$ there exists some proof $\mathcal{P}'$: $x \overset{*}{\rightarrow} \cdot \overset{*}{\leftarrow} y$ such that $w(\mathcal{P}) \geq w(\mathcal{P}')$.

It is clear that if a weighted reduction system A is weight decreasing joinable then A is Church-Rosser. We will now show a sufficient condition for the weight decreasing joinability.

Lemma 6. *Let A be a weighted reduction system. Then A is weight decreasing joinable if for any* $x, y \in D$ *and any proof* $\mathcal{P}$: $x \leftarrow \cdot \rightarrow y$ *one of the following conditions holds:*

(i) *there exists a proof* $\mathcal{P}'$: $x \overset{*}{\leftrightarrow} y$ *such that* $w(\mathcal{P}) > w(\mathcal{P}')$, *or*
(ii) *there exist proofs* $\mathcal{P}'$: $x \rightarrow \cdot \overset{*}{\leftrightarrow} y$ *and* $\mathcal{P}''$: $x \overset{*}{\leftrightarrow} \cdot \leftarrow y$ *such that* $w(\mathcal{P}) \geq w(\mathcal{P}')$ *and* $w(\mathcal{P}) \geq w(\mathcal{P}'')$, *or*

(iii) *there exists a proof $\mathcal{P}' : x \to y$ (or $x \leftarrow y$) such that $w(\mathcal{P}) \geq w(\mathcal{P}')$.*

Proof. By induction on the weight $w(\mathcal{Q})$ of a proof $\mathcal{Q} : x \overset{*}{\leftrightarrow} y$, we prove that there exists a proof $\mathcal{Q}'$: $x \overset{*}{\to} \cdot \overset{*}{\leftarrow} y$ such that $w(\mathcal{Q}) \geq w(\mathcal{Q}')$. *Base step* ($w(\mathcal{Q}) = 0$) is trivial. *Induction step*: Let $\mathcal{Q} : x \leftrightarrow x' \overset{*}{\leftrightarrow} y$ and let $\mathcal{S} : x' \overset{*}{\leftrightarrow} y$ be the subproof of $\mathcal{Q}$. From induction hypothesis, there exists a proof $\mathcal{S}'$: $x' \overset{*}{\to} \cdot \overset{*}{\leftarrow} y$ such that $w(\mathcal{S}) \geq w(\mathcal{S}')$. Thus, if $x \to x'$ then we have $\mathcal{Q}'$: $x \to x' \overset{*}{\to} \cdot \overset{*}{\leftarrow} y$ such that $w(\mathcal{Q}) \geq w(\mathcal{Q}')$. Otherwise we have a proof $\mathcal{Q}''$: $x \leftarrow x' \overset{n}{\to} \cdot \overset{*}{\leftarrow} y$ such that $w(\mathcal{Q}) \geq w(\mathcal{Q}'')$, where $\overset{n}{\to}$ denotes a reduction of n ($n \geq 0$) steps. By induction on n we will prove that $\mathcal{Q}'$ exists. The case $n = 0$ is trivial. Let $\mathcal{Q}''$: $x \leftarrow x' \to z \overset{n-1}{\longrightarrow} \cdot \overset{*}{\leftarrow} y$ and let $\mathcal{P}$: $x \leftarrow x' \to z$ be the subproof of $\mathcal{Q}''$. Then $\mathcal{P}$ can be replaced with $\mathcal{P}'$ satisfying one of the above conditions (i), (ii), or (iii).

Case (i). $\mathcal{P}' : x \overset{*}{\leftrightarrow} z$ and $w(\mathcal{P}) > w(\mathcal{P}')$. Then we have $\hat{\mathcal{Q}}$: $x \overset{*}{\leftrightarrow} z \overset{n-1}{\longrightarrow} \cdot \overset{*}{\leftarrow} y$ such that $w(\mathcal{Q}'') > w(\hat{\mathcal{Q}})$. Thus, by using induction hypothesis concerning the weight $w(\mathcal{Q})$, we obtain $\mathcal{Q}'$ from $\hat{\mathcal{Q}}$.

Case (ii). $\mathcal{P}'$: $x \to z' \overset{*}{\leftrightarrow} z$ and $w(\mathcal{P}) \geq w(\mathcal{P}')$. Then we have $\hat{\mathcal{Q}}$: $x \to z' \overset{*}{\leftrightarrow} z \overset{n-1}{\longrightarrow} \cdot \overset{*}{\leftarrow} y$ such that $w(\mathcal{Q}'') \geq w(\hat{\mathcal{Q}})$. Let $\hat{\mathcal{Q}}'$: $z' \overset{*}{\leftrightarrow} z \overset{n-1}{\longrightarrow} \cdot \overset{*}{\leftarrow} y$ be the subproof of $\hat{\mathcal{Q}}$. From induction hypothesis concerning the weight $w(\mathcal{Q})$ there exists a proof $\hat{\mathcal{Q}}''$: $z' \overset{*}{\to} \cdot \overset{*}{\leftarrow} y$ such that $w(\hat{\mathcal{Q}}') \geq w(\hat{\mathcal{Q}}'')$. Thus, by replacing $\hat{\mathcal{Q}}'$ of $\hat{\mathcal{Q}}$ with $\hat{\mathcal{Q}}''$, we have $\mathcal{Q}'$.

Case (iii). $\mathcal{P}' : x \leftarrow z$ and $w(\mathcal{P}) \geq w(\mathcal{P}')$. (If $\mathcal{P}' : x \to z$, the claim trivially holds.) Then we have $\hat{\mathcal{Q}}$: $x \leftarrow z \overset{n-1}{\longrightarrow} \cdot \overset{*}{\leftarrow} y$ such that $w(\mathcal{Q}'') \geq w(\hat{\mathcal{Q}})$. From induction hypothesis concerning the number n of reduction steps, we have $\mathcal{Q}'$. □

The following lemma is used to show the Church-Rosser property of non-left-linear systems in Section 7.

Lemma 7. *Let $A_0 = \langle D, \underset{0}{\to} \rangle$ and $A_1 = \langle D, \underset{1}{\to} \rangle$. Let $\mathcal{P}_i$: $x_i \overset{*}{\underset{1}{\leftrightarrow}} y$ ($i = 1, \cdots, n$) and let $\rho = \sum_{i=1}^{n} w(\mathcal{P}_i)$. Assume that for any $a, b \in D$ and any proof $\mathcal{P}$: $a \overset{*}{\underset{1}{\leftrightarrow}} b$ such that $w(\mathcal{P}) \leq \rho$ there exist proofs $\mathcal{P}'$: $a \overset{*}{\underset{1}{\to}} c \overset{*}{\underset{1}{\leftarrow}} b$ with $w(\mathcal{P}') \leq w(\mathcal{P})$ and $a \overset{*}{\underset{0}{\to}} c \overset{*}{\underset{0}{\leftarrow}} b$ for some $c \in D$. Then, there exist proofs $\mathcal{P}'_i$: $x_i \overset{*}{\underset{0}{\to}} z$ ($i = 1, \cdots n$) and $\mathcal{Q}$: $y \overset{*}{\underset{1}{\leftrightarrow}} z$ with $w(\mathcal{Q}) \leq \rho$ for some z (Figure 3.1).*

Proof. By induction on ρ. *Base step* ($\rho = 0$) is trivial. *Induction step*: From induction hypothesis, we have proofs $\tilde{\mathcal{P}}_i$: $x_i \overset{*}{\underset{0}{\to}} z'$ ($i = 1, \cdots n-1$) and $\tilde{\mathcal{Q}}$: $y \overset{*}{\underset{1}{\leftrightarrow}} z'$ for some z' such that $\sum_{i=1}^{n-1} w(\mathcal{P}_i) \geq w(\tilde{\mathcal{Q}})$. By connecting the proofs $\tilde{\mathcal{Q}}$ and $\mathcal{P}_n$ we have a proof $\hat{\mathcal{P}}$: $z' \overset{*}{\underset{1}{\leftrightarrow}} y \overset{*}{\underset{1}{\leftrightarrow}} x_n$. Since $\sum_{i=1}^{n-1} w(\mathcal{P}_i) \geq w(\tilde{\mathcal{Q}})$ and $w(\hat{\mathcal{P}}) = w(\tilde{\mathcal{Q}}) + w(\mathcal{P}_n)$, it follows that $\rho \geq w(\hat{\mathcal{P}})$. By the assumption, we have proofs $\tilde{\mathcal{P}}$: $z' \overset{*}{\underset{1}{\to}} z \overset{*}{\underset{1}{\leftarrow}} x_n$ with $\rho \geq w(\hat{\mathcal{P}}) \geq w(\tilde{\mathcal{P}})$ and $z' \overset{*}{\underset{0}{\to}} z \overset{*}{\underset{0}{\leftarrow}} x_n$ for some z. Thus we obtain proofs $\mathcal{P}'_i$: $x_i \overset{*}{\underset{0}{\to}} z$ ($i = 1, \cdots, n$).

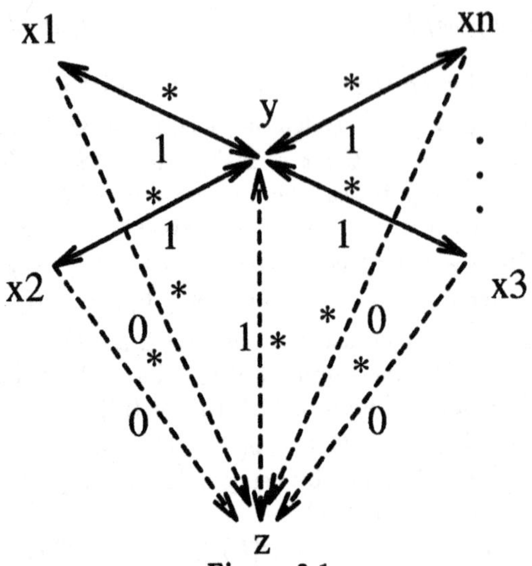

Figure 3.1

By combining subproofs of $\hat{\mathcal{P}}$: $z' \underset{1}{\overset{*}{\leftrightarrow}} y \underset{1}{\overset{*}{\leftrightarrow}} x_n$ and $\tilde{\mathcal{P}}$: $z' \underset{1}{\overset{*}{\rightarrow}} z \underset{1}{\overset{*}{\leftarrow}} x_n$, we can make $\mathcal{Q}'$: $y \underset{1}{\overset{*}{\leftrightarrow}} z' \underset{1}{\overset{*}{\rightarrow}} z$ and $\mathcal{Q}''$: $y \underset{1}{\overset{*}{\leftrightarrow}} x_n \underset{1}{\overset{*}{\rightarrow}} z$. Note that $\rho + \rho \geq w(\hat{\mathcal{P}}) + w(\tilde{\mathcal{P}}) = w(\mathcal{Q}') + w(\mathcal{Q}'')$. Thus $\rho \geq w(\mathcal{Q}')$ or $\rho \geq w(\mathcal{Q}'')$. Take $\mathcal{Q}'$ as $\mathcal{Q}$ if $\rho \geq w(\mathcal{Q}')$; otherwise, take $\mathcal{Q}''$ as $\mathcal{Q}$. □

4 Term Rewriting Systems

In the following sections, we briefly explain the basic notions and definitions concerning term rewriting systems [3, 5].

Let $\mathcal{F}$ be an enumerable set of function symbols denoted by $f, g, h, \cdots$, and let $\mathcal{V}$ be an enumerable set of variable symbols denoted by $x, y, z, \cdots$ where $\mathcal{F} \cap \mathcal{V} = \phi$. By $T(\mathcal{F}, \mathcal{V})$, we denote the set of terms constructed from $\mathcal{F}$ and $\mathcal{V}$. The term set $T(\mathcal{F}, \mathcal{V})$ is sometimes denoted by T. $V(t)$ denotes the set of variables occurring in a term t.

A substitution θ is a mapping from a term set $T(\mathcal{F}, \mathcal{V})$ to $T(\mathcal{F}, \mathcal{V})$ such that for a term t, $\theta(t)$ is completely determined by its values on the variable symbols occurring in t. Following common usage, we write this as $t\theta$ instead of $\theta(t)$.

Consider an extra constant □ called a hole and the set $T(\mathcal{F} \cup \{\Box\}, \mathcal{V})$. Then $C \in T(\mathcal{F} \cup \{\Box\}, \mathcal{V})$ is called a context on $\mathcal{F}$. We use the notation $C[\ ,\ldots,\ ]$ for the context containing n holes ($n \geq 0$), and if $t_1, \ldots, t_n \in T(\mathcal{F}, \mathcal{V})$, then $C[t_1, \ldots, t_n]$ denotes the result of placing $t_1, \ldots, t_n$ in the holes of $C[\ ,\ldots,\ ]$ from left to right. In particular, $C[\ ]$ denotes a context containing precisely one hole. s is called a subterm of t if $t \equiv C[s]$. If s is a subterm occurrence of t, then we write $s \subseteq t$. If a term t has an occurrence of some (function or variable) symbol e, we write $e \in t$. The variable occurrences $z_1, \cdots, z_n$ of $C[z_1, \cdots, z_n]$ are fresh if $z_1, \cdots, z_n \notin C[\ , \cdots,\ ]$ and $z_i \not\equiv z_j$ $(i \neq j)$.

A rewrite rule is a pair $\langle l, r\rangle$ of terms such that $l \notin \mathcal{V}$ and any variable in r also occurs in l. We write $l \to r$ for $\langle l, r\rangle$. A redex is a term $l\theta$, where $l \to r$. In this case $r\theta$ is called a contractum of $l\theta$. The set of rewrite rules defines a reduction relation $\to$ on T as follows:

$t \to s$ iff $t \equiv C[l\theta]$, $s \equiv C[r\theta]$ for some rule $l \to r$, and some $C[\ ]$, θ.

When we want to specify the redex occurrence $\Delta \equiv l\theta$ of t in this reduction, we write $t \xrightarrow{\Delta} s$.

Definition 8. A term rewriting system R is a reduction system $R = \langle T(\mathcal{F}, \mathcal{V}), \to \rangle$ such that the reduction relation $\to$ on $T(\mathcal{F}, \mathcal{V})$ is defined by a set of rewrite rules. When we want to specify the term rewriting system R in the reduction relation $\to$, we write $\underset{R}{\to}$. If R has $l \to r$ as a rewrite rule, we write $l \to r \in R$.

We say that R is left-linear if for any $l \to r \in R$, l is linear (i.e., every variable in l occurs only once). If R has a critical pair then we say that R is overlapping: otherwise non-overlapping [3, 5]. A rewrite rule $l \to r$ is duplicating if r contains more occurrences of some variable than l; otherwise, $l \to r$ is non-duplicating. We say that R is non-duplicating if every $l \to r \in R$ is non-duplicating [5].

5 Left-Right Separated Conditional Systems

In this section we introduce a new conditional term rewriting system R in which l and r of any rewrite rule $l \to r \Leftarrow x_1 = y_1, \cdots, x_n = y_n$ do not share the same variable; every variable y_i in r is connected to some variable x_i in l through the equational conditions $x_1 = y_1, \cdots, x_n = y_n$. A decidable sufficient condition for the Church-Rosser property of R is presented.

Definition 9. A left-right separated conditional term rewriting system is a conditional term rewriting system with extra variables in which every conditional rewrite rule has the form:

$l \to r \Leftarrow x_1 = y_1, \cdots, x_n = y_n$

with $l, r \in T(\mathcal{F}, \mathcal{V})$, $V(l) = \{x_1, \cdots, x_n\}$ and $V(r) \subseteq \{y_1, \cdots, y_n\}$ $(n \geq 0)$ such that:

(i) $l \notin V$ is linear,
(ii) $\{x_1, \cdots, x_n\} \cap \{y_1, \cdots, y_n\} = \phi$,
(iii) $x_i \not\equiv x_j$ if $i \neq j$,
(iv) no variable has more occurrences in r than it has in the conditional part "$x_1 = y_1, \cdots, x_n = y_n$".

Note. In the above conditional rewrite rule, the left-hand side l and the right-hand side r have separate variables, i.e., $V(l) \cap V(r) = \phi$, because of (ii). Since every variable y_i in r is connected to some variable x_i in l through the equational condition, it holds that $V(r\theta) \subseteq V(l\theta)$ for the substitution $\theta = [x_1 := y_1, \cdots, x_m := y_m]$. Thus, $l\theta \to r\theta$ is an unconditional rewrite rule, and it is non-duplicating due to (iv).

Example 1. The following R is a left-right separated conditional term rewriting system:

$$R \quad \begin{cases} f(x,x') \to g(y,y) \Leftarrow x = y, x' = y \\ h(x,x',x'') \to c \Leftarrow x = y, x' = y, x'' = y \end{cases}$$

The following R' is however not a left-right separated conditional term rewriting system since the condition (iv) does not hold:

$$R' \quad \{ f(x,x') \to h(y,y,y) \Leftarrow x = y, x' = y$$

□

Definition 10. Let R be a left-right separated conditional term rewriting system. We inductively define reduction relations $\xrightarrow[R_i]{}$ for $i \geq 0$ as follows: $\xrightarrow[R_0]{} = \phi$, $\xrightarrow[R_{i+1}]{} = \{\langle C[l\theta], C[r\theta]\rangle \mid l \to r \Leftarrow x_1 = y_1, \cdots, x_n = y_n \in R$ and $x_j\theta \overset{*}{\underset{R_i}{\longleftrightarrow}} y_j\theta$ $(j = 1, \cdots, n)\}$. Note that $\xrightarrow[R_i]{} \subseteq \xrightarrow[R_{i+1}]{}$ for all $i \geq 0$. $s \to t$ iff $s \xrightarrow[R_i]{} t$ for some i.

The weight $w(\mathcal{P})$ of a proof $\mathcal{P}$ of a left-right separated conditional term rewriting system R is defined as the total redctuton steps appearing in the recursive structure of $\mathcal{P}$.

Definition 11. A proof $\mathcal{P}$ and its weight $w(\mathcal{P})$ are inductively defined as follows:

(i) The empty sequence λ is a (0 step) proof of $t \overset{*}{\underset{R_n}{\longleftrightarrow}} t$ $(n \geq 0)$ and $w(\lambda) = 0$.

(ii) An expression $\mathcal{P} : s \xrightarrow[\mathbf{[r},C[\,],\mathcal{P}_1,\cdots,\mathcal{P}_m]]{} t$ (resp. $t \xleftarrow[\mathbf{[r},C[\,],\mathcal{P}_1,\cdots,\mathcal{P}_m]]{} s$) is a proof of $s \xrightarrow[R_n]{} t$ (resp. $t \xleftarrow[R_n]{} s$) $(n \geq 1)$, where $\mathbf{r}$ is a rewrite rule $l \to r \Leftarrow x_1 = y_1, \cdots, x_m = y_m \in R$ and $C[\,]$ is a context such that for some substitution θ, $t \equiv C[l\theta]$, $s \equiv C[r\theta]$, and $\mathcal{P}_i$ is a proof of $x_i\theta \overset{*}{\underset{R_{n-1}}{\longleftrightarrow}} y_i\theta$ $(i = 1, \cdots, m)$. $w(\mathcal{P}) = 1 + \sum_{i=1}^{m} w(\mathcal{P}_i)$. $\mathcal{P}_1, \cdots, \mathcal{P}_m$ are subproofs associated with the proof $\mathcal{P}$.

(iii) A finite sequence $\mathcal{P} : \mathcal{P}_1 \cdots \mathcal{P}_m$ $(m \geq 1)$ of proofs is a proof of $t_0 \overset{*}{\underset{R_n}{\longleftrightarrow}} t_m$ $(n \geq 1)$, where $\mathcal{P}_i$ $(i = 1, \cdots, m)$ is a proof of $t_{i-1} \xrightarrow[R_n]{} t_i$ or $t_{i-1} \xleftarrow[R_n]{} t_i$. $w(\mathcal{P}) = \sum_{i=1}^{m} w(\mathcal{P}_i)$.

$\mathcal{P}$ is a proof of $s \overset{*}{\leftrightarrow} t$ if it is a proof of $s \overset{*}{\underset{R_n}{\longleftrightarrow}} t$ for some n. For convenience, we often use the abbreviations introduced in Section 3; i.e., we abbreviate a proof $\mathcal{P}$ of $s \overset{*}{\leftrightarrow} t$ by $\mathcal{P}$: $s \overset{*}{\leftrightarrow} t$, and the form of a proof is indicated by writing, for example, $\mathcal{P}$: $s \overset{*}{\rightarrow} \cdot \overset{*}{\leftarrow} t$, $\mathcal{P}'$: $s \leftarrow \cdot \overset{*}{\rightarrow} \cdot \leftarrow t$, etc. We use the symbols $\mathcal{P}, \mathcal{Q}, \cdots$ for proofs.

Let $l \rightarrow r \Leftarrow x_1 = y_1, \cdots, x_m = y_m$ and $l' \rightarrow r' \Leftarrow x'_1 = y'_1, \cdots, x'_n = y'_n$ be two rules in a left-right separated conditional term rewriting system R. Assume that we have renamed the variables appropriately, so that two rules share no variables. Assume that $s \notin V$ is a subterm occurrence in l, i.e., $l \equiv C[s]$, such that s and l' are unifiable, i.e., $s\theta \equiv l'\theta$, with the most general unifier θ. Note that $r\theta \equiv r$, $r'\theta \equiv r'$, $y_i\theta \equiv y_i$ $(i = 1, \cdots, m)$ and $y'_j\theta \equiv y'_j$ $(j = 1, \cdots, n)$ as $\{x_1, \cdots, x_m\} \cap \{y_1, \cdots, y_m\} = \phi$ and $\{x'_1, \cdots, x'_n\} \cap \{y'_1, \cdots, y'_n\} = \phi$. Since $l \equiv C[s]$ is linear and the domain of θ is contained in $V(s)$, $C[s]\theta \equiv C[s\theta]$. Thus, from $l\theta \equiv C[s]\theta \equiv C[l'\theta]$, two reductions starting with $l\theta$, i.e., $l\theta \rightarrow C[r']$ and $l\theta \rightarrow r$, can be obtained by using $l \rightarrow r \Leftarrow x_1 = y_1, \cdots, x_m = y_m$ and $l' \rightarrow r' \Leftarrow x'_1 = y'_1, \cdots, x'_n = y'_n$ if we assume the equations $x_1\theta = y_1, \cdots, x_m\theta = y_m$ and $x'_1\theta = y'_1, \cdots, x'_n\theta = y'_n$. Then we say that $l \rightarrow r \Leftarrow x_1 = y_1, \cdots, x_m = y_m$ and $l' \rightarrow r' \Leftarrow x'_1 = y'_1, \cdots, x'_n = y'_n$ are overlapping, and $E \vdash \langle C[r'], r\rangle$ is a conditional critical pair associated with the multiset of equations $E = [x_1\theta = y_1, \cdots, x_m\theta = y_m, x'_1\theta = y'_1, \cdots, x'_n\theta = y'_n]$ in R. We may choose $l \rightarrow r \Leftarrow x_1 = y_1, \cdots, x_m = y_m$ and $l' \rightarrow r' \Leftarrow x'_1 = y'_1, \cdots, x'_n = y'_n$ to be the same rule, but in this case we shall not consider the case $s \equiv l$. If R has no critical pair, then we say that R is non-overlapping.

Example 2. Let R be the left-right separated conditional term rewriting system with the following rewrite rules:

$$R \quad \begin{cases} f(x', x'') \rightarrow g(x) \Leftarrow x' = x, x'' = x \\ f(y', h(y'')) \rightarrow g(y) \Leftarrow y' = y, y'' = y \end{cases}$$

Let $\theta = [x' := y', x'' := h(y'')]$ be the most general unifier of $f(x', x'')$ and $f(y', h(y''))$. By applying the substitution θ to the conditional parts "$x' = x, x'' = x$" and "$y' = y, y'' = y$" we have the multiset of equations $E = [y' = x, h(y'') = x, y' = y, y'' = y]$. Then, assuming the equations in E, $g(x) \leftarrow g(x)\theta \leftarrow f(x', x'')\theta \equiv f(y', h(y'')) \rightarrow g(y)$. Thus, we have a condtional critecal pair $E \vdash \langle g(x), g(y)\rangle$. □

Note that in a left-right separated conditional term rewriting system the application of the same rule at the same position does not imply the same result as the variables occurring in the left-hand side of a rule do not cover that in the right-hand side: See the following example.

Example 3. Let R be the left-right separated conditional term rewriting system with the following rewrite rules:

$$R \begin{cases} f(x) \rightarrow g(y) \Leftarrow x = y \\ a \rightarrow c \\ b \rightarrow c \end{cases}$$

It is obvious that R is non-overlapping. We have however two reductions $f(c) \rightarrow g(a)$ and $f(c) \rightarrow g(b)$, as $c \overset{*}{\leftrightarrow} a$ and $c \overset{*}{\leftrightarrow} b$. Thus the application of the first rule at the root position of $f(c)$ does not guarantee a unique result. □

We next discuss how to compare the weights of *abstract* proofs including the assumed equations of E. $E \sqcup E'$ denotes the union of multisets E and E'. We write $E \sqsubseteq E'$ if no elements in E occur more than E'.

Definition 12. Let E be a multiset of equations $t' = s'$ and a fresh constant $\bullet$. Then relations $t \underset{E}{\sim} s$ and $t \underset{E}{\sim\!\triangleright} s$ on terms is inductively defined as follows:

(i) $t \underset{[t=s]}{\sim} s$.
(ii) If $t \underset{E}{\sim} s$ then $s \underset{E}{\sim} t$.
(iii) If $t \underset{E}{\sim} r$ and $r \underset{E'}{\sim} s$ then $t \underset{E \sqcup E'}{\sim} s$.
(iv) If $t \underset{E}{\sim} s$ then $C[t] \underset{E}{\sim} C[s]$.
(v) If $l \rightarrow r \Leftarrow x_1 = y_1, \cdots, x_n = y_n \in R$ and $x_i\theta \underset{E_i}{\sim} y_i\theta$ $(i = 1, \cdots, n)$ then $C[l\theta] \underset{E}{\sim\!\triangleright} C[r\theta]$ where $E = E_1 \sqcup \cdots \sqcup E_n$.
(vi) If $t \underset{E}{\sim\!\triangleright} s$ then $t \underset{E \sqcup [\bullet]}{\sim} s$.

In the above definition the fresh constant $\bullet$ keeps in E the number of concrete rewriting steps appearing in an *abstract* proof. We write $t \underset{E}{\triangleleft\!\sim} s$ if $s \underset{E}{\sim\!\triangleright} t$.

Lemma 13. *Let $E = [p_1 = q_1, \cdots, p_m = q_m, \bullet, \cdots, \bullet]$ be a multiset in which $\bullet$ occurs k times $(k \geq 0)$, and let $\mathcal{P}_i$: $p_i\theta \overset{*}{\leftrightarrow} q_i\theta$ $(i = 1, \cdots, m)$.*

(1) *If $t \underset{E}{\sim} s$ then there exists a proof $\mathcal{Q}$: $t\theta \overset{*}{\leftrightarrow} s\theta$ with $w(\mathcal{Q}) \leq \sum_{i=1}^{m} w(\mathcal{P}_i) + k$.*
(2) *If $t \underset{E}{\sim\!\triangleright} s$ then there exists a proof $\mathcal{Q}'$: $t\theta \rightarrow s\theta$ with $w(\mathcal{Q}') \leq \sum_{i=1}^{m} w(\mathcal{P}_i) + k + 1$.*

Proof. By induction on the construction of $t \underset{E}{\sim} s$ and $t \underset{E}{\sim\!\triangleright} s$ in Definition 12, we prove (1) and (2) simultaneously. *Base Step*: Trivial as (i) $t \underset{[t=s]}{\sim} s$ of Definition 12. *Induction Step*: If we have $t \underset{E}{\sim} s$ by (ii) (iii) (iv) and $t \underset{E}{\sim\!\triangleright} s$ by (vi) of Definition 12, then from the induction hypothesis (1) and (2) clearly follow. Assume that $t \underset{E}{\sim\!\triangleright} s$ by (v) of Definition 12. Then we have a rule $l \rightarrow r \Leftarrow x_1 = y_1, \cdots, x_n = y_n$ such that $t \equiv C[l\theta']$, $s \equiv C[r\theta']$, $x_i\theta' \underset{E_i}{\sim} y_i\theta'$ $(i = 1, \cdots, n)$ for some θ' and $E = E_1 \sqcup \cdots \sqcup E_n$. From the induction hypothesis and $E = E_1 \sqcup \cdots \sqcup E_n$, it can be easily shown that $\mathcal{Q}_i$: $x_i\theta'\theta \overset{*}{\leftrightarrow} x_i\theta'\theta$ $(i = 1, \cdots, n)$ and $\sum_{i=1}^{n} w(\mathcal{Q}_i) \leq \sum_{i=1}^{m} w(\mathcal{P}_i) + k$. Therefore we have a proof $\mathcal{Q}'$: $t\theta \rightarrow s\theta$ with $w(\mathcal{Q}') \leq \sum_{i=1}^{m} w(\mathcal{P}_i) + k + 1$. □

Theorem 14. *Let R be a left-right separated conditional term rewriting system. Then R is weight decreasing joinable if for any conditional critical pair $E \vdash \langle q, q' \rangle$ one of the following conditions holds:*

(i) $q \underset{E'}{\sim} q'$ *for some E' such that $E' \sqsubseteq E \sqcup [\bullet]$, or*

(ii) $q \underset{E_1}{\sim\triangleright} \cdot \underset{E_2}{\sim} q'$ *and* $q \underset{E_1'}{\sim} \cdot \underset{E_2'}{\triangleleft\sim} q'$ *for some E_1, E_2, E_1', and E_2' such that $E_1 \sqcup E_2 \sqsubseteq E \sqcup [\bullet]$ and $E_1' \sqcup E_2' \sqsubseteq E \sqcup [\bullet]$, or*

(iii) $q \underset{E'}{\sim\triangleright} q'$ *(or* $q \underset{E'}{\triangleleft\sim} q'$*) for some E' such that $E' \sqsubseteq E \sqcup [\bullet]$.*

Note. If R has finitely many rewrite rules then R has finitely many conditional critical pairs. For each $E \vdash \langle q, q' \rangle$, it is decidable whether one of the above conditions (i), (ii), or (iii) holds since each relation between q and q' is restricted by an upper bound $E \sqcup [\bullet]$. Thus, the theorem presents a decidable sufficient condition for guaranteeing the Church-Rosser property of R having finte rewrite rules.

Proof. The theorem follows from Lemma 6 if for any $\mathcal{P}$: $t \leftarrow p \rightarrow s$ $(t \not\equiv s)$ one of the following conditions holds: (i) there exists a proof $\mathcal{Q}$: $t \overset{*}{\leftrightarrow} s$ $w(\mathcal{P}) > w(\mathcal{Q})$, or (ii) there exist proofs $\mathcal{Q}_1$: $t \rightarrow \cdot \overset{*}{\leftrightarrow} s$ and $\mathcal{Q}_2$: $t \overset{*}{\leftrightarrow} \cdot \leftarrow s$ such that $w(\mathcal{P}) \geq w(\mathcal{Q}_1)$ and $w(\mathcal{P}) \geq w(\mathcal{Q}_2)$, or (iii) there exists a proof $\mathcal{Q} : t \rightarrow s$ (or $t \leftarrow s$) such that $w(\mathcal{P}) \geq w(\mathcal{Q})$. Hence we will show that one of (i), (ii), or (iii) holds for a given proof $\mathcal{P}$: $t \leftarrow p \rightarrow s$.

Let $\mathcal{P}$: $t \overset{\Delta}{\leftarrow} p \overset{\Delta'}{\rightarrow} s$ where two redexes $\Delta \equiv l\theta$ and $\Delta' \equiv l'\theta'$ are associated with two rules $\mathbf{r}_1$: $l \rightarrow r \Leftarrow x_1 = y_1, \cdots, x_m = y_m$ and $\mathbf{r}_2$: $l' \rightarrow r' \Leftarrow x_1' = y_1', \cdots, x_{m'}' = y_{m'}'$ respectively.

Case 1. Δ and Δ' are disjoint. Then $p \equiv C[\Delta, \Delta']$ for some context $C[\ ,\ ]$ and $\mathcal{P}$: $t \equiv C[t', \Delta'] \overset{\Delta}{\leftarrow} C[\Delta, \Delta'] \overset{\Delta'}{\rightarrow} C[\Delta, s'] \equiv s$ for some t' and s'. Since we can take $\mathcal{Q}_1 = \mathcal{Q}_2$: $t \equiv C[t', \Delta'] \overset{\Delta'}{\rightarrow} C[t', s'] \overset{\Delta}{\leftarrow} C[\Delta, s'] \equiv s$ with $w(\mathcal{Q}_1) = w(\mathcal{Q}_2) = w(\mathcal{P})$, (ii) holds.

Case 2. Δ' occurs in θ of $\Delta \equiv l\theta$ (i.e., Δ' occurs below the pattern l). Without loss of generality we may assume that $\mathbf{r}_1$: $C_L[x_1, \cdots, x_m] \rightarrow C_R[y_1, \cdots, y_n] \Leftarrow x_1 = y_1, \cdots, x_m = y_m$ (all the variable occurrences are displayed), $\mathcal{P}'$: $p \equiv C[C_L[p_1, \cdots, p_m]] \overset{\Delta}{\rightarrow} t \equiv C[C_R[t_1, \cdots, t_n]]$ with subproofs $\mathcal{P}_i$: $p_i \overset{*}{\leftrightarrow} t_i$ $(i = 1, \cdots, m)$, and $\mathcal{P}''$: $p \equiv C[C_L[p_1, p_2, \cdots, p_m]] \overset{\Delta'}{\rightarrow} s \equiv C[C_L[p_1', p_2, \cdots, p_m]]$ by $p_1 \overset{\Delta'}{\rightarrow} p_1'$. Thus $w(\mathcal{P}) = w(\mathcal{P}') + w(\mathcal{P}'')$ and $w(\mathcal{P}') = 1 + \sum_{i=1}^{m} w(\mathcal{P}_i)$. Since we have a proof $\mathcal{Q}'$: $p_1' \overset{\Delta'}{\leftarrow} p_1 \overset{*}{\leftrightarrow} t_1$ with $w(\mathcal{Q}') = w(\mathcal{P}'') + w(\mathcal{P}_1)$, we can apply $\mathbf{r}_1$ to $s \equiv C[C_L[p_1', p_2, \cdots, p_m]]$ too. Then, we have a proof $\mathcal{Q}$: $s \equiv C[C_L[p_1', \cdots, p_m]] \rightarrow t \equiv C[C_R[t_1, \cdots, t_n]]$ with $w(\mathcal{Q}) = 1 + w(\mathcal{Q}') + \sum_{i=2}^{m} w(\mathcal{P}_i) = w(\mathcal{P})$. Thus, (iii) follows.

Case 3. Δ and Δ' coincide by the application of the same rule, i.e., $\mathbf{r} = \mathbf{r}_1 = \mathbf{r}_2$. (We mentioned in Example 3 that in a left-right separated conditional term rewriting system the application of the same rule at the same position does not imply the same result as the variables occurring in the left-hand side of a rule do not cover that in the right-hand side. Thus this case is necessary even if the system is non-overlapping.) Let the rule applied to Δ and Δ' be $\mathbf{r}$: $C_L[x_1,\cdots,x_m] \to C_R[y_1,\cdots,y_n] \Leftarrow x_1 = y_1,\cdots,x_m = y_m$ (all the variable occurrences are displayed, and $m \geq n$ by the condition (iv) of Definition 9), and let $\mathcal{P}'$: $p \equiv C[C_L[p_1,\cdots,p_m]] \xrightarrow{\Delta} t \equiv C[C_R[t_1,\cdots,t_n]]$ with subproofs $\mathcal{P}'_i$: $p_i \overset{*}{\leftrightarrow} t_i$ $(i = 1,\cdots,m)$ and $\mathcal{P}''$: $p \equiv C[C_L[p_1,\cdots,p_m]] \xrightarrow{\Delta'} s \equiv C[C_R[s_1,\cdots,s_n]]$ with subproofs $\mathcal{P}''_i$: $p_i \overset{*}{\leftrightarrow} s_i$ $(i = 1,\cdots,m)$. Here $w(\mathcal{P}) = w(\mathcal{P}') + w(\mathcal{P}'') = 2 + \sum_{i=1}^m w(\mathcal{P}'_i) + \sum_{i=1}^m w(\mathcal{P}''_i)$. Then, we have a proof $\mathcal{Q}$: $t \equiv C[C_R[t_1,\cdots, t_n]] \overset{*}{\leftrightarrow} C[C_R[p_1,\cdots,p_n]] \overset{*}{\leftrightarrow} C[C_R[s_1,\cdots,s_n]] \equiv s$ with $w(\mathcal{Q}) = \sum_{i=1}^n w(\mathcal{P}'_i) + \sum_{i=1}^n w(\mathcal{P}''_i) < 2 + \sum_{i=1}^m w(\mathcal{P}'_i) + \sum_{i=1}^m w(\mathcal{P}''_i) = w(\mathcal{P})$. (Note that $m \geq n$ is necessary to guarantee $w(\mathcal{Q}) < w(\mathcal{P})$.) Hence (i) holds.

Case 4. Δ' occurs in Δ but neither Case 2 nor Case 3 (i.e., Δ' overlaps with the pattern l of $\Delta \equiv l\theta$). Then, there exists a conditional critical pair $[p_1 = q_1,\cdots,p_m = q_m] \vdash \langle q,q'\rangle$ between $\mathbf{r}_1$ and $\mathbf{r}_2$, and we can write $\mathcal{P}$: $t \equiv C[q\theta] \xleftarrow{\Delta} p \equiv C[\Delta] \xrightarrow{\Delta'} s \equiv C[q'\theta]$ with subproofs $\mathcal{P}_i$: $p_i\theta \overset{*}{\leftrightarrow} q_i\theta$ $(i = 1,\cdots,m)$. Thus $w(\mathcal{P}) = \sum_{i=1}^m w(\mathcal{P}_i) + 2$. From the assumption about critical pairs the possible relations between q and q' are give in the following subcases.

Subcase 4.1. $q \underset{E'}{\sim} q'$ for some E' such that $E' \sqsubseteq E \sqcup [\bullet]$. By Lemma 13 and $E' \sqsubseteq E \sqcup [\bullet]$, we have a proof $\mathcal{Q}'$: $q\theta \overset{*}{\leftrightarrow} q'\theta$ with $w(\mathcal{Q}') \leq \sum_{i=1}^m w(\mathcal{P}_i) + 1 < w(\mathcal{P})$. Hence it is obtained that $\mathcal{Q}$: $t \equiv C[q\theta] \overset{*}{\leftrightarrow} s \equiv C[q'\theta]$ with $w(\mathcal{Q}) < w(\mathcal{P})$. Thus, (i) holds.

Subcase 4.2. $q \underset{E_1}{\sim\!\triangleright} \cdot \underset{E_2}{\sim} q'$ and $q \underset{E'_1}{\sim} \cdot \underset{E'_2}{\triangleleft\!\sim} q'$ for some E_1, E_2, E'_1, and E'_2 such that $E_1 \sqcup E_2 \sqsubseteq E \sqcup [\bullet]$ and $E'_1 \sqcup E'_2 \sqsubseteq E \sqcup [\bullet]$. By Lemma 13 and $E_1 \sqcup E_2 \sqsubseteq E \sqcup [\bullet]$, we have a proof $\mathcal{Q}'$: $q\theta \to \cdot \overset{*}{\leftrightarrow} q'\theta$ with $w(\mathcal{Q}') \leq \sum_{i=1}^m w(\mathcal{P}_i) + 2 = w(\mathcal{P})$. Hence we can take $\mathcal{Q}_1$: $t \equiv C[q\theta] \to \cdot \overset{*}{\leftrightarrow} s \equiv C[q'\theta]$ with $w(\mathcal{Q}_1) \leq w(\mathcal{P})$. Simiraly we have $\mathcal{Q}_2$: $t \equiv C[q\theta] \overset{*}{\leftrightarrow} \cdot \leftarrow s \equiv C[q'\theta]$ with $w(\mathcal{Q}_2) \leq w(\mathcal{P})$. Thus, (ii) follows.

Subcase 4.3. $q \underset{E'}{\sim\!\triangleright} q'$ (or $q \underset{E'}{\triangleleft\!\sim} q'$) and $E' \sqsubseteq E \sqcup [\bullet]$. By Lemma 13 and $E' \sqsubseteq E \sqcup [\bullet]$, we have a proof $\mathcal{Q}'$: $q\theta \to q'\theta$ with $w(\mathcal{Q}') \leq \sum_{i=1}^m w(\mathcal{P}_i) + 2 = w(\mathcal{P})$. Hence we obtain $\mathcal{Q}$: $t \equiv C[q\theta] \to s \equiv C[q'\theta]$ with $w(\mathcal{Q}) \leq w(\mathcal{P})$. For the case of $q \underset{E'}{\triangleleft\!\sim} q'$ we can obtain $\mathcal{Q}$: $s \leftarrow t$ with $w(\mathcal{Q}) \leq w(\mathcal{P})$ similarly. Thus, (iii) holds. □

Corollary 15. *Let R be a left-right separated conditional term rewriting system. Then R is weight decreasing joinable if R is non-overlapping.*

Example 4. Let R be the left-right separated conditional term rewriting system with the following rewrite rules:

$$R \quad \begin{cases} f(x',x'') \to h(x,f(x,b)) \Leftarrow x' = x, x'' = x \\ f(g(y'),y'') \to h(y,f(g(y),a)) \Leftarrow y' = y, y'' = y \\ a \to b \end{cases}$$

Here, R has the conditional critical pair
$[g(y') = x, y'' = x, y' = y, y'' = y] \vdash \langle h(x,f(x,b)), h(y,f(g(y),a))\rangle$.

Since $h(x, f(x,b)) \underset{[y''=x]}{\sim} h(y'', f(x,b)) \underset{[g(y')=x]}{\sim} h(y'', f(g(y'),b)) \underset{[y''=y,y'=y]}{\sim}$ $h(y, f(g(y),b)) \underset{[\bullet]}{\sim} h(y, f(g(y),a))$, we have $h(x, f(x,b)) \underset{E'}{\sim} h(y, f(g(y),a))$ where $E' = [g(y') = x, y'' = x, y'' = y, y' = y, \bullet]$. Thus, from Theorem 14 it follows that R is weight decreasing joinable. □

We say that $E = [p_1 = q_1, \cdots, p_m = q_m]$ is satisfiable (in R) if there exist proofs $\mathcal{P}_i$: $p_i\theta \overset{*}{\leftrightarrow} q_i\theta$ $(i = 1, \cdots, m)$ for some θ; otherewise E is unsatisfiable. Note that the satisfiability of E is generally undecidable. Theorem 14 requests that every conditional critical pair $E \vdash \langle q, q' \rangle$ satisfies (i), (ii) or (iii). However, it is clear that we can ignore conditional critical pairs having unsatisfiable E. Thus, we can strengthen Theorem 14 as follows.

Corollary 16. *Let R be a left-right separated conditional term rewriting system. Then R is weight decreasing joinable if any conditional critical pair $E \vdash \langle q, q' \rangle$ such that E is satisfiable in R satisfies (i), (ii) or (iii) in Theorem 14.*

6 Conditional Linearization

The original idea of the conditional linearization of non-left-linear term rewriting systems was introduced by De Vrijer [4], Klop and De Vrijer [6] for giving a simpler proof of Chew's theorem [2, 9]. In this section, we introduce a new conditional linearization based on left-right separated conditional term rewriting systems. The point of our linearization is that by replacing traditional conditional systems with left-right separated conditional systems we can easily relax the non-overlapping limitation.

Now we explain a new linearization of non-left-linear rules. For instance, let consider a non-duplicating non-left-linear rule $f(x,x,x,y,y,z) \to g(x,x,x,z)$. Then, by replacing all the variable occurrences x,x,x,y,y,z from left to right in the left handside with distinct fresh variable occurrences x',x'',x''',y',y'',z' respectively and connecting every fresh variable to corresponding original one with equation, we can make a left-right separated conditional rule $f(x',x'',x''',y',y'',z') \to g(x,x,x,z) \Leftarrow x' = x, x'' = x, x''' = x, y' = y, y' = y, z' = z$. More formally we have the following definition, the framework of which originates essentially from De Vrijer [4], Klop and De Vrijer [6].

Definition 17. **(i)** If $\mathbf{r}$ is a non-duplicating rewrite rule $l \to r$ and $l \equiv C[y_1, \cdots, y_m]$ (all the variable occurences of l are displayed), then the (left-right separated) conditional linearization of $\mathbf{r}$ is a left-right separated conditional rewrite rule $\mathbf{r}_L$: $l' \to r \Leftarrow x_1 = y_1, \cdots, x_m = y_m$ where $l' \equiv C[x_1, \cdots, x_m]$ and $x_1, \cdots, x_m$ are distinct fresh variables. Note that $l'\theta \equiv l$ for the substitution $\theta = [x_1 := y_1, \cdots, x_m := y_m]$.

(ii) If R is a non-duplicating term rewriting system, then R_L, the conditional linearization of R, is defined as the set of the rewrite rules $\{\mathbf{r}_L | \mathbf{r} \in R\}$.

Note. The non-duplicating limitation of R in the above definition is necessary to guarantee that R_L is a left-right separated conditional term rewriting system. Otherwise R_L does not satisfy the condition (iv) of Definition 9 in general.

The above conditional linearization is different from the original one by Klop and De Vrijer [4, 6] in which the left-linear version of a rewrite rule **r** is a traditional conditional rewrite rule without extra variables in the right handside and the conditional part. Hence, in the case **r** is already left-linear, Klop and De Vrijer [4, 6] can take **r** itself as its conditional linearization. On the other hand, in our definition we cannot take **r** itself as its conditional linearization since **r** is not a left-right separated rewrite rule.

Theorem 18. *If a conditional linearization R_L of a non-duplicating term rewriting system R is Church-Rosser, then R has unique normal forms.*

Proof. By Propsiton 3, similar to Klop and De Vrijer [4, 6]. □

Example 5. Let R be the non-duplicating term rewriting system with the following rewrite rules:

$$R \quad \begin{cases} f(x,x) \to h(x, f(x,b)) \\ f(g(y),y) \to h(y, f(g(y),a)) \\ a \to b \end{cases}$$

Note that R is non-left-linear and non-terminating. Then we have the following R_L as the linearization of R:

$$R_L \quad \begin{cases} f(x',x'') \to h(x, f(x,b)) \Leftarrow x' = x, x'' = x \\ f(g(y'),y'') \to h(y, f(g(y),a)) \Leftarrow y' = y, y'' = y \\ a \to b \end{cases}$$

In Example 4 the Church-Rosser property of R_L has already been shown. Thus, from Theorem 18 it follows that R has unique normal forms. □

7 Church-Rosser Property of Non-Duplicating Systems

In the previous section we have shown a general method based on the conditional linearization technique to prove the unique normal form property of non-left-linear overlapping non-duplicating term rewriting systems. In this section we show that the same conditional linearization technique can be used as a general method for proving the Church-Rosser property of some class of non-duplicating term rewriting systems.

Theorem 19. *Let R be a right-ground (i.e., no variables occur in the right hand-side of rewrite rules) term rewriting system. If the conditional linearization R_L of R is weight decreasing joinable then R is Church-Rosser.*

Proof. Let R and R_L have reduction relations $\rightarrow$ and $\underset{L}{\rightarrow}$ respectively. Since $\underset{L}{\rightarrow}$ extends $\rightarrow$ and R_L is weight decreasing joinable, the theorem clearly holds if we show the claim: For any t, s and $\mathcal{P}$: $t \overset{*}{\underset{L}{\leftrightarrow}} s$ there exist proofs $\mathcal{Q}$: $t \overset{*}{\underset{L}{\rightarrow}} r \overset{*}{\underset{L}{\leftarrow}} s$ with $w(\mathcal{P}) \geq w(\mathcal{Q})$ and $t \overset{*}{\rightarrow} r \overset{*}{\leftarrow} s$ for some term r. We will prove this claim by induction on $w(\mathcal{P})$. *Base Step* ($w(\mathcal{P}) = 0$) is trivial. *Induction Step*: Let $w(\mathcal{P}) = \rho > 0$. Form the weight decreasing joinability of R_L, we have a proof $\mathcal{P}'$: $t \overset{*}{\underset{L}{\rightarrow}} \cdot \overset{*}{\underset{L}{\leftarrow}} s$ with $\rho \geq w(\mathcal{P}')$. Let $\mathcal{P}'$ have the form $t \underset{L}{\rightarrow} s' \overset{*}{\underset{L}{\rightarrow}} \cdot \overset{*}{\underset{L}{\leftarrow}} s$. Without loss of generality we may assume that $C_L[x_1, \cdots, x_m, y_1, \cdots, y_n, \cdots, z_1, \cdots, z_p] \rightarrow C_R \Leftarrow x_1 = x, \cdots, x_m = x, y_1 = y, \cdots, y_n = y, \cdots, z_1 = z, \cdots, z_p = z$ (all the variable occurrences are displayed) is a linearization of $C_L[x, \cdots, x, y, \cdots, y, \cdots, z, \cdots, z] \rightarrow C_R$ and $\mathcal{P}''$: $t \equiv C[C_L[t_1^x, \cdots, t_m^x, t_1^y, \cdots, t_n^y, \cdots, t_1^z, \cdots, t_p^z]] \underset{L}{\rightarrow} s' \equiv C[C_R]$ with subproofs $\mathcal{P}_i^x$: $t_i^x \overset{*}{\underset{L}{\leftrightarrow}} t'_x$ $(i = 1, \cdots, m)$, $\mathcal{P}_j^y$: $t_j^y \overset{*}{\underset{L}{\leftrightarrow}} t'_y$ $(j = 1, \cdots, n)$, $\cdots$, $\mathcal{P}_k^z$: $t_k^z \overset{*}{\underset{L}{\leftrightarrow}} t'_z$ $(k = 1, \cdots, p)$ for some t'_x, t'_y, $\cdots$, t'_z. Then, from Lemma 7 and the induction hypothesis we have proofs $t_i^x \overset{*}{\rightarrow} t''_x$ $(i = 1, \cdots, m)$, $t_j^y \overset{*}{\rightarrow} t''_y$ $(j = 1, \cdots, n)$, $\cdots$, $t_k^z \overset{*}{\rightarrow} t''_z$ $(k = 1, \cdots, p)$. Hence we can take the reduction $t \equiv C[C_L[t_1^x, \cdots, t_m^x, t_1^y, \cdots, t_n^y, \cdots, t_1^z, \cdots, t_p^z]] \overset{*}{\rightarrow} C[C_L[t''_x, \cdots, t''_x, t''_y, \cdots, t''_y, \cdots, t''_z, \cdots, t''_z]] \rightarrow s' \equiv C[C_R]$. Let $\hat{\mathcal{P}}$: $s' \overset{*}{\underset{L}{\rightarrow}} \cdot \overset{*}{\underset{L}{\leftarrow}} s$ be the subproof of P'. From $\rho > w(\hat{\mathcal{P}})$ and induction hypothesis, we have $\hat{\mathcal{Q}}$: $s' \overset{*}{\underset{L}{\rightarrow}} r \overset{*}{\underset{L}{\leftarrow}} s$ with $w(\hat{\mathcal{P}}) \geq w(\hat{\mathcal{Q}})$ and $s' \overset{*}{\rightarrow} r \overset{*}{\leftarrow} s$ for some r. Thus, the theorem follows. □

The following corollary was originally proven by Oyamaguchi and Ohta [7].

Corollary 20. *[Oyamaguchi] Let R be a right-ground term rewriting system having a non-overlapping conditional linearization R_L. Then R is Church-Rosser.*

We next relax the right-ground limitation of R in Theorem 19.

Theorem 21. *Let R be a term rewriting system in which every rewrite rule $l \rightarrow r$ is right-linear (i.e., r is linear) and no non-linear variables in l occur in r. If the conditional linearization R_L of R is weight decreasing joinable then R is Church-Rosser.*

Proof. The proof is similar to that of Theorem 19. Let R and R_L have reduction relations $\rightarrow$ and $\underset{L}{\rightarrow}$ respectively. Since $\underset{L}{\rightarrow}$ extends $\rightarrow$ and R_L is weight decreasing joinable, the theorem clearly holds if we show the claim: for any t, s and $\mathcal{P}$: $t \overset{*}{\underset{L}{\leftrightarrow}} s$ there exist proofs $\mathcal{Q}$: $t \overset{*}{\underset{L}{\rightarrow}} r \overset{*}{\underset{L}{\leftarrow}} s$ with $w(\mathcal{P}) \geq w(\mathcal{Q})$ and $t \overset{*}{\rightarrow} r \overset{*}{\leftarrow} s$

for some term r. We will prove this claim by induction on $w(\mathcal{P})$. *Base Step* ($w(\mathcal{P}) = 0$) is trivial. *Induction Step*: Let $w(\mathcal{P}) = \rho > 0$. Form the weight decreasing joinability of R_L, we have a proof $\mathcal{P}'$: $t \xrightarrow[L]{*} \cdot \xleftarrow[L]{*} s$ with $\rho \geq w(\mathcal{P}')$. Let $\mathcal{P}'$ have the form $t \xrightarrow[L]{} \hat{s} \xrightarrow[L]{*} \cdot \xleftarrow[L]{*} s$. Without loss of generality we may assume that $C_L[x_1, \cdots, x_m, y_1, \cdots, y_n, \cdots, z_1, \cdots, z_p, v_1, \cdots, w_1] \rightarrow C_R[v, \cdots, w] \Leftarrow x_1 = x, \cdots, x_m = x, y_1 = y, \cdots, y_n = y, \cdots, z_1 = z, \cdots, z_p = z, v_1 = v, \cdots, w_1 = w$ (all the variable occurrences are displayed) is a linearization of a right-linear rewrite rule $C_L[x, \cdots, x, y, \cdots, y, \cdots, z, \cdots, z, v, \cdots, w] \rightarrow C_R[v, \cdots, w]$ and $t \equiv C[C_L[t_1^x, \cdots, t_m^x, t_1^y, \cdots, t_n^y, \cdots, t_1^z, \cdots, t_p^z, p_1, \cdots, q_1]] \xrightarrow[L]{} \hat{s} \equiv C[C_R[p, \cdots, q]]$ with subproofs $\mathcal{P}_i^x$: $t_i^x \overset{*}{\underset{L}{\leftrightarrow}} t'_x$ $(i = 1, \cdots, m)$, $\mathcal{P}_j^y$: $t_j^y \overset{*}{\underset{L}{\leftrightarrow}} t'_y$ $(j = 1, \cdots, n)$, $\cdots$, $\mathcal{P}_k^z$: $t_k^z \overset{*}{\underset{L}{\leftrightarrow}} t'_z$ $(k = 1, \cdots, p)$ for some t'_x, t'_y, $\cdots$, t'_z, and $\mathcal{P}^v$: $p_1 \overset{*}{\underset{L}{\leftrightarrow}} p$, $\cdots$, $\mathcal{P}^w$: $q_1 \overset{*}{\underset{L}{\leftrightarrow}} q$. Then, we can take $t \equiv C[C_L[t_1^x, \cdots, t_m^x, t_1^y, \cdots, t_n^y, \cdots, t_1^z, \cdots, t_p^z, p_1, \cdots, q_1]] \xrightarrow[L]{} s' \equiv C[C_R[p_1, \cdots, q_1]] \overset{*}{\underset{L}{\leftrightarrow}} \hat{s} \equiv C[C_R[p, \cdots, q]] \xrightarrow[L]{*} \cdot \xleftarrow[L]{*} s$ with the weight $w(\mathcal{P}')$. Let $\mathcal{P}''$: $t \equiv C[C_L[t_1^x, \cdots, t_m^x, t_1^y, \cdots, t_n^y, \cdots, t_1^z, \cdots, t_p^z, p_1, \cdots, q_1]] \xrightarrow[L]{} s' \equiv C[C_R[p_1, \cdots, q_1]]$. Then, from Lemma 7 and the induction hypothesis we have proofs $t_i^x \xrightarrow{*} t''_x$ $(i = 1, \cdots, m)$, $t_j^y \xrightarrow{*} t''_y$ $(j = 1, \cdots, n)$, $\cdots$, $t_k^z \xrightarrow{*} t''_z$ $(k = 1, \cdots, p)$. Hence we can take the reduction $t \equiv C[C_L[t_1^x, \cdots, t_m^x, t_1^y, \cdots, t_n^y, \cdots, t_1^z, \cdots, t_p^z, p_1, \cdots, q_1]] \xrightarrow{*} C[C_L[t''_x, \cdots, t''_x, t''_y, \cdots, t''_y, \cdots, t''_z, \cdots, t''_z, p_1, \cdots, q_1]] \rightarrow s' \equiv C[C_R[p_1 \cdots, q_1]]$. Let $\hat{\mathcal{P}}$: $s' \overset{*}{\underset{L}{\leftrightarrow}} \hat{s} \xrightarrow[L]{*} \cdot \xleftarrow[L]{*} s$. From $\rho > w(\hat{\mathcal{P}})$ and induction hypothesis, we have $\hat{\mathcal{Q}}$: $s' \xrightarrow[L]{*} r \xleftarrow[L]{*} s$ with $w(\hat{\mathcal{P}}) \geq w(\hat{\mathcal{Q}})$ and $s' \xrightarrow{*} r \xleftarrow{*} s$ for some r. Thus, the theorem follows. □

Corollary 22. *Let R be a term rewriting system in which every rewrite rule $l \rightarrow r$ is right-linear and no non-linear variables in l occur in r. If the conditional linearization R_L of R is non-overlapping then R is Church-Rosser.*

Example 6. Let R be the term rewriting system with the following rewrite rules:

$$R \quad \begin{cases} f(x, x, y) \rightarrow h(y, c) \\ g(x) \rightarrow f(x, c, g(c)) \\ c \rightarrow h(c, c) \end{cases}$$

Note that R is non-left-linear and non-terminating. Then we have the following R_L as the linearization of R: w

$$R_L \quad \begin{cases} f(x', x'', y') \rightarrow h(y, c) \Leftarrow x' = x, x'' = x, y' = y \\ g(x') \rightarrow f(x, c, g(c)) \Leftarrow x' = x \\ c \rightarrow h(c, c) \end{cases}$$

From Corollary 15, R_L is Church-Rosser. Thus, from Corollary 22 it follows that R is Church-Rosser. □

References

1. J. A. Bergstra and J. W. Klop, Conditional rewrite rules: Confluence and termination, *J. Comput. and Syst. Sci. 32* (1986) 323-362.
2. P. Chew, Unique normal forms in term rewriting systems with repeated variables, *Proc. 13th STOC* (1981) 7-18.
3. N. Dershowitz and J. P. Jouannaud, Rewrite Systems, in: J. V. Leeuwen, ed., *Handbook of Theoretical Computer Science B* (North-Holland, 1990) 244-320.
4. R. C. de Vrijer, Unique normal forms for combinatory logic with parallel conditional, a case study in conditional rewriting, *Techn. Report Free University Amsterdam* (1990)
5. J. W. Klop, Term rewriting systems, in: S. Abramsky, D. Gabbay and T. Maibaum, eds., *Handbook of Logic in Computer Science 2* (Oxford University Press, 1992) 1-116.
6. J. W. Klop and R. C. de Vrijer, Unique normal forms for lambda calculas with surjective pairing. *Information and Computation 80* (1989) 97-113.
7. M. Oyamaguchi and Y. Ohta, On the confluent property of right-ground term rewriting systems, *Trans. IEICE Japan, J76-D-I* (1993) 39-45, *in Japanese.*
8. M. Oyamaguchi, On the Church-Rosser property of nonlinear and nonterminating term rewriting systems, *Unpublished note, LA Symp (Summer)* (1992).
9. M. Ogawa, Chew's theorem revisited: uniquely normalizing property of nonlinear term rewriting systems, *Lecture Notes in Comput. Sci. 650* (Springer-Verlag, 1992) 309-318.

The Transformation of Term Rewriting Systems Based on Well-formedness Preserving Mappings

Jan C. Verheul and Peter G. Kluit

Delft University of Technology, TWI/TI, POB 356, 2600 AJ Delft, The Netherlands,
verheul@twi.tudelft.nl

Abstract. The paper describes a formal method to derive a simulating term rewriting system from an original system using six elementary syntactical transformations of terms. The derived (or transformed) rewrite systems allow one to mimic the operation of the original system, intended for a certain term domain, in another term domain. The paper presents formalizations for the notions term domain, domain mapping, simulation, transformation and composition of transformations. Formal proofs of the achieved results are presented when appropriate. Examples are given to illustrate the usefulness of the various transformations.

1 Introduction

In this paper we present the description of the application of the transformational approach to problem solving on term rewriting systems. The transformational approach will informally be defined as: construct a new input-output mapping, for a new domain, based on an existing input-output mapping for an existing domain. Determination of an output for a particular input can be done by transforming the input to the new domain, determining the output in the new domain by the new mapping and subsequently transforming the output back from the new domain to the original domain.

The transformational approach is very well-known in the field of differential equations and in the calculation of output signals of electronic circuits which can be described by differential equations. With problems from these areas it is customary to transform them from the time domain to the frequency domain and calculate with transformed signals and with transformed IO-mappings. The advantage of calculating with the composed mapping with two domain transformations and a transformed IO-mapping is in the case of Fourier transformations and harmonic signals a tremendous reduction of complexity. IO-mapping becomes just a matter of multiplication and in spite of the two additional domain transformations, the composed system with transformed IO-mapping is conceptually much simpler than the original system [KSS91].

A basic formalism is provided to underly the theory of transforming term rewriting systems: *retrenched vocabularies*. The relation between two term domains is formalized by the notion *well-formedness preserving mappings* and the notion of equivalence between a straightforward mapping and a composed mapping over a new domain is formalized by the notion *simulation*. The derivation of a simulating rewrite system is called *transformation*. A graphical illustration of the situation:

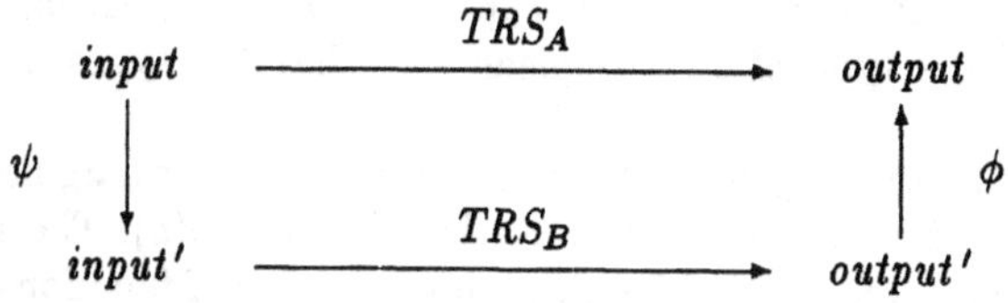

figure 1: transformational approach for term rewriting systems

The benefits of transforming term rewriting systems are manifold. It is possible to reduce the size of terms (the number of constructors) and thereby CPU and memory load in concrete implementations of rewrite systems. It is possible to reduce the number of rewrite rules in a rewrite system and thereby again CPU and memory load in practical implementations. Complexity reductions have, self evidently, also theoretical importance. It is also possible to transform rewrite systems to get an equivalent one with certain desired properties. It is for example possible to transform a rewrite system for an order sorted signature into one for a many sorted signature, without having to introduce coercion operators, like in [Gan91, GJM85]. The notion of simulation is defined in such a general way that terms in the original domain can be represented by classes of terms in the new domain. This concept is closely related to the notion of rewriting modulo a congruence [DJ90, BD89]. The theory that will be presented supports the transformation of rules with variables, but only for ground rewriting.

The paper is organized as follows. In section 2 the basic formalism, the formalism of retrenched vocabularies, is introduced. In section 3 the relation between retrenched vocabularies and other formalisms: the order sorted signature and the context free grammar formalism, is explored. In section 4 the mapping between term domains is treated with the definition of well-formedness preserving mappings. In section 5 the simulation concept is defined formally. In section 6 a number of useful elementary syntactic transformations are introduced. The effect of these transformations on the term domains: retrenched vocabularies, is described. Section 7 is devoted to the derivation of simulating rewrite systems from certain given rewrite systems. The derivation schemes are given in the form of imperative programming style algorithms and correctness of these algorithms is proven. In section 8 some examples are presented to illustrate the utility of the various transformations. Section 9 finally states some concluding remarks. Notational conventions for the rest of this paper:

- The symbol $\perp$ denotes the special value *undefined*.
- About set comprehension and its notation: $\{e(x_1, ..., x_n) \mid P(x_1, ..., x_n)\}$ denotes the smallest subset V_s of a not further to be specified universe U such that: $\forall x_1, ..., x_n \in U[P(x_1, ..., x_n) \Rightarrow e(x_1, ..., x_n) \in V_s]$. If the restriction to

some known subset V_r of U is a component of the predicate, this may be expressed left of the vertical bar, i.e. $\{e(x_1, ..., x_n) \in V_r \mid P(x_1, ..., x_n)\}$.
- The construction $\{a_1 \leftarrow v_1, ..., a_n \leftarrow v_n\}$ denotes the function that maps a_1 to v_1, ..., a_n to v_n and all other argument values to $\perp$.
- The conditional expression *if* P *then* T *else* E has its usual meaning, provided that the type of P is boolean and that the types of T and E are equal.
- The construction $f \dagger g$ denotes the function that maps argument value a to *if* $g(a) = \perp\!\!\!\perp$ *then* $\perp$ *else if* $g(a) = \perp$ *then* $f(a)$ *else* $g(a)$. The symbol $\perp\!\!\!\perp$ is a specially dedicated symbol for shrinking domains of functions.
- Functional composition is denoted by the infix operator $\circ$ and $g \circ f$ is the function defined by: $(g \circ f)(x) =$ *if* $f(x) = \perp$ *then* $\perp$ *else* $g(f(x))$.
- The number of elements in a set S is denoted by $|S|$.
- The powerset of a set S (the set of all distinct subsets of S) is denoted by $\mathcal{P}(S)$.

2 Retrenched Vocabularies

The theory of TRS-transformations is based on a formalism which is essentially a stripped down version of the order sorted signature formalism. It is called the *retrenched vocabulary* formalism and it is defined as:

Definition 1. A *retrenched vocabulary* is a 6-tuple $\langle C, V, \alpha, \rho, \mu, S\rangle$, with:
C : the set of constructors,
V : the set of variables,
$\alpha : C \rightarrow \mathcal{N}$, the arity function,
$\rho : C \times \mathcal{N}_+ \rightarrow \mathcal{P}(C)$, the restriction function,
$\mu : V \rightarrow \mathcal{P}(C) - \emptyset$, the match function,
S: the root restriction set;
for which conditions hold which are defined in terms of the components of the defining 6-tuple and in terms of the following quantities that can be derived from the defining 6-tuple:
$YC = \{\rho(c, i) \mid c \in C, 1 \leq i \leq \alpha(c)\}$: the set of *represented types* in the *constructor set*,
$YV = \{\mu(v) \mid v \in V\}$: the set of *represented types* in the *variable set*,
$Y = YC \cup YV \cup \{S\}$: the set of *represented types*,
$K = \{c \in C \mid \alpha(c) = 0\}$: the subset of zero arity constructors, also called *constants*,
$\rho^{-1} : C \rightarrow \mathcal{P}(C \times \mathcal{N})$ with $\rho^{-1}(c) = \{\langle c', i\rangle \mid c \in \rho(c', i)\}$: the *inverse restriction function*.
The conditions:

- $\forall c \in C, \forall 1 \leq n \leq \alpha(c)[\rho(c, n) \subseteq C] \wedge S \subseteq C$: a constructor restriction is defined to be a subset of the set of constructors,
- $\forall c \in C, \forall n > \alpha(c)[\rho(c, n) = \perp]$: the restriction function is not defined for values of its second argument greater than the arity in question,

- $S \neq \emptyset$: with $S = \emptyset$ it is not possible to construct well-formed terms, which is not very useful in practice,
- $YC \subseteq YV$: for each represented type that acts as function value of ρ or that equals S there is an associated variable,
- $\forall yv \in YV [\exists yc \in YC [yv \subseteq yc] \vee yv \subseteq S]$: all variables fit somewhere in well-formed terms: definition 6,
- for each represented type (each element of Y) there are countably infinite variables.

The derived sets, functions and the primary elements of the tuple will be indexed with a vocabulary specifier whenever ambiguity can arise, so for instance Y_A and α_A pertain to vocabulary $A = \langle C, V, \alpha, \rho, \mu, S\rangle$. □

Definition 2. The set T_A of *terms* over vocabulary $A = \langle C, V, \alpha, \rho, \mu, S\rangle$ is defined inductively as the smallest set such that: $K \subset T_A$; $V \subset T_A$; $\forall f \in C - K, \forall t_1, ..., t_{\alpha(f)} \in T_A [f(t_1, ..., t_{\alpha(f)}) \in T_A]$. □

Definition 3. The set TG_A of *ground terms* over vocabulary $A = \langle C, V, \alpha, \rho, \mu, S\rangle$ is defined inductively as the smallest set such that: $K \subset TG_A$; $\forall f \in C - K, \forall t_1, ..., t_{\alpha(f)} \in TG_A [f(t_1, ..., t_{\alpha(f)}) \in TG_A]$. □

Definition 4. The function $Root : T_A \rightarrow C \uplus V$ returns the root constructor of a given term. Definition by case: $\forall k \in K[Root(k) = k]$; $\forall v \in V[Root(v) = v]$; $\forall f \in C - K, \forall t_1, ..., t_{\alpha(f)} \in T_A[Root(f(t_1, ..., t_{\alpha(f)})) = f]$. □

Definition 5. The set $\mathcal{R}_A$ of *restricted terms* over vocabulary $A = \langle C, V, \alpha, \rho, \mu, S\rangle$ is defined inductively as the smallest set such that: $K \subset \mathcal{R}_A$; $V \subset \mathcal{R}_A$; $\forall f \in C - K, \forall t_1, ..., t_{\alpha(f)} \in \mathcal{R}_A [\forall 1 \leq i \leq \alpha(f) [\textit{if } t_i \in V \textit{ then } \mu(t_i) \subseteq \rho(f, i) \textit{ else } Root(t_i) \in \rho(f, i)] \Rightarrow f(t_1, ..., t_{\alpha(f)}) \in \mathcal{R}_A]$. □

Definition 6. The set $\mathcal{T}_A$ of *well-formed terms* over vocabulary $A = \langle C, V, \alpha, \rho, \mu, S\rangle$ is defined as: $\{t \in \mathcal{R}_A \mid \textit{if } t \in V \textit{ then } \mu(t) \subseteq S \textit{ else } Root(t) \in S\}$. □

Definition 7. The set $\mathcal{TG}_A$ of *well-formed ground terms* over vocabulary $A = \langle C, V, \alpha, \rho, \mu, S\rangle$ is defined as: $\mathcal{TG}_A = \mathcal{T}_A \cap TG_A$. □

Definition 8. The function $Var : T_A \rightarrow \mathcal{P}(V)$ returns the set of variables that occur in a given term. Definition with induction on term structure: $\forall k \in K[Var(k) = \emptyset]$; $\forall v \in V[Var(v) = \{v\}]$; $\forall f \in C - K, \forall t_1, ..., t_{\alpha(f)} \in T_A[Var(c(t_1, ..., t_{\alpha(f)})) = \bigcup_{i=1}^{\alpha(f)} Var(t_i)$. □

Definition 9. A variable v is called *maximally typed* in a term $t \in \mathcal{T}$ if and only if there is no substitution $\sigma = \{v' \leftarrow v\}$ with $v' \in V$ and $\mu(v') \supset \mu(v)$ (proper inclusion) and not a term $t' \in \mathcal{T}$ such that $t = t'^{\sigma}$. □

It is possible to identify a number of useful properties for retrenched vocabularies. The interesting fact about the following list of properties is that each of them can be established for vocabularies that do not already possess them by the transformations to be introduced in this paper.

Definition 10. Properties for retrenched vocabularies:

1. $\forall c \in C, \exists yc \in (YC \cup \{S\}) [c \in yc]$,
   If certain constructors do not occur as value of the restriction function or the root restriction set they cannot appear in a term. Retrenched vocabularies for which this condition holds will be called *integrated*.
2. $\emptyset \notin YC$,
   Empty-set valued restriction functions will prohibit the construction of ground terms, as at certain positions no constructors are allowed to occur. Retrenched vocabularies for which this condition holds will be called *satisfiable*.
3. $\forall yc \in YC [|yc| > 1]$,
   If only one constructor is allowed under a certain parent constructor then it is not necessary to discriminate between this particular parent and child, taking this parent and child together as a constructor with arity $m+n-1$ (assuming $\rho(c_{pa}, i) = \{c_{ch}\}$ for some i and $\alpha(c_{pa}) = m$ and $\alpha(c_{ch}) = n$) will result in an equally powerful system. The transformation is an instance of contraction (definition 28). Retrenched vocabularies for which this condition holds will be called *trivial combination free*.
4. $\forall y_1, y_2 \in Y[y_1 = y_2 \vee y_1 \cap y_2 = \emptyset]$,
   The ρ- and μ-values, subsets of C, should in some situations be confined to a (small) set of standard restriction sets, this suits better our human intuition and also eases formal reasoning about the restrictions considerably. Moreover: many sorted signatures are equivalent with retrenched vocabularies for which this condition holds. Retrenched vocabularies for which this condition holds will be called *normalized*.
5. $\forall c \in C, \forall 1 \leq i \leq \alpha(c)[S \cap \rho(c,i) = \emptyset]$.
   Sometimes it is useful to enforce the restriction that constructors which are allowed as start constructor are not allowed inside terms. Retrenched vocabularies for which this condition holds will be called *uniquely starting*.

□

3 Relations with Other Formalisms

The retrenched vocabulary formalism is in fact a stripped down version of the order sorted signature formalism. It is easy to derive a retrenched vocabulary from an order sorted signature and vice versa. Two general schemes will be presented for these derivation processes. It is also easy to derive a retrenched vocabulary from a context free grammar. The close relationship between sorted signatures and context free grammars is a known fact [HHKR89] and retrenched vocabularies are in fact an intermediate form between these two formalisms (the formalisms are equivalent with regard to their ability to define tree-like structures, they are of course different with regard to their ability to accept formal languages). The derivation schemes for order sorted signatures are based on the following perception of order sorted signatures, taken from [SNGM89].

Definition 11. An *order sorted signature* is a quadruple: $\langle S, <, \Sigma, V\rangle$ with:

- $S = \{s_1, ..., s_n\}$: the set of sorts (we assume S to be finite, this is not necessary however),

- $<$: a partial ordering on S ($\leq$ denotes the reflexive closure of $<$),
- $\Sigma = \{\Sigma_{\omega,s} \mid \omega \in S^* \wedge s \in S\}$: a family of typed operator sets, if there are no operators from sorts $s_1 ... s_n = \omega$ to sort s then $\Sigma_{\omega,s}$ is considered empty, all $\Sigma_{\omega,s}$ are assumed to be disjunct,
- $V = \{V_s \mid s \in S\}$: a family of typed variable sets, each V_s is assumed to contain countably infinite typed variables, all V_s are assumed to be disjunct.

□

Scheme 12. A retrenched vocabulary $A = \langle C, V, \alpha, \rho, \mu, S\rangle$ can be derived from an order sorted signature $Sig = \langle S, <, \Sigma, V\rangle$ as follows (symbols left of the equal signs pertain to A, symbols right of the equal signs pertain to Sig, unless specified otherwise):

$$C = \bigcup_{\omega \in S^*, s \in S} \Sigma_{\omega,s},$$
$$V = \bigcup_{s \in S} V_s,$$
$$\alpha = \{f \leftarrow |\omega| \mid \exists s \in S[f \in \Sigma_{\omega,s}]\} \cup \{v \leftarrow 0 \mid \exists s \in S[v \in V_s]\},$$
$$\rho = \{\langle f, i\rangle \leftarrow \{f' \in \Sigma_{\omega,s} \mid \omega \in S^* \wedge s \leq s_i\} \mid f \in \Sigma_{s_1 ... s_n, s_0} \wedge 1 \leq i \leq \alpha(f)\},$$
$$\mu = \{v \leftarrow \{f \in \Sigma_{\omega,s} \mid \omega \in S^*\} \mid v \in V_s\},$$
$$S = C_A.$$
□

Scheme 13. An order sorted signature $Sig = \langle S, <, \Sigma, V\rangle$ can be derived from a retrenched vocabulary $A = \langle C, V, \alpha, \rho, \mu, S\rangle$ as follows (symbols left of the equal signs pertain to Sig, symbols right of the equal signs pertain to A, unless specified otherwise, the meaning of Y can be found in definition 1):

$$S = \{\overline{\bigcap Y_s} \mid Y_s \subseteq Y \wedge Y_s \neq \emptyset \wedge \bigcap Y_s \neq \emptyset\} \cup \{\overline{C - \bigcup Y}\},$$
$$< = \{\langle \overline{C_l}, \overline{C_r}\rangle \mid \overline{C_l}, \overline{C_r} \in S_{Sig} \wedge C_l \subset C_s\},$$
$$\Sigma_{s_1 ... s_n, s_0} = \{f \in C \mid \alpha(f) = n \wedge \forall 1 \leq i \leq n[\overline{\rho(f,i)} = s_i] \wedge \exists C_s \subseteq C [f \in C_s \wedge \overline{C_s} = s_0 \wedge \neg \exists C'_s \subset C[C'_s \subset C_s]]\},$$
$$V_s = \{v \in V \mid \overline{\mu(v)} = s\},$$

in which the overbar (over C_s for instance) denotes an operator that hides the set structure of its argument (the only property this operator has to have is: $\forall S_1, S_2 \subseteq C[S_1 = S_2 \Leftrightarrow \overline{S_1} = \overline{S_2}]$). □

The main differences between the sorted signature and the vocabulary formalism are:

- a shortcut between operand and result types of operators, resulting in a restriction function for constructors,
- a restriction of the set of constructors which is allowed to act as root constructor of well-formed terms.

The first item is only a superficial, syntactical difference. It is possible to state the theory about transformations in terms of conventional operators with operand types and result types but this will turn out to be very awkward because each transformation will require a major restructuring of the type system. The shortcut of types is only for convenience of the presentation of our theory. The second item represents a structural difference between the conventional and the new formalism. Our theory requires the ability to exclude certain constructors from acting as root constructors of well-formed terms. This functionality can be added to the conventional sorted signature formalism by extending the formalism with a specially dedicated root-type which prescribes the valid result type of root operators of well-formed terms. Scheme 13 works only well if there is no restriction for root constructors in the vocabulary, i.e. if $S = C$.

The general scheme for the derivation of retrenched vocabularies from context free grammars is based on the following perception of context free grammars (taken from [ASU86]).

Definition 14. A *context free grammar* is a quadruple: $\langle S, N, T, P\rangle$ for which:

- $S \notin N, S \notin T$,
- $N \cap T = \emptyset$,
- $P \subset (N \uplus \{S\}) \times (N \uplus T)^*$,

and in which: S the specially designated start symbol of the grammar, N the set of non-terminals, T the set of terminals and P the set of productions. Each production consists of a non-terminal and a string of grammar symbols: non-terminals or terminals. Productions will be denoted by pairs consisting of a left hand side non-terminal and a right hand side symbol string. S is in fact also a non-terminal, be it a special one. S is not allowed to occur in right hand sides of productions. □

Scheme 15. A retrenched vocabulary $A = \langle C, V, \alpha, \rho, \mu, S\rangle$ can be derived from a context free grammar $G = \langle S, N, T, P\rangle$ as follows (symbols left of the equal signs pertain to A, symbols right of the equal signs pertain to G, unless specified otherwise):

$C = \{\overline{t} \mid t \in T\} \;\cup\; \{\overline{p} \mid p \in P\}$,
V, μ: not relevant,
$\alpha = \{\overline{t} \leftarrow 0 \mid t \in T\} \;\cup\; \{\overline{\langle n,\omega\rangle} \leftarrow |\omega| \mid \langle n,\omega\rangle \in P\}$,
$\rho = \{\langle\overline{\langle n, s_1 \ldots s_m\rangle}, i\rangle \leftarrow \{\overline{\langle n',\omega'\rangle} \in C_A \mid n' = s_i\} \cup \{\overline{t} \in C_A \mid t = s_i\} \mid$
$\qquad 1 \leq i \leq m \wedge \langle n, s_1 \ldots s_m\rangle \in P\}$,
$S = \{\overline{\langle n,\omega\rangle} \in C_A \mid n = S\}$. □

The context free grammar formalism and the retrenched vocabulary formalism are also not entirely equivalent. The problems are the absence of the variable concept in the CFG-formalism and the identification of ε-productions and terminals. Neither of them is a very serious deficiency. Variables play a relatively independent role in retrenched vocabularies. They can simply be added to the derived vocabulary. In the rule for the derivation of α can be seen that both terminals and ε-productions are mapped to constants. This is because both terminals an ε-productions represent leaf positions in parse trees. However, there is a difference between terminals and ε-productions. In the "yield" of the parse tree: the concatenation of all leaves from left to right, the ε-productions play no role because they represent the empty string. Terminals play a crucial role in the formation of the yield.

The domain concept of the transformational approach with term rewriting systems is based on retrenched vocabularies. A retrenched vocabulary A defines unambiguously a term domain: the set of well-formed terms $\mathcal{T}_A$. Transformation of terms and entire rewrite systems from domain $\mathcal{T}_A$ to another domain, associated with another vocabulary, say B, is based on the representation of terms from $\mathcal{T}_A$ by terms of $\mathcal{T}_B$. For a correct simulation of the A-domain by the B-domain it is required that each term of the A-domain has at least one representative in the B-domain. Each element of the A-domain is in general represented by an equivalence class of elements of the B-domain. Another requirement for a correct simulation is that each element of the B-domain is a representative of exactly one element of the A-domain.

4 Well-formedness Preserving Mappings

The most convenient starting point for the description of the various transformation actions between term domains is the transformation of terms of the simulating domain to terms of the simulated domain. The function ϕ that describes this transformation will be called a *well-formedness preserving mapping* (abbreviated WPM) and WPMs between two term domains are defined as:

Definition 16. A *well-formedness preserving mapping* between two term sets is a function $\phi : TG_B \rightarrow TG_A$ for which the following holds:

- $\forall t \in TG_B[t \in \mathcal{TG}_B \Leftrightarrow \phi(t) \in \mathcal{TG}_A]$,
- $\forall t \in \mathcal{TG}_A\ \exists t' \in \mathcal{TG}_B[\phi(t') = t]$. □

Because of the close correspondence between term sets and vocabularies WPMs will sometimes be referred to as mappings from vocabularies to vocabularies. WPMs are defined for the subset of ground terms. The incorporation of the variable concept poses a number of problems, more about this in section 9. The name WPM is derived from the first condition. The required surjectivity on the well-formed subsets (second condition), as opposed to bijectivity, allows sets of terms (classes) in the B-domain to represent terms of the A-domain. The function that transforms terms from the simulated domain to the simulating domain can be defined as a right inverse of the corresponding WPM ϕ, with regard to functional composition: a function ψ such that $\phi \circ \psi = id$. A few important properties of WPMs:

Theorem 17. The identity function *id* is a WPM from each vocabulary to itself. *Proof*: an easy verification of the conditions of definition 16. □

Theorem 18. If ϕ_2 is a WPM from vocabulary C to B and ϕ_1 a WPM from vocabulary B to A then $\phi_1 \circ \phi_2$ is a WPM from vocabulary C to vocabulary A. *Proof*: an easy verification of the conditions of definition 16. □

Theorem 19. If ϕ is a WPM from vocabulary B to A and ϕ is bijective on the well-formed subsets $\mathcal{TG}_A$ and $\mathcal{TG}_B$ then any right inverse ψ of ϕ is a WPM from vocabulary A to vocabulary B. *Proof*: an easy verification of the conditions of definition 16. □

The only reason to care about the non-wellformed terms is the self-evident desire to be able to define WPMs with rewrite systems. WPMs are mappings between two different term domains. The rewrite rules of the WPM should therefore be untyped themselves. This means that non-wellformed terms are also mapped to something. This "something" should not be a well-formed term otherwise there may occur problems with the determination of a right inverse of ϕ, necessary in the simulation process that will be described in next section.

5 The Step-simulation Relation

We are going to transform term rewriting systems for one term domain to equivalent term rewriting systems for another term domain. Because of the importance of the domain concept within this context, the notation for term rewriting systems will always be indexed with a vocabulary specifier, so TRS_A denotes a term rewriting system for terms from $\mathcal{T}_A$ with A a specific vocabulary.

Taking the composed route over the transformed rewrite system TRS_B in order to achieve the same effects as with TRS_A (see figure 1) will be called *simulating* TRS_A. The details of a simulation process for a TRS_A can be captured completely by a pair $\langle TRS_B, \phi \rangle$ in which TRS_B: the simulating term rewrite system and ϕ: a WPM from vocabulary B to vocabulary A. The simulation processes that will be described in this paper all follow the simulated process rewrite-step by rewrite-step. This particular class of simulations will be called *step-simulations*. The formal definition reads:

Definition 20. A pair $\langle TRS_B, \phi \rangle$ is called a *step-simulator* of TRS_A if and only if:

- ϕ is a WPM from B to A,
- $\forall t_1, t_2 \in \mathcal{TG}_A[t_1 \rightarrow_{TRS_A} t_2 \Rightarrow \forall t'_1 \in \Phi^{-1}(t_1), \exists t'_2 \in \Phi^{-1}(t_2)[t'_1 \rightarrow_{TRS_B} t'_2]$
  $t_1 \rightarrow_{TRS_A} t_2 \Leftarrow \exists t'_1 \in \Phi^{-1}(t_1), t'_2 \in \Phi^{-1}(t_2)[t'_1 \rightarrow_{TRS_B} t'_2]]$.

With Φ^{-1} the generalized inverse of ϕ: $\Phi^{-1}(x) = \{y \mid x = \phi(y)\}$. If ϕ is bijective on the well-formed subsets then the latter condition reduces to:

$$\forall t_1, t_2 \in \mathcal{TG}_B[t_1 \rightarrow_{TRS_B} t_2 \Leftrightarrow \phi(t_1) \rightarrow_{TRS_A} \phi(t_2)]. \qquad \square$$

The transformations of term rewriting systems that will be described in detail in this paper are all relatively simple transformations that bring about a slight change in the way in which terms are represented. These transformations can be seen as elementary transformations however, and large transformations can be composed from elementary ones. The concept of composition of transformations is formally supported with some convenient theorems.

Theorem 21. The pair $\langle TRS_A, id \rangle$ is a step-simulator of TRS_A for each vocabulary A and each term rewriting system TRS_A. *Proof*: an easy verification of the conditions of definition 20. □

Theorem 22. If $\langle TRS_C, \phi_2 \rangle$ is a step-simulator of TRS_B and $\langle TRS_B, \phi_1 \rangle$ is a step-simulator of TRS_A then $\langle TRS_C, \phi_1 \circ \phi_2 \rangle$ is a step-simulator of TRS_A (for each A, B, C and each TRS_A, TRS_B, TRS_C). *Proof*: an easy verification of the conditions of definition 20. □

Theorem 23. If $\langle TRS_B, \phi\rangle$ is a step-simulator of TRS_A and ϕ is bijective from $\mathcal{TG}_B$ to $\mathcal{TG}_A$ then $\langle TRS_A, \psi\rangle$ is a step-simulator of TRS_B for each A, B, each TRS_A, TRS_B and each right inverse ψ of ϕ. *Proof*: an easy verification of the conditions of definition 20. □

It is easy to see that TRS-properties like termination (strong normalization) and confluence are preserved from simulated to simulating system if ϕ is bijective on the well-formed subsets. Each rewrite step in the simulated system corresponds to exactly one rewrite step in the simulating system and each term in the simulated domain corresponds to exactly one term in the simulating domain. If ϕ is not bijective on the well-formed subsets then confluence is only preserved if ϕ is also taken in consideration. The problem is that two diverging rewrite sequences are in general only joinable in the sense that there is a common reduction *class*, i.e. two different terms t_1, t_2 for which $\phi(t_1) = \phi(t_2)$,

6 Six Elementary Syntactic Transformations

The six elementary transformations that will be elaborated upon in this section will be categorized in four *depth transformations* and two *breadth-depth transformations*. The transformations are elementary in the sense that larger, more complicate, transformations can be composed from them, while it is obvious that the elementary transformations themselves cannot be viewed as composed. Completeness of this set of six, a question closely related to the intended composable class of transformations, is discussed in section 9. The four depth-transformations are:

- leaf side distinguishing,
- root side distinguishing,
- leaf side identification,
- root side identification.

The two breadth-depth transformations are:

- contraction,
- expansion.

First an informal description of the transformations. Leaf side distinguishing is the renaming of different occurrences of one constructor with two different new names, depending on the leaf side context of that constructor, i.e. depending on which constructor occurs at a certain operand position of that constructor. Leaf side distinguishing is based on a bi-partition of the set of allowed constructors at a certain operand position. Root side distinguishing is the renaming of different occurrences of one constructor with two different new names, depending on the root side context of that constructor, i.e. depending on below which constructor and what operand position of that constructor the constructor to be distinguished occurs. Root side distinguishing is based on a bi-partition of the set of possible immediate root side contexts (which consist of a pair of a constructor and an integer, indicating an operand position) of the constructor to be distinguished.

The identification transformations are the inverses of the distinguishing transformations. Leaf side distinguishing is the renaming of occurrences of two different constructors with one new name and the leaf side context as decision criterion for the origin of the new constructor. Leaf side identification context requires an equal arity and disjunctness of the set of allowed constructors on the specific operand position of the constructors to be identified. Root side identification is the renaming of occurrences of two different constructors with one new name and the (immediate) root side context as decision criterion for the origin of the new constructor. Root side identification requires an equal arity and disjunctness of the set of possible immediate root-side contexts (which consists of a pair of a constructor and an integer, indicating an operand position) of the constructors to be identified.

Contraction is the taking together of two vertically related constructor occurrences, i.e. a constructor and one of its immediate descendants, into one single constructor. A graphical illustration of the situation:

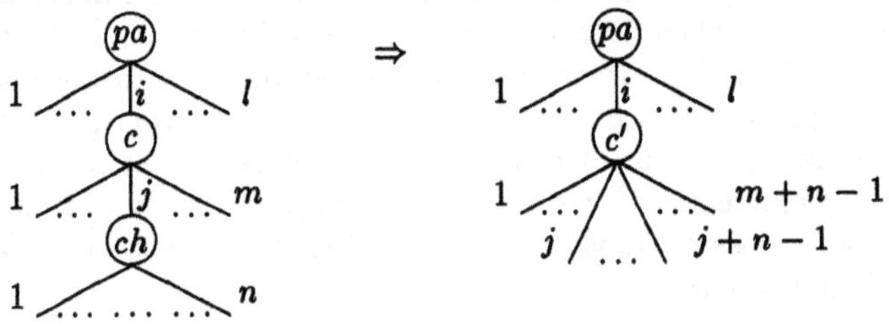

figure 2: Contraction breadth-depth transformation

Expansion is the inverse of the contraction transformation. It is the splitting of one constructor occurrence into two vertically related constructors.

In the rest of this section the transformations will be defined formally by expressing the shape of the new vocabulary in terms of the original one and by stating the conditions under which a certain vocabulary is transformable. For the sake of space only two depth and one breadth-depth transformation will be treated in detail. For these three transformations also a WPM from the new vocabulary to the original one will be given. The WPMs are meant only for ground terms (definition 16). The modifications of V and μ in definitions 24, 26, 28 are such that transformations of rules with variables (theorems 31, 32, 33) are possible. Since variables are only needed in rules, the particular identity of variables is irrelevant; systematic renaming of some or all variables in the rule leads to rules with equal functionality. We will therefore only describe the set of represented types which has to be supported with variables. The set of represented types in the variable set was a derived quantity and named YV in definition 1. We will give a derivation of the new YV from the original YV

instead of a derivation of the new V and μ because with YV fixed and with the last condition of definition 1, V and μ are determined up to renaming. The derivation of the new ρ is too complicated for a single expression. The entire expression will be split op in two parts, ρ'' defined in terms of ρ' and ρ' in terms of ρ.

The rewrite systems that will be presented for the definition of WPMs are untyped rewrite systems, defined by sets of rules, with as constructor domain the union of the constructor domains of both involved vocabularies.

Definition 24. The transformation *root side distinguishing* is defined formally as follows. The form of the new vocabulary $\langle C'', V'', \alpha'', \rho'', \mu'', S''\rangle$ is expressed in terms of the original vocabulary $\langle C, V, \alpha, \rho, \mu, S\rangle$, with YV as representative of V and μ:

- $C'' = (C - \{f\}) \cup \{f_1, f_2\}$;
- $\alpha'' = \alpha \dagger \{f \leftarrow \perp\!\!\!\perp, f_1 \leftarrow \alpha(f), f_2 \leftarrow \alpha(f)\}$;
- $\rho' = \rho \dagger \{\langle c, m\rangle \leftarrow (\rho(c,m) - \{f\}) \cup \{f_k\} \mid k \in \{1,2\}, \langle c, m\rangle \in C_{pa,k}\}$;
- $\rho'' = \rho' \dagger \{\langle f, i\rangle \leftarrow \perp\!\!\!\perp, \langle f_k, i\rangle \leftarrow \rho'(f,i) \mid 1 \leq i \leq \alpha(f), k \in \{1,2\}\}$;
- $S'' = S$;
- $YV'' = \bigcup_{yv \in YV} \textit{if } f \in yv \textit{ then } \{(yv - \{f\}) \cup \{f_1\}, (yv - \{f\}) \cup \{f_2\}\} \textit{ else } \{yv\}$.

if the following assumptions and conditions hold:

- f is the constructor to be distinguished into f_1 and f_2,
- $\{C_{pa,1}, C_{pa,2}\}$ is a partition of $C_{pa} = \rho^{-1}(f)$,
- $f \notin S$. □

Theorem 25. A WPM in the form of a term rewriting system for the transformation root side distinguishing looks as follows (same assumptions as in definition 24):

$$\begin{aligned}\{&c_{1,i}(v_1, ..., v_{p_i-1}, f_1(v_{p_i,1}, ..., v_{p_i,\alpha(f_1)}), v_{p_i+1}, ..., v_{\alpha(c_{1,i})}) \rightarrow \\ &\qquad c_{1,i}(v_1, ..., v_{p_i-1}, f(v_{p_i,1}, ..., v_{p_i,\alpha(f_1)}), v_{p_i+1}, ..., v_{\alpha(c_{1,i})}), \\ &c_{2,i}(v_1, ..., v_{p_i-1}, f_2(v_{p_i,1}, ..., v_{p_i,\alpha(f_2)}), v_{p_i+1}, ..., v_{\alpha(c_{2,i})}) \rightarrow \\ &\qquad c_{2,i}(v_1, ..., v_{p_i-1}, f(v_{p_i,1}, ..., v_{p_i,\alpha(f_2)}), v_{p_i+1}, ..., v_{\alpha(c_{2,i})}) \mid \\ &\langle c_{1,i}, p_i\rangle \in C_{pa,1}, \langle c_{2,i}, p_i\rangle \in C_{pa,2}, v_1, ..., v_{\alpha(c_{k,i})} \text{ are distinct variables } \}.\end{aligned}$$

The scheme for the rewrite rules looks complicated because of its generality. Parameters of this scheme are: the arity of the constructor f to be distinguished, the arity of possible parent constructors of f and the particular partition of $\rho^{-1}(f)$.

Proof. The transformation was the renaming of f with f_1 of f_2, depending on which constructor (and what operand position) occurred directly above f. The WPM is a mapping from the new domain to the old domain, so f_1 and f_2 have to be replaced by f. However, according to the definition of WPMs, well-formed terms have to be replaced by well-formed terms and non-well-formed terms by non-well-formed terms. A term with an occurrence of f_1 at the i-th position of c is well-formed only if $\langle c, i\rangle \in C_{pa,1}$, so combinations of $c(...f_1(...)...)$ with f_1 at i have to be replaced by $c(...f(...)...)$ only if $\langle c, i\rangle \in C_{pa,1}$. Likewise remarks for

$c(...f_2(...)...)$ if $\langle c, i\rangle \in C_{pa,2}$. If $c(...f_1(...)...)$ is not well-formed ($\langle c, i\rangle \notin C_{pa,1}$) then f_1 will not be replaced by f and the resulting term is not well-formed with regard to the target vocabulary, which is in accordance with the requirements for WPMs. □

Definition 26. The transformation *leaf side identification* is defined formally as follows. The form of the new vocabulary $\langle C'', V'', \alpha'', \rho'', \mu'', S''\rangle$ is expressed in terms of the original vocabulary $\langle C, V, \alpha, \rho, \mu, S\rangle$, with YV as representative of V and μ:

- $C'' = (C - \{f_1, f_2\}) \cap \{f\}$;
- $\alpha'' = \alpha \dagger \{f_1 \leftarrow \perp\!\!\perp, f_2 \leftarrow \perp\!\!\perp, f \leftarrow \alpha(f_1)\}$;
- $\rho' = \rho \dagger \{\langle f_k, i\rangle \leftarrow \perp\!\!\perp,$
  $\quad \langle f, i\rangle \leftarrow \rho(f_1, i) \cup \rho(f_2, i) \mid$
  $\quad\quad k \in \{1, 2\},$
  $\quad\quad 1 \leq i \leq \alpha(f_1)\}$;
- $\rho'' = \rho \dagger \{\langle c, m\rangle \leftarrow (\rho'(c, m) - \{f_1, f_2\}) \cup \{f\} \mid \rho'(c, m) \cap \{f_1, f_2\} \neq \emptyset\}$;
- $S'' = \mathit{if}\ f_1 \in S \vee f_2 \in S\ \mathit{then}\ S - \{f_1, f_2\} \cup \{f\}\ \mathit{else}\ S$;
- $YV'' = \{\mathit{if}\ f_1 \in yv \vee f_2 \in yv\ \mathit{then}\ yv - \{f_1, f_2\} \cup \{f\}\ \mathit{else}\ yv \mid yv \in YV\}$.

if the following conditions and assumptions hold:

- f_1 and f_2 are the constructors to be identified into f,
- $\alpha(f_1) = \alpha(f_2) > 0$,
- $\rho(f_1, p) \cap \rho(f_2, p) = \emptyset$, for p, the identification position,
- $\rho(f_1, i) = \rho(f_2, i)$, for all $1 \leq i \leq \alpha(f_1), i \neq p$,
- $\rho^{-1}(f_1) = \rho^{-1}(f_2)$. □

Theorem 27. A WPM in the form of a term rewriting system for the transformation leaf side identification looks as follows (same assumptions as in definition 26):

$$\begin{aligned}\{&f(v_1, ..., v_{p-1}, c_{1,i}(v_{p,1}, ..., v_{p,\alpha(c_{1,i})}), v_{p+1}, ..., v_{\alpha(f)}) \rightarrow \\ &\quad f_1(v_1, ..., v_{p-1}, c_{1,i}(v_{p,1}, ..., v_{p,\alpha(c_{1,i})}), v_{p+1}, ..., v_{\alpha(f)}), \\ &f(v_1, ..., v_{p-1}, c_{2,i}(v_{p,1}, ..., v_{p,\alpha(c_{2,i})}), v_{p+1}, ..., v_{\alpha(f)}) \rightarrow \\ &\quad f_2(v_1, ..., v_{p-1}, c_{2,i}(v_{p,1}, ..., v_{p,\alpha(c_{2,i})}), v_{p+1}, ..., v_{\alpha(f)}) \mid \\ &c_{1,i} \in \rho(f_1, p), c_{2,i} \in \rho(f_2, p), v_1, ..., v_{\alpha(f)} \text{ are distinct variables }\}.\end{aligned}$$

Proof: similar to the proof of theorem 25. □

Definition 28. The transformation *contraction* is defined formally as follows. The form of the new vocabulary $\langle C'', V'', \alpha'', \rho'', \mu'', S''\rangle$ is expressed in terms of the original vocabulary $\langle C, V, \alpha, \rho, \mu, S\rangle$, with YV the representative of V and μ:

- $C'' = C \cup \{c'\}$;
- $\alpha'' = \alpha \cup \{c' \leftarrow \alpha(c) + \alpha(ch) - 1\}$;
- $\rho'' = \rho \dagger \{\langle c, p\rangle \leftarrow \rho(c, p) - \{ch\}, \langle pa, i\rangle \leftarrow \rho(pa, i) \cup \{c'\} \mid c \in \rho(pa, i)\}$;
- $S'' = \mathit{if}\, c \in S\ \mathit{then}\ S \cup \{c'\}\ \mathit{else}\ S$;
- $YV'' = \{\mathit{if}\ c \in yv\ \mathit{then}\ yv \cup \{c'\}\ \mathit{else}\ yv \mid yv \in YV\}$.

if the following conditions and assumptions hold:

- P is an abbreviation for the predicate $|\rho(c,p)| = 1$,
- c and ch are the constructors to be contracted into c' and, p is the contraction position of c,
- $c \neq ch$.

The transformation may have introduced non-integratedness (definition 10.1) or unsatisfiability (definition 10.2). In some circumstances the constructors c and ch can no longer be used in well-formed terms because of this. The conditions for this to happen are simple but it has consequences for each element of the defining tuple, so the compact scheme for the derivation of the new vocabulary would become needlessly complicated if they were incorporated. The consequences for the vocabulary are always the same for a constructor c_n which has become non-integrated or unsatisfiable. Theorem 29 gives a scheme for the disposal of one single non-integrated or unsatisfiable constructor.

In case of the contracting breadth-depth transformation the constructors c and ch possibly become non-integrated or unsatisfiable. The conditions are:

- c becomes non-integrated if and only if $|\rho(c,p)| = 1$ (c is singleton set ρ-valued at p in the original vocabulary, and of course $\rho(c,p) = \{ch\}$ in that case),
- ch becomes non-integrated if and only if $ch \notin S$ and $c \in \rho(c_x, p_y) \Rightarrow c_x = c \wedge p_y = p$,
- c becomes unsatisfiable if and only if $ch \notin S$ and $c \in \rho(c_x, p_y) \Rightarrow c_x = c \wedge p_y = p$. □

Theorem 29. The following scheme can be used to make a vocabulary integrated if c_n is the only superfluous constructor. If there are more such constructors the scheme has to be applied repeatedly. The modified vocabulary $\langle C'', V'', \alpha'', \rho'', \mu'', S''\rangle$ is expressed in terms of the original vocabulary $\langle C, V, \alpha, \rho, \mu, S\rangle$:

$C'' = C - \{c_n\}$,
$V'' = V - \{v \in V \mid c_n \in \mu(v)\}$,
$\alpha'' = \{(c \leftarrow n) \in \alpha \mid c \in C''\}$,
$\rho'' = \{(\langle c, i\rangle \leftarrow S) \in \rho \mid c \in C''\}$,
$\mu'' = \{(v \leftarrow S) \in \mu \mid v \in V''\}$,
$S'' = S$.

Proof: straightforward. □

Theorem 30. A WPM in the form of a term rewriting system for the transformation contraction (same assumptions as in definition 28):

$$\{c_{new}(v_1, \ldots, v_{m+n-1}) \rightarrow c(v_1, \ldots, v_{j-1}, c_{ch}(v_j, \ldots, v_{j+n-1}), v_{j+n}, \ldots, v_{m+n-1}) \mid v_1, \ldots, v_{m+n-1} \text{ are distinct variables } \}$$

Proof: similar to the proof of theorem 25. □

7 Systematic Derivation of Simulating Rewrite Systems

Theorem 31. The following transformation scheme, stated in the form of an imperative programming style algorithm, derives step-simulating rewrite systems from given ones for the transformation *root side distinguishing.* Input parameters for the algorithm are: the rewrite system to be transformed, referred to by R, the particular constructor to be distinguished, referred to by f, and the non-trivial partition of $\rho^{-1}(f)$, referred to by $\{C_{pa,1}, C_{pa,2}\}$.

For each rule $h = \langle l, r\rangle$ in R Do:
  Replace all occurrences of f in both l and r at non-root positions by constructor f_1 or f_2, depending on if the direct parent-position pair is in $C_{pa,1}$ or $C_{pa,2}$;
  Replace all variables the μ-value of which contains f with variables with same μ-value except that f has been exchanged for f_1 or f_2, depending on if the direct parent-position pair is in $C_{pa,1}$ or in $C_{pa,2}$.
End of For-iteration over rules;
For each rule $h = \langle l, r\rangle$ in R Do:
  If $Root(l) = f \wedge Root(r) = f$ Then:
    Replace h by two rules which are derived from h, one by substituting f_1 for both roots, and one by substituting f_2 for both roots.
  End of If-Then;
  If $Root(l) = f \wedge Root(r) \neq f$ Then:
    Replace h by two rules which are derived from h, one by substituting f_1 for the root of l and one by substituting f_2 for the root of l.
  End of If-Then;
  If $Root(l) \neq f \wedge Root(r) = f$ Then:
    If $\rho^{-1}(Root(l)) \subseteq C_{pa,1}$ Then:
      Replace h by a rule which is derived from h by substituting f_1 for the root of r.
    End of If-Then;
    If $\rho^{-1}(Root(l)) \subseteq C_{pa,2}$ Then:
      Replace h by a rule which is derived from h by substituting f_2 for the root of r.
    End of If-Then;
    If $\rho^{-1}(Root(l)) \cap C_{pa,1} \neq \emptyset \wedge \rho^{-1}(Root(l)) \cap C_{pa,2} \neq \emptyset$ Then:
      Let r_1 be r in which the root node f has been replaced by f_1 and let r_2 be r in which the root node f has been replaced by f_2. Replace h by the two rule sets:

$$c_i(v_1, \ldots, v_{p-1}, l, v_{p+1} \ldots, v_{\alpha(c_i)}) \to c_i(v_1, \ldots, v_{p-1}, r_1, v_{p+1} \ldots, v_{\alpha(c_i)})$$

      for all $\langle c_i, p\rangle \in C_{pa,1}$ with $\{v_1, \ldots, v_{p-1}, v_{p+1}, \ldots, v_{\alpha(c_i)}\}$: distinct fresh variables, and each v_i such that it is maximally typed,

$$c_i(v_1, \ldots, v_{p-1}, l, v_{p+1} \ldots, v_{\alpha(c_i)}) \to c_i(v_1, \ldots, v_{p-1}, r_2, v_{p+1} \ldots, v_{\alpha(c_i)})$$

      for all $\langle c_i, p\rangle \in C_{pa,2}$, with $\{v_1, \ldots, v_{p-1}, v_{p+1}, \ldots, v_{\alpha(c_i)}\}$: distinct fresh variables, and each v_i such that it is maximally typed.
    End of If-Then.
  End of If-Then.

End of For-iteration over rules.

Proof of step-simulation: The transformation root side distinguishing corresponds to a WPM which is bijective on the well-formed subsets, so the fact to be proved is:

$\forall t_1, t_2 \in \mathcal{TG}_B[t_1 \rightarrow_{TRS_B} t_2 \Leftrightarrow \phi(t_1) \rightarrow_{TRS_A} \phi(t_2)]$ (definition 20).

Recall that TRS_A is the original rewrite system and TRS_B is a rewrite system, derived by the just presented algorithm.

Proof of $\Rightarrow$. Because of the bijectivity of ϕ on the well-formed subsets, the $\mathcal{TG}_A$-domain is completely covered by $\phi(\mathcal{TG}_B)$ and to each $t_{\phi,1}, t_{\phi,2} \in \mathcal{TG}_A$ correspond unambiguously two terms $t_1, t_2 \in \mathcal{TG}_B$ such that $t_{\phi,1} = \phi(t_1)$ and $t_{\phi,2} = \phi(t_2)$. It suffices to show that for all $t_{\phi,1}, t_{\phi,2}$ with corresponding t_1, t_2:

$t_{\phi,1} \rightarrow_{TRS_A} t_{\phi,2} \Rightarrow t_1 \rightarrow_{TRS_B} t_2$.

Let $t_{\phi,1} \rightarrow_{TRS_A} t_{\phi,2}$ and $h = \langle l, r \rangle$ the particular rule that was active in this rewrite operation. The shape of the rule h can be used to show with the different clauses of the algorithm that the implication holds. This is trivial for the case of the constructor to be distinguished not occurring in the root of l or r, and easy for the remaining cases. Note that the constructor to be distinguished can occur simultaneously at root positions of both rule patterns and at one or more internal positions of the rule patterns. The internal occurrences are covered by the first iteration of the algorithm, which replaces internal occurrences of the constructor to be distinguished and occurrences of this constructor in the μ-values of variables. If the constructor to be distinguished occurs at the root position of one or both rule patterns, things are more complicated. It can be seen easily that the cases which are distinguished in the second iteration of the algorithm are disjoint and together cover all situations of root-occurrences of f. Summarily:

1. $Root(l) = f \wedge Root(r) = f$,
2. $Root(l) = f \wedge Root(r) \neq f$,
3. $Root(l) \neq f \wedge Root(r) = f \wedge \rho^{-1}(Root(l)) \subseteq C_{s,1}$,
4. $Root(l) \neq f \wedge Root(r) = f \wedge \rho^{-1}(Root(l)) \subseteq C_{s,2}$,
5. $Root(l) \neq f \wedge Root(r) = f \wedge \rho^{-1}(Root(l)) \cap C_{s,1} \neq \emptyset \wedge \rho^{-1}(Root(l)) \cap C_{s,2} \neq \emptyset$.

For the cases 1. and 2. the implication to be proved holds because in the original domain the rule can match with root of l against f below a position from $C_{s,1}$ or against f below a position from $C_{s,2}$. For both cases there is a rule in the transformed rule-set. It is also easy to see that the results of the rewrite operations (the contracta $t_{\phi,2}$ and t_2) correspond, i.e. $t_{\phi,2} = \phi(t_2)$.

For the cases 3. and 4. there is only one simulating rule necessary because the root of the left hand side can only match against a constructor either below some positions of $C_{s,1}$ or below some positions of $C_{s,2}$. Again it is easy to see that the results of the rewrite operations correspond: $t_{\phi,2} = \phi(t_2)$.

For case 5. the added constructor above the roots of both rule patterns is actually to make the implication $\Leftarrow$ hold, i.e. it is to achieve that in the new domain not more rewrite operations are possible than in the original domain. The various different added root constructors (from $C_{s,1}$ and $C_{s,2}$) are necessary to keep the implication $\Rightarrow$ hold in all circumstances. If

$$\begin{aligned}&Root(l) \neq f \wedge\\ &Root(r) = f \wedge\\ &\rho^{-1}(Root(l)) \cap C_{s,1} \neq \emptyset \wedge\\ &\rho^{-1}(Root(l)) \cap C_{s,2} \neq \emptyset\end{aligned}$$

then the root of l can match against any of the positions in $\rho^{-1}(Root(l)) \cap C_{s,1}$ and against any of the positions in $\rho^{-1}(Root(l)) \cap C_{s,2}$. The rule-set which is the result of the transformation in this situation ensures that the implication holds for this situation. Again it is easy to see that the results of the rewrite operations correspond: $t_{\phi,2} = \phi(t_2)$.

Proof of $\Leftarrow$. If $t_1 \rightarrow_{TRS_B} t_2$ with $h = \langle l, r\rangle$ the particular rewrite rule that was active in this rewrite operation, then it is easy to show, with case analysis, that $\phi(t_1) \rightarrow_{h'} \phi(t_2)$, with h' the rule that acted as original of h in the transformation process of the rewrite system. Therefore $\phi(t_1) \rightarrow_{TRS_A} \phi(t_2)$. Problematic in the proof of $\Leftarrow$ is the fifth clause of the algorithm:

$$\begin{aligned}&Root(l) \neq f \wedge\\ &Root(r) = f \wedge\\ &\rho^{-1}(Root(l)) \cap C_{s,1} \neq \emptyset \wedge\\ &\rho^{-1}(Root(l)) \cap C_{s,2} \neq \emptyset.\end{aligned}$$

In all situations, except that of the fifth clause, it is possible to determine unambiguously the rule that was the source of the transformed rule. If, for instance, the transformed rule has f_1 as left hand side root, the original rule had f as left hand side root. In the case of the fifth clause the distinguished constructor (f_1 or f_2) is no longer root constructor of the right hand side of the transformed rule because an additional constructor has been fixed on top of it. So the transformed rule may have had a rule with f as root of right or a rule with f as constructor just below the root of right. However it is easy so see that always at least one of these possible origins was in the original rule-set (if both possible origins were in the original set, the original set contained a superfluous rule i.e. $c(\ldots, l, \ldots) \rightarrow c(\ldots, f(\ldots), \ldots)$ which was implied by: $l \rightarrow f(\ldots)$). □

Theorem 32. The following transformation scheme, stated in the form of an imperative programming style algorithm, derives step-simulating rewrite systems from given ones for the transformation ***leaf side identification***. Input parameters for the algorithm are: the rewrite system to be transformed, referred to by R and the particular constructors to be identified, referred to by f_1, f_2.

For each rule $h = \langle l, r\rangle$ in R Do:
 Replace all occurrences of f_1 and f_2 in both l and r by an occurrence of the constructor f;

Replace all variables in both l and r the μ-value of which contains f_1 and/or f_2, with one with same μ-value except that f_1 and/or f_2 has been exchanged for f.

End of For-iteration over rules.

Proof of step-simulation: similar with proof of theorem 31. Easier because of the lesser complexity of the algorithm, as compared with the algorithm of theorem 31. □

The relative simplicity of the algorithm of theorem 32 is due to the fact that variables can restrict the set of constructors which is allowed to act as root of terms to be matched against them further than the restriction of ρ for that operand position.

Theorem 33. The following transformation scheme, stated in the form of an imperative programming style algorithm, derives step-simulating rewrite systems from given ones for the transformation *contraction*. Input parameters for the algorithm are: the rewrite system to be transformed, referred to by R and the particular constructors to be contracted, referred to by c and ch. The *workarea* and *finishedarea* are variables to which sets of rules can be bound, *finishedarea* is assumed to be empty at the start of the algorithm.

For each rule $h = \langle l, r \rangle$ in R Do:
    Replace all combined occurrences of c and ch in both l and r by an occurrence of c';
    Replace all variables in both l and r the μ-value of which contains c with one with same μ-value except that c has been exchanged for c'.
End of For-iteration over rules;
For each rule $h = \langle l, r \rangle$ in R Do:
    Place a copy of h in an emptied *workarea*;
    For each variable v in $Var(l)$ Do:
        For each rule $h' = \langle l', r' \rangle$ in the *workarea* Do:
            If v occurs below c at position j Then
                If $\neg \exists c_x \in C[c_x \neq c \wedge c_x \in \mu(v)]$ Then
                    Remove h' from the *workarea*.
                End If-Then;
                If $ch \in \mu(v)$ Then
                    Add a rule to the *workarea* which is derived from h' by replacing occurrences $c(t_1, ..., t_{j-1}, v, t_{j+1}, ..., t_m)$ of v below c at j by $c'(t_1, ..., t_{j-1}, v_1, ..., v_n, t_{j+1}, ..., t_m)$ and other occurrences of v by: $ch(v_1, ..., v_n)$ with $\{v_1, ..., v_n\}$: distinct fresh variables which are maximally typed.
                End of If-Then.
            End of If-Then.
        End of For-iteration over rules in *workarea*.
    End of For-iteration over variables in $Var(l)$;
    For each rule $h' = \langle c_l(t_1, ..., t_{\alpha(cl)}), c_r(u_1, ..., u_{\alpha(cr)}) \rangle$ in the *workarea* Do:

If $c_l = ch \vee c_r = ch$ Then

If $ch \notin S \wedge \forall c_x \in C, 1 \leq i \leq \alpha(c_x)[ch \in \rho(c_x, i) \Rightarrow c_x = c \wedge i = j]$ Then

Remove h' from the *workarea*.

End of If-Then;

Add a rule to the *workarea* which is derived from h' by replacing the pattern $c_x(t_1, ..., t_{\alpha(cx)})$ ($x = l$ or $x = r$) by:

$c'(v_1, ..., v_{j-1}, t_1, ..., t_{\alpha(cx)}, v_{j+1}, ..., v_m)$

if $c_x = ch$ and by:

$c(v_1, ..., v_{j-1}, c_x(t_1, ..., t_{\alpha(cl)}), v_{j+1}, ..., v_m)$

if $c_x \neq ch$, with $\{v_1, ..., v_{j-1}, v_{j+1}, ..., v_m\}$: distinct fresh variables which are maximally typed.

End of If-Then.

End of For-iteration over rules in the *workarea*;

Add rules in *workarea* to *finishedarea*.

End of For-iteration over rules in R.

Proof of step-simulation: similar with proof of theorem 31. □

8 Examples

Some examples to illustrate the utility of the various transformations. Consider the following two rewrite rules, intended for typed rewriting:

$$f_1(g(a)) \rightarrow f_1(h(b)),$$
$$f_2(g(a)) \rightarrow f_2(h(b)).$$

If the type system of the underlying signature is such that f_1 and f_2 are the only two constructors that can occur directly above g and if $g \notin S$ then these two rules can be replaced by the single rule:

$$g(a) \rightarrow h(b).$$

Opportunities for this kind of generalization will not occur often in practice because the precondition combined with the required form of the rules is rather strong. With distinguishing and identification transformations the scope of applicability of this kind of generalization can be enlarged substantially. If, in the above two rules, f_1 and f_2 are not the only constructors that can occur above g then it is possible to perform root side distinguishing on g and for instance to call g: g_1 when it occurs below f_1 or f_2 and g_2 in other circumstances. After this distinguishing renaming the two rules (in which g has become g_1) can be replaced by:

$$g_1(a) \rightarrow h(b).$$

If the root constructors of both patterns of one of the rules are not identical, for instance:

$$f_l(g(a)) \rightarrow f_r(h(b)),$$
$$f_2(g(a)) \rightarrow f_2(h(b)),$$

then it is sometimes possible to identify these root constructors. If the type system is such that the sets of constructors which are allowed to occur below f_l and f_r are disjoint then f_l and f_r can be identified and for instance be called f_1. After this identification a generalization has become possible.

The rather strong preconditions for the identification transformation: equal arity and disjunct sets of allowed constructors for the specific identification position, can also be made to hold by applications of one or more of the other transformations. The arity of constructors can be enlarged by the expansion transformation. An n-ary constructor f can be transformed into an n+1-ary constructor f' by expanding f into f' with one fixed constant at the n+1-th operand position. The transformation is trivial but the developed theory offers an integrated formal framework for this transformation. The disjunctness requirement for identification can be made to hold by subsequent distinguishing. Consider the vocabulary with unary constructors f_1, f_2, g_1, g_2, g_3 and a type system such that $\rho(f_1, 1) = \{g_1, g_2\}$ and $\rho(f_2, 1) = \{g_2, g_3\}$. Identification of f_1 and f_2 is not allowed because $\{g_1, g_2\} \cap \{g_2, g_3\} \neq \emptyset$. Root side distinguishing g_2 will make the identification possible. If g_2 is called g_{21} below f_1 and g_{22} below f_2 then the transformed vocabulary will have the constructors $f, g_1, g_{21}, g_{22}, g_3$ and the type system will be such that: $\rho(f, 1) = \{g_1, g_{21}, g_{22}, g_3\}$.

9 Concluding Remarks

Although the retrenched vocabulary formalism has a larger distance to algebraic semantics of terms than the order sorted signature formalism [EM85], the presented elementary transformations can easily be harmonized with algebraic semantics. If each operator corresponds to a typed function (or operation) then constructor distinguishing corresponds to *specialization*, constructor identification to *generalization* and contraction to *functional composition*. It is also easy to interpret the modifications of the type system with the various transformations in terms of operations on values. More about the semantical aspects of the transformations and also more detailed treatment of all six transformations, mentioned in this document, can be found in [Ver94].

The problem with the incorporation of the variable concept in the WPM-notion is that WPMs will become set-valued functions in some circumstances. There are no fundamental objectives against the extension of the WPM concept to variables however. Additional research is necessary to work things out properly. Additional research is also necessary for the development of optimizing transformation strategies, i.e. strategies which determine for a given vocabulary (signature) and rewrite system a sequence of transformations to reach a minimal number of rules or minimal sized terms.

About completeness of the proposed set of elementary transformations: it is not possible to cover the space of all step simulating transformations with a finite set of elementary transformations. Not even if the domain mapping is bijective. The number of transformations which can be composed with a finite set of elementary transformations is at most countably infinite, which is less than the number of bijective functions between two countably infinite term sets. However, it is possible to describe mathematically the class of transformations which is generated by the proposed set of six transformations and to prove completeness with regard to this class. A paper on this subject is forthcoming.

References

[ASU86] Alfred V. Aho, Ravi Sethi, and Jeffrey D. Ullman. *Compilers, Principles, Techniques and Tools.* Addison-Wesley Publishing Company, 1986. ISBN: 0-201-10194-7.

[BD89] Leo Bachmair and Nachum Dershowitz. Completion for rewriting modulo a congruence. *Theoretical Computer Science*, 67:173–201, 1989.

[DJ90] N. Dershowitz and J.P. Jouannaud. Rewrite systems. In J. van Leeuwen, editor, *Handbook of Theoretical Computer Science*, volume B, pages 243–320. Elsevier Science Publishers B.V., 1990.

[EM85] H. Ehrig and B. Mahr. *Fundamentals of Algebraic Specification*, volume 1. Springer, Berlin, 1985.

[Gan91] H. Ganzinger. Order-sorted completion: the many-sorted way. *Theoretical Computer Science*, (89):3–32, 1991.

[GJM85] J.A. Goguen, J.P. Jouannaud, and J. Meseguer. Operational semantics for order-sorted algebra. In *Proc. 12th Internat. Conf. on Automata, Languages and Programming*, number 194 in LNCS, pages 221–231. Springer, 1985.

[HHKR89] J. Heering, P.R.H. Hendriks, P. Klint, and J. Rekers. The syntax definition formalism sdf - reference manual. In *SIGPLAN Notices*, volume 24(11), pages 43–75, 1989.

[KSS91] H. Kwakernaak, R. Sivan, and R.C.W. Strijbos. *Modern Signals and Systems.* Prentice-Hall, Englewood Cliffs, 1991.

[SNGM89] W. Smolka, W. Nutt, J.A. Goguen, and J. Meseguer. *Resolution of Equations in Algebraic Structures*, volume 2, chapter 10: Order-Sorted Equational Computing, pages 297–367. Academic Press, 1989.

[Ver94] J.C. Verheul. Well-formedness preserving mappings, a useful relation between retrenched vocabularies. Technical report, Delft University of Technology, Holland, 1994. to appear.

Abstract Notions and Inference Systems for Proofs by Mathematical Induction

Claus-Peter Wirth* and Klaus Becker

Fb. Informatik, Universität Kaiserslautern, D-67663, Germany
{wirth, klbecker}@informatik.uni-kl.de

Abstract. Soundness of inference systems for inductive proofs is sometimes shown ad hoc and a posteriori, lacking modularization and interface notions. As a consequence, these soundness proofs tend to be clumsy, difficult to understand and maintain, and error prone with difficult to localize errors. Furthermore, common properties of the inference rules are often hidden, and the comparison with similar systems is difficult. To overcome these problems we propose to develop soundness proofs systematically by presenting an abstract frame inference system a priori and then to design each concrete inference rule locally as a sub-rule of some frame inference rule and to show its soundness by a small local proof establishing this sub-rule relationship. We present a frame inference system and two approaches to show its soundness, discuss an alternative, and briefly classify the literature. In an appendix we give an example and briefly discuss failure recognition and refutational completeness.

1 Motivation

Given some set of first-order axioms 'R', one is often not only interested in those properties 'Γ' which are logical consequences of 'R', i. e. which hold in all models of 'R': " $\mathrm{R} \models \Gamma$ "; but also in properties which are only required to hold in some specific sub-class of the class of models of 'R'. Instead of restricting the class of models by some required property, one may also ask for those 'Γ' for which (instead of " $\mathrm{R} \models \Gamma$ ") only " $\mathrm{R} \models \Gamma\tau$ " must hold for all τ taken from a specific set of (e. g. ground) substitutions. Notions of validity resulting from combinations of possible restrictions of these two kinds are usually called *inductive validity* and the inductively valid properties are called *inductive theorems*. In Wirth & Gramlich (1994) we discussed the most important of these notions in a unified framework on the basis of positive/negative-conditional equational specifications as introduced in Wirth & Gramlich (1993). These theorems are called "inductive" since (finite) proofs for most of them require mathematical induction. Inductive reasoning extends deductive reasoning by capturing infinite deductive proofs in a finite cyclic representation, e. g. capturing

* supported by the Deutsche Forschungsgemeinschaft, SFB 314 (D4-Projekt)

$$\frac{\Gamma(x_0)}{\Gamma(0) \quad \dfrac{\Gamma(\mathsf{s}(x_1))}{\Gamma(\mathsf{s}(0)) \quad \dfrac{\Gamma(\mathsf{s}(\mathsf{s}(x_2)))}{\Gamma(\mathsf{s}(\mathsf{s}(0))) \quad \ldots}}} \qquad \text{(using “}x_i = 0 \vee \exists x_{i+1}\colon x_i = \mathsf{s}(x_{i+1})\text{”) in something like} \qquad \frac{\Gamma(x_0)}{\Gamma(0) \quad \dfrac{\Gamma(\mathsf{s}(x_1))}{\dfrac{\Gamma(x_1)}{\text{(back to top)}}}}$$

(where the formulas below each line imply the formula above). For this kind of cyclic reasoning to be sound, the deductive reasoning must terminate for each instantiation of the theorem. This can be guaranteed when one requires for each cyclic reasoning the preconditions (usually called *induction hypotheses*) (e. g. "$\Gamma(x_1)$") to be smaller than the *"induction" conclusion* (e. g. "$\Gamma(\mathsf{s}(x_1))$") w. r. t. some wellfounded ordering, called *induction ordering*.

In Walther (1994) we can read the following about proving inductive theorems: "Research on automated induction these days is based on two competing paradigms: *Implicit induction* (also termed *inductive completion*, *inductionless induction*, or, less confusingly, *proof by consistency*) evolved from the *Knuth-Bendix Completion Procedure* .... The other research paradigm ... is called *explicit induction* and resembles the more familiar idea of induction theorem proving using induction axioms."

While the two paradigms are not uniformly defined in the research community, we call the latter paradigm "explicit" because in the underlying inference systems each cyclic reasoning is made explicit in a single inference step which brings together induction hypotheses and conclusions in a set of *induction base* and *induction step* formulas and explicitly guarantees the termination of their cycles with the help of a sub-proof or -mechanism for the wellfoundedness of the induction ordering resulting from the step formulas.

The inference systems for implicit induction, however, permit us to spread the cyclic reasoning as well as the termination control over several inference steps. To rediscover the inductive cycles in the reasoning we usually have to inspect several inference steps instead of a single one that explicitly does the induction. Possibly since this seemed to be somewhat difficult compared to the older and well-known explicit induction, the paradox "inductionless induction" became a name for implicit induction. Another reason for this name might be the emphasis the researchers in the field of implicit induction gave to the refutational completeness (cf. Appendix B) of their inference systems: In general the set of inductively valid theorems is not enumerable for all reasonable and interesting notions of inductive validity; therefore refutational completeness is highly appreciated for an inference system for inductive theorem proving as it is an optimal theoretical quality. Refutational completeness, however, does not help to find finite proofs for inductively valid formulas (whereas the ability of an inductive theorem prover to detect invalid formulas is most important under a practical aspect (cf. Appendix A), especially for *inductive* theorem proving because of the important role generalizations play in it, where an invalid formula can result from a valid input theorem due to over-generalization).

To succeed in proving an inductive theorem in finite time, implicit inductive theorem provers have to solve exactly the same problem as explicit inductive theorem provers, namely to find a finite cyclic representation for an infinite deductive proof as well as an induction ordering guaranteeing the termination of its cycles. Therefore, if a theorem prover with sufficient deductive power fails to show an inductive theorem, then either it fails to construct the proper reasoning cycles or its mechanisms for bookkeeping of ordering information (cf. sect. 2) or for satisfying ordering constraints are too weak. While inference systems for implicit induction usually have sufficient potentiality to construct the proper reasoning cycles, their bookkeeping of relevant ordering information may be insufficient for certain proofs (cf. Appendix C for an example), even though their powerful orderings for satisfying the ordering constraints partially compensate for this insufficiency. Inference systems for explicit induction on the other hand do not require any bookkeeping of ordering information since the ordering information is only of local importance within single explicit induction steps. Explicit inductive theorem provers usually use rather simple semantic orderings which have turned out to be powerful enough for almost all practical applications. These provers, however, usually do not find more sophistically structured reasoning cycles, e. g. in mutually recursive domains or in the case (which may be of more importance in practice) that the required instantiations of the induction hypotheses are difficult to be guessed when the step formulas are synthesized and do not become obvious before the induction hypotheses can be applied to the induction conclusions (cf. Protzen (1994) for a simple example). From an abstract point of view and beyond the technicalities usually involved, however, all inductive proofs of our intuition can be formulated according to each of the two paradigms without changing the reasoning cycles.

Note, however, that the scope of mathematical induction (i. e. reasoning in terminating cycles) goes beyond inductive theorem proving (w. r. t. our definition above). A field of application of mathematical induction which is very similar to inductive theorem proving w. r. t. data and problem structures is to prove (ground) confluence for first-order clauses, cf. Becker (1993) and Becker (1994).

Therefore, in our opinion, while arguing for the one and against the other paradigm should be overcome, a unifying representation of all these induction-based approaches is not only theoretically possible, but also strongly required from a practical point of view: The possible combination of insight, methods, techniques, and heuristics based on results of research in all these fields will be beneficial for the search for proofs by mathematical induction in practice.

The abstract notions and the frame inference system we present in this paper are proposed as a top level step towards such a unification. This frame inference system is necessarily more similar to inference systems for implicit than for explicit induction, since, from a top level view and according to our definition of the terms, explicit induction is a form of implicit induction where the induction is restricted to be done explicitly within single inference steps. The results of the field of explicit induction, however, are most important for developing concrete inference systems consisting of practically useful sub-rules of our frame inference rules. While the construction of such concrete inference systems as well as a comparison and a combination of implicit and explicit induction are supported by the abstract framework of this paper, a proper treatment of these subjects cannot be given here but must be elaborated in several future papers.

2 Prover States

Prover states are intended to represent the state of the prover, i. e. to record which sub-tasks of a proof have been successfully established, which goals remain to be proved, etc.. Technically, the prover states are the field of our inference relation '$\vdash$'. For *deductive* reasoning we may start with a set 'G' of *goals* which contains the theorems we want to prove, transform them and finally delete them after they have become trivial tautologies. Such an inference relation, starting from the theorems to be proved and reducing them until some termination criterion is satisfied, is called *analytic*, cf. e. g. Bibel & Eder (1993). These theorems must belong to some set 'Form' which contains the formulas the inference system can treat. If '$\vdash$' permits only sound transformation and deletion steps, then from "$G \vdash^{\bullet} \emptyset$" (where '$\vdash^{\bullet}$' denotes the reflexive and transitive closure of '$\vdash$') we may conclude that 'G' is valid. For practical reasons, we would like to have a set 'L' of *lemmas* at hand. Into this 'L' inference steps can store axioms of the specification, already proved lemmas or the results of any theorem prover called for solving special tasks which might be helpful in our inference process. We can then use 'L' for transformations of 'G' that are only known to be sound relative to 'L'. Thus our prover states should be pairs of sets '(L, G)' such that "$(\emptyset, G) \vdash^{\bullet} (L, \emptyset)$" implies validity of '$G$' and '$L$'. For *inductive* reasoning, we additionally need a set 'H' of *induction hypotheses*. Similar to 'L', the set 'H' may be built up during the inference process and used for transformations of 'G' which are *founded* on 'H' in the sense that they are only known to be sound relative to 'H'. Similar to the treatment of lemmas, our prover states should be triples of sets '(L, H, G)' such that "$(\emptyset, \emptyset, G) \vdash^{\bullet} (L, H, \emptyset)$" implies validity of '$G$', '$L$', and '$H$'. Unlike the lemmas in 'L', however, the hypotheses in 'H' are not known to be valid before the whole induction proof is done (i. e. "$G = \emptyset$"). Here we are running the risk of being caught in a cyclic kind of reasoning like: "The goal can be deleted, since it is valid due to the hypothesis, which is valid, if the goal can be deleted, ...". On the other hand, some kind of cyclic reasoning is really required for successful inductive proofs of finite length. We only have to make sure that this cyclic reasoning terminates. This can be achieved by equipping each formula in 'H' or 'G' with a weight and allowing a hypothesis to transform a goal only if the weight of the hypothesis is smaller than the weight of the goal w. r. t. a *wellfounded quasi-ordering* '$\lesssim$' (i. e. some reflexive and transitive relation $\lesssim$, whose *ordering* $< := \lesssim \setminus \gtrsim$ does not allow infinite descending sequences $\gimel_0 > \gimel_1 > \gimel_2 > \dots$), which we will call the *induction ordering* in what follows.

We would like to point out that we distinguish between the weight of a formula and the actual formula itself: For explicit induction, weights are not needed on the inference system level because each inductive reasoning cycle is encapsuled in a single inference step which combines induction conclusions with induction hypotheses into step formulas. For implicit induction, however, induction conclusions and hypotheses are not joined from the beginning. Instead, the conclusions are taken as goals and transformed until it becomes obvious which hypotheses will be useful for proving them; an idea which is just now coming into view of the researchers in the field of explicit induction, cf. Protzen (1994). At this point, when hypotheses are to be applied to the transformed goals, their weights are needed to transfer ordering information from the original goals that generated the hypotheses to the transformed goals. Roughly speaking, the goals must store the weights of hypotheses for which they carry the proof work. (This still permits mutual induction.) A possible weight for a formula, which is so natural that there are hardly any other weights in the literature on implicit induction, is (the value of a measure applied to) the formula itself. However, if we require the weight of a goal to be determined by the formula alone, then the chance to transform or delete this goal by means of some fixed hypothesis (which must be smaller w. r. t. '$<$') gets smaller with each transformation of the goal into other goals which are smaller w. r. t. '$<$'. Such transformation of goals is usually called *simplification*. While simplification of a goal is an important[2] heuristic, the weight of the goal should not change during simplification. This can be stated more generally: For concrete inference rules it is very important that a goal can transmit its weight unchanged to the goals which it is transformed into. This would be generally impossible if the weight of a formula were restricted to be the formula itself. In our approach, therefore, each element '$\beth$' of "$H \cup G$" is some *syntactic construct* from a set 'SynCons'. Besides its formula 'form($\beth$)', '$\beth$' may have some additional contents describing its weight. Theoretically, one can consider each syntactic construct to be a pair made up of a formula expressing some property and a weight carrying the ordering information for avoiding non-terminating cycles in the use of inductive arguments. For the description of concrete inference systems within our abstract framework, however, it is more convenient not to restrict the syntactic constructs to this form because in some of these inference systems the formulas share some structure with the weights: E. g., in Bachmair (1988) the formulas are the weights, and in Becker (1994) the weights restrict the semantics of the formulas by the so-called "reference compatibility". The distinction between formulas and syntactic constructs (i. e. formulas augmented with weights) has the following advantages compared to other inference systems for implicit induction:

Easier Design of Inference Rules: The design of concrete inference rules (e. g. as sub-rules of the abstract inference rules of sect. 5 below) becomes simpler because a transformation of the actual formula does not necessarily include an appropriate transformation of its weight (into a smaller one) and is thus not restricted by ordering constraints. For an illustrating example cf. Appendix C.

No Global Ordering Restriction: The design of inference steps becomes possible which transform the formula of a goal into another one which is bigger w. r. t. the induction ordering, cf. Gramlich (1989).

High Quality of Ordering Information: The loss of ordering information during simplification of a goal (as described above) is avoided, which (as far as we know) was first described in Wirth (1991), exemplified by a failure

[2] when the induction ordering contains the evaluation ordering of the functional definitions of the specification, cf. Walther (1994)

of a formal induction proof just caused by this loss of ordering information. There it is also sketched how to store the weight of the goal to avoid this information loss, — an idea which is also to be found in Becker (1993). For an illustrating example cf. Appendix C.

Focus on Relevant Ordering Information: Some induction proofs are only possible if we do not measure the whole formula (as often is the case when a clause is measured as the multi-set of *all* its literals) but only some subformula, subterm or variable of it. Focusing on certain variables is common for human beings (speaking of "induction on variable x", e. g.). While for the mechanisms usually applied inside the induction rule of explicit induction focusing on certain variables (called "*measured variables*" in Boyer & Moore (1979) and Walther (1994)) is standard, for focusing in implicit induction a marking concept was introduced in Wirth (1991) due to practical necessity, exhibited by an example. The more general focusing that can be achieved with the syntactic constructs here, permits us not to measure those parts of formulas which (due to unsatisfiable ordering constraints) block the application of useful hypotheses, thereby permitting us to focus on the literals (or even terms, variables) that do get smaller on their way from the induction conclusion to the induction hypothesis. This permits additional applications of induction hypotheses. For an illustrating example cf. Appendix C.

All in all, the inference relation '$\vdash$' should operate on prover states which are triples " (L, H, G) " of finite sets such that 'L' contains formulas (from 'Form') and 'H' and 'G' contain syntactic constructs (from 'SynCons') whose formulas may be accessed via the function " form: SynCons $\rightarrow$ Form ". While prover states being trees of syntactic constructs may be more useful in practice, the simpler data structure presented here suffices for the purposes of this paper.

3 Counterexamples and Validity

For powerful inductive reasoning we must be able to restrict the test for the weight of a hypothesis to be smaller than the weight of a goal (which must be satisfied for the permission to apply the hypothesis to the goal) to the special case semantically described by their formulas. This can be achieved by considering only such instances of their weights that result from ground substitutions describing invalid instances of their formulas. A syntactic construct augmented with such a substitution providing extra information on the invalidity of its formula is called a ***counterexample***. A syntactic construct whose formula is valid thus has no counterexamples. We assume the existence of some set 'Info' describing this extra information and require the induction ordering '$\precsim$' to be a wellfounded quasi-ordering not simply on " SynCons " but actually on " SynCons$\times$Info ". Furthermore, we require "being a counterexample" to be a well-defined basic property which must be either true or false for each $(\daleth, J) \in$ SynCons$\times$Info. Finally, in order to formally express the relation between counterexamples and our abstract notion of validity, we require for each syntactic construct $\daleth \in$ SynCons that 'form($\daleth$)' is valid iff there is no $J \in$ Info such that $(\daleth, J)$ is a counterexample.

Let us consider this final requirement for two instantiations of our abstract notion of validity of formulas:

For the case of inductive validity of a formula 'Γ' given as indicated in sect. 1 (i. e. iff $\models_{\mathcal{A}} \Gamma\tau$ for all "inductive substitutions" τ and all algebras $\mathcal{A}$ belonging to the class 'K' of those models which satisfy some additional property required for this kind of inductive validity) an appropriate way of satisfying the requirement is to define a formula to be a clause, to define a syntactic construct to be a pair (Γ, f) of a formula Γ and a weight f (described in terms of the variables of the formula), to define " $\text{form}((\Gamma, f)) := \Gamma$ ", to define the elements of 'Info' to be triples "$(\tau, \mathcal{A}, \kappa)$" with $\mathcal{A} \in \text{K}$ and κ valuating the remaining free variables of " $\Gamma\tau$ " to elements of the universe (or carrier) of the algebra $\mathcal{A}$, and then to say that " $((\Gamma, f), (\tau, \mathcal{A}, \kappa)) \in \text{SynCons}\times\text{Info}$ is a counterexample " if τ is an inductive substitution and $\mathcal{A}_\kappa$ evaluates " $\Gamma\tau$ " to false.

For the case of ground joinability (as defined in Becker (1993) or Becker (1994)), an appropriate way is to define a formula to be either a clause C (where validity means joinability of all ground instances $C\tau$) or a pair of clauses (C, D) (where validity means joinability of all those ground instances $C\tau$ for which $((C, D), \tau)$ is reference compatible), to define syntactic constructs to be pairs of clauses, 'form' to be the identity function, 'Info' to be the set of substitutions, and then to say that " $((C, D), \tau) \in \text{SynCons}\times\text{Info}$ is a counterexample " if τ is ground, $((C, D), \tau)$ is reference compatible, and " $C\tau$ " is not joinable.

Note that our notion of "counterexample" is a semantic one contrary to the notion of "inconsistency proof" used in Bachmair (1988). Generally speaking, an abstract frame inference system that is to be fixed prior to the design of concrete inference rules has to be sufficiently stable and therefore its notions should not rely on our changeable ideas on formal proofs.

Finally note that even with our emphasis here on proving valid formulas positively (instead of being refutationally complete), the somewhat negative kind of argumentation with counterexamples is handier, somewhat less operationally restricted, and more convenient for defining and proving properties of practically useful inference systems than the less local *formal proofs* used in the positive proving approach of Gramlich (1989) or Reddy (1990).

4 Foundedness

In this section we move from counterexamples to an even higher level of abstraction which allows the reader to forget about ' $\lesssim$ ', 'Info', and counterexamples from sect. 5 on. We use the notion of counterexamples to lift ' $\lesssim$ ' from "SynCons×Info" to subsets of 'SynCons' by explaining what we mean by saying that a set H of hypotheses is *founded* on a set G of goals (written $H \curvearrowright G$) or by saying that a set G of goals is *strictly founded* on a set H of hypotheses (written $G \searrow H$ or $H \swarrow G$). Roughly speaking, $H \curvearrowright G$ indicates that the hypotheses are known to be valid if a *final* prover state (i. e. one with an empty set of goals) can be entailed. $H \swarrow G$ indicates that the goals in G can be deleted by the application of smaller hypotheses from H.

Definition 1 Foundedness. Let $M, H, G \subseteq$ SynCons. Let '$\searrow/\!\curvearrowright$' be a symbol for a single relation. Now M is said to be *strict/quasi-founded* on (H, G) (denoted by $M \searrow/\!\curvearrowright (H, G)$) if

$$\forall \aleph \in M\colon \forall I \in \text{Info:}\; \left(\begin{array}{l} ((\aleph, I) \text{ is a counterexample}) \;\Rightarrow \\ \left(\begin{array}{l} \exists \beth \in H\colon \exists J \in \text{Info:} \left(\begin{array}{l} ((\beth, J) \text{ is a counterexample}) \\ \wedge\, (\aleph, I) > (\beth, J) \end{array} \right) \\ \vee\, \exists \beth \in G\colon \exists J \in \text{Info:} \left(\begin{array}{l} ((\beth, J) \text{ is a counterexample}) \\ \wedge\, (\aleph, I) \gtrsim (\beth, J) \end{array} \right) \end{array} \right) \end{array} \right).$$

M is said to be *strictly founded* on H (denoted by $M{\searrow}H$) if $M \searrow/\!\curvearrowright (H, \emptyset)$.
M is said to be (quasi-) *founded* on G (denoted by $M{\curvearrowright}G$) if $M \searrow/\!\curvearrowright (\emptyset, G)$.
Note that (for $\aleph \in$ SynCons) the expressive power of " $\{\aleph\} \searrow/\!\curvearrowright \ldots$ " is higher than that of " $\{\aleph\}{\searrow}\ldots$ " and " $\{\aleph\}{\curvearrowright}\ldots$ " together, since " $\{\aleph\}{\searrow}H \vee \{\aleph\}{\curvearrowright}G$ " implies " $\{\aleph\} \searrow/\!\curvearrowright (H, G)$ ", but the converse does not hold in general.

Corollary 2. *Let* $H \subseteq$ SynCons. *Now each of the following seven properties is logically equivalent to validity of* form$[H]$*:*

(1) $H{\curvearrowright}\emptyset$	(4) $\forall G{\subseteq}$SynCons: $H{\curvearrowright}G$
(2) $H{\searrow}\emptyset$	(5) $\forall G{\subseteq}$SynCons: $H{\searrow}G$
(3) $H{\searrow}H$ *(due to*	(6) $\exists G{\subseteq}$SynCons: ((form$[G]$ *is valid*) $\wedge$ $H{\curvearrowright}G$)
wellfoundedness of '>')	(7) $\exists G{\subseteq}$SynCons: ((form$[G]$ *is valid*) $\wedge$ $H{\searrow}G$)

Corollary 3. *Let* $H \subseteq G \subseteq$ SynCons. *Now:* $\emptyset{\searrow}H{\curvearrowright}G$.

Corollary 4. *The following inclusion-properties hold:* $\searrow \subseteq \curvearrowright$. $\searrow\circ\curvearrowright \subseteq \searrow$. $\curvearrowright\circ\searrow \subseteq \searrow$.

Corollary 5.

(1)	$M_1 \curvearrowright N_1 \;\wedge\; M_2 \curvearrowright N_2$	$\Rightarrow$	$M_1{\cup}M_2 \curvearrowright N_1{\cup}N_2$
(2)	$M_1 \searrow N_1 \;\wedge\; M_2 \searrow N_2$	$\Rightarrow$	$M_1{\cup}M_2 \searrow N_1{\cup}N_2$
(3)	$M_1{\cup}M_2 \curvearrowright N_1$	$\Rightarrow$	$M_1 \curvearrowright N_1{\cup}N_2$
(4)	$M_1{\cup}M_2 \searrow N_1$	$\Rightarrow$	$M_1 \searrow N_1{\cup}N_2$
(5)	$G \searrow H$	$\Rightarrow$	$G \searrow H{\setminus}G$

Note that the last item of the previous as well as the first item of the following corollary rely on the wellfoundedness of '>'.

Corollary 6.

(1)	$M \searrow/\!\curvearrowright (H, G) \;\wedge\; H \curvearrowright G{\cup}M$	$\Rightarrow$	$M \curvearrowright G$
(2)	$M \searrow/\!\curvearrowright (H, G) \;\wedge\; G \searrow N$	$\Rightarrow$	$M \searrow H{\cup}N$

Corollary 7. $\curvearrowright$ *is a quasi-ordering.*

Corollary 8.
$\searrow$ *is a transitive relation, which is neither irreflexive nor generally reflexive.*
Let $\forall i{\in}\mathbb{N}$: H_i, $G_i \subseteq$ SynCons. *Then (by the wellfoundedness of* $>$*)* $\forall i{\in}\mathbb{N}$: $H_i{\searrow}H_{i+1}$ *implies that* form$[H_i]$ *must be valid. More generally,* $\forall i{\in}\mathbb{N}$: $H_i \searrow/\!\curvearrowright (H_{i+1}, G_{i+1})$ *implies* $H_i \curvearrowright \bigcup_{j>i} G_j$. *Moreover, the restriction of* $\searrow$ *to those* $H \subseteq$ SynCons *with invalid* form$[H]$ *is a wellfounded ordering.*

5 The Frame Inference System

We now come to four abstract inference rules defining ' $\vdash$ '. Thus, in this and the following three sections, ' $\vdash$ ' will be restricted to application of one of the four following inference rules.

In what follows, let $\Gamma \in$ Form ; $\aleph, \gimel \in$ SynCons ; L, L' be finite subsets of 'Form'; and H, H', G, G', and M be finite subsets of 'SynCons':

Expansion:
$$\frac{(L \quad ,H \quad ,G \quad)}{(L \quad ,H \quad ,G \cup \{\gimel\})}$$

Hypothesizing:
$$\frac{(L \quad ,H \quad ,G \quad)}{(L \quad ,H \cup \{\aleph\} ,G \quad)}$$
if L is invalid or $\{\aleph\} \curvearrowright H \cup G$

Acquisition:
$$\frac{(L \quad ,H \quad ,G \quad)}{(L \cup \{\Gamma\} ,H \quad ,G \quad)}$$
if L is invalid or Γ is valid.

Deletion:
$$\frac{(L \quad ,H \quad ,G \cup \{\gimel\})}{(L \quad ,H \quad ,G \quad)}$$
if L is invalid or $\{\gimel\} \searrow\!/\!\curvearrowright (H, G)$

The Expansion rule has two typical applications. The first introduces sub-goals for a goal that is to be deleted, cf. the Transformation rule below. The second is the very difficult task of introducing new conjectures that are needed for the whole induction proof to work. The Hypothesizing rule makes a new hypothesis '$\aleph$' available for the proof. Since forward reasoning on hypotheses is hardly required, it can usually be restricted to the following sub-rule (cf. Corollary 3) which just stores the goals. This storing is necessary indeed because these goals usually have been transformed when they become useful for inductive reasoning.

Memorizing:
$$\frac{(L ,H \quad ,G \cup \{\aleph\})}{(L ,H \cup \{\aleph\} ,G \cup \{\aleph\})}$$

The Acquisition rule makes a new lemma 'L' available for the proof. The rule may be used to include axioms from the specification or formulas proved in other successful runs of the inference system or by any other sound prover which seems appropriate for some special purposes. The Deletion rule permits the deletion of a goal that is strictly founded on some hypotheses. While we cannot to go into details on how to find this out, the Deletion rule especially permits us to remove a goal if its formula is implied by the formula of an instance of a hypothesis and this instance is smaller than the goal in our induction ordering. More frequently, however, is the Deletion rule used in the following combination with several preceding Expansion steps:

Transformation: $\dfrac{(L\,,H\,,G\cup\{\text{℩}\})}{(L\,,H\,,G\cup M\quad)}$

if L is invalid or $\{\text{℩}\} \searrow\!/\!\curvearrowright (H, G{\cup}M)$

The Transformation rule replaces a goal '℩' with a (possibly empty) set 'M' of sub-goals whose completeness may rely on hypotheses from 'H' or lemmas from 'L'. It is the real working rule of the frame inference system. The intended design of concrete inference systems for specific kinds of validity mainly consists of finding corresponding sub-rules of the Transformation rule.

In the following sections we will present two alternative approaches for explaining why the above inference system implements the ideas presented in the beginning of sect. 2. The first is called "analytic" because it is based on an invariance property that holds for an initial state and is kept invariant by the analytic inference steps. The second is called "backwards" because it is based solely on an invariance property which holds for a final (i. e. successful) state and is kept invariant when one applies the inference rules in backward direction.

6 The Analytic Approach

The analytic approach was first formalized in Wirth (1991). In sect. 2 we indicated that if ' $\vdash$ ' permits sound steps only, then from " $(\emptyset, \emptyset, G) \vdash^{\bullet} (L, H, \emptyset)$ " we may conclude that 'G' is valid. This idea is formalized in:

Definition 9 Soundness of Inference Step.
The inference step " $(L, H, G) \vdash (L', H', G')$ " is called *sound* if validity of 'form[G']' implies validity of 'form[G]'.

Corollary 10. *If all inference steps in " $(L, H, G) \vdash^{\bullet} (L', H', \emptyset)$ " are sound, then* 'form[G]' *is valid.*

Besides the soundness of an inference *step* described above, it is also useful to know about invariant properties of a prover *state* because they can be used to justify why an inference step must be sound. The following such property is most natural, stating that all lemmas are valid and that the hypotheses are founded on the goals, i. e. that for each counterexample for a hypothesis there is a smaller counterexample for a goal.

Definition 11 Correctness of Prover State.
A prover state (L, H, G) is called *correct* if L is valid and $H \curvearrowright G$.

While the first part of this definition should be immediately clear, " $H \curvearrowright G$ " states that the goals carry the proof work (which has to be done for the hypotheses) in such a way that the transformation of a goal may make use of hypotheses which are smaller (w. r. t. our induction ordering $<$) than the goal itself since minimal counterexamples for goals cannot be deleted that way. While "correctness of prover states" obviously formulates this idea, it is not the only possible way to do it:

Definition 12 Weak Correctness of Prover State.
A prover state (L, H, G) is called *weakly correct* if
L is valid and $(H \swarrow G \Rightarrow (\text{form}[H] \text{ is valid}))$.

By the corollaries 4 and 2(3) we get:

Corollary 13. *If a prover state is correct, then it is weakly correct, too.*

As announced above, correctness of prover states really permits us to conclude that the inference steps of our frame inference system are sound:

Lemma 14 Soundness of Inference Steps.
If (L, H, G) is a [weakly] correct prover state, then an inference step " $(L, H, G) \vdash (L', H', G')$ " (with the above rules) is sound.

(For a proof cf. Appendix D.) Furthermore, correctness holds indeed for an initial state and is kept invariant by the frame inference system:

As a corollary of corollaries 3 and 13 we get:

Corollary 15 Initial State is Correct.
Let L be valid and $H \subseteq G$. Now (L, H, G) is [weakly] correct.

Lemma 16 Invariance of Correctness of Prover States.
If " $(L, H, G) \vdash (L', H', G')$ " (with the above rules) and the prover state (L, H, G) is [weakly] correct, then '(L', H', G')' is [weakly] correct, too.

(For a proof cf. Appendix D.) Finally, "correctness of prover states" as an invariance property is not only useful to conclude soundness of single steps, but also globally useful, which can be seen in the following corollary stating that the lemmas and hypotheses gathered in a final prover state are valid:

As a corollary of the corollaries 3, 2(1), and 13 we get:

Corollary 17 For Final State: Correctness means Validity.
$(L', H', \emptyset)$ is [weakly] correct iff " $L' \cup \text{form}[H']$ " is valid.

7 The Backwards Approach

The backwards approach was first formalized in Becker (1994).

Definition 18 Inductiveness and Inductive Soundness.
A prover state (L, H, G) is called *inductive* if $((L \text{ is valid}) \Rightarrow H \swarrow G)$.
The inference step " $(L, H, G) \vdash (L', H', G')$ " is called *inductively sound*[3] if inductiveness of (L', H', G') implies inductiveness of (L, H, G).

[3] In Becker (1994) this is called *preservation of* (inductive) *counterexamples*, but we cannot use this name here because the notion of "counterexample" is different there.

Inductiveness is a technical notion abstracted from inference systems similar to the frame inference system of sect. 5. Roughly speaking, inductiveness of a state means that an inductive proof of it is possible in the sense that a final (i. e. successful) prover state can be entailed. This is because the goals can be deleted, since there are either false lemmas (ex falso quodlibet) or false hypotheses below all invalid goals.

Inductive soundness can replace soundness of prover steps, by the following argumentation, which (just like soundness) requires to think ' $\vdash$ ' backwards, starting from a final prover state $(L', H', \emptyset)$, which must be inductive. Now, if the steps deriving a final state are inductively sound, then all states involved must be inductive. Finally, inductiveness of an initial state implies validity of the initial set of goals.

As a corollary of Corollary 3 we get:

Corollary 19 Final State Must Be Inductive. $(L', H', \emptyset)$ *is inductive.*

By Corollary 2(5) for the forward and by the corollaries 3, 4 and 2(3) for the backwards direction we get:

Corollary 20 For Initial State: Inductiveness means Validity of Goals. *Let* L *be valid and* $H \subseteq G$ *. Now,* 'form[G]' *is valid iff* (L, H, G) *is inductive.*

By the corollaries 19 and 20 we conclude:

Corollary 21. *If all inference steps in* " $(\emptyset, \emptyset, G) \vdash^{*} (L', H', \emptyset)$ " *are inductively sound, then* 'form[G]' *is valid.*

Unlike soundness, inductive soundness also captures the basic idea of our frame inference system which for the analytic approach had to be expressed by some correctness property, namely the idea that transformations of goals may make use of hypotheses which are smaller than the goal itself since minimal counterexamples for goals can never be deleted that way. Note, however, that "being not inductive" is no invariance property of ' $\vdash$ ' (like correctness is) because one never knows whether it holds for some state or not: If all steps are inductively sound, we only know that the property of "being not inductive" is never removed by an inference step, but this does not mean that it ever holds. Especially for successful proofs it never does, cf. Corollary 19. Instead, inductiveness (i. e. "being inductive") is an invariance property of ' $\dashv$ '.

Since inductive soundness captures the basic idea of our frame inference system, we get (cf. Appendix D for a proof):

Lemma 22 Inductive Soundness of Inference Steps.
An inference step with the above rules is inductively sound.

8 Discussion of the Two Approaches

While the relation between the two approaches of the previous two sections is not simple, both seem to be equally useful in capturing the ideas presented in the beginning of sect. 2 as well as in explaining the soundness of our inference system: The following is a corollary of 10, 14, 15, & 16, as well as a corollary of 21 & 22:

Corollary 23 Soundness of " $(\emptyset, \emptyset, G) \vdash^{\circledast} (L', H', \emptyset)$ ".
If " $(\emptyset, \emptyset, G) \vdash^{\circledast} (L', H', \emptyset)$ " (with the above rules), then " $\mathrm{form}[G]$ " is valid.

The analytic approach even permits a slightly stronger conclusion via Corollary 17:

Corollary 24. *If " $(\emptyset, \emptyset, G) \vdash^{\circledast} (L', H', \emptyset)$ " (with the above rules), then " $\mathrm{form}[G] \cup L' \cup \mathrm{form}[H']$ " is valid.*

Considering the design of concrete inference systems by presenting sub-rules of the frame inference rules, another advantage of the analytic approach could be that the additional assumption of correctness of the states could be essential for the sub-rule relationship.

Finally, we compare the analytic and the backwards approach independently of our frame inference system. Here, we consider one approach to be superior to the other, when it permits additional successful proofs, whereas we do not respect the fact that one notion may be more appropriate for effective concretion than the other. Invariance of correctness cannot be superior to invariance of weak correctness or to inductive soundness, since a step with the former and without one of the latter properties starts from an invalid set of goals and thus the required soundness of inference steps does not permit additional proofs. Invariance of weak correctness cannot be superior to invariance of correctness (or else to inductive soundness), since a step with the former and without one of the latter properties leads to (or else starts from) an invalid set of goals and thus the required soundness of inference steps does not permit additional proofs. Finally, inductive soundness is very unlikely to be superior to invariance of [weak] correctness, since a step with the former and without one of the latter properties leads to an invalid set of lemmas or hypotheses. Moreover, if we do not consider all proofs but only the existence of proofs, then (on our non-effective level!) all approaches are equivalent: Using the Deletion rule $|G|$ -times we get:

Corollary 25 Completeness of " $(\emptyset, \emptyset, G) \vdash^{\circledast} (L', H', \emptyset)$ ".
If " $\mathrm{form}[G]$ " is valid, then " $(\emptyset, \emptyset, G) \vdash^{\circledast} (\emptyset, \emptyset, \emptyset)$ " (with the above rules).

Note, however, that (as far as we know) for the construction of effective concrete inference systems based on rules which are no (effective) sub-rules of the rules of our frame inference system, each of the three approaches (i. e.: soundness and invariance of correctness; soundness and invariance of weak correctness; inductive soundness) may be superior to each of the others. The same may hold for generalizing them to inference systems on generalized prover states. E. g., for

Parallelization: $$\frac{(L, H, G \cup G')}{(L, H, G) \qquad (L, H, G')}$$

it is obvious how to generalize inductive soundness, whereas the two other approaches do not seem to permit an appropriate generalization based on local properties of triples (L, H, G).

9 The "Switched" Frame Inference System

In sect. 2 we pointed out that we have to avoid non-terminating reasoning cycles between hypotheses and goals. In our formalization we achieved this by founding a hypothesis '$\aleph$' on smaller or equal goals from 'G', i. e. " $\{\aleph\}\curvearrowright G$ " (cf. the condition of the Hypothesizing rule), and by applying to a goal '$\gimel$' only strictly smaller hypotheses from 'H', i. e. " $\{\gimel\}\searrow H$ " (cf. the condition of the Deletion rule). From a cyclic reasoning " $H\curvearrowright G\searrow H$ " we immediately get " $H\searrow H$ " and " $G\searrow G$ " by Corollary 4, and then 'form[$H\cup G$]' is valid by Corollary 2(3), which means that the reasoning cycle is sound. Now an alternative way to achieve this is the following: Instead of doing our quasi-decreasing step '$\curvearrowright$' from hypotheses to goals " $H\curvearrowright G$ " and our strictly decreasing step '$\searrow$' from goals to hypotheses " $G\searrow H$ ", we could go from hypotheses to goals with a strict and from goals to hypotheses with a quasi step. More precisely: The condition of the Hypothesizing rule would be changed into " if L is invalid or $\{\aleph\} \searrow/\curvearrowright (G,H)$ ", and the condition of the Deletion rule is changed into " if L is invalid or $\{\gimel\}\curvearrowright G\cup H$ ". The Expansion and the Acquisition rules remain unchanged. A Memorizing sub-rule cannot exist and the Transformation rule must be composed of several Expansions, an optional following Hypothesizing, and then a Deletion into one of the following forms:

Memorizing Switched Transformation: $$\frac{(L\,,H\qquad\quad ,G\cup\{\aleph\})}{(L\,,H\cup\{\aleph\}\,,G\cup M\quad)}$$

if L is invalid or $\{\aleph\} \searrow/\curvearrowright (G\cup M, H)$

Simple Switched Transformation: $$\frac{(L\,,H\qquad ,G\cup\{\aleph\})}{(L\,,H\qquad ,G\cup M\quad)}$$

if L is invalid or $\{\aleph\}\curvearrowright G\cup M\cup H$

When we then also switch '$\curvearrowright$' and '$\searrow$' (i. e. replace one by the other) in the definitions of "correctness" and "inductiveness" and when we require an initial state additionally to have an empty set of hypotheses, then we get the analogous results for the soundness of our *switched* frame inference system. The reasons why we prefer the non-switched version presented here are the following:

From a user's point of view, the non-switched version may be more convenient, because hypotheses become available earlier and easier via the Memorizing rule, which does not exist for the switched version.

From the inference system designer's point of view, the non-switched version is more convenient, due to the following argumentation: With both the switched and the non-switched version of the inference system, a proof can be thought to consist mainly of steps of the kind that a goal '$\aleph$' may become available as a hypothesis and is then transformed into sub-goals 'M'. For the non-switched case this can be achieved by an application of the Memorizing and then of the Transformation rule. For the switched inference system this is just a Memorizing Switched Transformation. One shortcoming of the switched version results from the fact that the transformation of a goal into sub-goals has to be strictly decreasing instead of quasi-decreasing (as required for the non-switched case). The design of quasi-decreasing transformations, however, is easier than that of strictly decreasing ones, for the same reason as exhibited in sect. 2 ("Easier Design of Inference Rules") and as illustrated in Appendix C. Therefore, the non-switched inference system allows for small grain inference steps which in the switched system must be replaced with a very big inference step bridging over all quasi-decreasing steps until a strictly decreasing step is reached. Another shortcoming of the Memorizing Switched Transformation is that each simplification step has to decrease the weight of the goal strictly, i. e. that the possibility to apply some fixed smaller hypothesis gets more and more unlikely with each simplification step, cf. sect. 2 ("High Quality of Ordering Information"). If we, however, use the Simple Switched Transformation instead, then the goal is not made available as a hypothesis. Thus, in order not to lose the possibility to apply hypotheses, simplification should not be done via inference steps of the switched inference system, but incorporated into the hypotheses applicability test. With the non-switched version, however, the full inference power of the whole inference system can be homogeneously used for simplification.

On the other hand, the comparison of weights for an applicability test of a hypothesis to a goal is simpler in the switched inference system because there the weight of the hypothesis is often equal to the weight of the goal, in which case the test is successful. While this does not allow for additional proofs with the switched inference system, it may allow to avoid possibly complicated reasoning on ordering properties.

10 Classifying Other Work

In this section we give an incomplete sketch of the literature on inference systems for implicit induction and briefly classify these inference systems according to our presentation here.

In Bachmair (1988) our sets of hypotheses and goals are not separated yet. A disadvantage of this is that a success of a proof due to an empty set of goals is more difficult to detect and that understanding the inference system gets more difficult without the concepts of hypotheses and goals. The missing separation into hypotheses and goals also requires both the foundedness step from hypotheses to goals (as in our switched system of sect. 9) and the step from goals to hypotheses (as in the non-switched system of sect. 5) to be strictly decreasing (i. e. '$\searrow$'), which means a combination of the disadvantages of both the switched and the non-switched approach. The soundness of a proof in Bachmair's inference system results from the fact that a fair derivation sequence $M_i \vdash M_{i+1}$ ($i \in \mathbb{N}$) always satisfies $M_i \curvearrowright M_{i+1}$ and has a sub-sequence such that $\forall i \in \mathbb{N}$: $M_{j_i} \searrow M_{j_{i+1}}$, which means that 'form[M_0]' must be valid, cf. corollaries 8, 7, and 2(6). The foundedness relations are defined by use of sizes of inconsistency proofs for the equations in M_i instead of the counterexamples themselves. As already mentioned in sect. 3, it is in general undesirable to base the notions of a frame inference system on formal proofs instead of semantic notions because the latter impose no operational restrictions and are likely to change less frequently. One of the operational restrictions in Bachmair (1988) is the confluence requirement for the specifying set of rules. That this restriction can be removed was noted in Gramlich (1989) by defining the foundedness relations by use of the sizes of positive proofs measured via the applications of equations from M_i.

The important separation between hypotheses and goals was introduced in Reddy (1990), where a frame inference system similar to our switched one in sect. 9 is used, the argumentation for soundness follows the analytic approach using operationally restricted versions of soundness and (switched) correctness, and the foundedness notions are still the operationally restricted ones of Gramlich (1989). In Wirth (1991) this operational restriction is overcome by using a semantic foundedness notion.

In Fraus (1993) we have found the first[4] argumentation for soundness following the backwards approach. While the inference system already is of the (superior) non-switched style, the foundedness relations are still operationally restricted (by measuring positive proofs in the natural deduction calculus). In Becker (1994) we finally find the backwards approach based on semantic foundedness notions as presented here.

11 Conclusion

We tried to give an intuitive understanding of proofs by mathematical induction, exhibited the essential requirements, and provided a simple data structure for prover states. To enable a clear understanding of the functions of inference systems for proofs by mathematical induction, we introduced the concept of "foundedness" which also has applications beyond this paper. We presented an abstract frame inference system, elaborated two approaches for explaining why this system is sound, and argued why we prefer our frame inference system to the switched one. We classified argumentation for soundness occurring in the literature according to our taxonomy. When practically appropriate concrete inference systems are designed as systems of sub-rules of the rules of the presented frame inference system, soundness of these systems is given immediately. While we did not present a concrete inference system in this paper, the example in Appendix C should be sufficient to make our intention obvious.

Appendix A Safe Steps and Failure Recognition

As already mentioned in sect. 1, the ability of an inductive theorem prover to detect invalid formulas is most important under a practical aspect, especially for *inductive* theorem proving because of the important role generalizations play in it, where an invalid formula can result from a valid input theorem due to over-generalization. Thus, suppose we have some failure predicate 'FAIL' defined on sets of formulas which is *correct* in the sense that $\forall F \in \mathrm{FAIL}$: ($F$ is invalid). Note that this failure predicate is defined on *sets of* formulas for operational

[4] This actually goes back to an unpublished manuscript of Alfons Geser (1988) at the University of Passau entitled "An inductive proof method based on narrowing".

reasons, namely in order to be able to recognize that one formula contradicts another one; whereas for a theoretical treatment it would be sufficient to define it on single formulas since one of those formulas must be invalid in a consistent specification; but to find out which formula it is is undecidable in general.

We define a prover state (L, H, G) to be a *failure state* if

$$(L \cup \text{form}[H \cup G]) \in \text{FAIL}.$$

Note that we have included L and H (instead of just testing G) because we want to be able to detect an invalid lemma or hypothesis when it has just been generated and do not want to have to wait until it will have been harmfully applied to a goal. One is tempted to argue that, in case of [weak] correctness of a prover state, an invalid hypothesis implies the existence of an invalid goal, but this argument again does not respect the operational aspect.

Now, when an inductive theorem prover has realized to be in a failure state, the following questions arise: How far do we have to backtrack to reach a state with valid formulas? Have some of our original input formulas been invalid? For answering these questions the following notion is useful:

An inference step " $(L, H, G) \vdash (L', H', G')$ " is called *safe* if

validity of " $L \cup \text{form}[H \cup G]$ " implies validity of " $L' \cup \text{form}[H' \cup G']$ ".

It is not reasonable to require all possible steps of an inductive theorem prover to be safe, since this property is undecidable for generalization steps which play a major role in inductive theorem proving. For concrete inference systems, however, it is usually possible to give interesting sufficient conditions for the application of an inference rule to be safe. Now, when the prover has found out that a prover state (L'', H'', G'') is a failure state and all steps in $(L', H', G') \vdash^{*} (L'', H'', G'')$ are known to be safe, then (L', H', G') must be a failure state, too. To recover from this failure we may iterate the following:

If this (L', H', G') is the original input state (with L' known to be valid and $H' \subseteq G'$), then we have refuted our original set of goals G' and should stop proving. Otherwise the step that yielded (L', H', G'), say $(L, H, G) \vdash (L', H', G')$, must be carefully inspected: If it is known to be safe we backtrack this step and reiterate. Otherwise it might be possible to find a (minimal) subset of $L'' \cup \text{form}[H'' \cup G'']$ for which the failure predicate FAIL is still known to hold and which also is (implied by) a subset of $L \cup \text{form}[H \cup G]$; in which case we also backtrack this step and reiterate. Otherwise, when $(L, H, G) \vdash (L', H', G')$ is likely to be an unsafe step which might have caused the failure, we backtrack this step and may try to go on with a hopefully safe inference step instead.

Appendix B Refutational Completeness

For achieving refutational completeness we need a wellfounded ordering $>_{\text{refut}}$ on finite sets of syntactic constructs. To be able to refute initial failure states we need the following property.

Definition 26 FAIL-Completeness.

The failure predicate FAIL is *complete w. r. t. '* $\vdash$ *' and '* $>_{\text{refut}}$ *'* if for all finite sets $L \subseteq$ Form; $H, G \subseteq$ SynCons; if form$[G]$ is invalid, but (L, H, G) is not a failure state, then there are finite sets L', H', G' with $(L,H,G) \vdash^{\oplus} (L',H',G')$ and $G >_{\mathrm{refut}} G'$.

By wellfoundedness of '$>_{\mathrm{refut}}$' we immediately get:

Corollary 27 Refutational Completeness.
Let $L \subseteq$ Form; $H, G \subseteq$ SynCons be finite sets. Assume either that '$\vdash$' is sound or that L is valid, $H \subseteq G$, and '$\vdash$' is inductively sound. Furthermore assume FAIL *to be complete w. r. t. '$\vdash$' and '$>_{\mathrm{refut}}$'. Now, if* form[G] *is invalid, then there is some failure state (L',H',G') with $(L,H,G) \vdash^{\oplus} (L',H',G')$.*

Definition 28 Fairness.
Let β be an ordinal number with $\beta \preceq \omega$. Let $L_i \subseteq$ Form; $H_i, G_i \subseteq$ SynCons for all $i \prec 1{+}\beta$. Consider the derivation $(L_i,H_i,G_i) \vdash (L_{i+1},H_{i+1},G_{i+1})$ $(i \prec \beta)$. It is called *fair* if $\beta \prec \omega \wedge (L_\beta,H_\beta,G_\beta) \notin \mathrm{dom}(\vdash)$ (i. e. no inference rule can be applied to $(L_\beta,H_\beta,G_\beta)$) or $\exists i \prec 1{+}\beta\colon G_i = \emptyset$ or $\forall i \prec 1{+}\beta$: ((form[$G_i$] invalid $\wedge$ (L_i,H_i,G_i) not a failure state) $\Rightarrow$ $\exists j \prec 1{+}\beta\colon G_i >_{\mathrm{refut}} G_j$).

Corollary 29. *Let β be an ordinal number with $\beta \preceq \omega$. Let $L_i \subseteq$ Form; $H_i, G_i \subseteq$ SynCons be finite sets for all $i \prec 1{+}\beta$. Let $(L_i,H_i,G_i) \vdash (L_{i+1},H_{i+1},G_{i+1})$ $(i \prec \beta)$ be a fair derivation. Assume either that '$\vdash$' is sound or that L_0 is valid, $H_0 \subseteq G_0$, and '$\vdash$' is inductively sound. Furthermore assume* FAIL *to be complete w. r. t. '$\vdash$' and '$>_{\mathrm{refut}}$'. Now, if* form[G] *is invalid, there must exist some $i \prec 1{+}\beta$ such that (L_i,H_i,G_i) is a failure state.*

Appendix C An Example

In this section we give an example to illustrate our abstract argumentation on the benefit of separating weights from formulas and of using non-switched inference systems. Consider the following specification of a member-predicate "mbp(x,l)" testing for x occurring in the list l, a delete-function "dl(x,l)" deleting all occurrences of x in the list l, a remove-copies-function "rc(x,l)" removing repeated occurrences of x in the list l, and a brushing function "br(k,l)" removing repeated occurrences in the list l for all elements of the list k:

$\mathsf{mbp}(x,\mathsf{nil})$	$=\mathsf{false}$		$\mathsf{dl}(x,\mathsf{cons}(y,l))$	$=\mathsf{cons}(y,\mathsf{dl}(x,l))$	**if** $x \neq y$
$\mathsf{mbp}(x,\mathsf{cons}(y,l))$	$=\mathsf{true}$	**if** $x = y$	$\mathsf{rc}(x,\mathsf{nil})$	$=\mathsf{nil}$	
$\mathsf{mbp}(x,\mathsf{cons}(y,l))$	$=\mathsf{mbp}(x,l)$	**if** $x \neq y$	$\mathsf{rc}(x,\mathsf{cons}(y,l))$	$=\mathsf{cons}(y,\mathsf{dl}(x,l))$	**if** $x = y$
$\mathsf{dl}(x,\mathsf{nil})$	$=\mathsf{nil}$		$\mathsf{rc}(x,\mathsf{cons}(y,l))$	$=\mathsf{cons}(y,\mathsf{rc}(x,l))$	**if** $x \neq y$
$\mathsf{dl}(x,\mathsf{cons}(y,l))$	$=\mathsf{dl}(x,l)$	**if** $x = y$	$\mathsf{br}(\mathsf{nil},l)$	$=l$	
			$\mathsf{br}(\mathsf{cons}(x,k),l)$	$=\mathsf{br}(k,\mathsf{rc}(x,l))$	

Suppose we want to show the following theorem (", " denotes "logical or"):

(0) $\mathsf{br}(k,\mathsf{cons}(x,l)) = \mathsf{cons}(x,\mathsf{br}(k,l)),\ \mathsf{mbp}(x,l) = \mathsf{true}$

saying that either x occurs in l or the wave-front $\mathsf{cons}(x,\ldots)$ can be rippled out. Applying a covering set of substitutions to (0) we get a base case for $\{k \mapsto \mathsf{nil}\}$ (which is trivial after two rewriting steps applying the first rule for br) and the following case for $\{k \mapsto \mathsf{cons}(y,k)\}$:

(1) $\mathsf{br}(\mathsf{cons}(y,k),\mathsf{cons}(x,l)) = \mathsf{cons}(x,\mathsf{br}(\mathsf{cons}(y,k),l)),\ \mathsf{mbp}(x,l) = \mathsf{true}$

After two rewriting steps applying the second rule for br we get:

(2) $\mathsf{br}(k, \mathsf{rc}(y, \mathsf{cons}(x, l))) = \mathsf{cons}(x, \mathsf{br}(k, \mathsf{rc}(y, l))),\ \ \mathsf{mbp}(x, l) = \mathsf{true}$

A rewriting step applying the third rule for rc in the left-hand side of the first equation yields

(3) $x = y,\ \ \mathsf{br}(k, \mathsf{cons}(x, \mathsf{rc}(y, l))) = \mathsf{cons}(x, \mathsf{br}(k, \mathsf{rc}(y, l))),\ \ \mathsf{mbp}(x, l) = \mathsf{true}$

as well as the goal

(G) $x \neq y,\ \ \mathsf{br}(k, \mathsf{rc}(y, \mathsf{cons}(x, l))) = \mathsf{cons}(x, \mathsf{br}(k, \mathsf{rc}(y, l))),\ \ \mathsf{mbp}(x, l) = \mathsf{true}$

whose proof we do not treat here.

Applying our induction hypothesis (0) instantiated by $\{\, l \mapsto \mathsf{rc}(y, l) \,\}$, i. e.

(I) $\mathsf{br}(k, \mathsf{cons}(x, \mathsf{rc}(y, l))) = \mathsf{cons}(x, \mathsf{br}(k, \mathsf{rc}(y, l))),\ \ \mathsf{mbp}(x, \mathsf{rc}(y, l)) = \mathsf{true}$

to (3) we get:

(4) $\mathsf{mbp}(x, \mathsf{rc}(y, l)) \neq \mathsf{true},$

$x = y,\ \ \mathsf{br}(k, \mathsf{cons}(x, \mathsf{rc}(y, l))) = \mathsf{cons}(x, \mathsf{br}(k, \mathsf{rc}(y, l))),\ \ \mathsf{mbp}(x, l) = \mathsf{true}$

Rewriting with the lemma $\mathsf{mbp}(x, \mathsf{rc}(y, l)) = \mathsf{mbp}(x, l)$ finally yields a tautology.

For this induction proof to be sound we have to find a wellfounded ordering in which the instance (I) of our hypotheses is (strictly) smaller than the goal (3) to which it is applied. When we consider the weight of a formula to be the formula itself (considered as the multi-set of (its literals considered as the multi-sets of) its terms), this cannot be achieved with a simplification ordering, whereas the evaluation ordering of the specification is a sub-relation of a simplification ordering, namely the lexicographic path ordering given by the following precedence on function symbols: $\mathsf{mbp} \succsim \mathsf{true}, \mathsf{false}$; $\mathsf{br} \succsim \mathsf{rc} \succsim \mathsf{dl} \succ \mathsf{cons}$. The first thing we can do is to use weight pointers to avoid the deterioration of ordering information on the way from (1) to (3): Initially the weight pointer of (0) points to (0) itself, i. e. the weight of this formula is this formula itself. The same holds for (1) because when applying the substitution the weight is instantiated as well as the formula. During the simplification steps yielding (2) and then (3) the weight remains unchanged, i. e. the weight of the formula (3) is still the formula (1). Now (I) is indeed strictly smaller than (1) in the lexicographic path ordering given by: $\mathsf{br} \succ \mathsf{cons}$; $\mathsf{br} \succsim \mathsf{rc}$; $\mathsf{br} \succ \mathsf{mbp}$. Thus the high quality of ordering information preserved by separating the formula from its weight when simplifying (1) into (3) now permits us to justify the application of (I) to (3) with a simplification ordering, i. e. we can now realize that our (partial) proof is sound. This success, however, is not very convincing because the last pair in the above precedence (i. e. $\mathsf{br} \succ \mathsf{mbp}$) is not all motivated by the evaluation ordering of our specification. After all, our example proof is nothing but a structural induction and thus should work out without such a sophisticated ordering. If we have a closer look on the derivation from (0) (or (1)) to (3), then we notice that the first literal as usual does get smaller in the evaluation ordering on the way from our original goal (over the goal the hypothesis is applied to) to the instantiated hypothesis (I), but the second literal does not. Thus, if we focus on the first literal by setting our weight to it, then we only have to show that the first literal of (I) is smaller than the first literal of (1) which does not require the unmotivated precedence of $\mathsf{br} \succ \mathsf{mbp}$. Going one step further, setting the weight of (0) to be the variable k, the weight of (1) and (3) becomes $\mathsf{cons}(y, k)$ which is trivially greater than the weight k of (I).

The proof of (G) gets easier when we have the following lemma:
(L) $\mathsf{dl}(x,l)=l,\ \mathsf{mbp}(x,l)=\mathsf{true}$
We will now use this lemma to show why the design of inference rules gets easier when we separate the weight of a formula from the formula itself and when we do not use the switched inference system of sect. 9 instead of the non-switched one of sect. 5. Suppose we want to apply (L) in the proof of a formula
(10) $\Gamma[\mathsf{dl}(x,l)]$ containing $\mathsf{dl}(x,l)$ as a subterm.
Rewriting with (L) yields the following two sub-goals:
(11) $\mathsf{mbp}(x,l)=\mathsf{true},\ \Gamma[l]$ (12) $\mathsf{mbp}(x,l)\neq\mathsf{true},\ \Gamma[\mathsf{dl}(x,l)]$
When we restrict the weight of a formula to be the formula itself, then the sub-goals (11) and (12) must be smaller than (10) (without focusing on $\Gamma[\mathsf{dl}(x,l)]$). When we choose the switched inference system and want to make (10) available as an induction hypotheses, then the weights of the sub-goals (11) and (12) must be strictly smaller than the weight (10) in a wellfounded ordering. For (11) this might be achieved again by the unmotivated trick of extending the precedence on function symbols with $\mathsf{dl}\succ\mathsf{mbp}$ (additionally to $\mathsf{mbp}\succsim\mathsf{true}$); for (12), however, this does not seem to be reasonably possible in general. Thus the design of an inference rule applying (L) in the intended form to (10) must be very difficult to develop without separated weights or for a switched frame inference system, because the step from (10) to (12) must be replaced with a very big inference step bridging over all steps following in the proof of (12) until all branches of this proof have reached a smaller weight. Separating weights from formulas and using a non-switched frame inference system, however, the design of such an inference rule is very easy: One just sets the weight of (11) and (12) to the the original weight of formula (10).

Appendix D The Proofs

Proof of Lemma 14: The only rule whose soundness is non-trivial is the Deletion rule for form$[G]$ being valid. By Corollary 2(2) we get $G\searrow\emptyset$. Thus by Corollary 6(2) and the condition of the rule we get $\{ℷ\}\searrow H$. By Corollary 13 $(L,H,G\cup\{ℷ\})$ is weakly correct. Thus (since $G\cup\{ℷ\}\searrow H$ due to Corollary 5(2)) we can conclude that form$[H]$ must be valid. By Corollary 2(7) this means that form$(ℷ)$ is valid, i. e. that form$[G\cup\{ℷ\}]$ is valid.

Proof of Lemma 16: Assume L to be valid. We show invariance of weak correctness first: Expansion: Assume $G\cup\{ℷ\}\searrow H$. By Corollary 5(4) we get $G\searrow H$. Now form$[H]$ must be valid due to weak correctness of (L,H,G). Hypothesizing: Assume $G\searrow H\cup\{ℵ\}$. By Lemma 22 we get $G\searrow H$. By weak correctness of (L,H,G), form$[H]$ must be valid, which by Corollary 2(7) means that form$[G\cup H]$ is valid. By the condition of the rule and Corollary 2(6), we know that form$(ℵ)$ is valid. Therefore form$[H\cup\{ℵ\}]$ is valid, too. Acquisition: Trivial. Deletion: Assume $G\searrow H$. By Lemma 22 we get $G\cup\{ℷ\}\searrow H$. By weak correctness of $(L,H,G\cup\{ℷ\})$, form$[H]$ must be valid.

Finally we show invariance of correctness: Expansion: By Corollary 5(3). Hypothesizing: By Corollary 7 we get $G\curvearrowright G$. If we assume $H\curvearrowright G$, we get

$H \cup G \curvearrowright G$ by Corollary 5(1). By the condition of the rule and Corollary 7 this means $\{\aleph\} \curvearrowright G$. By our assumption we now get $H \cup \{\aleph\} \curvearrowright G$ via Corollary 5(1). Acquisition: Trivial. Deletion: Assume $H \curvearrowright G \cup \{\gimel\}$. By Corollary 6(1) from the condition of the rule we get $\{\gimel\} \curvearrowright G$. By Corollary 7 we have $G \curvearrowright G$ and then by Corollary 5(1) $G \cup \{\gimel\} \curvearrowright G$. By the assumption and Corollary 7 this means $H \curvearrowright G$.

Proof of Lemma 22: Expansion: By Corollary 5(4). Hypothesizing: By Corollary 7 we get $H \curvearrowright H$ and then from the condition of the rule $H \cup \{\aleph\} \curvearrowright H \cup G$ by Corollary 5(1). If we now assume $G \searrow H \cup \{\aleph\}$, we get $G \searrow H \cup G$ by Corollary 3, and then by corollaries 5(5) and 5(4) $G \searrow H$. Acquisition: Trivial. Deletion: Assume $G \searrow H$. By the condition of the rule and Corollary 6(2) we get $\{\gimel\} \searrow H$. Thus by our assumption and Corollary 5(2) we get $G \cup \{\gimel\} \searrow H$.

Acknowledgements: We would like to thank Jürgen Avenhaus, Alfons Geser, Bernhard Gramlich, Ulrich Kühler, and Martin Protzen for useful hints.

References

Leo Bachmair (1988). *Proof by Consistency in Equational Theories.* 3rd IEEE symposium on Logic In Computer Science, pp. 228-233.

Klaus Becker (1993). *Proving Ground Confluence and Inductive Validity in Constructor Based Equational Specifications.* TAPSOFT 1993, LNCS 668, pp. 46-60, Springer.

Klaus Becker (1994). *Rewrite Operationalization of Clausal Specifications with Predefined Structures.* PhD thesis, Fachbereich Informatik, Universität Kaiserslautern.

Wolfgang Bibel, E. Eder (1993). *Methods and Calculi for Deduction.* In: Dov M. Gabbay, C. J. Hogger, J. A. Robinson (eds.). *Handbook of Logic in Artificial Intelligence and Logic Programming.* Vol. 1, pp. 67-182, Clarendon.

Robert S. Boyer, J Strother Moore (1979). *A Computational Logic.* Academic Press.

Ulrich Fraus (1993). *A Calculus for Conditional Inductive Theorem Proving.* 3rd CTRS 1992, LNCS 656, pp. 357-362, Springer.

Bernhard Gramlich (1989). *Inductive Theorem Proving Using Refined Unfailing Completion Techniques.* SEKI-Report SR-89-14 (SFB), Fachbereich Informatik, Universität Kaiserslautern. *Short version in:* 9th ECAI 1990, pp. 314-319, Pitman.

Martin Protzen (1994). *Lazy Generation of Induction Hypotheses.* 12th CADE 1994, LNAI 814, pp. 42-56, Springer.

Uday S. Reddy (1990). *Term Rewriting Induction.* 10th CADE 1990, LNAI 449, pp. 162-177, Springer.

Christoph Walther (1994). *Mathematical Induction.* In: *Handbook of Logic in Artificial Intelligence and Logic Programming.* eds. cf. above. Vol. 2, pp. 127-228, Clarendon.

Claus-Peter Wirth (1991). *Inductive Theorem Proving in Theories specified by Positive/Negative Conditional Equations.* Diplomarbeit, Fachbereich Informatik, Universität Kaiserslautern.

Claus-Peter Wirth, Bernhard Gramlich (1993). *A Constructor-Based Approach for Positive/Negative-Conditional Equational Specifications.* 3rd CTRS 1992, LNCS 656, pp. 198-212, Springer. Revised and extended version in J. Symbolic Computation (1994) **17**, pp. 51-90, Academic Press.

Claus-Peter Wirth, Bernhard Gramlich (1994). *On Notions of Inductive Validity for First-Order Equational Clauses.* 12th CADE 1994, LNAI 814, pp. 162-176, Springer.

Author Index

Bachmair, Leo 1
Balbiani, Philippe 31
Basin, David A. 15
Becker, Klaus 353
Bockmayr, Alexander 51
Bündgen, Reinhard 71
Dershowitz, Nachum 89
Ferreira, M. C. F. 106
Ganzinger, Harald 1
Gélis, Jean-Michel 124
Glauert, John 144
Gramlich, Bernhard 166
Hintermeier, Claus 186
Hoot, Charles 206
Inverardi, Paola 223
Jouannaud, Jean-Pierre 235
Kennaway, Richard 247
Khasidashvili, Zurab 144
Kluit, Peter G. 332
Kucherov, Gregory 262
Matsumoto, Yoshihiro 276
Mori, Akira 276
Ohlebusch, Enno 296
Oyamaguchi, Michio 316
Rusinowitch, Michaël 262
Sadfi, Walid 235
Toyama, Yoshihito 316
Verheul, Jan C. 332
Walsh, Toby 15
Werner, Andreas 51
Wirth, Claus-Peter 353
Zantema, Hans 106

Springer-Verlag and the Environment

We at Springer-Verlag firmly believe that an international science publisher has a special obligation to the environment, and our corporate policies consistently reflect this conviction.

We also expect our business partners – paper mills, printers, packaging manufacturers, etc. – to commit themselves to using environmentally friendly materials and production processes.

The paper in this book is made from low- or no-chlorine pulp and is acid free, in conformance with international standards for paper permanency.

Lecture Notes in Computer Science

For information about Vols. 1–912

please contact your bookseller or Springer-Verlag

Vol. 913: W. Schäfer (Ed.), Software Process Technology. Proceedings, 1995. IX, 261 pages. 1995.

Vol. 914: J. Hsiang (Ed.), Rewriting Techniques and Applications. Proceedings, 1995. XII, 473 pages. 1995.

Vol. 915: P. D. Mosses, M. Nielsen, M. I. Schwartzbach (Eds.), TAPSOFT '95: Theory and Practice of Software Development. Proceedings, 1995. XV, 810 pages. 1995.

Vol. 916: N. R. Adam, B. K. Bhargava, Y. Yesha (Eds.), Digital Libraries. Proceedings, 1994. XIII, 321 pages. 1995.

Vol. 917: J. Pieprzyk, R. Safavi-Naini (Eds.), Advances in Cryptology - ASIACRYPT '94. Proceedings, 1994. XII, 431 pages. 1995.

Vol. 918: P. Baumgartner, R. Hähnle, J. Posegga (Eds.), Theorem Proving with Analytic Tableaux and Related Methods. Proceedings, 1995. X, 352 pages. 1995. (Subseries LNAI).

Vol. 919: B. Hertzberger, G. Serazzi (Eds.), High-Performance Computing and Networking. Proceedings, 1995. XXIV, 957 pages. 1995.

Vol. 920: E. Balas, J. Clausen (Eds.), Integer Programming and Combinatorial Optimization. Proceedings, 1995. IX, 436 pages. 1995.

Vol. 921: L. C. Guillou, J.-J. Quisquater (Eds.), Advances in Cryptology – EUROCRYPT '95. Proceedings, 1995. XIV, 417 pages. 1995.

Vol. 922: H. Dörr, Efficient Graph Rewriting and Its Implementation. IX, 266 pages. 1995.

Vol. 923: M. Meyer (Ed.), Constraint Processing. IV, 289 pages. 1995.

Vol. 924: P. Ciancarini, O. Nierstrasz, A. Yonezawa (Eds.), Object-Based Models and Languages for Concurrent Systems. Proceedings, 1994. VII, 193 pages. 1995.

Vol. 925: J. Jeuring, E. Meijer (Eds.), Advanced Functional Programming. Proceedings, 1995. VII, 331 pages. 1995.

Vol. 926: P. Nesi (Ed.), Objective Software Quality. Proceedings, 1995. VIII, 249 pages. 1995.

Vol. 927: J. Dix, L. Moniz Pereira, T. C. Przymusinski (Eds.), Non-Monotonic Extensions of Logic Programming. Proceedings, 1994. IX, 229 pages. 1995. (Subseries LNAI).

Vol. 928: V.W. Marek, A. Nerode, M. Truszczynski (Eds.), Logic Programming and Nonmonotonic Reasoning. Proceedings, 1995. VIII, 417 pages. 1995. (Subseries LNAI).

Vol. 929: F. Morán, A. Moreno, J.J. Merelo, P. Chacón (Eds.), Advances in Artificial Life. Proceedings, 1995. XIII, 960 pages. 1995 (Subseries LNAI).

Vol. 930: J. Mira, F. Sandoval (Eds.), From Natural to Artificial Neural Computation. Proceedings, 1995. XVIII, 1150 pages. 1995.

Vol. 931: P.J. Braspenning, F. Thuijsman, A.J.M.M. Weijters (Eds.), Artificial Neural Networks. IX, 295 pages. 1995.

Vol. 932: J. Iivari, K. Lyytinen, M. Rossi (Eds.), Advanced Information Systems Engineering. Proceedings, 1995. XI, 388 pages. 1995.

Vol. 933: L. Pacholski, J. Tiuryn (Eds.), Computer Science Logic. Proceedings, 1994. IX, 543 pages. 1995.

Vol. 934: P. Barahona, M. Stefanelli, J. Wyatt (Eds.), Artificial Intelligence in Medicine. Proceedings, 1995. XI, 449 pages. 1995. (Subseries LNAI).

Vol. 935: G. De Michelis, M. Diaz (Eds.), Application and Theory of Petri Nets 1995. Proceedings, 1995. VIII, 511 pages. 1995.

Vol. 936: V.S. Alagar, M. Nivat (Eds.), Algebraic Methodology and Software Technology. Proceedings, 1995. XIV, 591 pages. 1995.

Vol. 937: Z. Galil, E. Ukkonen (Eds.), Combinatorial Pattern Matching. Proceedings, 1995. VIII, 409 pages. 1995.

Vol. 938: K.P. Birman, F. Mattern, A. Schiper (Eds.), Theory and Practice in Distributed Systems. Proceedings,1994. X, 263 pages. 1995.

Vol. 939: P. Wolper (Ed.), Computer Aided Verification. Proceedings, 1995. X, 451 pages. 1995.

Vol. 940: C. Goble, J. Keane (Eds.), Advances in Databases. Proceedings, 1995. X, 277 pages. 1995.

Vol. 941: M. Cadoli, Tractable Reasoning in Artificial Intelligence. XVII, 247 pages. 1995. (Subseries LNAI).

Vol. 942: G. Böckle, Exploitation of Fine-Grain Parallelism. IX, 188 pages. 1995.

Vol. 943: W. Klas, M. Schrefl, Metaclasses and Their Application. IX, 201 pages. 1995.

Vol. 944: Z. Fülöp, F. Gécseg (Eds.), Automata, Languages and Programming. Proceedings, 1995. XIII, 686 pages. 1995.

Vol. 945: B. Bouchon-Meunier, R.R. Yager, L.A. Zadeh (Eds.), Advances in Intelligent Computing - IPMU '94. Proceedings, 1994. XII, 628 pages.1995.

Vol. 946: C. Froidevaux, J. Kohlas (Eds.), Symbolic and Quantitative Approaches to Reasoning and Uncertainty. Proceedings, 1995. X, 420 pages. 1995. (Subseries LNAI).

Vol. 947: B. Möller (Ed.), Mathematics of Program Construction. Proceedings, 1995. VIII, 472 pages. 1995.

Vol. 948: G. Cohen, M. Giusti, T. Mora (Eds.), Applied Algebra, Algebraic Algorithms and Error-Correcting Codes. Proceedings, 1995. XI, 485 pages. 1995.

Vol. 949: D.G. Feitelson, L. Rudolph (Eds.), Job Scheduling Strategies for Parallel Processing. Proceedings, 1995. VIII, 361 pages. 1995.

Vol. 950: A. De Santis (Ed.), Advances in Cryptology - EUROCRYPT '94. Proceedings, 1994. XIII, 473 pages. 1995.

Vol. 951: M.J. Egenhofer, J.R. Herring (Eds.), Advances in Spatial Databases. Proceedings, 1995. XI, 405 pages. 1995.

Vol. 952: W. Olthoff (Ed.), ECOOP '95 - Object-Oriented Programming. Proceedings, 1995. XI, 471 pages. 1995.

Vol. 953: D. Pitt, D.E. Rydeheard, P. Johnstone (Eds.), Category Theory and Computer Science. Proceedings, 1995. VII, 252 pages. 1995.

Vol. 954: G. Ellis, R. Levinson, W. Rich. J.F. Sowa (Eds.), Conceptual Structures: Applications, Implementation and Theory. Proceedings, 1995. IX, 353 pages. 1995. (Subseries LNAI).

VOL. 955: S.G. Akl, F. Dehne, J.-R. Sack, N. Santoro (Eds.), Algorithms and Data Structures. Proceedings, 1995. IX, 519 pages. 1995.

Vol. 956: X. Yao (Ed.), Progress in Evolutionary Computation. Proceedings, 1993, 1994. VIII, 314 pages. 1995. (Subseries LNAI).

Vol. 957: C. Castelfranchi, J.-P. Müller (Eds.), From Reaction to Cognition. Proceedings, 1993. VI, 252 pages. 1995. (Subseries LNAI).

Vol. 958: J. Calmet, J.A. Campbell (Eds.), Integrating Symbolic Mathematical Computation and Artificial Intelligence. Proceedings, 1994. X, 275 pages. 1995.

Vol. 959: D.-Z. Du, M. Li (Eds.), Computing and Combinatorics. Proceedings, 1995. XIII, 654 pages. 1995.

Vol. 960: D. Leivant (Ed.), Logic and Computational Complexity. Proceedings, 1994. VIII, 514 pages. 1995.

Vol. 961: K.P. Jantke, S. Lange (Eds.), Algorithmic Learning for Knowledge-Based Systems. X, 511 pages. 1995. (Subseries LNAI).

Vol. 962: I. Lee, S.A. Smolka (Eds.), CONCUR '95: Concurrency Theory. Proceedings, 1995. X, 547 pages. 1995.

Vol. 963: D. Coppersmith (Ed.), Advances in Cryptology - CRYPTO '95. Proceedings, 1995. XII, 467 pages. 1995.

Vol. 964: V. Malyshkin (Ed.), Parallel Computing Technologies. Proceedings, 1995. XII, 497 pages. 1995.

Vol. 965: H. Reichel (Ed.), Fundamentals of Computation Theory. Proceedings, 1995. IX, 433 pages. 1995.

Vol. 966: S. Haridi, K. Ali, P. Magnusson (Eds.), EURO-PAR '95 Parallel Processing. Proceedings, 1995. XV, 734 pages. 1995.

Vol. 967: J.P. Bowen, M.G. Hinchey (Eds.), ZUM '95: The Z Formal Specification Notation. Proceedings, 1995. XI, 571 pages. 1995.

Vol. 968: N. Dershowitz, N. Lindenstrauss (Eds.), Conditional and Typed Rewriting Systems. Proceedings, 1994. VIII, 375 pages. 1995.

Vol. 969: J. Wiedermann, P. Hájek (Eds.), Mathematical Foundations of Computer Science 1995. Proceedings, 1995. XIII, 588 pages. 1995.

Vol. 970: V. Hlaváč, R. Šára (Eds.), Computer Analysis of Images and Patterns. Proceedings, 1995. XVIII, 960 pages. 1995.

Vol. 971: E.T. Schubert, P.J. Windley, J. Alves-Foss (Eds.), Higher Order Logic Theorem Proving and Its Applications. Proceedings, 1995. VIII, 400 pages. 1995.

Vol. 972: J.-M. Hélary, M. Raynal (Eds.), Distributed Algorithms. Proceedings, 1995. XI, 333 pages. 1995.

Vol. 973: H.H. Adelsberger, J. Lažanský, V. Mařík (Eds.), Information Management in Computer Integrated Manufacturing. IX, 665 pages. 1995.

Vol. 974: C. Braccini, L. DeFloriani, G. Vernazza (Eds.), Image Analysis and Processing. Proceedings, 1995. XIX, 757 pages. 1995.

Vol. 975: W. Moore, W. Luk (Eds.), Field-Programmable Logic and Applications. Proceedings, 1995. XI, 448 pages. 1995.

Vol. 976: U. Montanari, F. Rossi (Eds.), Principles and Practice of Constraint Programming — CP '95. Proceedings, 1995. XIII, 651 pages. 1995.

Vol. 977: H. Beilner, F. Bause (Eds.), Quantitative Evaluation of Computing and Communication Systems. Proceedings, 1995. X, 415 pages. 1995.

Vol. 978: N. Revell, A M. Tjoa (Eds.), Database and Expert Systems Applications. Proceedings, 1995. XV, 654 pages. 1995.

Vol. 979: P. Spirakis (Ed.), Algorithms — ESA '95. Proceedings, 1995. XII, 598 pages. 1995.

Vol. 980: A. Ferreira, J. Rolim (Eds.), Parallel Algorithms for Irregularly Structured Problems. Proceedings, 1995. IX, 409 pages. 1995.

Vol. 981: I. Wachsmuth, C.-R. Rollinger, W. Brauer (Eds.), KI-95: Advances in Artificial Intelligence. Proceedings, 1995. XII, 269 pages. (Subseries LNAI).

Vol. 982: S. Doaitse Swierstra, M. Hermenegildo (Eds.), Programming Languages: Implementations, Logics and Programs. Proceedings, 1995. XI, 467 pages. 1995.

Vol. 983: A. Mycroft (Ed.), Static Analysis. Proceedings, 1995. VIII, 423 pages. 1995.

Vol. 985: T. Sellis (Ed.), Rules in Database Systems. Proceedings, 1995. VIII, 373 pages. 1995.

Vol. 986: Henry G. Baker (Ed.), Memory Management. Proceedings, 1995. XII, 417 pages. 1995.

Vol. 987: P.E. Camurati, H. Eveking (Eds.), Correct Hardware Design and Verification Methods. Proceedings, 1995. VIII, 342 pages. 1995.

Vol. 988: A.U. Frank, W. Kuhn (Eds.), Spatial Information Theory. Proceedings, 1995. XIII, 571 pages. 1995.

Vol. 989: W. Schäfer, P. Botella (Eds.), Software Engineering — ESEC '95. Proceedings, 1995. XII, 519 pages. 1995.

Vol. 990: C. Pinto-Ferreira, N.J. Mamede (Eds.), Progress in Artificial Intelligence. Proceedings, 1995. XIV, 487 pages. 1995. (Subseries LNAI).